BEIHEFT 70 ZUR NOVA HEDWIGIA

BEIHEFTE ZUR
NOVA HEDWIGIA

HEFT 70

The Marine Algae and Coastal Environment of Tropical West Africa

by

G.W. LAWSON

and

D. M. JOHN

with 65 plates and 14 figures

1982 · J. CRAMER
In der A.R. Gantner Verlag Kommanditgesellschaft
FL-9490 VADUZ

Department of Botany
University of Ghana
Legon, Ghana

Present address:
D.M. John
Department of Botany
British Museum (Natural History)
Cromwell Road,
London SW75BD

Present address:
G.W. Lawson
Department of Biological Sciences
University of Lagos
Nigeria

Printed in Germany
by Strauss & Cramer GmbH, 6945 Hirschberg 2
ISBN 3-7682-5470-4

Contents

Preface

This is the first time a comprehensive account of the marine algae of tropical West Africa has been attempted and in writing it we have tried to put on record all that we have discovered and learnt about seaweeds in this part of Africa during the past quarter of a century. We have been able to study at first hand shore habitats and to make collections of seaweeds from nearly all the maritime countries of West Africa, and these along with other collections and descriptions of shore ecology in the region, both older ones and those of comparatively recent date, have been taken into account in the preparation of this work. In addition all published information on the marine algae of tropical West Africa - information which is often very widely scattered in the literature - has been brought together and assessed.

The result will, we hope, be useful to the student of West African algae in providing him with a general background to those features of the environment in the tropical parts of the region which affect the biology of the sea shore as well as giving him a ready means of identifying any specimen that may come to hand. The specialist will also find in these pages a great deal of previously unpublished data on the distribution, ecology, taxonomy and morphology of the seaweeds we have dealt with. Most of the species are illustrated and the great majority of the illustrations have been drawn by ourselves, or directly under our supervision, from material we have often personally collected.

In a region such as tropical West Africa, which has even now been so relatively little studied phycologically, a marine flora such as this must be regarded as a pioneer work. It must be considered as a synthesis or summary of our present but as yet incomplete knowledge of the region. New records will be added for many years, new information about known species is being constantly compiled, and plant name changes will continue to be made. Though we believe the majority of the algae have been named with a fair degree of certainty, nevertheless, as pointed out in the relevant parts of the text, there are groups in which determinations must be regarded as uncertain until such time as they have undergone a global revision. There is also some algal material from the region which, for one reason or another, cannot at the present time be ascribed to a genus, let alone to a species. In this context it is relevant to refer to Dixon's (Dixon, 1963) general comment on marine algal floras: "It is doubtful if more than about 70% can be identified accurately in any flora or taxonomic treatise, for any part of the world."

We would like to extend thanks to the many individuals who have, in various ways, enabled us to complete this work. The following specialists have confirmed or, in some cases, identified material for us: — Dr. I.A. Abbott, Professor M. Bodard, Professor P.S. Dixon, Dr. F. Drouet, Dr. W.F. Farnham, Dr. R.L. Fletcher, Professor P. Gayral, Dr. M.D. Guiry, Professor C. van den Hoek, Mrs L. Irvine, Dr. H.W. Johansen, Dr. D.F. Kapraun, Dr. J.T. Koster, Dr. G. Kraft, Dr. T.A. Norton, Dr. E. Post, Mr J.H. Price, Dr. Y. Saito, Dr. P.C. Silva, Mr R.H. Simons, Mrs M. Steentoft, Professor W.R. Taylor, Dr. E. Woelkerling, Dr. E.M. Wollaston, Professor H.B.S. Womersley, and also the late Dr. H. Blackler, Professor P. Dangeard, Dr. E.Y. Dawson and Professor G.F. Papenfuss. Some of these specialists, along with others including Mme F. Ardré, Professor M. Doty, Mme G. Feldmann, the late Professor J. Feldmann, Mme P. Lemoine, Dr. W.F. Prud'homme van Reine and Professor R. Sourie have, through correspondence, given us useful advice on certain difficult groups, or helped us to locate collections. Special thanks are due to Mr. J.H. Price of the British Museum (Natural History) for giving advice both on literature and on many of our taxonomic decisions, also in providing the account of the genus *Callithamnion*. We are indebted to those who have helped us with the illustrations namely Mr S.K. Avumatsodo, Mr J.O.K. Ansah, Mrs J. Mitchelmore, Mr E. Ofei and Miss C. Townsend; to Dr. M.D. Guiry for providing the drawings of our *Cryptonemia* material; to Mr F.C. Reynolds for preparing the photographs; and to Mrs J.A. Moore, Mr A.P. Moore and Miss J. Whiting for invalvable assistance in the final stages of preparation of this work. Thanks also go to a number of our former colleagues namely Mr J.B. Hall, Dr. J.M. Lock and Mr W. Pople, for advice during the writing of this manuscript, for critically reading certain sections, and in some instances for help with the literature. We wish to record our appreciation to Professor E. Laing and Professor G.C. Clerk for providing facilities in the Department of Botany, Legon, and to Professor J.O. De Graft Hanson and Mr J.B. Hall for providing the latin diagnoses. Finally, special mention should be made of our assistant over many years, Mr F.O.K. Seku, for his diligent and enthusiastic participation in all phases of this work. In a work of this nature, whose preparation has covered a number of years, many individuals have provided assistance in one way or another and we extend our thanks to all of those we may have not mentioned by name.

We would like also to extend our thanks to the Directors and Curators of the following institutions for allowing us work space or the loan of reference material and collections: — Herbarium of the Allan Hancock Foundation, University of Southern California; Farlow Herbarium of Cryptogamic Botany, Harvard University; Universitets Botaniske Museum, Copenhagen; Botanisches Museum, Berlin; Instituto de Botânica "Dr. Gonçalo Sampaio", Universidade do Porto; Botanical Institut of the University of Coimbra; Muséum National d'Histoire Naturelle, Paris; Museum of the Royal Norwegian Society, Trondheim; British Museum (Natural History), London.

This whole enterprise would not have been made possible but for the generous financial support given to us by the University of Ghana, Legon in the form of special grants and incidental expense. UNESCO also kindly provided us with funds covering many of our collecting trips along the West African coast.

G.W.L.
D.M.J.

INTRODUCTION

The coastline of tropical West Africa, for phytogeographical reasons which will become apparent in the section on "Marine Phytogeography", is taken as extending for about 3,800 km (ca 2,356 miles) from the northern border of Gambia southwards to the Equator (Fig. 1). In an extensive indentation of the coast, now known as the Bight of Benin, lie a chain of islands including Bioko (formerly Macias Nguema Biyogo and Fernando Póo), Príncipe and São Tomé whilst off Guinea-Bissau are many small islands forming the Archipel de Bijagos. The benthic algae occurring in the many lagoon systems, salt pans and estuaries found in this the Gulf of Guinea region of West Africa are considered along with those growing under the more strictly marine conditions of open rocky shores.

An Historical Review

From the phycological point of view tropical West Africa has until recently been a comparatively little-known region. The scattered literature relating to benthic marine algae found along this coast, though by no means extensive, is not always easy to locate. Some of the earlier records are in obscure sources or buried in large general works and thus easily overlooked. Systematic lists of algae begin to appear towards the turn of the nineteenth century although some published records go back almost 200 years. Most of the ecological surveys of the shore algae along large stretches of the coast have been made only in the last two and a half decades. A number of these surveys have been made by ourselves, and our observations, together with those from other recent accounts and from earlier literature, have been used in the preparation of this book.

Some of the earlier records of marine algae from the region must be regarded as doubtful. For example, in 1805 the large brown alga *Saccorhiza polyschides* (Lightfoot) Batters was described under the name of *Ulva bulbosa* by Palisot de Beauvois from plants supposedly found growing on rocks "à Shama, sur la côte de Guinée" (now Ghana). Despite the fact that his illustration, somewhat misleadingly labelled *Ulva tuberosus*, is unmistakably of this well-known seaweed, and even though there exists an authenticated specimen collected by Beauvois in the Hornemann collection in Paris (Norton and Burrows, 1969), we consider it an exceedingly remote possibility that the type locality of *Ulva bulbosus* is Shama in Ghana. *Saccorhiza polyschides* is a plant normally found in the colder waters around Europe with North West Africa the southernmost extension of its present range. Its most southerly mention is by Primo (1953) who reports finding one of the characteristic holdfasts on rocks at Villa Cisneros (now Dakhla), about on the Tropic of Cancer in former Spanish Sahara. Furthermore we have been unable to find it after careful search of the original locality in Ghana, and *Saccorhiza polyschides* is a plant of such a size that it would be difficult to overlook. It is of course just possible that *Saccorhiza* did once grow further south in tropical West Africa but now no longer occurs along this coast. There is no evidence for a change of water temperature during the past 200 years or so that might account for such a disappearance.

Prior to the recent account of the shore ecology of Gambia by John and Lawson (1977b), only a number of scattered records of marine algae existed in the literature for the short coastline (ca 85 km) of this small country. Of these only 4 are specifically for Gambia; another 29 records, mostly in the older literature, refer to "Senegambia". Gambia is unusual in being virtually an enclave within Sénégal, and it is not clear what meaning the term Senegambia had when used in the past. By far the greatest amount of collecting in the region, however, has been carried out to the north of Gambia around the Cap Vert peninsula in Sénégal. Therefore, it seems quite probable that the majority of the algal records for Senegambia are based on plants actually collected to the north of Gambia. Since Gambia is taken as the northernmost limit of the truly tropical marine flora, then the Senegambia records have been discounted unless the species were rediscovered in our recent survey of this country.

To the south of Gambia are the mostly sandy or muddy and largely mangrove-fringed coasts of the Casamance region of Sénégal, Guinea-Bissau and Guinée, about which little is known phycologically. Bodard (1966a) reports 4 marine algae collected at Cap Skiring in southern Sénégal whilst Chevalier (1920) lists 16 species acknowledged as having been identified by P. Hariot from this Casamance region. A brief account is given by Sourie (1954) and Marchal (1960) of the plants and animals found on the rocky shore around Conakry, the capital of Guinée. These accounts are principally concerned with the zoobenthos and many of the more common algae are simply referred to genus. Some records of *Bostrychia* spp. are given in Schnell (1950) and in various publications by Post (1936, 1955a, 1955b, 1959, 1963a). We can find no mention of any marine algae from Guinea-Bissau or from the many offshore islands forming the Archipel de Bijagos.

Our knowledge of the algal flora of Sierra Leone is of recent date although botanical investigations in that country extend back over 200 years. The first brief list of marine algae was published by Lawson (1954b) and was followed by a short account of the distribution of plants and animals on the rocky shores of the Sierra Leone peinsula on the northern side of which stands the capital, Freetown (Lawson, 1957a). A few algae are mentioned by Longhurst (1958) in a paper dealing mainly with the zonation of the zoobenthos along the estuary of the Sierra Leone River. There is one very early mention of the brown alga *Sargassum vulgare* C. Agardh (as *S. cymosum* C. Agardh form *latifolium* C. Agardh) by Agardh (1821) based on material collected by A. Afzelius, a Swedish agricultural adviser stationed at Freetown in 1792. A recent survey made by us (John and Lawson, 1977a) has added a further 48 records of marine algae to the 25 species (excluding blue-green algae) previously reported from Sierra Leone. Aleem (1978) has also recently listed 85 species of algae identified from shore collections and from the occasional dredge sample taken at depths of 10 to 20 m along the continental shelf of the Sierra Leone peninsula. Fifty two of these species are new records for Sierra Leone and about two-thirds of this number

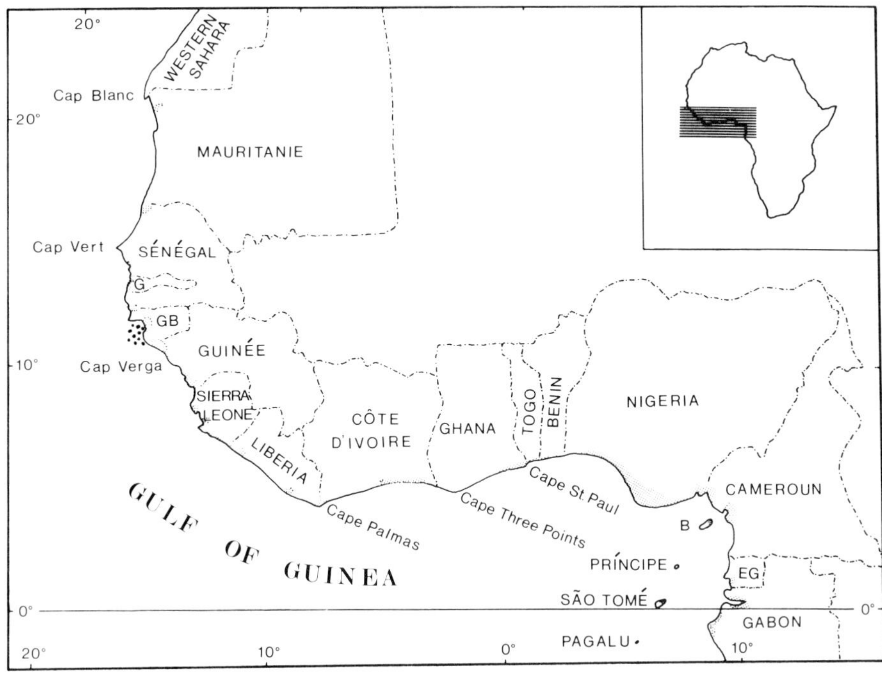

Fig. 1. Map of West Africa including the coastal region (Gambia to Equator) and offshore islands covered by this work. The principal areas of mangrove (after Diels *et al.*, 1963; Keay, 1959) are indicated by stippling. G, Gambia; GB, Guinea-Bissau; EG, Equatorial Guinea; B, Bioko. Inset shows the region in relation to the rest of the African continent.

have not previously been reported from the Gulf of Guinea. Doubt must attach to many of these new regional records as the species have often not been previously recorded from the north-eastern seaboard of the Atlantic, and the only two floras mentioned by Aleem (1978, p. 397) cover the eastern tropical and subtropical coasts of the Americas (Taylor, 1960) and the Canary Islands (Børgesen, 1925-1930). Several algae are mentioned by Aleem (1980b) in a paper on marine fungi in Sierra Leone and he lists 36 species in another (Aleem, 1980a) dealing with marine blue-green algae from this country.

The paucity of algological records from Liberia and Côte d'Ivoire may partly be explained by the relative inaccessibility until recent times of many of the more rocky areas of coastline. The first general account of the shore ecology of Côte d'Ivoire and the southernmost corner of Liberia (Cape Palmas), together with an annotated check-list of the marine algae, have only recently been published (John, 1972c, 1977b). A description of shore zonation and an annotated list of 88 species of marine algae found along almost the entire Liberian coast has also appeared (De May *et al.*, 1977). In the earlier literature only a few and often single records are given by Amossé (1970), Bodard (1966a, 1966c), Fox (1957), Lawson and Price (1969), Price *et al.* (1978) and Post (1963b, 1966a, 1966b) for Côte d'Ivoire, and Askenasy (1888) for

Monrovia in Liberia.

Ghana (formerly the Gold Coast), of all the countries bordering the Gulf of Guinea region of West Africa, has received not only some of the earliest but also the most detailed attention from phycologists. A few years after the somewhat doubtful record of a brown alga by the Frenchman, Baron Palisot de Beauvois already mentioned (p. 9), a list of plants including marine algae was published by Hornemann (1819) from "Danish Guinea", the coastal part of present-day Ghana, collected by P. E. Isert who was a German-born physician. Almost a century later an encrusting red alga was reported by Foslie (1909) from "Fort Brandenburg" (now Princes' Town Fort) having been collected by Dr H. Brauns in 1892. The first systematic accounts of marine algae from Ghana were provided by Dickinson (1951, 1952) and Dickinson and Foote (1950, 1951) from collections made by Miss Vera Foote, a teacher at Achimota College and later a lecturer at the then University College of the Gold Coast. In recent years the number of marine algae known from Ghana has more than trebled, many of the additional records mentioned by John and Lawson (1972a) have come from subtidal collections made by SCUBA diving. Lists have also been published by Lawson (1960c, 1965) which include algae growing both in brackish-water as well as freshwater habitats. General accounts of the ecology of shore algae are to be found in a series of papers by Lawson (1953, 1956, 1957b) whilst subtidal ecology is discussed by John *et al.* (1977) and Lieberman *et al.* (1979). Various anomalous distribution patterns found on harbour breakwaters in Ghana and adjacent countries are described and discussed by John and Pople (1973).

The combined coastlines of Togo and Benin (formerly Dahomey) stretch for about 170 km (ca 105 miles) and consist almost entirely of sandy palm-fringed beaches often backed by extensive lagoon systems. These beaches of unconsolidated material are unsuitable for the growth of benthic algae. Before the coastal survey of these two countries by John and Lawson (1972b) just 3 species had been reported (Pilger, 1911; Steentoft, 1967); these algae were probably collected in the drift. It seems that the only extensive surfaces for the attachment of shore organisms have appeared over the past 20 years as a result of the construction of harbours, jetties and other marine structures. For the first time areas of rock have been uncovered by local coastal erosion caused by the interference of such structures with the normal west-east movement of sand along the shore (Pl. I). A total of 52 algal species were discovered in 1971 by John and Lawson (1972b) along the shore of Togo and Benin, and this number might be expected to further increase with time.

Plate I (after John & Lawson, 1972b): A. An aerial view showing the effect of interference by the breakwaters of Lomé Harbour (Togo) on the west to east longshore drift of sand. On the western side of the longest breakwater sand has accumulated whilst on the eastern side it has been lost resulting in the uncovering of rocky platforms. B. A view to the east of the breakwaters showing a recently uncovered rocky platform, a low sandy cliff and fallen palm trees resulting from the erosion of the original sandy beach.

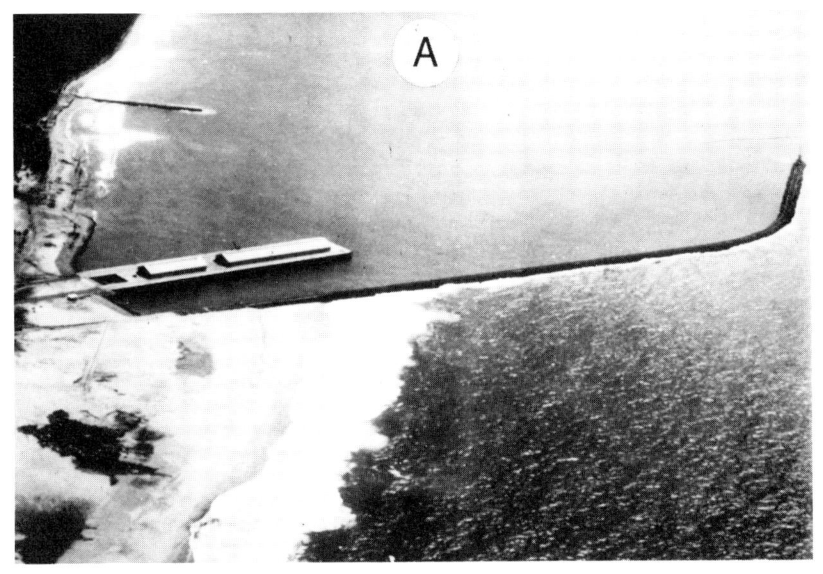
A

B

Nigeria has a somewhat similar coastline to that of Togo and Benin, although towards the east of the country are extensive muddy deposits forming the delta of the Niger River. The earliest published work on Nigerian algae was by Mills (1932) and dealt solely with marine and freshwater diatoms around Warri on the western side of the delta. Man-made constructions such as the breakwaters of Lagos Harbour have provided suitable surfaces for the development of benthic algal vegetation and this has been described by Fox (1957) and Steentoft Nielsen (1958). It is likely that the offshore oil drilling platforms standing off the eastern part of the coast may also provide surfaces for the growth of sessile marine organisms including algae; these platforms have yet to be studied.

Cameroun borders the Bight of Benin where the West African coast, after running almost parallel to the Equator, turns suddenly southward. The first annotated list and ecological account of the marine algae of this country was published by Pilger (1911) based on collections and observations made in 1908 at six localities by C. L. Ledermann, a horticulturalist at the Victoria Botanical Gardens in what was then a German colony. Almost 50 years elapsed before Lawson made a brief visit to what was then called British Cameroons and gave a first hand account of the distribution of shore algae (Lawson, 1955). Detailed re-examination of Lawson's collections, and further ones made by us during a visit in February 1974, has led to the discovery of a number of algae not previously reported from Cameroun.

There is nothing known regarding the marine algal flora of Equatorial Guinea (formerly Rio Muni), a small country sandwiched between Cameroun to the north and Gabon to the south. Bioko, which is also under the same administration, is the largest of the offshore islands in the Bight of Benin and lies just about 30 km southwest of Victoria on the coast of Cameroun. About 30 species of algae are now known from the shores of this island, and prior to the survey carried out by one of us in December 1981 the only published records were of the brown alga *Hydroclathrus clathratus* (Bory) Howe and 3 encrusting calcareous red algae mentioned by Schmidt and Gerloff (1957).

Gabon straddles the Equator and nearly all of the algal records from here appear to be based on collections made along the northern part of its coast, usually in the vicinity of the capital, Libreville. Much of our early knowledge of the marine algae of Gabon rests on the identifications of Hariot (1896) of plants collected by F.-R. Thollon and Pierre in the estuary of the Gabon River. A total of twenty three other papers have been traced, many of them also dating back to the last century, that make mention of single and sometimes secondary citations of Gabonese algae. These records are often based on plants collected by the Frenchman, E. Jardin in the middle part of the nineteenth century. The only account of the shore ecology of Gabon is of recent date (John and Lawson, 1974a) and contains over 70 new records of marine algae.

Some of the earliest published records of marine algae from tropical West Africa are of collections from two of the offshore islands. The famous Austrian-born botanist and zoologist F. M. J. Welwitsch collected algae from the islands of Príncipe and São Tomé in 1853 and 1860 respectively while en route and returning from an expedition to Angola. Barton (1897) lists just 4 algae collected by Welwitsch whilst Hariot (1908) mentions 38 species and varieties from these islands. In the 1880's the

Portuguese took a direct interest in the floras of these islands and Professor J. Henriques published a series of papers (Henriques, 1885, 1886a, 1886b, 1887) on São Tomé plants collected by F. Newton, A.F. Moller and F. J. Diaz Quintas of which 35 are species of algae. Two of these works (1886a, 1887) are simply extracts of information published in the earlier ones (1885, 1886b) with mistakes in the original text corrected and some new ones introduced. In a later paper Henriques (1917) considered both plants and animals from São Tomé but all the algal records are secondary citations. Much of our knowledge of crustose members of the family Corallinaceae in the region comes from the identifications and descriptions by Foslie (1897, 1900a, 1907, 1909) and Foslie and Printz (1929) of material collected from São Tomé. Rodriques (1960) describes and illustrates 7 of the brown algae from this island whilst Tandy (1944) has contributed a short list of algae collected by A. W. Exell in November 1932 during an expedition to islands in the Gulf of Guinea. Marine and freshwater blue-green algae from São Tomé and Príncipe have been reported by Sampaio (1958, 1962a, 1962b, 1963) largely based on specimens collected by Professor A. Roseira between September and November 1954. Carpine, who was on the French research vessel "Calypso" that visited the Gulf of Guinea in 1956, gives the only detailed account of the ecology and distribution of marine organisms on these two islands (Carpine, 1959). His algae were identified by Mme P. Huvé and this is the only large collection of algae from the region we have been unable to examine; this collection may now be lost (Mme Huvé, *pers. comm.*). Steentoft (1967) provides determinations of marine algae collected in 1956 from São Tomé by C.A. Thorold and re-appraises all the earlier records from this island and Príncipe, recognising in all 101 species and varieties. She reviews the history of algal collecting on the islands and also considers some of the environmental factors influencing shore life in this region of West Africa.

Bioko (formerly Fernando Póo, more recently known as Macias Nguema Biyogo) is the largest island in the Gulf of Guinea. The algal flora of this island seems relatively depauperate (ca 30 spp.) and was virtually unknown until very recently. Pagalu (formerly Annobon) is the smallest and outermost of the chain of islands and lies just south of the Equator. Some of the 35 algae reported from its shore are as yet unknown in the region of West Africa covered by this work though they are mentioned in footnotes.

Many of the earlier publications cited in this review contain but few observations on the ecology of the marine algae found in tropical West Africa, and are often simply names of algae sometimes included in more general lists of plants. The principal publications dealing with shore ecology are cited in the relevant section. The tropical coast of West Africa is also included in a review article entitled "The Littoral Ecology of West Africa" by Lawson (1966). A critical assessment of the marine algae of the West African coast and adjacent islands, including a comprehensive bibliography, is being currently undertaken and the parts concerning the Chlorophyta and Xanthophyta, Phaeophyta, and the Rhodophyta have or are about to be published (Lawson and Price, 1969; Price *et al.*, 1978, 1983; John *et al.*, 1979). These papers include largely secondary citations for the tropical region of West Africa and can be used in conjunction with this book in providing the detailed background to certain of our taxonomic decisions.

The Physical Environment

Description of the Coast

It is currently believed that the rifting of the Pre-Cambrian shield of a supercontinent led to the opening of the Atlantic Ocean beginning with the rotation of Africa and South America away from North America about 180 million years B.P. (see Raven and Axelrod, 1974, for references). The original coastline of tropical West Africa was formed by the separation and spreading apart of Africa and South America commencing 125 - 130 million years B.P. Fracture zones known under the names of St. Paul, Romanche, Chain and Guinea were important features during this process and determined the shape of the coastline (Arens *et al.*, 1971). The final separation of the continents probably took place less than 100 million years B.P., and by the end of the Cretaceous period (ca 65 million years B.P.) they were about 800 km apart at their closest point and were still linked by many islands existing along the mid-Atlantic ridge and its flanks.

Many of the rocky headlands along the coast of tropical West Africa consist of the old Pre-Cambrian formations whilst between these are found later sedimentary basins and river deltas (Hosper, 1971; Martin, 1971). Rocky shores and cliffs of hard igneous rock occur particularly in the central part of the region whereas in the north-west, conglomerates, limestones, soft sandstones and shales predominate. Both the coastal plain and the sublittoral show evidence of former shore lines notched out during the Quaternary transgressions and regressions of the sea (Allen and Wells, 1962; McMaster *et al.*, 1970). According to Webb (1958, 1960), sand deposition is responsible for the gradual extension of the coastline in a seaward direction in the central parts of the region. Mount Cameroun and the principal islands in tropical West Africa are evidently of volcanic origin and the shores of the islands are largely of basaltic and ferruginous rocks, black sand, as well as scoria.

Today much of the central part of the West African coast runs almost parallel to the Equator (ca 05° N) and consists for long stretches of wave-beaten beaches fringed with palm trees and backed by extensive lagoon systems (see Fig. 1). These lagoons have been classified by Boughey (1957) as "open" or "closed" lagoons on the basis that the former always retain a connection with the sea whereas the latter are usually closed by a sand bar but may, annually or less frequently, temporarily break through to the sea during the rainy season. Bernard (1937) and De Rouville (1946) have recognised two types of lagoon. One type is narrow and brackish, lies immediately behind a sand barrier, has many widely separated temporary openings, and is not associated with large river systems. An example of this type of lagoon is the Lac Nokoué or Benin lagoon (Girault and Kimpe, 1967) and it corresponds to the "closed" lagoon type of Boughey. Such lagoons may be in contact with the sea for much of the year due to their channels being kept open artificially, or else the sand bars closing them may be deliberately breached by the local people who fish the lagoons.

Those lagoons having an artificially restricted opening such as the Sakumo lagoon in Ghana, where contact with the sea is maintained by a culvert under the Accra to Tema coast road, have been called by Pauly (1975) "semi-closed" lagoons. The other type of lagoon is fed by a large river and usually forms an extensively branched freshwater estuary. The lagoon systems at the mouth of the Volta and Niger Rivers in Ghana and Nigeria respectively are examples of this type. The closed lagoons are flooded seasonally rather than daily and consequently show extremes of salinity over the year. In contrast, the open lagoons are tidal and in their inlet channels fluctuations take place in salinity in relation to the "wet" and "dry" season, nonetheless, salinity variations are not so extreme as in closed lagoons (see Figs 12, 13). Where mangroves grow in or near the lagoon openings they afford suitable surfaces for the attachment of some of the smaller benthic algae provided marine influences are strong.

Cameroun, Gabon, Ghana, Liberia, Sierra Leone, and to a lesser extent Côte d'Ivoire and Gambia, are countries that have fairly extensive rocky shores. Many of the offshore islands, as has already been indicated, are volcanic in origin and so are almost wholly rocky, often with high cliffs. In addition to rocky shores there are man-made marine structures including harbour breakwaters, jetties, outflow pipes, and oil drilling platforms that may provide artificial surfaces for the development of benthic organisms.

Our knowledge of the nature of the seabed along the continental margin of tropical West Africa is of recent date and still very incomplete. It would appear that there are a number of rocky areas in comparatively shallow water just offshore. For instance, there are between Côte d'Ivoire and Nigeria a series of narrow rocky banks stretching almost unbroken for considerable distances and running almost parallel to the coast (Allen, 1965; Martin, 1971). These banks consist of coarse shelly sandstone and are believed to be beach rocks formed during regressions of the sea. Between such outcrops may be found wide areas of sand, gravel or mud overlain in some places with shells or shell fragments. Extensive areas of cobbles or nodules, consisting largely of calcareous red algae, also occur. These areas, often misleadingly called "coral banks", have been reported off Ghana (Buchanan, 1958; John and Lawson, 1972a; John *et al.*, 1977; Lieberman *et al.*, 1979), Sierra Leone (Longhurst, 1958; McMaster *et al.*, 1970), Côte d'Ivoire and Togo (Barbey, 1968), as well as from the offshore islands of Príncipe and São Tomé (Carpine, 1959). The absence of coral reefs along this coast has been attributed to the periodic upwelling of cooler water at certain times of the year, to lack of suitable bedrock, and to high turbidity (Buchanan, 1954; Ekman, 1953). Only in a few shallow-water and protected coves, where solar heating of the water may limit the influence of cold upwelled water, has any concretionary phenomenon been observed (Laborel, 1974).

Climatic Factors

Within the Tropics two peaks of solar radiation occur during the year. At the Equator these coincide with the equinoxes on 21 March and 21 September when the sun passes directly overhead at midday. In the northern hemisphere any point between

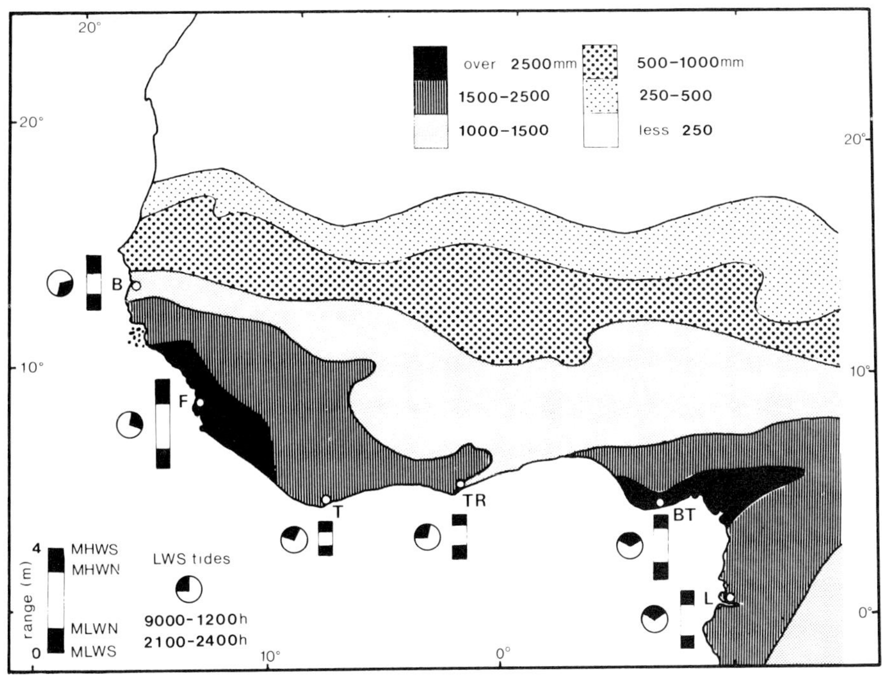

Fig. 2. The mean annual rainfall and tidal data for West Africa. Variation in the mean range or amplitude (MHW - mean high water, MLW - mean low water) of spring (S) and neap (N) tides and the approximate time o'clock of low water spring tides (LWS) are shown at six stations. B, Banjul (Gambia); BT, Bonney Town (Nigeria); F, Freetown (Sierra Leone); L, Libreville (Gabon); T, Tabou (Côte d'Ivoire); TR, Takoradi (Ghana).

the Equator and the Tropic of Cancer (23° 27'N), for example, will have a peak in solar radiation between March and June during the sun's apparent northward migration and another between June and September during its apparent southward migration. This bimodal radiation cycle regulates the overall climate of West Africa by influencing two main air masses that are in opposition to one another.

To the south is a mass of moist equatorial air influenced by maritime conditions, and to the north is a hot and dry, stable subtropical continental air mass over the Sahara Desert (23° - 30° N). Where the two air masses meet and interact there is a zone of climatic instability known as the intertropical or monsoon front. This front lies roughly in an east-west direction, and the division of West Africa into a number of rainfall belts (Fig. 2) is principally caused by its movement and those of the air masses it divides northwards and southward. Between December and March the intertropical front is south of about 8° to 9° N, and the dry body of continental air is the dominant influence of this the period of the major 'dry' season for much of West Africa. Only near the Equator is this the time of the 'wet' or 'rainy' season. In association with the sun's apparent movement, this intertropical front gradually migrates northward at least 10° reaching 17° to 21° N by July or August. From July or

August to September the front recedes once again southwards. The rains usually occur 320 to 480 km to the south of the front where the humid air reaches an altitude of at least 915 to 1,220 m before precipitation occurs.

Along the central coastal region of West Africa from Guinée to western Nigeria there is a distinct four-season climatic regime: a major 'dry' season (November/December to February/March) when temperatures are generally highest and rainfall and relative humidity least, a two peak 'wet' or 'rainy' season (March/April to July/August, September/November) associated with lower temperatures and higher relative humidity and rainfall, and a short 'dry' season (August/September) between the two rainfall peaks. The break in the rains resulting in the short dry season occurs when the main rains are further north. To the north of 8° to 9° latitude the two wet seasons gradually merge so the year becomes divided into a two-season climatic regime of one wet season and only one dry season. The wet season also becomes shorter the further it is away from the Equator. Just north of the Equator (e.g. Libreville) the wet season coincides with the northern hemisphere winter and the short dry season occurs between May and September when the intertropical front is furthest north.

The actual amount of rainfall as well as the exact periods of the rains may not be related solely to the position of the intertropical front but is influenced also by differences in the winds crossing the coast, the presence of coastal currents, and by coastal orientation and interior relief (see Fig. 2). In general, the wet season becomes shorter and the rainfall decreases from south to north in West Africa. Neverthless, there is a relatively dry climate from southeastern Ghana eastwards through Togo and Benin due in part to the prevailing wind blowing almost parallel to the coast without releasing its moisture. Other factors believed to be responsible for this drier central part of the tropical West African coast include the penetration to the coast of the dry Harmattan wind and the presence of cold coastal water. In contrast, the coastal region extending from the southern border of Liberia northward to the Casamance region of Sénégal has a very heavy rainfall as it receives during the northern hemisphere summer the full force of the moisture laden south-westerly winds. Rainfall is also heavier and the pattern modified where there exists high internal relief such as along the Liberian-Guinean mountain range and around Mount Cameroun.

Winds are produced in the centre of the dry subtropical mass of continental air and these are the prevailing easterlies or trade winds. In the southern hemisphere the dominant winds are the south-westerlies and these penetrate well north of the Equator following the northward migration of the intertropical front. Thus as the front moves northwards in April the wind begins to veer around so that the dominant wind direction is no longer easterly but south-westerly. When the strength of the prevailing wind is weak, then a predominant feature of the coastal region may be a diurnal land-sea breeze. The significance of the change in direction of the wind, which normally blows during the day from the south-west for much of the year along the central part of the tropical West Africa coast, as well as fluctuations in its strength, are discussed in relation to water loss from seaweeds by Jeník and Lawson (1967).

The general features of the African climate are given in Jackson (1961), van Chi-Bonnardel (1973) and Thompson (1965) whilst a summary of climatological conditions in West Africa between latitudes 4° to 20°N is provided by Toupet (1968).

Oceanographic Features

A conspicuous feature of the surface currents in the Atlantic Ocean is the presence of two large gyral systems, one on each side of the Equator. The water in the North Atlantic gyral shows a clockwise circulation whilst in the South Atlantic the circulation is counter-clockwise (Fig. 3). This circulation pattern has the effect of bringing widespread warm water, and hence a tropical marine climate, to the western shores of the Atlantic. On the eastern side cold water influences result in a more extensive temperate marine climate and a narrower region of warm water conditions. Also along the West African coast the periodic upwelling of subsurface water helps to maintain the cooling effect far into the Tropics.

The cold Canary Current, flowing from north to south, is part of the northern gyral circulation and influences the North West African coast. On the other hand, the principal current affecting the South West African coast is the Benguela Current flowing from south to north and belonging to the South Atlantic gyral system. These two ocean currents become warmer in the lower latitudes and then turn westward as they approach the Equator forming the North and South Equatorial Currents.

On the older maps of surface currents the Canary Current is shown to send a branch into the Gulf of Guinea. More recently the origin of this so-called Guinea Current has been given a number of different interpretations. One such interpretation is that there is an equalising current known as the Equatorial Counter Current in the calm doldrum zone between the large Atlantic gyrals. When this eastwardly flowing current reaches Africa it is called the Guinea Current. Again, Berrit (1969) has interpreted the Guinea Current as a series of small inshore gyrals between the coast and the Equator. The confusion regarding this current is understandable as many of the details of its characteristics are as yet uncertain. From time to time the Guinea Current, which is predominantly eastward flowing, reverses its direction and this behaviour has also not been satisfactorily explained. It is apparent that the real situation is much more complex than many of the maps of water circulation in the area would suggest. The statement of Jones (1969) that "The Equatorial Atlantic Current system is a complicated and yet incompletely understood complex of dynamic water bodies of varying magnitude and direction, originating from several diverse regions", is still a correct summary of our present limited appreciation of what really happens.

As far as the ecology of the West African coast is concerned the general picture is one of a relatively narrow zone of warm water of low salinity ($T > 24°C$, $S < 35‰$ in the region of the Equator, squeezed as it were, between colder waters of higher salinity from the north and south. This warmer water occupying the central region of the Gulf of Guinea is sometimes referred to as "Guinea Water" (Berrit, 1961, 1962a, 1962b, 1969). It varies seasonally having a minimum temperature be-

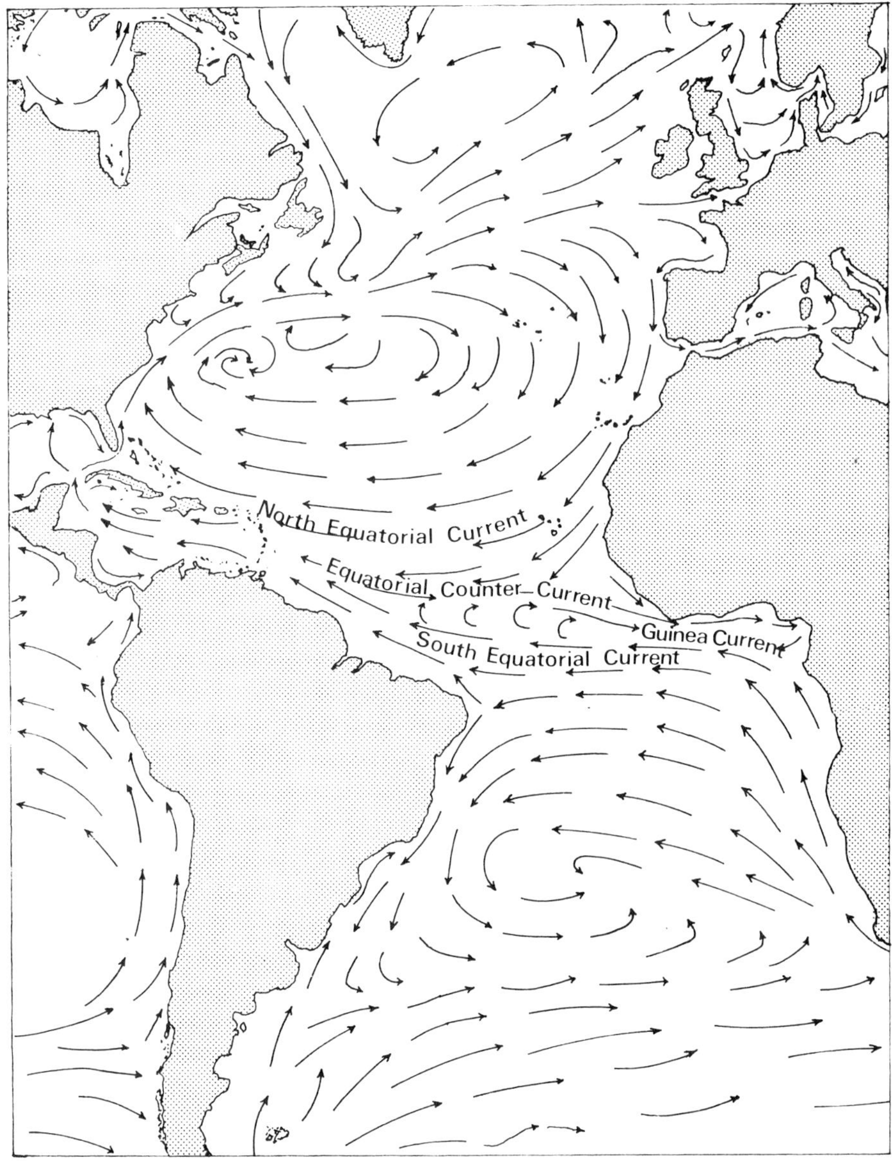

Fig. 3. The circulation of the surface waters in the Atlantic Ocean (modified after Carruther, 1961).

tween July and October (Fig. 4) when the ambient air temperature is low, rainfall and cloud cover are at a maximum, and solar radiation is minimal. Furthermore it is not constant in position but moves north during the northern summer and south during the southern summer. This means that for those parts bordering the coast of tropical West Africa to the north and south there is a regime of alternating colder and warmer surface waters at different times of the year.

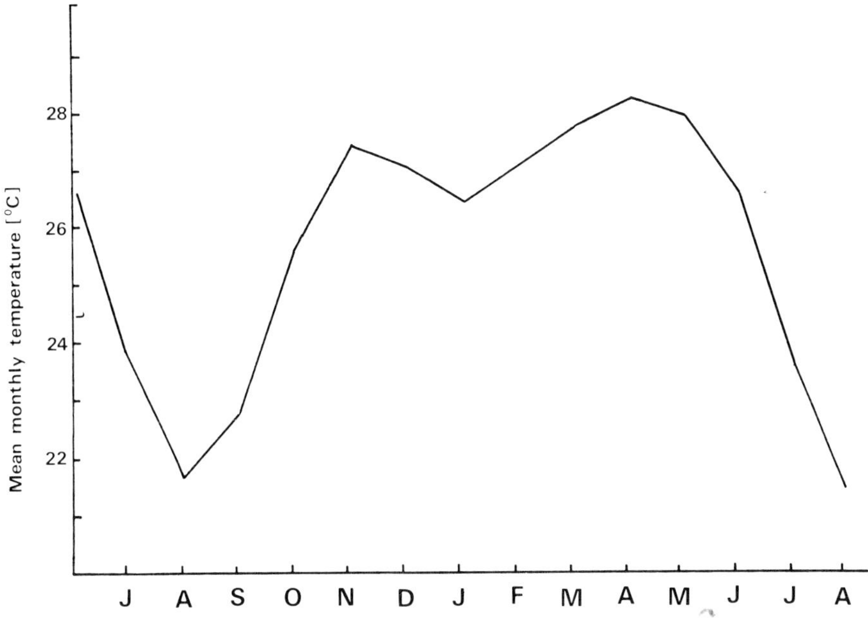

Fig. 4. The annual temperature fluctuations of the surface waters off the coast of Ghana (after Ofori-Adu, 1975).

The situation is further complicated, however, by the occurrence of periodic "upwelling" of colder and more saline water (T<24° C, S>35‰) that take place between Cape Palmas in Liberia and just to the east of Lagos in Nigeria. Usually two periods of upwelling take place in the year, the principal one between late June and October, whereas weaker and more limited upwellings may occur for short periods between December and March. Various theories have been propounded to account for upwelling but so far none gives a completely satisfactory explanation of the phenomenon. In many of the studies on upwelling along this part of the West African coast (Longhurst, 1964; Ingham, 1970; Morliere, 1970; Morliere and Rebert, 1972; among others) it has been assumed that it is due to wind driven divergence of the surface water. However, from work at the Ghana Fisheries Research Unit, Tema (1970) and by Bakun *et al.* (1973) and Houghton (1976), it is becoming increasingly evident that the upwelling event and subsequent changes in hydrography show a poor correlation with local winds. Pople and Mensah (1971) suggested that the upwelling is a local phenomenon resulting from the interplay of meteorological factors with thermocline convection in the sea. They put forward the view that evaporation causes an increase in surface salinity and that sinking of this denser water leads to a turnover with the less dense colder water rising to the surface. Houghton (1973) has questioned this suggested mechanism on the grounds that evaporation is insufficient to increase salinity enough to account for this water circulation. It seems that there is no one explanation for coastal upwelling in the region although shelf waves may in some way be implicated (Houghton, 1976;

Houghton and Beer, 1975). A consequence of the upwelling is an increase in nutrient levels leading to a rise in phytoplankton productivity (Anang, 1979; Longhurst, 1964; Reyssac, 1970) and probably contributing to the high turbidity of the surface water over this period.

Besides the sudden changes that occur in salinity and temperature due to upwelling there is also considerable decrease in surface salinity in those parts of tropical West Africa that receive much inflow of freshwater during the rainy season. Thus, in the Bight of Benin the surface water salinity may drop below 28‰ during the rainy season (Berrit, 1969) whilst salinities less than 20‰ have been reported off Sierra Leone (Watts, 1958) and Guinée (Marchal, 1960).

Taking into account the oceanographic features that have been outlined above, various attempts have been made to classify the West African coast into a number of different areas or sectors. Postel (1968) considers the region to be divisible into three areas: (a) a northern area of cool water, with high salinity (north of Cap Blanc which is the border between Mauritanie and former Spanish Sahara), (b) a transition area where the water is alternately warm and cool, without any marked drops in salinity (Cap Blanc to Cap Verga in Guinée), and (c) a southern area with warm or cool water and slight to marked temporary drops in salinity (Cap Verga to the Equator). He considers this third area to be further divided into three sectors - a Guinean Sector of warm water, with considerable temporary drops in salinity (Cap Verga to Cape Palmas in Liberia), a Côte d'Ivoire Sector of alternating warm and cool water, with slight temporary drops in salinity (Cape Palmas to Cape St. Paul in eastern Ghana), and a Cameroun Sector of warm water, with marked temporary drops in salinity (Cape St. Paul to the Equator). The Côte d'Ivoire Sector should perhaps be more correctly extended further eastwards since cold upwellings are known to take place as far as Cape Formosa in Nigeria. The surface waters to the south of the Equator are somewhat cooler and more saline than those found in the Gulf of Guinea region of tropical West Africa but this area is not included in Postel's scheme.

Steentoft (1967) has pointed out that for shore organisms not only the marine season but the climatic season is important and the two must be considered together. Thus, although marine conditions along the part of the coast of tropical West Africa from Sierra Leone to Cameroun are similar to those of the islands of Príncipe and São Tomé they differ considerably in climate. The mainland is for a longer or shorter period hot and dry during the northern hemisphere winter, and cool and wet for the rest of the year, whereas the islands have a hot wet climate during the northern winter and a cool dry climate during the northern summer.

The coast of tropical West Africa is subjected to a more or less continuous pounding by Atlantic waves, with the roughest seas occurring most frequently from about May to October which also coincides with the time of the rainy season. These heavy seas together with discharge from rivers and lagoons result in considerable inshore turbidity and silting. The sand bars closing many of the lagoons do not become breached until well into the wet season and may continue discharging silt-laden water beyond the moderation of the weather in October. Thus, several factors contribute to an inshore band of turbid water which, according to Longhurst (1964), extends from 1 to 6 km from the coast in the central part of the region. The high turbidity together with heavy seas inhibits the collecting of subtidal algae over the rainy season. A special case of inshore turbidity is exhibited along parts of the coasts of Togo and Benin where a turbid yellow water zone is apparent. This is the result of the continuous discharge at Kpémé in Togo of washings from a phosphate factory.

Tides

The coastline of tropical West Africa is relatively straight and, as in other tropical areas, the tidal range is small varying from about 0.4 to 0.9 m at neap tides and 0.6 to 1.8 m at spring tides (see Fig. 2). Nevertheless, an increased tidal amplitude is found in river estuaries where the water is channelled between converging shores. Thus Freetown, which lies on the southern bank of the estuary of the Sierra Leone River, has a tidal range of almost twice that reported from many other places along the African coast. Marine influences often extend well above the theoretical limit of the high water due to the effects of waves and surf which may be considerable along much of the coast.

Along the African coast the tides are of the semi-diurnal type with two low and two high waters occurring in any 24 hour period. The actual time of the lowest spring tides in each month varies from place to place along the coast (see Fig. 2). Of the two high tides in any day one is always higher than the other and similarly of the low tides one is always lower. Thus there is a higher high water, a lower high water, a higher low water, and a lower low water each day. The two low and two high tides each day differ in amplitude by as much as 0.3 m though the amount of this difference follows the astronomic tidal cycle of 18.6 years. In general the lower low waters occur in the day and the higher low waters in the night when the sun is south of the Equator (September - March), whereas the position becomes reversed after the equinox when the sun is in the northern hemisphere. The ecological implications of this phenomenon are fully discussed in the section on shore ecology. For a full explanation of the very complex behaviour of the tides refer to Doodson and Warburg (1941) and Dronkers (1964).

Information on predicted tidal levels and times for the West African coast are given in the annual publication of the British Admiralty on tides and tidal streams in the Atlantic and Indian Oceans (Vol. 2).

MARINE PHYTOGEOGRAPHY

General Considerations

Sound taxonomic work is a precondition for all types of biogeographical studies, and lists from different areas must be strictly comparable if conclusions are to be drawn with certainty. If, for instance, a single taxon goes under different names in two different areas or two closely related but separate species from such areas are misidentified and placed under the same name, then some distortion of the apparent distribution patterns is inevitable. We are fully aware that many new records will be made for marine algae on the coast of tropical West Africa and that numerous taxonomic problems remain unsolved in our region and, indeed, in other tropical areas or regions with which comparisons need to be made. Nevertheless, we believe that enough is now known for a preliminary analysis to be made and for an attempt to be undertaken to relate the observed distribution patterns to the climatic, oceanographical, and perhaps historical features, which might be responsible for them.

The West African Coast

The distinctive nature of the tropical West African marine algal flora is indicated clearly in a recent study of the warm temperate and tropical flora of the Atlantic as a whole using the technique of ordination (Lawson, 1978). The algal flora of a total of 87 countries and islands with a combined species composition of over 1500 species were ordinated using reciprocal averaging (Hill, 1973). An example of the type of results obtained is shown in Fig. 5 in which are plotted the first two axes of the ordination.

This ordination diagram shows a number of interesting features. In the first place the floras of the two sides of the Atlantic are clearly distinguished from each other, the American localities appearing as a group to the left of the diagram and the African to the centre and right. It is interesting to note that Ascension Island (27), lying midway across the Atlantic, appears to have affinities with the American rather than the African side. The countries of tropical West Africa form a very close knit grouping with only Côte d'Ivoire showing some divergence from the remainder.

The algal floras of both Gambia and Gabon are very clearly associated with the tropical West African group of algae and so must be regarded as belonging to this floristic region. Thus Gambia may be taken to represent the northernmost boundary of the flora of the region. In fact just two of the species found in Gambia *(Hypnea arbuscula, Ulva popenguinensis)* have not been found in countries to the south (John and Lawson, 1977b). On the other hand nearly 30% of the total marine algal flora (excluding blue-green algae) found in the Gulf of Guinea region of West Africa are not present further north than Gambia.

The southern phytogeographical boundary of tropical West Africa cannot be so clearly defined. The reason for this is that the shores of southern Gabon, Congo Republic, Cabinda and Zaîre are still so little-known phycologically. As pointed out above, however, the affinities of the flora of Gabon are overwhelmingly with that of the Gulf of Guinea. But collections from Gabon have almost all been made in the vicinity of Libreville which lies almost on the Equator (John and Lawson, 1974a). For this reason we have taken the Equator as the southern limit of tropical West Africa for the purposes of this book, though we recognise the possibility that this limit may be placed further south after more exploration has been carried out.

At the extreme upper right of the ordination diagram (Fig. 5) the Salvage Islands (1) and former Spanish Sahara (3) are shown as completely unrelated to the Gulf of Guinea group of species but form a loose group together with other colder-water localities namely the Canaries, Cape Verde Islands, St. Helena, and Mauritanie. Between these two groups, but with more affinity to the tropical group, lies an intermediate cluster consisting of Angola (25), Sénégal (5) and two of the Gulf of Guinea islands namely Príncipe (18) and São Tomé (19) together with another island Pagalu (29), which lies just outside tropical West Africa as defined above. Congo Republic (22) and Zaîre (24), lying together but some distance away, also probably belong to this group although some doubt attaches to their exact placement as the records from these two countries are few, usually more than a century old, and in need of revision.

It is interesting to see that the islands of Príncipe, São Tomé and Pagalu come in this transition group rather than, as might be expected, in the tropical group which their geographical position might suggest. Their affinity with the floras of Sénégal and Angola is perhaps an indication that the water temperature is somewhat lower around these islands than along the mainland coast of tropical West Africa. Steentoft (1967) has indicated that there are also differences in the climatic seasons between these offshore islands and the mainland although climate is more likely to effect the zonation of littoral communities rather than their composition.

The proximity of Sénégal and Angola in the ordination diagram, despite their wide spatial separation, emphasises the feature they possess in common, namely that they each lie adjacent to the tropical region of West Africa - one in the northern hemisphere and the other in the southern hemisphere. There are a number of algal species including *Bryopsis balbisiana* Lamouroux ex C. Agardh, *B. corymbosa* J. Agardh, *Halimeda tuna* (Ellis & Solander) Lamouroux, and *Gelidium versicolor* (Gmelin) Lamouroux which are found in both Angola and Sénégal, but which are absent from tropical West Africa. The transitional nature of the floras along the coasts of Sénégal and Angola have been commented upon by Sourie (1954) and Lawson *et. al.* (1975) respectively.

In the ordination diagram Mauritanie lies equidistant from former Spanish Sahara and Sénégal. Probably the affinities of the main part of the Mauritanian coast are essentially with the latter. The species that Mauritanie shares with former Spanish Sahara are largely confined to a small area on the western tip of the Cap Blanc peninsula whilst several tropical species reach their northern limit on the warmer

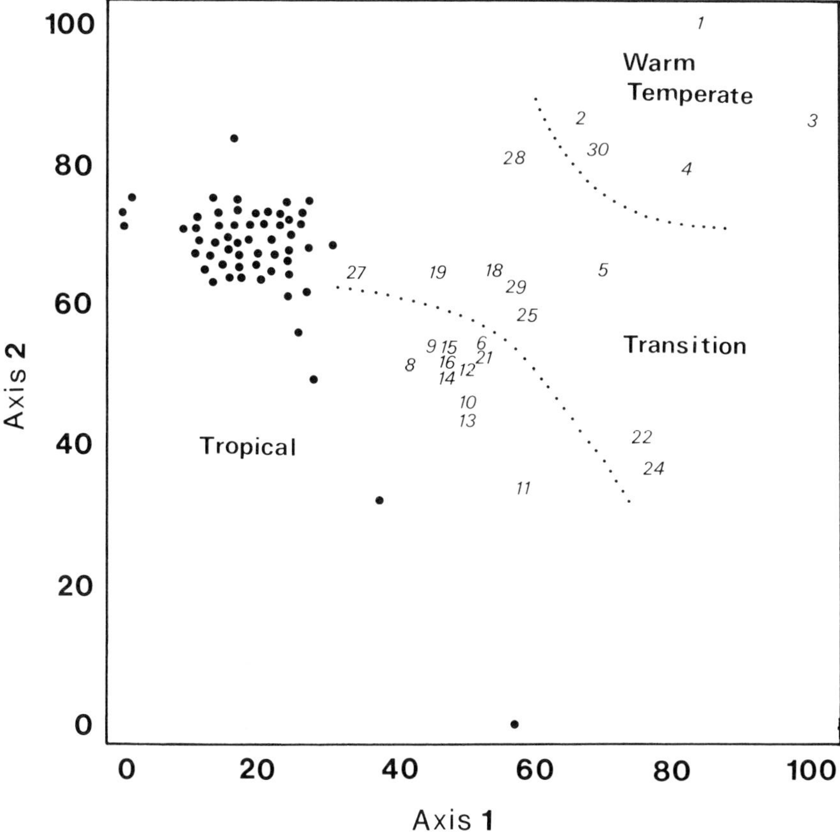

Fig. 5. Ordination of the benthic marine algal floras of the tropical and subtropical Atlantic Ocean. The position of countries and islands on the western side of the Atlantic are indicated by closed circles whilst those on or near the West African coast are numbered (see Fig. 6 for key to numbers).

Baie du Lévrier side of this peninsula. Thus Cap Blanc appears to represent a boundary between the warm temperate seaweed flora of North West Africa and the subtropical transition flora of Mauritanie and the northern part of Sénégal (Lawson and John, 1977).

South West Africa is not included in the ordination diagram as its algal flora bears little resemblance to those found elsewhere in the Atlantic. There is a very restricted penetration of colder water species (e.g. *Acrosorium maculatum* (Kützing) Papenfuss, *Plocamium beckeri* Simons and *P. suhrii* Kützing) into Angola from the south, these being largely confined to the southern half of the country. On the other hand many large and distinctly cold water algae such as *Iridaea capensis* J. Agardh and

Laminaria schinzii Foslie are found at Swakopmund in South West Africa but do not extend northwards into Angola. This suggests that, at least as far as the algae are concerned, the transition area between Ekman's (Ekman, 1953) tropic and boreal-antiboreal littoral provinces takes place in the northernmost part of the South West African coast (John *et al.*, 1981; Lawson *et al.*, 1975). Penrith and Kensley (1970a, 1970b) and Kensley and Penrith (1973) have come to the same conclusion when considering the littoral fauna. According to the map in Hedgpeth (1957), showing the littoral faunal provinces of the world, this is the only part of the world in which the boreal-antiboreal zone is not separated from the tropic by a warm temperate zone.

The groupings described above are used in compiling the map in Fig. 6 showing the phytogeographical regions of the West African coast. It is apparent that Postel's (Postel, 1968) divisions of the coast into a number of areas or sectors based solely on oceanographic considerations may have little phytogeographical significance. For example, he puts the southern boundary of his "transition area" at Cap Verga in Guinée whereas, as indicated above, there is phycological evidence to show that the tropical flora extends much further north than this Cape, in fact to Gambia, and that as far as algae are concerned the transition zone begins just south of the Cap Vert peninsula in Sénégal and extends northwards as far as the Baie du Lévrier in Mauritanie. Again, his division of the tropical West African coast into three sectors has no meaning phytogeographically since the algal floras of the countries bordering the region appear to form a very homogeneous group. Nevertheless, it must be pointed out that many differences in diversity and abundance are to be found along this coast which probably relate to strictly local environmental circumstances. For instance, the low diversity and stunted growth of algae along the shores of Cameroun (Lawson, 1955), Gambia (John and Lawson, 1977b) and Sierra Leone (John and Lawson, 1977a; Lawson, 1957a), might reflect local conditions of reduced inshore salinity due to heavy precipitation over much of the year and the presence of large river systems. Again Bassindale (1961) has reported that less than 50% of the animals and plants are common to the eastern and western parts of the Ghanaian coast. Recent collecting has revealed that this difference is not nearly as dramatic as he supposed but it does appear to be real. Formerly these differences were believed to be related to changes in the nature and orientation of the shore that occur in the vicinity of Cape Three Points. Recently Ofori-Adu (1975) has found higher inshore water temperatures and lowered salinities along the western part of the coast, so making it possible to divide it into an eastern and western sector based solely on hydrographic considerations.

Fig. 6. Map of West Africa showing the marine phytogeographical regions based on the ordination (see Fig. 5). Key to the numbers: 1, Salvage Islands; 2, Canary Islands; 3, Western Sahara (former Spanish Sahara); 4, Mauritanie; 5, Sénégal; 6, Gambia; 7, Guinea-Bissau (former Portuguese Guinea); 8, Guinée; 9, Sierra Leone; 10, Liberia; 11, Côte d'Ivoire; 12, Ghana; 13, Togo; 14, Benin (former Dahomey); 15, Nigeria; 16, Cameroun; 17, Bioko (former Fernando Póo, Macias Nguema Biyogo); 18, Príncipe; 19, São Tomé; 20, Republic of the Congo (Brazzaville); 23, Cabinda; 24, Zaîre (former Congo Republic); 25, Angola; 26, South West Africa (Namibia); 27, Ascension Island; 28, Saint Helena; 29, Pagalu (former Annobon); 30, Cape Verde Islands (off the map).

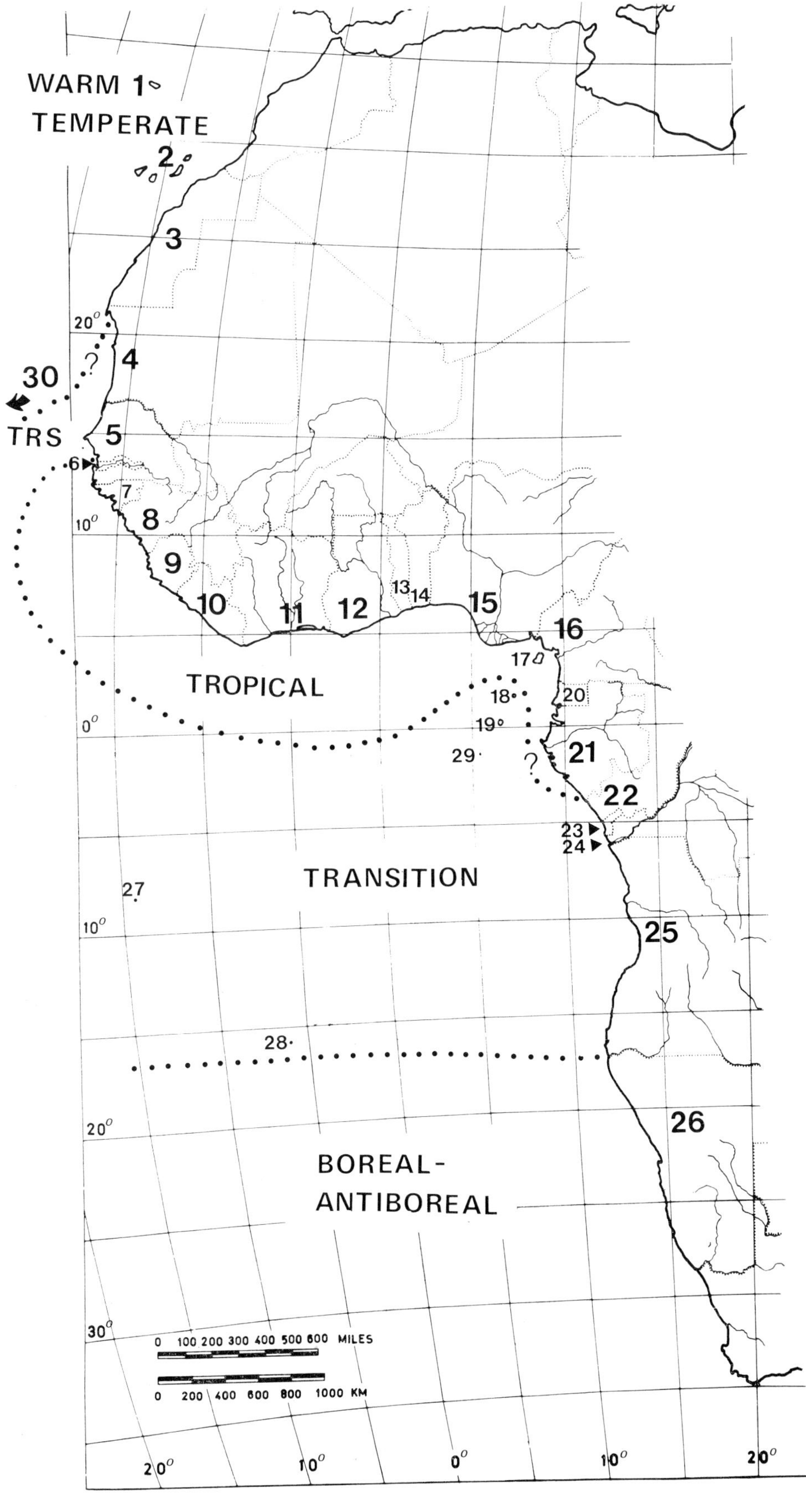

WARM 1
TEMPERATE
2
3
20°
4
30
TRS
5
6
7
8
10°
9
10
11
12
13
14
15
16
17
TROPICAL
18
20
0°
19
21
29
22
23
24
TRANSITION
27
10°
25
28
20°
26
BOREAL-
ANTIBOREAL
30°
0 100 200 300 400 500 600 MILES
0 200 400 600 800 1000 KM
20°
10°
0°
10°
20°

Affinities with Other Regions

In the previous section it has been shown that ordination of the algal floras of the tropical Atlantic produces two distinct groupings, one for the Caribbean region and another for West Africa. Despite this apparently clear separation, however, the two regions have a great many species in common. The tropical western Atlantic has a very rich marine algal flora; Taylor (1960) lists about 760 species although the area he deals with is very wide and his list includes many subtropical species. Subsequent discoveries have added at least another 100 species to the area covered by Taylor's marine flora. By way of contrast the marine algal flora of the eastern tropical Atlantic is a relatively impoverished one. Bearing in mind that tropical West Africa is a much more limited and strictly tropical area than that covered by Taylor's flora for the Americas, the fact that this present work deals with only about 300 species (excluding Cyanophyta) or about two fifths as many species as his region is undoubtedly significant. Of these West African species about two thirds are common to both sides of the Atlantic and some, for example, *Dohrniella antillarum, Halymenia duchassaingii, Sargassum filipendula* and *Waldoia antillanum* are apparently confined to these two regions. Again, the degree of endemism is much greater on the western (about 55%) than the eastern (about 7%) side of the Atlantic Ocean.

Various theories such as the relative inhospitability of the tropical West African coast have been put forward to account for these facts and perhaps one of the more plausible of these is an historical one. It was suggested (see Hoek, 1975) on the basis of paleontological and oxygen isotope evidence that the reduced sea temperatures during the Pleistocene glaciations would have been sufficient to shift the 20°C winter isotherm latitudinally by about 15 to 20° towards the Equator in the eastern Atlantic. This would have been sufficient to wipe out completely the tropical marine flora of West Africa while leaving that of the western Atlantic relatively unaffected. Such a catastrophic occurrence would adequately explain the relative poverty, low endemism, and the similarities with the Caribbean flora of the present tropical West African flora. Thus, recolonization of West Africa would have occurred mainly from the tropical western Atlantic subsequent to the glaciations with relatively little time for new speciation to take place (Hoek, 1975). More recent work, however, has indicated that during the glaciations tropical sea temperatures even in the eastern Atlantic were relatively unaffected (Climap, 1976; Gates, 1976). If this was so then another explanation will have to be sought for the facts given above. For example, the greater diversity of marine habitats in the Caribbean area may be related in part to the presence of fringing coral reefs and associated seagrass-dominated lagoon communities.

Though tropical West Africa has more species in common with the western tropical Atlantic than with any other region, it also shares a great many with both the Indian and Pacific Oceans, 56% and 58% respectively. In other words there is a very large pantropical element. A few West African species are found in one or other of these oceans but not in the western Atlantic. For example, *Halymenia actinophysa* is known only from Ghana and the Pacific coast of America whereas *Corynomorpha prismatica* is known only from the Gulf of Guinea and the Indian Ocean.

MARINE ECOLOGY

Factors Affecting Algal Distribution

Although much of the West African coast consists of sandy beaches, the larger attached algae with which this book is concerned are virtually absent from such unstable areas and are restricted to places where there is rock or other hard stable surfaces for attachment. Where the substratum is suitable they grow between the tide marks extending upwards if there is some spray or splash and downwards to varying depths depending on the clarity of the water. The intertidal or littoral region of the shore is of special interest as the organisms that occupy it have to be able to live for part of the time underwater and for the remainder in air. At lower levels more time is spent underwater than in air, and at upper levels the reverse is true. Thus there is a gradient of conditions under which plants and animals grow. What, in fact, is observed is that the seaweeds occur in more or less definite belts or zones which may be wide or narrow, and overlapping or more or less exclusive. In West Africa, where the range of the tides is not great, the belts are relatively narrow but may be widened vertically by heavy wave action or horizontally by a gentle sloping shore.

It might be thought that the belts are present because the particular algae in them are adapted to live at the levels at which they occur. But experiments both in West Africa (Lawson, 1966) and elsewhere (Lewis, 1964; Lodge, 1948) have shown that when rocks are artificially denuded of all living organisms, the initial recolonization of those bare areas is often by plants normally restricted to other levels. It is only after some considerable time that the original belts are restored and the recolonized area becomes indistinguishable from the rest of the shore.

It is certainly true that different algae vary in their response to environmental factors. When they are exposed to air, for example, they lose water at different rates. In Fig. 7, it can be seen that *Sargassum vulgare* loses water more rapidly than *Bryocladia thyrsigera* which in turn loses water at a faster rate than *Ulva fasciata.* It can be said that *Ulva* is better adapted to resist water loss than *Bryocladia* which itself is more adapted to do so than *Sargassum*, and they in fact occupy zones in these relative positions on the shore. This simple relationship between resistance to desiccation and height on the shore does not always hold, however, and other factors must be involved.

From these two sets of observations, namely recolonization experiments where algae are found outside their normal belts, and experiments that show different species to possess varying resistance to environmental stresses, it can perhaps be concluded that the definition of the belts is largely due to competitive factors. In other words because one species is better adapted to withstand conditions prevailing at one par-

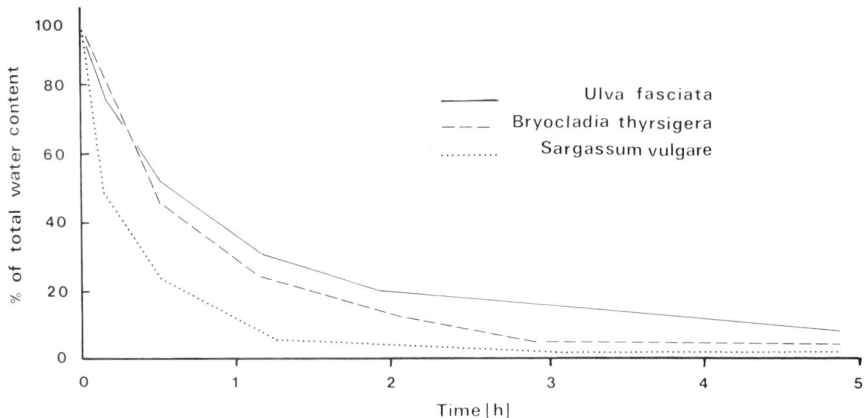

Fig. 7. The desiccation curves for three species of algae when exposed on the shore during the daytime (after Jeník and Lawson, 1967).

ticular level than other species, it is able to compete successfully and exclude them from that level. What probably happens is that spores of many species are constantly being deposited randomly at all levels on the shore but due to direct physical conditions and to competitive factors operating in the germling stages most are killed off except where they are best fitted to survive.

If algal zonation is due to differences in the tolerance of species to environmental factors, and the advantages or disadvantages that ensue, it follows that the pattern of zonation will be altered when the environmental factors themselves vary. Zonation must therefore be regarded not as a static but as a dynamic process in which the actual level a particular organism occupies may be influenced by seasonal changes in climate, tides and oceanographic conditions.

On the Ghanaian coast it has been demonstrated that there is a regular seasonal invasion of higher levels on the shore by many algae during the months of the northern hemisphere summer and a corresponding "retreat" during the northern winter when they become bleached and die (Pl. IIA, B). Analysis of the possible factors responsible for these "migrations" suggest that desiccation operating through the

Plate II: A. An area of shore at Komenda (Ghana) photographed in January showing the mat of algae and the crustose lithothamnia having become bleached white in the upper part of the lower eulittoral subzone. B. A close-up of one of the rocky outcrops at Komenda showing bleached algae including *Gracilaria dentata* and other red algae. C. A log stranded high up on the beach at Sassandra (Côte d'Ivoire) with the lower part covered by a variegated mat of algae characterising the littoral fringe (after John, 1972c). D. A close-up of the mat on the log showing lighter patches of the green alga *Rhizoclonium ambiguum* surrounded by a darker felty covering of the red alga *Lophosiphonia reptabunda* and in some bare areas a few individuals of the small limpet *Siphonaria pectinata* (after John, 1972c).

A
B
D
C

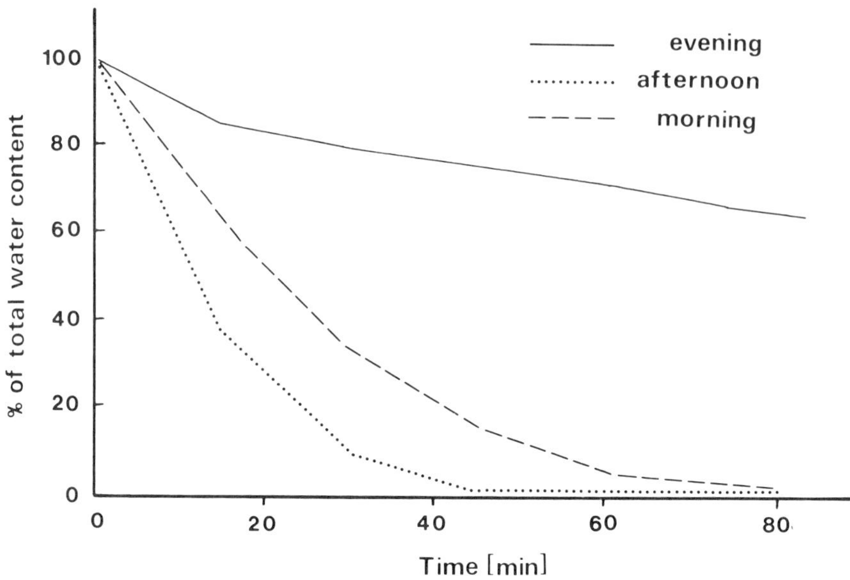

Fig. 8. The desiccation curve for *Gracilaria dentata* when exposed on the shore at different periods during the day and night (unpublished data from S.O. Asare).

agency of the tides is responsible. After the equinox in September the lowest of the two low waters in any period of 24 hours occurs during the daytime, whereas in March the position is reversed and the lowest low water takes place during the night with only the higher low water occurring during the daytime. Fig. 8 shows how much greater is the water loss from *Gracilaria dentata*, for example, during the day than in the evening. It is undoubtedly much more damaging for the lower zone algae to be exposed to air during the day than during the night, and their seasonal changes in level and percentage cover correlate well with this tidal factor (Fig. 9). The fact that the period when lower low waters occur during the day is the hot and dry season in Ghana undoubtedly reinforces this effect. On the other hand in such places as São Tomé and Gabon, where it is wet and humid between October and April, the effect on zonation of lower low waters occurring during the daytime might be expected to be mitigated at least to some extent.

Local variations in the zonation pattern, often caused by the absence or reduced abundance of certain characteristic shore organisms, may not always find a ready explanation. The recent past history of an area of shore may be important since newly exposed or recently denuded substrata may have been colonised only very recently and the community present may represent a stage in a successional sequence. Three phases of recolonization of cleared rock surfaces have been recognised in Ghana. Ignoring micro-organisms, the first phase is a complete cover of the green algae *Ulva* and *Enteromorpha*. Soon small spots of calcareous encrusting red algae appear and after a few weeks completely dominate the previously denuded area.

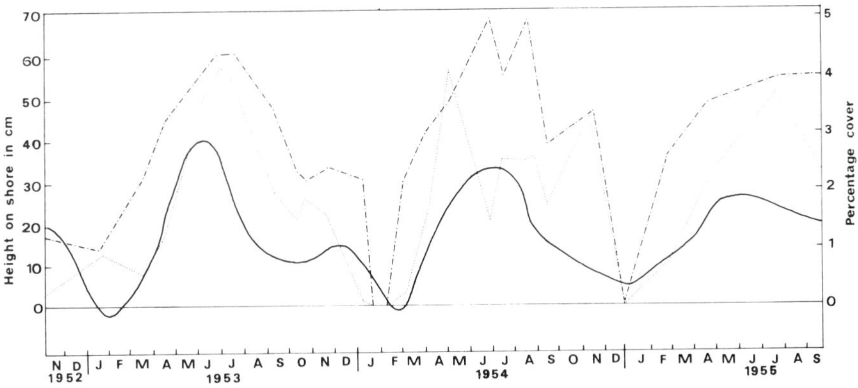

Fig. 9. The diagram shows the relationship between seasonal tidal changes and the seasonal variations in zonation and cover of *Hypnea musciformis.* The continuous line is the tidal curve produced by joining together the heights of the lowest of the lower low waters which occurred in each month over a period of nearly three years but excluding those which took place at night i.e. 06.00 to 18.00 h. The broken line is the upper limit of *Hypnea* and the dotted line the percentage cover of the alga (after Lawson, 1957b).

Later a mixture of algae, including some of the original species present and others not normally occurring in the belt, cover the area. The final phase is a slow one and six or more years may elapse before the area becomes indistinguishable from those surrounding it. A somewhat similar series of events has been observed on plexiglass plates fixed to the sea bed at a depth of about 13 m. The main difference between these artificial plates and cleared rock surfaces in the littoral zone was the absence of an initial cover of green algae; instead, filamentous brown algae were some of the first colonizers together with lithothamnia*. After a year sessile animals became dominant on the plates but the continual death and slouring-off of these organisms resulted in new primary surfaces becoming available for colonization. It is believed that the changes observed on these artificial surfaces are similar to those that take place on the nearby submarine banks when new surfaces become exposed. Sometimes stability may never be attained and the community can be considered then to be a pre-climax held in check by perhaps periodic sand-scouring or sand-burial.

Many algae found on old harbour breakwaters along the Gulf of Guinea coast are absent from those recently built in Benin and Togo and from nearby rocky platforms only exposed following the interference of these artificial structures with the west-east longshore drift of sand (see Pl. I). It is probable that the algal flora has not had time to reach stability and the wide variation in the zonation patterns observed on these new surfaces supports this view. It appears that the available space at the appropriate level is often occupied by opportunist species rather than the more usual species which may take longer to become established. There are, moreover, no nearby rock outcrops to provide a reservoir of spore-parents for colonization. With the exception of the barnacle subzone in Benin it appears that colonization by the

*This term with a small "L" and unitalicized denotes encrusting calcareous red algae.

characteristic zone-forming organisms is much more rapid for animals than for plants.

Below the level of the lowest spring tides conditions for life become much more uniform since the organisms are never exposed to air. As a result of this there is usually no sharp and obvious zonation of algae subtidally and the changes that occur with increased depth are nearly always gradual ones. Many of the shallow water algae simply become smaller, less abundant and eventually disappear altogether in deeper water. This effect is related to changes that take place in the quality and quantity of light associated with increase in depth. Nevertheless, the restriction to deeper water of certain algal species found off the coast of Ghana may be due either to intolerance to water movement or to an inability to compete with the abundant shallow-water plants. Some of those species that were found to occur over a wide depth range were larger and better developed in deeper water. This might reflect the longer growing season in deeper water, possibly related to reduced turbidity found further offshore and to lessened water movement associated with depth.

The general distribution of subtidal algae is not only governed by light attenuation but also by the nature of the available substrata. Sandy, muddy, shell or cobble bottoms afford unstable surfaces for the long-term attachment of benthic algae in habitats subject to considerable water movement. The absence of algae or phanerogams anchored in sand or mud is a reflection of the instability of such bottoms under West African conditions, particularly in shallow water. Again, the powerful surge effect produced on the bottom during the heavy seas of the rainy season leads to agitation and rolling of the cobbles strewn over the sandy seabed. This agitation results in abrasion of the cobble surfaces causing the removal of most algae other than crustose growth forms. Such action probably accounts for the great number of plants washed ashore at the start of the rainy season or after storms during the major dry season. Some of the plants found growing on the cobbles during the relatively stable months of the dry season may have survived the unfavourable season in situ as spores, germlings or juveniles, or else arose from persistent holdfasts or stoloniferous bases. Many algae must survive the period of unstable weather conditions on the rocky banks, and these may act as spore parents for those that subsequently develop on the cobbles with the return of stable conditions.

Rocky platforms are also affected by movement of sand and may be partly buried and re-exposed repeatedly during the period of increased water movement. One of the few plants adapted to growing on rocks intermittently covered to a variable depth by shifting sand is the red alga *Dictyurus fenestratus;* this has a creeping stoloniferous base giving rise to erect stipes bearing net-like branches. Another alga commonly found growing on such sand-covered rocks is *Sargassum filipendula,* but this plant can survive and regenerate only if the rock surfaces are not permanently buried. At depths of below 30 m off the Ghanaian coast, where the sandy or muddy bottom is more stable than at shallower depths, algae are found growing on the long conical shells of *Turritella.* These shells almost invariably contain hermit crabs whose presence might ensure that they do not become buried but remain exposed as a surface for algal attachment.

The low turbidity, reduced turbulence and minimal cloud cover during the major dry season suggests that this is the most favourable time of the year for the growth and development of subtidal algae. This favourable period varies off Ghana from less than four to just over seven months but may be considerably shorter in other parts of the region, for example, off Cameroun. In contrast, as has been mentioned earlier, littoral algae occur in reduced abundance and occupy lower levels on the shore during the dry season of the year in Ghana. Thus, the growing seasons for littoral and sublittoral algae appear to be out of phase. This might permit the development, on the return of favourable growing conditions to the disturbed habitat, of spores from plants growing in the favourable habitat. In fact, about two-thirds of the sublittoral algal species collected off Tema, Ghana, also grow in the littoral zone.

There is a conspicuous absence of larger algae along wave-sheltered parts of boulder beaches, rough stone breakwaters, or where rocky platforms are much gullied and broken-up into ledges and holes. Large populations of algal-feeding reef fishes are a feature common to all such areas. It has been shown experimentally (John and Pople, 1973) that the dominance of lithothamnia, articulated coralline algae, and low-growing cushion or mat-forming species along parts of the breakwater system of Tema Harbour in Ghana is the result of grazing and browsing by fish. The absence of herbivorous fish elsewhere, and hence the usual development of a large and diversified algal vegetation, appears to be due to the absence of apertures and ledges which afford refuge for the herbivorous fish from larger predator fish.

In many wave-exposed situations the rocks in the sublittoral fringe and below may be covered by a dense population of the sea urchin *Echinometra lucunter* Linnaeus whilst lithothamnia are the only conspicuous algae present. The rôle played by sea urchins appear to be similar to that of the fish, as demonstrated by the development of a dense algal turf following their experimental removal from a large lower eulittoral zone tidepool at Matrakni Point in the western region of Ghana.

The only study on plant/animal interactions on open littoral rocks in the Gulf of Guinea has involved the experimental removal of the small limpet *Siphonaria pectinata* Linnaeus (Gauld and Buchanan, 1959). In areas of rock cleared of this limpet there was a rapid and prolific growth of blue-green algae suggesting that grazing by this animal may be an important factor governing the distribution of such algae on the shore. Therefore grazing pressure by fish and various invertebrates may be of prime importance in determining the abundance and composition of algal vegetation along many parts of the tropical West African coast.

The earlier studies on recolonization of artificially denuded rock surfaces in Ghana suggest that, as in other parts of the world, several years can elapse before a stable climax vegetation develops. It seems that along the coast of tropical West Africa instability of sand and cobble bottoms, as well as the local influence of fish and invertebrate grazing, may persistently interfere with the development of true climax vegetation.

Zonation on Rocky Shores

Introduction

Because the restriction of certain groups of plants and animals to definite belts or zones is a universal phenomenon of rocky shores the pattern of littoral zonation forms a convenient basis on which to describe and compare them. The scheme advanced by T.A. and A. Stephenson (1949) for describing rocky shore zonation was the first which aimed at world-wide applicability and has in fact been used to describe rocky shores in many parts of the world (see Stephenson and Stephenson, 1972). Their scheme is based on the fact that certain types of organism characterise approximately the same level on rocky shores throughout the world, though the actual species differ widely. They maintain that zones should be defined in terms of the organisms composing them rather than by physical characteristics such as average or extreme tide levels. The terminology of this scheme has been modified and improved by Lewis (1961, 1964). The system for describing shore zonation by the French biologists Molinier and Picard (1953) and Pérès and Picard (1956) is based on similar criteria to those outlined below.

According to the scheme proposed by Lewis a rocky shore can be divided into a littoral zone and an upper sublittoral zone or fringe. The littoral zone may be considered to be that part of the shore where the organisms are adapted to, or have a requirement for, alternate emersion from and immersion in seawater, or wave splash and spray. The upper limit of the littoral zone is marked by the disappearance of marine organisms. The littoral zone is divided into a littoral fringe, the upper parts of which may be beyond the tidal range and only influenced by seawater spray and splash, and a eulittoral zone in which there is a more or less regular daily tidal immersion and emersion. The littoral fringe on the Gulf of Guinea coast extends upwards from the upper limit of barnacles in quantity to the disappearance of small snails. The eulittoral may itself be subdivided into an upper barnacle subzone and a lower subzone dominated by lithothamnia. The upper part of the sublittoral zone is uncovered only at extreme low waters of spring tides (ELWS). In some places a sublittoral fringe may be recognised if there is sufficient biological justification i.e. if it contains characteristic organisms not found either in the littoral zone or in the sublittoral proper. The junction between the littoral and the sublittoral zone is usually well-defined in colder waters, where it is taken as the upper limit of the large brown laminarian seaweeds or kelps, whereas in the tropical parts of West Africa it is rather ill-defined but may be indicated by the upper limit of *Sargassum, Dictyopteris* or sea urchins.

Tropical West African Coast

Most ecological accounts of rocky shore zonation in the region have tried, as far as possible, to follow one or other of the schemes mentioned. The only publication using the scheme of Molinier and Picard (1953) and Pérès and Picard (1956), based on experience in the Mediterranean, is that dealing with the distribution of marine organisms on two of the offshore islands (Carpine, 1959). In this book we have adopted the more widely used scheme of Stephenson and Stephenson, modified by Lewis, as a framework for summarising the patterns of littoral zonation of attached organisms growing along the West African coast and adjacent islands.

The main features of rocky shore zonation in Ghana have been intensively studied both for plants and animals by Lawson (1953, 1956, 1957b) and John and Pople (1973), while Bassindale (1961) and Gauld and Buchanan (1959) have confined their observations largely to the animals. Accounts exist for Cameroun (Lawson, 1955; Pilger, 1911), Côte d'Ivoire (John, 1972c), Liberia (De May *et al.*, 1977; 1972c), Gabon (John and Lawson, 1974a), Gambia (John and Lawson, 1977b), Nigeria (Steentoft Nielsen, 1958), Togo and Benin (John and Lawson, 1972b), Sierra Leone (Lawson, 1957a), and the offshore islands of Príncipe and São Tomé (Carpine, 1959). Thus the littoral ecology of most of the countries bordering the Gulf of Guinea region of West Africa has received some attention.

Workers mainly interested in animals, such as Bassindale (1961) and Gauld and Buchanan (1959), have usually distinguished between rocky shores and those that consist largely of boulders. Where the boulders are unstable the littoral organisms are poorly represented, but even when relatively stable the algal diversity is often low and much of the fauna is confined to the undersurface of the rocks. In fact, it is often difficult to interpret the zonation pattern on boulder surfaces and on the four-armed concrete blocks which are sometimes used in the construction of harbour breakwaters. Most of the remarks on littoral zonation that follow refer to the patterns generally found on more or less continuous rock surfaces.

Undoubtedly the most important factor modifying the pattern of zonation along the rocky parts of tropical West Africa is the amount of wave action. This is a very complex factor varying with the slope and aspect of the rock surfaces. Increasing wave action modifies the belts in two ways: firstly, increasing their vertical extent, and secondly altering the species composition. A diagrammatic representation of zonation of certain dominant shore organisms in relation to wave-exposure on open rocky shores in Ghana is given in Fig. 10. It shows that with progressive increase in wave action the belts become wider and extend upwards and also that certain organisms are found only over a limited range of exposure. With some variations this general pattern is repeated along many of the rocky parts of the coast. The zonation encountered under conditions of moderate to strong wave action must be taken as the most usual type in the region as only few places may be regarded as being truly sheltered.

Littoral fringe:

This is the uppermost zone on the shore under the influence of the sea, and is probably the one showing the most constant composition despite wide variations in the amount of wave action which might be found at different localities. The most characteristic organisms at this level on the shore are small snails (littorines) such as

Littorina punctata Gmelin and *Nodilittorina miliaris* Quoy & Gaimard.When these two snails occur together the latter usually occupies the somewhat higher levels. *Littorina punctata* is uncommon in the offshore islands and in Cameroun and Gabon; in these countries *Nodilittorina* is often the dominant snail. *Littorina cingulifera* Dunker may sometimes be present but in relatively low numbers. The zone is comparatively barren of algae except for cushion-like forms or small mats found in cracks, crevices, beneath rock overhangs or on open rocks shaded by trees; more extensive mats sometimes occur on the lower side of stranded logs (Pl. IIC). The most common plants present are often somewhat bleached specimens of red algae such as *Bostrychia* spp. or *Lophosiphonia reptabunda;* occasionally entangled with these are the green strands of *Rhizoclonium,* and epiphytic patches of filamentous blue-green algae may also occur (Pl. IID). In such situations the small isopod *Ligia gracilipes* Budde-Lund may be in some abundance. The rock surfaces may be covered in places by a blackish powdery penetrating layer composed of the blue-green alga *Entophysalis deusta.*

Eulittoral zone:

This zone may be conveniently divided into two subzones: an upper one, which becomes more or less completely dried-out during low water, usually characterized by the barnacle *Chthamalus dentatus* Krauss, and a lower one often remaining moist even at low water and usually dominated by lithothamnia. The gastropod *Siphonaria pectinata* may on occasion replace the barnacles as the dominant organism in very wave-sheltered situations such as on the leeward side of harbour breakwaters. It is sometimes under these circumstances associated with the oyster *Ostrea cucullata* Born. One feature common to many localities along the coast of tropical West Africa is the presence of an encrusting brown algal belt in what might be considered as the lower part of the upper eulittoral subzone where barnacles are still present but in low numbers. This brown algal belt is best developed when barnacles are in low abundance throughout the subzone as is found on the hard granite boulders in the western region of Ghana. The main encrusting brown alga is *Basispora africana* (Pl. IIID) and this is sometimes accompanied by another member of the Ralfsiaceae of very similar appearance, namely *Ralfsia expansa.* Algae are not generally well represented in this upper and drier subzone although occasionally extensive patches of blue-green algae may be present. Other algae found in this eulittoral subzone include green clumps of *Enteromorpha* and especially *Ulva* together with brown tufts of *Ectocarpus, Giffordia* or *Sphacelaria,* and often bleached tufts of the red alga *Centroceras clavulatum.* The gastropod *Nerita atrata* Gmelin is found on more wave-sheltered rocks at this level and in the subzone below. A belt of oysters may occasionally occur on nearly horizontal rocks in this upper subzone and in the one below but cannot be regarded as a characteristic feature of open rocks in the region since in wave-exposed situations they are largely confined to shallow tidepools. On sand-scoured beach rocks the green algae are often dominant, although on gently sloping surfaces *Bachelotia antillarum* may form extensive brown mats if wave action is not very great.

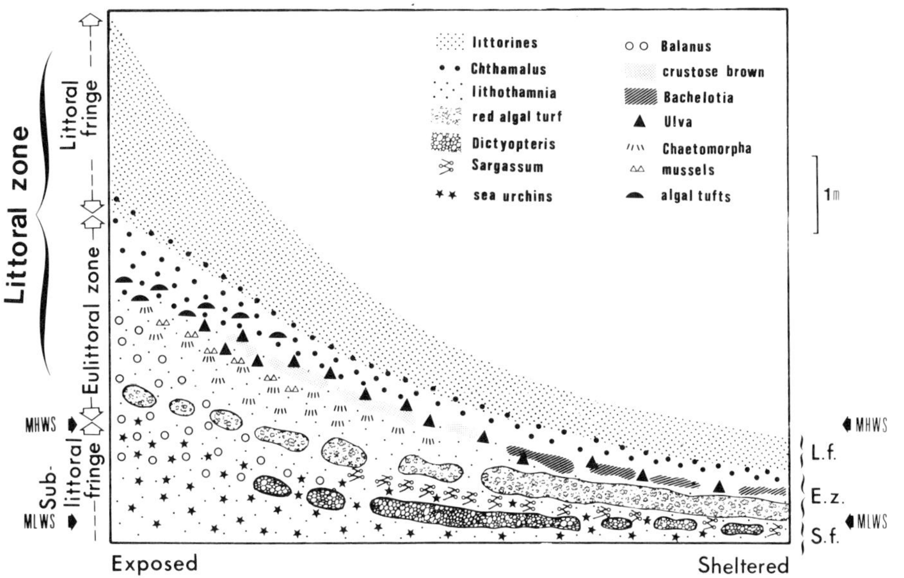

Fig. 10. A diagrammatic representation of the zonation of the dominant littoral organisms on rocky shores in West Africa in relation to exposure to wave action.

The lower subzone, in contrast to the one above, hardly ever dries out completely and even at low water is often kept moist by wave splash or spray and perhaps, in some cases, by the water holding capacity of the algal turf. This subzone is usually characterised by lithothamnia; no other algae seem so constant to it as are some of the animals. On wave-exposed rocks the mussel *Perna perna* Linnaeus occurs together with limpets such as *Fissurella nubecula* Linnaeus and *Patella safiana* Lamark, and the carnivorous snails *Thais nodosa* Linnaeus and *T. haemostoma* Linnaeus. Where there is considerable wave action the large barnacle *Balanus tintinnabulum* Linnaeus is often abundant and the shore crab *Grapsus grapsus* Linnaeus appears in large numbers though often ranging over the whole of the littoral zone. The small mussel *Brachydontes puniceus* Gmelin is present in a few wave-sheltered places especially where there is some brackish-water influence, whilst in such situations a continuous crust of worm tubes covered by lithothamnia may also be found. On smooth, steeply sloping and moderately wave-sheltered rocks large areas are often covered by the slimy greyish to brownish-green zonanthid *Palythoa* (Pl. IIID). The great majority of the shore algae are confined to this lower subzone, although in heavily wave-exposed situations only crustose growth forms may be found (Pl. IIIA). More usually, however, the crusts are accompanied or overgrown by algae which, although they may occur as large erect forms elsewhere, are here in the form of low cushion-like tufts, ring-like patches, or as closely adherent felted mats. Most conspicuous in such places are stiff green tufts of *Chaetomorpha antennina*, whilst also common and usually growing directly on the lithothamnia are species of *Centroceras*, *Gelidium*, *Herposiphonia*, *Laurencia* and *Taenioma*. With increasing shelter many of these algae still persist and the patches become thicker and denser, eventually coalescing to form a variegated algal turf or mat so that the lithothamnia appear much less conspicuous (Pl. IIIB, C). Other algae which now become abundant include species of *Bryopsis, Caulerpa, Colpomenia, Gracilaria, Hypnea, Padina,* and articulated corallines such as *Amphiroa, Corallina* and *Jania* (Pl. IVA,B). These latter three plants often form a carpet on gently sloping rocks at the outflows of tidepools or where there is a certain amount of surge action. On beach rocks at about this level, particularly where there is some sand present, there are frequently found loose carpets of *Bryocladia thyrsigera* occasionally accompanied by finely divided plants of *Grateloupia filicina* or *Gracilaria verrucosa.*

Sublittoral fringe:

This upper part of the sublittoral zone has defied precise definition in West Africa because it lacks a reliable indicator. The brown algae *Dictyopteris delicatula* and *Sargassum vulgare* have in the past been regarded as defining the upper limit of the fringe along the Gulf of Guinea coast but they are not always a constant feature (Pl. IV C). *Sargassum* is usually absent from moderately wave-exposed rocks and *Dictyopteris* disappears when wave action is severe (Pl. IIIA) whilst neither is found, even in wave-sheltered situations, when there are large populations of herbivorous fish. Sea urchins are present on most shores in the region, being particularly abundant on open wave-exposed rocks, and therefore are perhaps the best indicators of the upper limit of the sublittoral fringe (see Pl. IIIA, Pl. IVC). The most common sea urchins present are *Echinometra lucunter* and, on sheltered harbour breakwaters, the long-spined *Diadema antillarum* Philippi; a less conspicuous urchin *Arbacia lixula* Troschel is sometimes also to be found. In the upper eulittoral zone sea urchins occur only in tidepools (Pl. IVD) or in small circular holes which retain water during low tide.

Plate III: A. A very wave-exposed cliff face at Busua Beach (Ghana) showing the little-developed barnacle subzone, followed by the clearly demarcated lithothamnia subzone devoid of any erect algae, and finally the sea urchin *Echinometra lucunter* characterising the sublittoral fringe. B. A sloping and moderately wave-sheltered rocky shore at Cape Shilling (Sierra Leone) showing greyish patches of barnacles characterising the upper eulittoral subzone and the lithothamnia of the lower eulittoral subzone almost completely covered by a carpet of erect algae. C. A moderately wave-sheltered and steeply sloping rocky surface at Sassandra (Côte d'Ivoire) having a dense carpet largely of articulated coralline algae covering the lower eulittoral subzone. D. A moderately wave-sheltered cliff face at Matrakni Point (Ghana) having a greyish coloured species of *Palythoa* almost completely covering the lower eulittoral subzone, above this zoanthid follows blackish crusts of *Basispora africana,* and finally in the littoral fringe occur small snails largely confined to fissures and crevices.

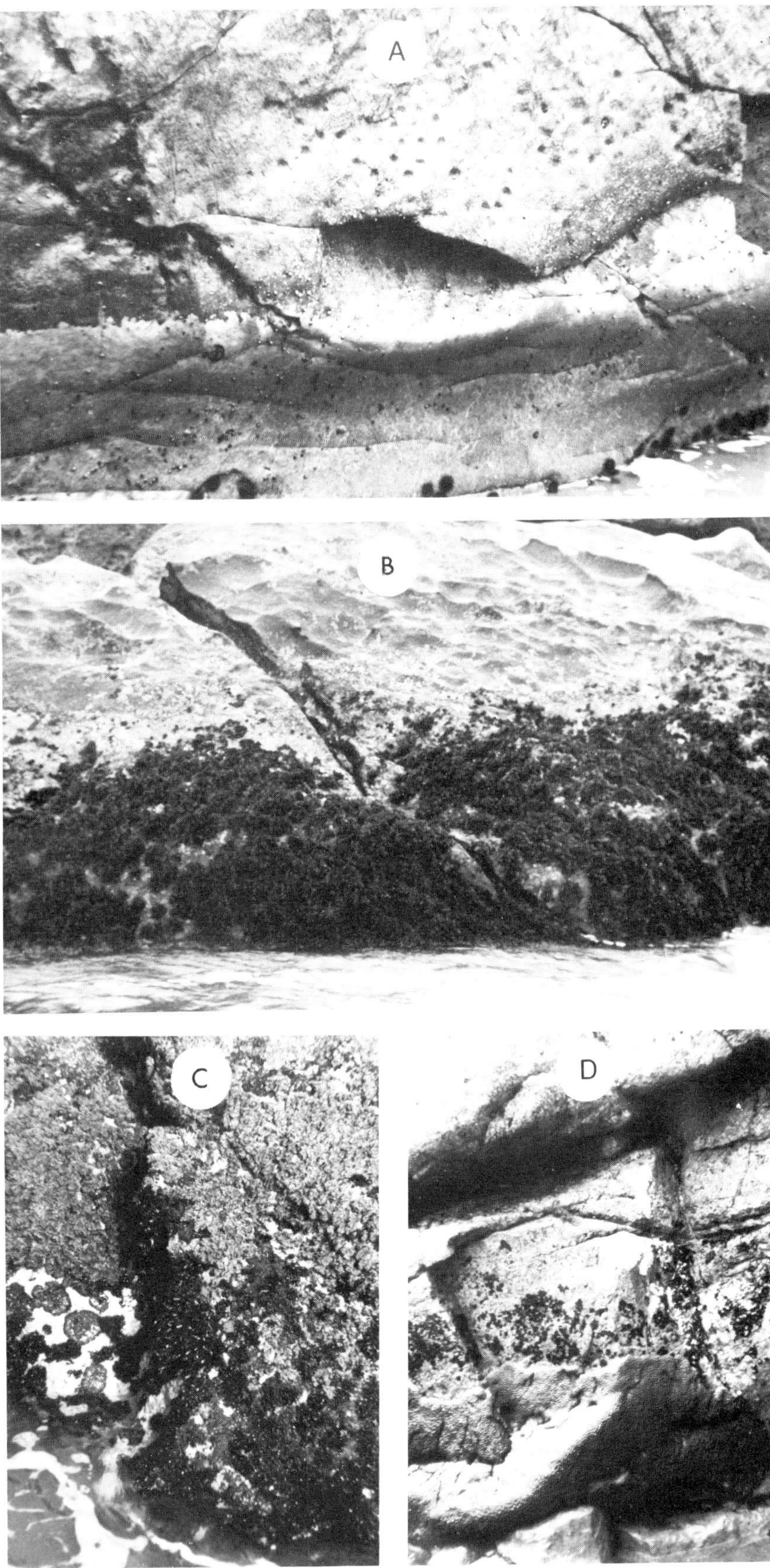
A
B
C
D

Sublittoral Habitats

Inshore rocks

Where wave action is severe, the larger softer algae including *Sargassum vulgare* and *Dictyopteris delicatula* are absent from the sublittoral fringe and shallow sublittoral. In such wave-exposed situations the dominant organisms are lithothamnia and the black sea urchin *Echinometra lucunter*. When *Sargassum* and *Dictyopteris* do occur, they often extend less than half a metre below the low water of spring tides only to give way to lithothamnia and sea urchins. If *Sargassum vulgare* extends to greater depths, as in many wave-sheltered habitats, it is then gradually replaced by *S. filipendula*. This takes place, for example, in the immediate sublittoral near the entrance to the Gabon River and at a much greater depth off the Ghana coast. A well-developed and often diversified community of epilithic and epiphytic algae occurs in some wave-sheltered bays or inlets where may be found dense beds of *Sargassum*. Nevertheless, in such sheltered habitats where the rocky substratum is much divided up into holes, ledges and gullies, most of the larger algae are absent and the common forms are crustose, low cushion-like or mat-forming species, articulated corallines, and sometimes clumps of *Hypnea cervicornis*. Such is the situation found in Miemia Bay in the western region of Ghana where there are large populations of fish known to be algal-feeders. Somewhat similar is the structure and composition of the plant and animal communities occurring along the moderately wave-sheltered parts of many of the rough stone breakwaters in the region.

Offshore rock, cobble and shell bottoms

The absence of coral reefs along the West African coast has been attributed to the upwelling of colder water at certain times of the year, to high turbidity, and to the lack of suitable bedrock in shallow water (Buchanan, 1954; Ekman, 1953). Nevertheless, as mentioned previously, banks of lithified beach sediment occur offshore, particularly between Côte d'Ivoire and Nigeria. Strewn over sandy bottoms may also be found roughly hemispherical cobbles formed by the accretion of encrusting calcareous red algae. These cobbles range in size from a few centimetres to one or two decimetres and are irregularly pitted so affording a good substratum for benthic algae. Cobble bottoms are reported off a number of West African countries having

Plate IV: A. A close-up of a carpet of algae at Komenda (Ghana) showing various algae including the fan-shaped fronds of *Padina* (lower half), the lanceolate erect fronds of *Caulerpa taxifolia* (top left), a clump of *Gracilaria dentata* (top centre), and the foliaceous fronds of a species of *Dictyota* fringing the tidepool (bottom). B. A close-up of a carpet of algae at Komenda including a much branched clump of *Hypnea musciformis* and the lighter coloured branches of the articulated coralline alga *Amphiroa*. C. A close-up of the sublittoral fringe at Sassandra (Côte d'Ivoire) showing clumps of *Sargassum*, a variegated mat of *Dictyopteris delicatula*, and the sea urchin *Echinometra lucunter* usually occupying areas devoid of soft and erect algae. D. A large tidepool on Matrakni Point (Ghana) containing large numbers of the sea urchin *Echinometra* and devoid of most algae other than crustose forms.

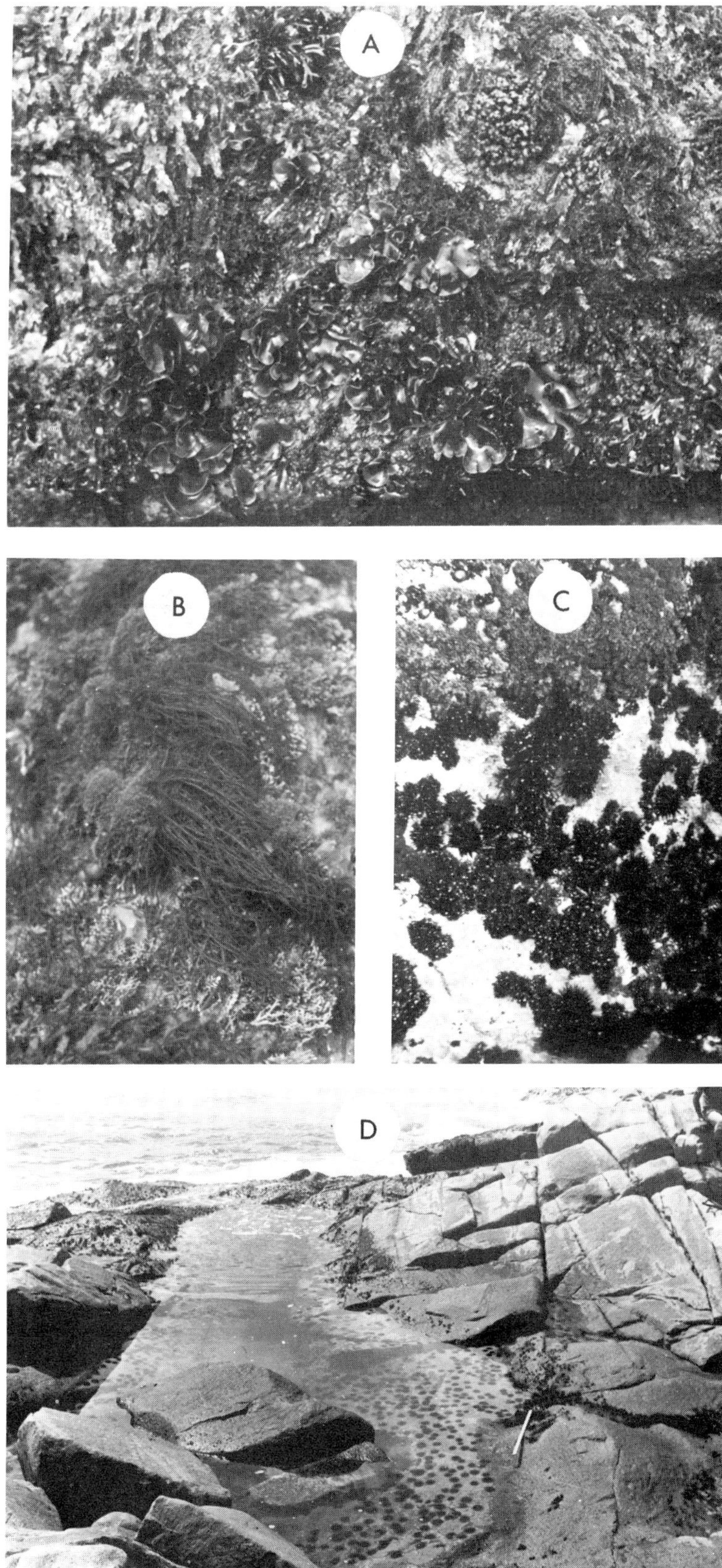

A
B
C
D

a sea board, as well as from the offshore islands of Príncipe and São Tomé. Another suitable surface for the attachment of deeper water algae is formed by the shells of the gastropod *Turritella.* These shells are fairly abundant off the coast of Ghana particularly where there are areas of grey mud at depths in excess of 30 m. The remarks that follow refer to the ecology of subtidal algae found off the Ghanaian coast, which is the only part of tropical West Africa that has been the subject of a detailed sublittoral survey (John *et al.*, 1977). With local variations, the general features of submarine communities may well be similar in other parts of the region.

One of the most common algae on low and undulating parts of the offshore rocky banks lying adjacent to sandy bottoms at a depth of about 8 to 14 m is *Dictyurus fenestratus*, a beautiful red net-like alga which forms patches sometimes covering several square metres. It always grows attached to rock although its extensive stoloniferous attaching organ may be covered to a variable depth by sand. *Sargassum filipendula*, sometimes found in the same general area, is probably the largest subtidal alga in the region often reaching a metre or more in height. It appears to be rather easily detached as large numbers of plants may frequently be found floating just off the bottom in the vicinity of attached individuals of the same species. Other algae commonly growing together with these two species include *Dictyopteris delicatula, Gelidiopsis variabilis, Corynomorpha prismatica, Thamnoclonium claviferum*, and species of the genus *Galaxaura.*

Where the rocky banks are raised to form low cliffs, gullies and ledges the larger algae are much less common. In such areas lithothamnia are often common with the softer algae now forming an inconspicuous mat less than a cm in height, only on occasion are there to be found larger clumps of *Dictyopteris delicatula, Hypnea cervicornis, Laurencia majuscula, Thamnoclonium claviferum*, and purplish-red patches of filamentous blue-green algae. Large gorgonians or soft corals sometimes reaching over a metre in height are common particularly below 20 m. Demersal algal-feeding fishes are found in very large concentrations where these shallow (8 - 30 m) rocky banks are much broken-up (Pl.VA). They include pomacentrids such as *Abudefduf saxatilis* Linnaeus (sergeant major fish) and the territorial *Stegastes imbricata* Jenyns, as well as *Pseudoscarus hoefleri* Steindachner (parrot fish) and *Acanthurus monroviae* Steindachner (surgeon fish). The long-spined and voracious black sea urchin *Diadema antillarum* Philippi is commonly found during the day under rock ledges and in holes along such parts of the rocky banks (Pl. VA).

Plate V: A. The broken-up edge of a rocky platform at a depth of 12 m off the coast of Ghana showing the long-spined sea urchin *Diadema antillarum* and the mostly algal-feeding reef fish commonly known as the sergeant major fish (*Abudefduf saxatilis*). B. Cobbles lying over the sandy seabed close to the rocky platform and photographed in the dry season when they are covered by a diversified carpet of soft and erect algae. C. Two cobbles showing soft and erect algae growing on their pitted surface, the filamentous alga *Spyridia* and the membranaceous alga *Hymenena*. D. Two cobbles showing one of the most common alga growing on them, the membranaceous alga *Halymenia actinophysa*.

A

B

C

D

The algal vegetation that develops on the calcareous cobbles is very uniform in appearance, with many species present but no one species dominant (Pl.VB-D). There is an overall decrease with depth in the number of algal species growing on the cobbles though many of those that occur over a wide depth range are generally larger and better developed below about 18 m. This is particularly true in the case of *Botryocladia guineensis. Caulerpa taxifolia, Champia vieillardii, Halymenia actinophysa, Rhodophyllis gracilarioides* and *Waldoia antillanum*, which are some of the more common plants in these rather featureless cobble areas.

A few algae, including species of *Codium, Halymenia* and *Spyridia*, have been dredged from depths as great as 40 m. These plants were probably growing on small rocky outcrops or on cobbles. Other algae from such depths include *Gracilaria foliifera* and a foliaceous red alga (aff. *Hymenena* sp.) dredged from areas of unstable bottom where they are always found growing attached to *Turritella* shells.

Other Marine Habitats

Tidepools

Irregularities on rocky shores often result in the formation of littoral tidepools of varying size and topography. The abundance of plants in these pools is often determined by the nature of the bottom and when the floor is permanently lined with sand the algae may be confined to the steep sides. Larger and deeper pools, particularly those having a restricted outlet to the sea at low water, have conditions similar to the open sea and so many of the organisms growing in them are also found in the shallow sublittoral. The most common plants in such pools include *Sargassum vulgare, Dictyopteris delicatula* and species of *Dictyota, Galaxaura,* and *Padina*. In some wave-sheltered localities, however, where the pools are lined with boulders, they may be filled by dense but species-poor mats of algae even though such pools occur at low levels on the shore and maintain a connection with the sea at low water. Blue-green algae are common in these pools sometimes growing over large bushy plants of *Hypnea cervicornis* and clumps of *Dictyota* spp. A notable feature of such pools is the presence of large populations of algal-feeding fish including especially pomacentrids such as *Abudefduf hamyi* Rochebrune and *Microspathodon frontatus* Emery.

Tidepools in rocky areas away from beaches, and therefore relatively free from sand, frequently have a lining of lithothamnia accompanied by such encrusting brown algae as *Lobophora variegata* and *Ralfsia expansa*. The lithothamnia crust may be elevated at the sides of the pools as branched and anastomosing structures which form small shelf-like platforms level with the water surface; these are often full of holes due to the activities of burrowing animals. Much of the bottom of some pools may be covered by green or greyish-brown zoanthids such as species of *Palythoa*. In some very wave-exposed situations, such as the tip of Matrakni Point in western Ghana (Pl. IVD), the eulittoral pools are almost devoid of algae other than lithothamnia but contain very dense populations of the sea urchin *Echinometra lucunter*, often over fifty to the square metre.

Large changes in temperature, salinity, oxygen tension and hydrogen-ion concentration may take place during exposure to air in small pools at high levels on the shore, and may govern the types of organisms growing in them. The diversity of algae in such pools is often very low; for example, very shallow pools in the upper eulittoral subzone may contain only mats of *Bachelotia antillarum* often associated with a good deal of sand. Purely rocky pools at this same level may be lined with lithothamnia together with plants of *Ulva* or *Enteromorpha*. A more diverse algal flora may be found in small pools at this level if they are shaded by rock overhangs or trees. The fauna of these pools is also sparse, consisting often of juvenile *Littorina* and *Nerita* with hermit crabs, or brownish colonies of the reef coral *Siderastrea radians* Pallas where there is shelter.

The tidepools in the littoral fringe, unlike those of the eulittoral, are not replenished with fresh seawater by the rise and fall of the tides, and depend for their existence on salt spray, wave-splash and rainwater. Partly shaded pools at this level studied at Cape Palmas in Liberia contained bushy *Cladophora* plants heavily epiphytised by pennate diatoms, whereas the more usual unshaded pools generally had a carpet of blue-green algae growing over a layer of sediment. Such lining carpets of blue-green algae may be best developed in pools contaminated by bird droppings and other organic matter, and the water of such pools may be quite green in colour due to large populations of planktonic algae. Pools at higher levels in the littoral fringe are frequently filled by dense floating scums of blue-green algae most commonly consisting of *Microcoleus lyngbyaceus* and *Anacystis* spp.

Artificial harbours

Increased commercial activity in West Africa in recent years has led to the construction of many new harbours consisting of rough stone or concrete block breakwater systems. As might be expected, the distribution of littoral and sublittoral organisms on such structures is often similar to that found on natural rocky areas. Nevertheless, there are significant differences between sheltered or moderately wave-sheltered parts of such breakwaters, and similarly exposed areas of natural rock (Fig. 11). Many of the larger and more characteristic algae of the littoral zone and shallow sublittoral are entirely absent or else in low abundance along the moderately wave-sheltered parts of rough stone breakwaters in Ghana, Côte d'Ivoire, Benin and Togo. The dominant algae present include lithothamnia, cushion-like or mat-forming species, articulated corallines, and occasionally bushy clumps of *Hypnea cervicornis*. In such situations there is always an abundance of algal-feeding reef fishes such as pomacentrids, parrot fish, and surgeon fish that probably find protection against attack from larger predators in the many holes and ledges that the breakwaters provide. There are similar large populations of such reef fishes, together with a low abundance of primary producers, in sheltered bays and inlets having boulder or broken-up rocky shores and bottoms.

The upper levels of the shore in the very wave-sheltered inner parts of the harbours differ from most similar areas of natural rock in the scarcity both of littorinids and gastropods. In such situations a mat of blue-green algae, often principally consisting of *Microcoleus lyngbyaceus*, usually grows over the mixture of oil and sediment which covers most hard surfaces. Larger algae are absent and there are no organisms which can be used to define the sublittoral fringe. Thus, pollution and heavy sedimentation appear to be important factors adversely affecting the development of benthic organisms inside harbours, as well as in sheltered inlets and bays used as natural harbours.

Lagoons and Estuaries

A rather special type of environment for benthic algae is that found in the many lagoons and estuaries occurring along the West African coast. These habitats differ from the open shore in that salinity is much more variable, wave action is minimal, and mud brought down by rivers colours the water and often gives a muddy substratum. Such conditions are unfavourable for the majority of marine algae and are no doubt responsible for the absence of most open shore algae from this type of environment. At first sight it might appear that algae are completely absent from such areas, but closer examination shows them to be present although very stunted and inconspicuous. Some algae such as *Catenella impudica* and *Caloglossa* spp. are largely confined to brackish-water habitats and appear well adapted to them.

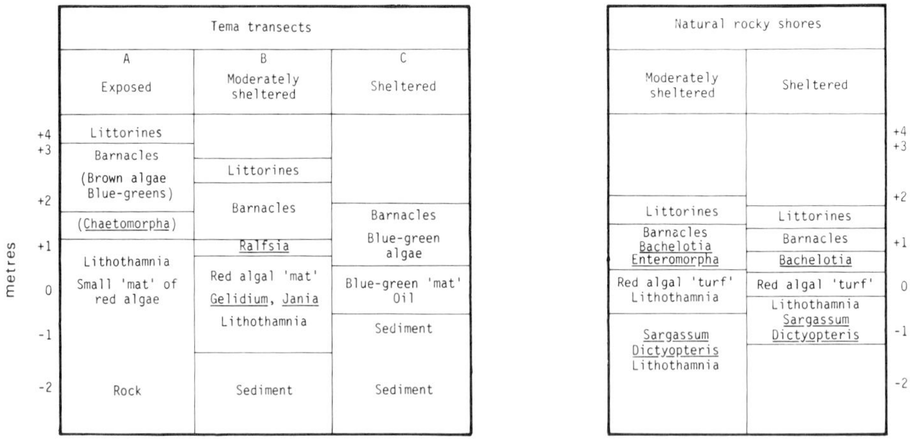

metres	Tema transects: A Exposed	Tema transects: B Moderately sheltered	Tema transects: C Sheltered
+4	Littorines		
+3	Barnacles	Littorines	
+2	(Brown algae Blue-greens)	Barnacles	Barnacles
	(Chaetomorpha)		Blue-green algae
+1	Lithothamnia	Ralfsia	
0	Small 'mat' of red algae	Red algal 'mat' Gelidium, Jania	Blue-green 'mat' Oil
-1		Lithothamnia	Sediment
-2	Rock	Sediment	Sediment

Natural rocky shores: Moderately sheltered	Natural rocky shores: Sheltered	metres
		+4
		+3
Littorines	Littorines	+2
Barnacles Bachelotia Enteromorpha	Barnacles	+1
	Bachelotia	
Red algal 'turf' Lithothamnia	Red algal 'turf'	0
Sargassum Dictyopteris Lithothamnia	Lithothamnia Sargassum Dictyopteris	-1
		-2

Fig. 11. The distribution of the dominant organisms along transects on the Tema breakwater system in Ghana compared with the pattern on local rocky shores. Parentheses around names indicate that the organism is only sometimes common. Heights are given above or below Chart Datum (after John and Pople, 1973).

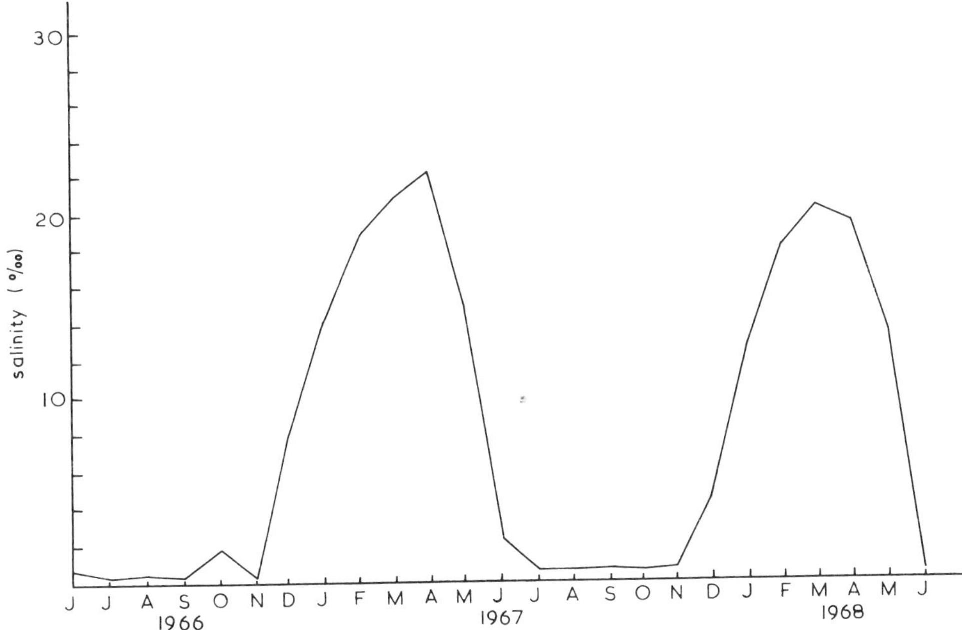

Fig. 12. The annual salinity fluctuations at high tide in the Lagos lagoon, Nigeria which forms part of an "open" lagoon system. The salinity never rises above that of seawater due to tidal influences and falls markedly during the heavy rains of the wet season (after Fagade and Olaniyan, 1974).

Fluctuations in salinity are least in estuaries, and in those lagoons that retain a permanent connection with the sea. Their salinity never rises above that of the sea itself although it may drop almost to zero after heavy rains (Fig. 12). The diurnal tidal rise and fall in such "open" lagoons allows for the development of a genuine littoral zone. On the other hand many lagoons may become cut-off from the sea for part of the year by the development of a sand bar; in these "closed" lagoons salinity fluctuations may be extreme (Fig. 13). After the lagoon is cut-off by the bar the trapped seawater may become diluted by the addition of rainwater, but on the cessation of the rains evaporation may cause the salinity to rise to more than twice that of seawater, and some parts of the lagoon may dry out completely. When this happens an almost black crust of blue-green algae may be left on the cracked surface of the mud whilst a floating scum is common on the surface of small and very saline pools left in depressions. Blue-green algal species found in such habitats include *Microcoleus chthonoplastes*, *Agmenellum thermale* and *Anacystis dimidiata*, but sometimes larger algae such as *Chara zeylanica* Klein ex Willdenow may occur.

The muddy bottom of many lagoons is a considerable barrier to the establishment of algae except for a number of microscopic forms especially blue-greens, though animals such as fiddler crabs may be present in large numbers. It is the mangroves

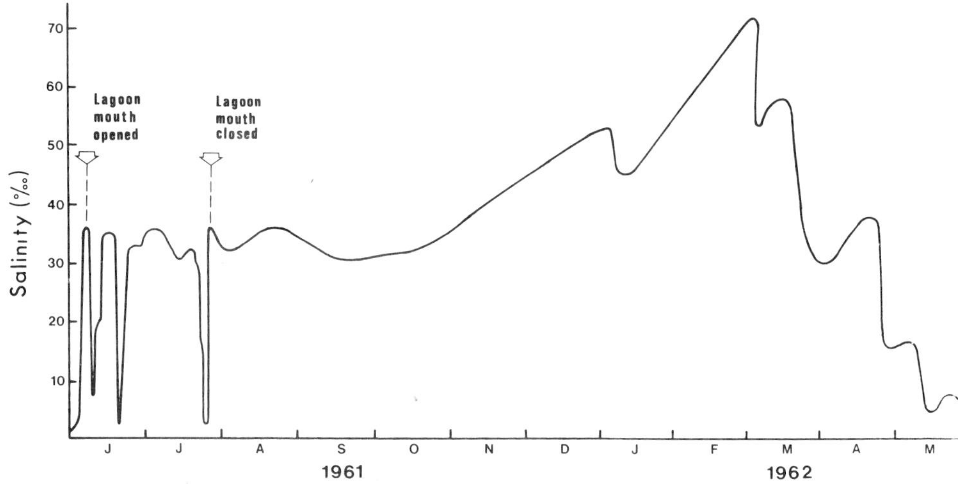

Fig. 13. The annual salinity fluctuations in a small "closed" lagoon in Ghana. The sand bar at the lagoon mouth became breached in June as a result of a rise in lagoon water level due to the heavy rains, allowing seawater to enter the lagoon. A complex interplay of floods and spring tides brought about wide fluctuations in salinity until the lagoon mouth closed once more in August. The September rains prevented any rise in salinity and only after the end of the rains in October did the salinity rise owing to evaporation. Transient falls in salinity were caused by occasional rains and a marked fall did not occur until the heavy rains began in late April (after Pople, in Ewer and Hall, 1972).

that grow in these lagoons that usually provide the hard surfaces for the attachment of littoral organisms. In the case of *Avicennia* it is the upright breathing roots or pneumatophores spreading some distance round the tree that are often colonized. It should be noted, however, that *Avicennia* is found in both open and closed lagoons and it is only in the former that there is any appreciable development of algae on the pneumatophores (Pl. VI A,B). *Rhizophora*, the other main genus of mangrove in West Africa, occurs mainly in open lagoons where its arching stilt roots may become covered by marine organisms (Pl. VI C,D).

Because of the tidal rise and fall over the mangrove roots, attaching organisms become arranged in a series of belts, although the vertical extent of these is often very restricted because of the reduced wave action in the sheltered conditions that prevail. The littoral zone on mangroves can sometimes be divided into a littoral fringe and a eulittoral zone although the number of subzones in the latter is somewhat variable. The fact that mangroves are usually rooted in mud means that there is no hard substratum for the development of sublittoral or sublittoral fringe organisms.

Plate VI: A. The breathing roots or pneumatophores of the black mangrove *Avicennia africana* growing through mud at the mouth of the River Densu (Ghana) and only exposed at low water (after Lawson, 1960c). B. A close-up of some of the breathing roots showing them mainly covered by a thick felty mat of brackish-water algae, principally *Bostrychia radicans*. C. The stilt roots of the red mangrove *Rhizophora* growing into mud at the mouth of the River Volta in Ghana (after Ewer and Hall, 1972). D. A close-up of a stilt root of a *Rhizophora* tree near Douala (Cameroun) showing dark patches of blue-green algae.

A
B
D
C

The littoral fringe, if represented at all, is usually recognised by the presence of the small snail *Littorina scabra* Lamark (Fig. 14). The eulittoral zone may in some places be divided into three subzones: an upper subzone dominated by species of *Chthamalus* or occasionally *Balanus amphitrite* Darwin, a middle oyster subzone of *Ostrea tulipa* and, only present on occasion, a lower algal subzone. The lower subzone of algae when present forms a greyish furry mat consisting of one or more species of *Bostrychia* together with other small red algae such as species of *Caloglossa, Catenella,* and *Polysiphonia,* and filaments of various green algae such as *Cladophora, Chaetomorpha* and *Rhizoclonium.* When green algae are present by themselves they may occupy higher levels corresponding to the upper eulittoral or even the littoral fringe. The small snail *Littorina scabra* may also extend downwards into the barnacle subzone along with a species of *Thais.*

Although the shores of lagoons are always muddy or sandy, some estuaries, on the other hand, may have appreciable amounts of rock, laterite or even stranded logs which provide a stable substratum allowing the development of a distinct zonation of littoral organisms. In the case of wide estuaries, such as that of the Sierra Leone River, the pattern of zonation near the mouth is almost indistinguishable from that of open rocky coasts. It is only some distance up such estuaries where tidal influences, salinity and wave action are reduced that changes become apparent.

In the littoral fringe the littorinids characteristic of the open coast, namely *Littorina punctata* and *Nodilittorina miliaris,* gradually give way up the estuary to *Littorina cingulifera* as they become reduced in vertical extent before finally disappearing. The barnacles forming the upper part of the eulittoral zone similarly do not extend so far up the shore but are more persistent than the snails and reach further up the estuaries. A low cover of algae dominated by *Bostrychia radicans* may replace the barnacles in many places. In marked contrast to open shore habitats a distinctive subzone of the oyster *Ostrea tulipa* is developed in the middle part of the eulittoral where it is usually accompanied by the small limpet *Siphonaria pectinata,* a species also found on open coasts. Again, further up the estuary *Siphonaria* drops out and the oysters may become gradually replaced by the small mussel *Brachydontes puniceus.* Characteristically a dirty grey felt or low turf of algae is found covering the lower parts of the eulittoral zone. It contains varying amounts of species of *Bostrychia, Caloglossa, Cladophora, Gelidium, Enteromorpha, Murrayella,* and several blue-green algae. Sometimes this lower subzone of the eulittoral may be dominated instead by animals such as serpulid worms (*Pomatoleis* spp.) accompanied by *Balanus amphitrite* and sponges. The position of the sublittoral fringe is often occupied by mud flats from which algae are usually absent, but at the mouths of some smaller rivers rocks occur on which algae grow. At the mouth of the Ankobra River in Ghana, such rocks bear well-developed plants of *Grateloupia filicina* in an almost pure stand. Similarly a mat of *Grateloupia* also occurs at the entrance to the Tabou River in Côte d'Ivoire, but here it is accompanied by *Bryocladia thyrsigera* and occupies the lower eulittoral subzone rather than the sublittoral fringe just below a belt of bushy plants of *Gelidium corneum.*

The patterns of littoral zonation observed in estuarine conditions are often complicated by other factors. For example, where there is a good deal of shade cast by trees or the river bank many of the red algae from the lower eulittoral subzone are

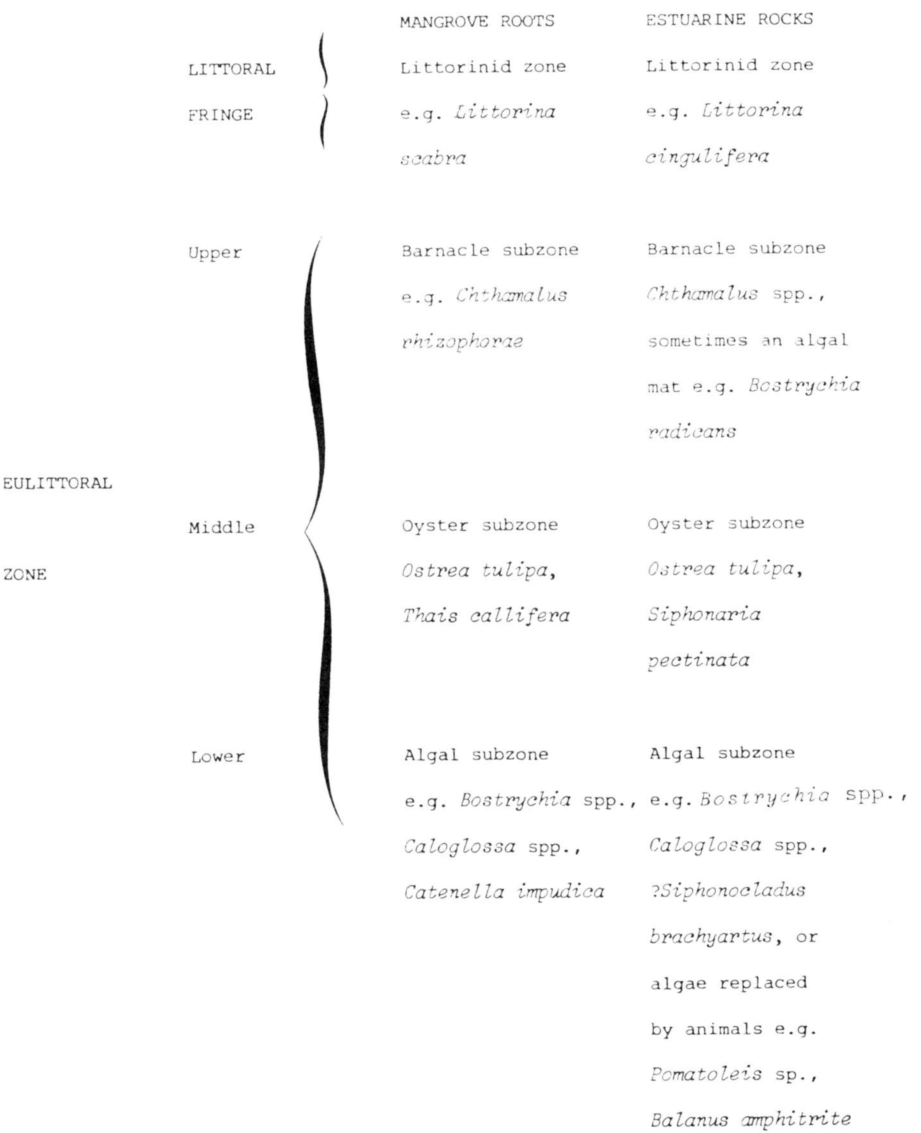

		MANGROVE ROOTS	ESTUARINE ROCKS
	LITTORAL FRINGE	Littorinid zone e.g. *Littorina scabra*	Littorinid zone e.g. *Littorina cingulifera*
EULITTORAL ZONE	Upper	Barnacle subzone e.g. *Chthamalus rhizophorae*	Barnacle subzone *Chthamalus* spp., sometimes an algal mat e.g. *Bostrychia radicans*
	Middle	Oyster subzone *Ostrea tulipa*, *Thais callifera*	Oyster subzone *Ostrea tulipa*, *Siphonaria pectinata*
	Lower	Algal subzone e.g. *Bostrychia* spp., *Caloglossa* spp., *Catenella impudica*	Algal subzone e.g. *Bostrychia* spp., *Caloglossa* spp., ?*Siphonocladus brachyartus*, or algae replaced by animals e.g. *Pomatoleis* sp., *Balanus amphitrite*

Fig. 14. The typical pattern of littoral zonation on unshaded substrata where tidal influences are still appreciable.

found at much higher levels and even extend into the littoral fringe. There may, in addition, be a considerable growth of *Rhizoclonium* in the littoral fringe sometimes accompanied by blue-green algae such as *Entophysalis deusta.*

Littoral zonation in estuaries is often confusing and difficult to interpret because of the many variables that may influence it, but it will be apparent that it bears a strong similarity to that found on mangroves. The most notable feature that both types of environment have in common is that eulittoral zone may be subdivided into an upper barnacle zone, a middle oyster zone, and a lower algal zone (see Fig. 14).

Salt Pans

Situated along the Gulf of Guinea coast are solar salt factories where seawater is evaporated down in shallow pans so as to obtain the salt. The pans are arranged in a series and there is a progressive increase in salinity until the water in the pans at the end of the series is supersaturated and the mud floors become lined with large salt crystals. It has been observed in Ghana that the water in some of the most saline pans is often turbid due to the presence of large populations of the planktonic alga *Dunaliella salina* (Dunal) Teodoresco. In general the water of the lower salinity pans is very clear and the mud floors are usually covered by a thick spongy mat of blue-green algae which occasionally break up and float as surface flocks. The angiosperm *Sesuvium portulacastrum* (Linnaeus) Linnaeus often fringes these pans and sometimes its branches grow into them. This halophyte also affords a surface for the growth of the blue-green algae. The most common blue-green algae in these shallow pans are *Coccochloris peniocystis, Gomphosphaeria aponina, Johannesbaptista pellucida, Schizothrix friesii and Spirulina subsalsa.*

SYSTEMATIC ACCOUNT

Introduction

The sequence of orders and families in the four divisions considered in this work follows, as far as possible, the arrangement given in the "Check-list of British Marine Algae" (Parke and Dixon, 1976). Genera are listed alphabetically under each family, and the species placed in alphabetical order under each of the genera. The only departure from this arrangement is in the family Corallinaceae where all the crustose genera are grouped separately from those that are articulated.

For each entry the complete author citation is given as follows: firstly, the reference to the author for the publication of the epithet or for making the now accepted generic assignment; secondly, if applicable, the first published name (basionym) and the authority. The nomenclature has been revised as far as possible and discussion on certain of our decisions is contained in the remarks following the species entry in question. Further details can often be obtained by referring to "Seaweeds of the West Coast of Tropical Africa and Adjacent Islands: a Critical Assessment". This is published in four parts as follows: Chlorophyta and Xanthophyta (Lawson and Price, 1969), Phaeophyta (Price *et al.*, 1978), class Bangiophyceae of the Rhodophyta (John *et al.*, 1979), and class Florideophyceae of the Rhodophyta (Price *et al.*, 1983). If a species is preceded by a question mark then there is some element of doubt regarding its inclusion; this is generally discussed in the remarks. When there is very strong evidence to suggest that a species or genus is incorrectly attributed to the region then it is given as a doubtful record. Footnotes often refer to species which might be expected to occur in tropical West Africa but have until now been reported just to the south of the region covered by this work. These are usually species found in the Congo Republic or the island of Pagalu (=Annobon) where tropical conditions are likely to prevail.

Descriptions of the orders and families in the Chlorophyta, Phaeophyta and Rhodophyta are not included but can be found in the following floras: "The Marine Algae of California" by Abbott and Hollenberg (1976) and "The Marine Algae of the Eastern Tropical and Subtropical Coasts of the Americas" by W.R. Taylor (1960). Family descriptions will also appear in the several volumes comprising "Seaweeds of the British Isles", the first published volume deals with the orders Nemaliales and Gigartinales in the Rhodophyta (Dixon and Irvine, 1977a). The families of the Cyanophyta are given in the various revisions of this group by Drouet (1968, 1973, 1978, 1981) and Drouet and Daily (1952).

The descriptions and dimensions of the species are largely based on specimens from tropical West Africa examined by ourselves. Various sources have been consulted for such information when we have been unable to locate a collection but have no good reason to doubt the validity of the published record. We have based our decriptions of the lithothamnia principally on the very detailed accounts of Foslie; these are often of specimens collected from the islands of Príncipe and São Tomé. In very polymorphic genera (e.g. *Cladophora*) and all members of the Cyanophyta we have taken cell dimensions from the literature, recognising that the often limited amount of material examined by us might not be fully representative of the range of variation found in the species.

The Cyanophyta are common shore algae in the tropics and so cannot be ignored by any serious collector in the region. Unfortunately there are special problems involving the taxonomy of the blue-green algae with specialists in the group holding wide and often very divergent views. For this reason all published and unpublished names of species from the region of West Africa covered by this present work are given. Many of the collections from this region have been determined by F. Drouet and so we have adopted his taxonomic system and used his names. According to Drouet many of the criteria traditionally used to define blue-green algal species are influenced by habitat conditions and so he considers many formerly recognised species to be no more than ecophenes*. Criticism of his approach has come from a number of algal specialists including Bourrelly (1957, 1970a, 1970b), Forest (1968), Geitler (1960), Komarek (1973) and Stanier *et al.* (1971). Traditionally the blue-green algae have been treated in much the same manner as other algal groups (see floras of Desikachary, 1959; Geitler, 1930-32), but it is generally accepted that such an approach is very artificial particularly below the family level. Some workers have recently advocated treating the blue-green algae in a similar manner to the bacteria and calling them the cyanobacteria.

We make reference to plates or figures in the works of others though most of the species are illustrated by line drawings. An attempt has been made wherever possible to emphasise in the drawings features of the plant which will aid in distinguishing between genera and species, whether they be of the general habit or of details of vegetative or reproductive structures. In a few instances illustrations made by other authors have been reproduced especially when the specimens drawn have not been available to us. The blue-green algae have been treated throughout in a somewhat different manner to the other groups and the only illustrations are those that assist in distinguishing between the terms used by Drouet (1968) to separate those filamentous species not containing heterocysts.

The ecology of the species is described in some detail when we have collected the plant personally, otherwise notes have been taken from exsiccatae collections or from ecological descriptions found in the literature for the region. The phytogeographical boundaries mentioned when referring to the world distribution of species and genera are those proposed by Ekman (1953) and are as follows: arctic-antarctic, boreal-antiboreal, warm temperate and tropic. These so-called littoral provinces are shown by Hedgpeth (1957, pl. 1) on a world map. The local distribution of each species in tropical West Africa is also given along with the reference(s) in which it is

*Forms representing the range of phenotypic variation found in a species.

cited. When such a reference is followed by the letters "p.p." (pro parte), it may be taken that reassessment has shown the material on which the original record was based to be heterogeneous. New records, whether they are derived from our own recent shore collecting or from an examination of herbarium material, appear with the word "unpublished" following the mention of the country. If the West African material shows some points of special taxonomic interest, or where it differs in some way from the usual description of the species, then this information is included in the remarks section under the entry in question.

DIVISION CHLOROPHYTA · CLASS CHLOROPHYCEAE

Order Ulotrichales · Family Ulotrichaceae

Urospora Areschoug 1866

Doubtful Record

Urospora sp.

To this genus has been tentatively referred a few short (<500 μm high), simple, narrow (10-16(-20) μm diameter) filaments (see Pl. 3, fig. 9) growing on the roots of mangroves or within a low algal mat on the Sierra Leone peninsula (John and Lawson, 1977a; Lawson, 1957a).

Family Monostromataceae

Gomontia Bornet et Flahault 1888

Doubtful Record

Gomontia polyrhiza (Lagerheim) Bornet & Flahault

This widely distributed species is reported perforating the shells of molluscs collected on the Sierra Leone peninsula (Aleem, 1978). The first report of this unicellular alga from the mainland coast of West Africa is considered doubtful until the determination can be verified.

Order Ulvales

Family Ulvaceae

Enteromorpha Link 1820

Plants free-floating or more commonly attached by a discoid holdfast, terete and tubular or if flattened above then only having the margins hollow, simple or often alternately divided, sometimes tapering and terminating in a single series of cells; chloroplasts parietal, covering all or only the upper part of a cell, with one to several pyrenoids; zoospores having two or four flagella; gametes biflagellate.

World Distribution: cosmopolitan.

The members of this genus possess a remarkable degree of morphological plasticity and can only be separated with any degree of confidence on the basis of the arrangement of the cells and on certain cellular components.

Key to the Species

1. Branches terete and tubular towards the base, becoming flattened above and hollow only at the margins; apices often forcipate or distinctly notched ..... *E. linza*
1. Branches terete to somewhat flattened and tubular throughout; apices often .... rounded 2

2. Branches in surface view showing the cells arranged in distinct longitudinal and sometimes transverse rows .... *E. lingulata*
2. Branches in surface view showing the cells either irregularly arranged throughout or only in the upper and older parts 3

3. Cells irregularly arranged throughout the plant . *E. intestinalis* subsp. *intestinalis*
3. Cells irregularly arranged only in the upper and older parts of the plant 4

4. Branches having the cells mostly quadrate and less than 10 µm in breadth, only arranged in longitudinal rows in the younger parts; pyrenoids rarely more than 1 per cell .... *E. prolifera* subsp. *prolifera*
4. Branches having the cells quadrate to distinctly rectangular, from 10-40 µm in size, often in conspicuous transverse and longitudinal rows in the younger parts; pyrenoids 1 to several per cell 5

5. Plants much branched throughout and forming hair-like tufts; branches terete and filiform, always less than 1 mm in diameter; cells containing from 3 to 12 pyrenoids.. *E. clathrata*
5. Plants simple or only branched towards the base; branches terete to somewhat flattened, often slightly inflated above and contorted, usually from 1-4 mm in width; cells containing 1 to 3 pyrenoids............. *E. flexuosa* subsp. *flexuosa*

Enteromorpha clathrata (Roth) Greville Pl. 1, figs 1, 2.

Greville, 1830; *Conferva clathrata* Roth, 1806.

Plants commonly forming hair-like tufts and occasionally as extensive mats, with individuals from 2-15 cm in length; branches terete and filiform, up to 1 mm in diameter, much divided and entangled, with successive orders shorter and tapering towards the apex, often spine-like but never terminating in a uniseriate row of cells; cells often quadrate to rectangular and occasionally polygonal, from 10-45 µm in size, arranged in distinct longitudinal and transverse rows only in the younger branches; pyrenoids from 3 to 12 in each cell.

For additional figures see Bliding, 1963, p. 107, 108, 110 to 112, figs 64-68.

Occurring on the upper parts of rock ledges and concrete surfaces in the eulittoral zone and growing over a wide range of exposure to wave action; occasionally forming a distinct belt in the barnacle subzone. Sometimes found in brackish-water habitats, such as on rocks in river estuaries.

Distribution. World: as for genus. West Africa: Cameroun: John 1977a. Côte d'Ivoire: John 1977a. Ghana: John 1977a, Lawson 1966, Lawson & Price 1969,

Townsend & Lawson 1972; *E. clathrata* f. *pumita*, Lawson & Price 1969; *E. prolifera* var. *tubulosa*, Lawson & Price 1969; *E. torta*, Lawson & Price 1969. Liberia: De May *et al.*, 1977. Nigeria: Fox 1957, John 1977a, Lawson & Price 1969, Steentoft Nielsen 1958, Steentoft 1967. São Tomé: John 1977a, Lawson & Price 1969, Steentoft 1967. Sierra Leone: Aleem 1978, Fox 1957, John & Lawson 1977a, Lawson & Price 1969. Togo: John 1977a, John & Lawson 1972b.

Enteromorpha flexuosa (Wulfen ex Roth) J. Agardh — Pl. 1, figs 5-7.

J. Agardh, 1882-1883; *Conferva flexuosa* Wulfen, in Roth, 1800.

subsp. **flexuosa** J. Agardh

J. Agardh, 1882-1883.

Plants forming tufts or often extensive mats, with individuals from 3-20 cm in length; branches terete or somewhat flattened above, sometimes inflated and contorted, usually from 1-4 mm in width, simple or more usually divided towards the base into similar and subequal branches; cells angular, quadrate or rectangular, from 9-28 μm in size, arranged in longitudinal and transverse rows only in the younger parts; chloroplast usually forming a band towards the centre of the cell; pyrenoids 1 to 3 per cell.

For additional figures see Bliding, 1963, p. 75 to 78, figs 38-41.

Occurs over a wide range of exposure to wave-action growing on rocks and stranded logs in the eulittoral zone; occasionally found in the littoral fringe as a low mat associated with a good deal of sand. Reported from Nigeria as growing entangled with *E. clathrata* on the stilt roots of *Rhizophora* in the brackish-water environment of a mangrove swamp.

Distribution. World: as for genus. West Africa: Cameroun: John 1977a; *E. compressa* (pro parte), Pilger 1911. Côte d'Ivoire: John 1972c, 1977a. Gabon: John 1977a, John & Lawson 1974a. Gambia: John & Lawson 1977b. Ghana: John 1977a, Lawson & Price 1969, Steentoft 1967, Townsend & Lawson 1972; *E. flexuosa*, Lawson 1960a; *E. compressa*, Lawson & Price 1969; *E. intermedia*, Fox 1957. Liberia: De May *et al.* 1977, John 1972c, 1977a. Nigeria: John 1977a, Lawson & Price 1969, Steentoft 1967; *E. intermedia*, Fox 1957. São Tomé: John 1977a, Lawson & Price 1969, Steentoft 1967; *E. prolifera*, Hariot 1908, Henriques 1917. Sierra Leone: Fox 1957, John & Lawson 1977a, Lawson & Price 1969, Steentoft 1967. Togo: John 1977a, John & Lawson 1972b.

Enteromorpha intestinalis (Linnaeus) Link — Pl. 1, fig. 9.

Link, 1820; *Ulva intestinalis* Linnaeus, 1753.

subsp. **intestinalis** De Silva & Burrows

De Silva and Burrows, 1973.

Plants about 10 cm or more in length; branches terete below and tubular, somewhat compressed or inflated above and often contorted, to about 1 cm in width, commonly simple and occasionally with just a few branches arising towards the base; cells in surface view angular or polygonal, from 12-20 μm in diameter, always irregularly arranged throughout; branches in section having the cells quadrate or subquadrate, with the lumen from 15-20 μm in height and the outer wall thicker than the lateral walls; chloroplast small, parietal and in the upper part of a cell; pyrenoids 1 per cell.

For additional figures see Bliding, 1963, p. 140, 142, 143, 145 to 147, figs 87-92.

There is no information on the ecology of this plant in the region but the Hornemann specimen in the Copenhagen Herbarium (C) is a simple plant, much inflated and contorted, and similar to forms Bliding (1963) found which growing in "brackish, eutrophic water with rise of temperature often get wrinkled and blown up.". Elsewhere in the Atlantic Ocean it is often found in tidepools high on the shore and in places where there is little wave action.

Distribution. World: cosmopolitan. West Africa: Ghana: *E. intestinalis*, Lawson & Price 1969, Townsend & Lawson 1972; *Ulva intestinalis* var. *ß. crispa*, Hornemann 1819.

Enteromorpha lingulata J. Agardh

J. Agardh, 1882-1883.

Plants as found in the Gulf of Guinea only a few cms in length though up to 30 cm in other areas, usually growing as tufts; branches terete to slightly flattened or inflated above, simple or many times divided near the base or throughout; cells quadrate to irregularly rectangular, from 10-30 μm in size, arranged in longitudinal and transverse rows throughout; chloroplast usually band-shaped; pyrenoids from 2 to 4 cell.

For figures see Kapraun, 1970, p. 249, 256, 261, 267, 273, 275, 281, 285, figs 9, 10, 39, 40, 55, 56, 75, 76, 98, 102, 134, 146-148.

Occurs on beach rocks in the lower eulittoral subzone which are moderately sheltered from wave action.

Plate 1: Figs 1, 2. *Enteromorpha clathrata:* 1. Plant forming a hair-like tuft of alternately divided filiform branches; 2. Portion of a branch showing the cells arranged in transverse and longitudinal rows. Figs 3, 4. *Enteromorpha linza:* 3. Plants divided only towards the base and the branches notched or forcipate at apex; 4. Section of an upper part of a branch showing the hollow margin. Figs 5-7. *Enteromorpha flexuosa* subspecies *flexuosa:* 5, 6. Variations in the degree of branching of plants; 7. Surface of a branch showing the regular arrangement of the cells and the presence of usually a single pyrenoid per cell. Fig. 8. *Enteromorpha prolifera* subspecies *prolifera:* surface of a branch showing the small quadrate cells. Fig. 9. *Enteromorpha intestinalis* subspecies *intestinalis:* surface of an older branch showing the irregularly arranged cells.

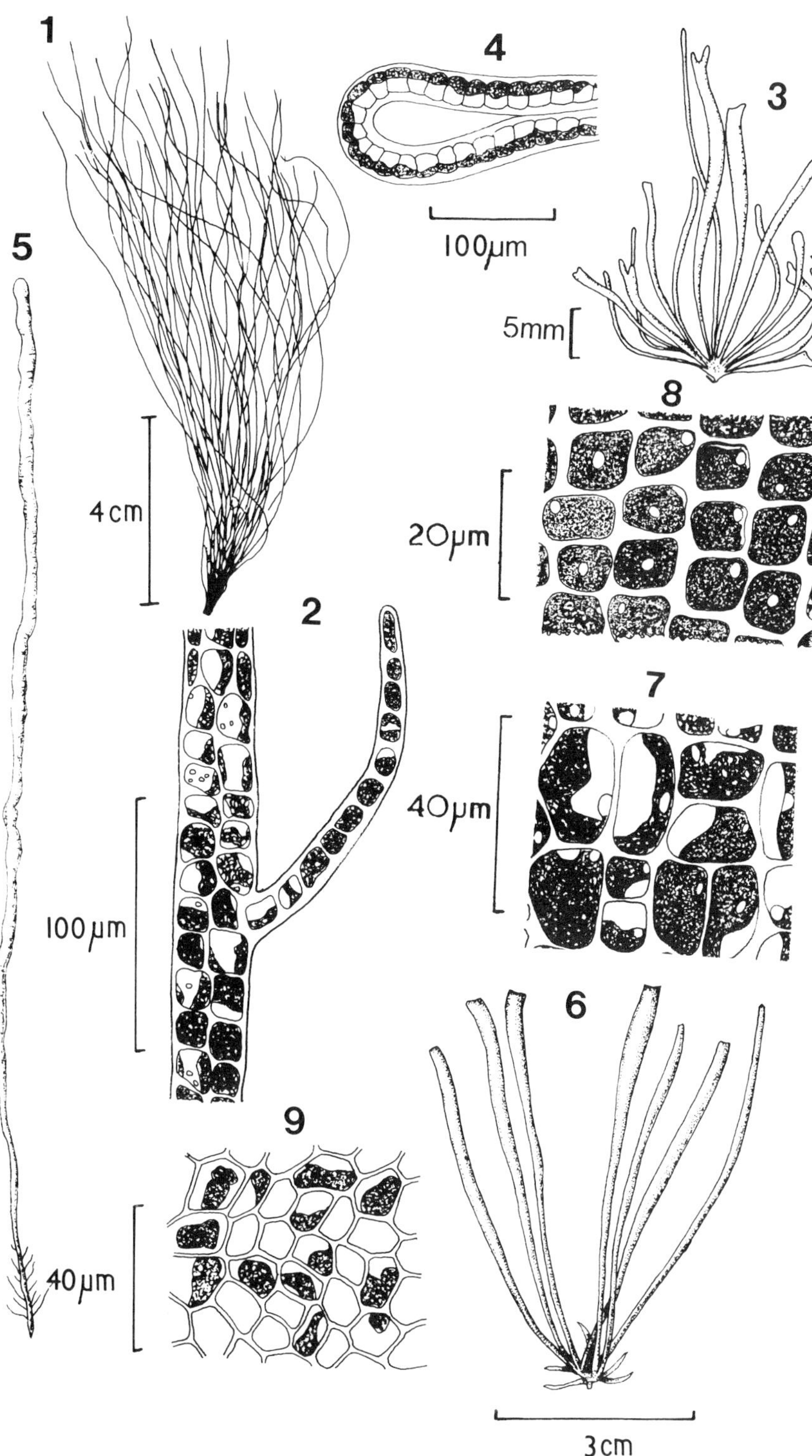
1
4
3
100µm
5
5mm
8
4cm
20µm
2
7
40µm
100µm
6
9
40µm
3cm

Distribution. World: in some warm temperate and tropical seas. West Africa: Cameroun: *E. compressa* (pro parte), Pilger 1911. Gabon: John & Lawson 1974a. Gambia: John & Lawson 1977b. Nigeria: *E. clathrata,* Fox 1957, Steentoft 1967, Steentoft Nielsen 1958.

Remarks. The Nigerian plants on subsequent examination we believe to be this species and not *E. clathrata.* Similarly the Ledermann plant (No. 368), identified by Brand for Pilger as *E. compressa,* is a mixed collection. Bliding (1963) regards the Australian and European plants he studied to be closely related to or perhaps even conspecific with *E. flexuosa* subsp. *flexuosa.* The doubtful status of this species is discussed by Lawson and Price (1969).

Enteromorpha linza (Linnaeus) J. Agardh — Pl. 1, figs 3, 4.

J. Agardh, 1882-1883; *Ulva linza* Linnaeus, 1753.

Plants from 2-6 cm in length, forming tufts or sometimes extensive mats; branches terete and tubular below, becoming flattened above and hollow only towards the margins, to about 4 mm in width, simple or on occasion West African plants proliferously divided towards the base, with the tips often characteristically notched or forcipate; cells in surface view in distinct longitudinal rows in the younger parts and irregularly arranged elsewhere; branches in section having the cells quadrate to rectangular, with the lumen from (6-)10-16 μm in diameter and from 10-26 μm in height, having the inner wall about 7 μm and the outer wall from 3-4 μm thick; chloroplasts parietal and lobed; pyrenoids from 1 to 2(-4) per cell.

For additional figures Bliding, 1963, p. 128, 129, 131, figs 79-81.

Common on beach rocks in the eulittoral zone in places where wave action is moderate and frequently associated with much sand; occasionally found on hard substrates in lagoons.

Distribution. World: widespread from boreal-antiboreal to tropical seas. West Africa: Gabon: John & Lawson 1974a. Ghana: Lawson & Price 1969, Townsend & Lawson 1972; *E. procera* f. *crainosa,* f. *opposita,* f. *subnuda,* Lawson & Price 1969; *Ulva lanceolata,* Hornemann 1819.

Enteromorpha prolifera (O.F. Müller) J. Agardh — Pl. 1, fig. 8.

J. Agardh, 1882-1883; *Ulva prolifera* O.F. Müller, 1778.

subsp. **prolifera** J. Agardh

J. Agardh, 1882-1883.

Plants from 2-10 cm in length, usually forming dense tufts; branches terete and tubular to compressed or inflated, with the different orders of branches varying considerably in length, often with numerous, simple or much divided proliferations;

cells usually small and quadrate, from 7-10 μm in size, arranged in longitudinal rows only in the younger parts; chloroplasts parietal and entire; pyrenoids large and usually 1 per cell.

For additional figures see Bliding, 1963, p. 47 to 49, 51, 53, figs 19-23.

Found on only one occasion in Ghana as a few individuals growing in a large tide-pool in the eulittoral zone.

Distribution. World: probably cosmopolitan. West Africa: Ghana: Lawson 1960a, Lawson & Price 1969, Townsend & Lawson 1972; *E. prolifera* var. *capillaris,* Chapman 1971. Sénégal (Casamance): Chevalier 1920.

Doubtful Record

Enteromorpha chaetomorphoides Børgesen

The report from Sierra Leone by Aleem (1978) is the first mention of this species from the eastern side of the Atlantic Ocean. We regard this West African record as doubtful until the material on which it is based is re-examined and the determination verified.

Ulva Linnaeus 1753

Plants membranaceous, consisting of a variously dissected and lobed frond, distromatic and rarely hollow, more or less stipitate and having a discoid holdfast; cells having a cup-shaped chloroplast and a single nucleus, with one to several pyrenoids; zoospores quadriflagellate and four to eight per cell; gametes biflagellate.

World Distribution: cosmopolitan.

Mention has been made of the difficulties to be found in distinguishing between *Ulva fasciata, U. lactuca* and *U. rigida* growing in Ghana (Lawson, 1956; Lawson and Price, 1969) due to the presence of apparent intermediates possibly representing hybrids.

Key to the Species

1. Plants often much divided into long, ribbon-like, lanceolate segments; frond in section often thicker in the middle than towards the margins; cells rectangular and to about 40 μm in height . . . . *U. fasciata*
1. Plants simple or little divided, ovate, lanceolate or orbicular; frond in section of almost constant thickness throughout; cells quadrate to rectangular and less than 30 μm in height 2
2. Fronds usually with a margin of small triangular teeth or spines. . . . . *U. popenguinensis*
2. Fronds normally smooth margined 3

3. Plants flaccid and with or without a distinct stipe; frond in section from 30-50 μm thick, with the cells nearly quadrate and to 20 μm in height.... *U. lactuca*
3. Plants firm and having a distinct stipe; frond in section from 50-70(-110) μm thick, with the cells rectangular and about 30 μm high .... *U. rigida*

Ulva fasciata Delile Pl. 2, figs 1, 2.

Delile, 1813.

Plants to 60 cm in length, irregularly divided into several more or less distinct ribbon-like, lanceolate segments, sometimes having irregularly or pinnately arranged marginal lobes, with the margins undulate or crenate, basally having a short and distinct stipe; frond in section 70-115 μm thick; cells at margin somewhat separated and nearly quadrate, elsewhere rectangular, from 10-16 μm in diameter and up to 40 μm in height, with thick walls.

For additional figures see Krishnamurthy and Joshi, 1969, p. 127, figs 3, 9, 15.

Occurs over a wide range of exposure to wave action both on rocks and occasionally on larger algae, and extends from the barnacle subzone down into the sublittoral fringe. Large and well-developed plants are to be found in moderately sheltered situations where there is some wave- or current-surge; more stunted individuals occur in places where wave-action is greater.

Distribution. World: pantropical. West Africa: Côte d'Ivoire: John 1972c, 1977a. Ghana: Dickinson & Foote 1951, Jeník & Lawson 1967, John 1977a, Lawson 1953, 1954a, 1956, 1957b, 1966, Lawson & Price 1969, Steentoft 1967, Stephenson & Stephenson 1972. Liberia: De May *et al.* 1977, John 1972c, 1977a. São Tomé: John 1977a, Lawson & Price 1969, Steentoft 1967, Tandy 1944.

Ulva lactuca Linnaeus Pl. 2, figs 3, 4.

Linnaeus, 1753.

Plants about 10 cm in length, usually broader than long, orbicular to ovate-lanceolate, often irregularly divided and sometimes becoming proliferous in older parts, with the margins undulate or folded, basally having an indeterminate stipe and an attachment disc; frond in section from 35-50 μm in thickness; cells rectangular, from 13-15 μm in diameter and up to 20 μm in height, with thin walls.

For additional figures see Bliding, 1968, p. 541, 543, 545, figs 3-5.

Plate 2: Figs 1, 2. *Ulva fasciata:* 1. Plant membranaceous and irregularly divided into ribbon-like lanceolate fronds; 2. Section through the middle of a frond showing the two layers of rectangular cells. Figs 3, 4. *Ulva lactuca:* 3. Plant membranaceous and irregularly divided into fronds that are broader than long; 4. Section showing two layers of almost quadrate cells. Figs 5, 6. *Ulva popenguinensis:* 5. Plant membranaceous and lobed or bifid; 6. Portion of a frond showing a marginal spine. Fig. 7. *Ulva rigida:* Section showing the two layers of rectangular cells.

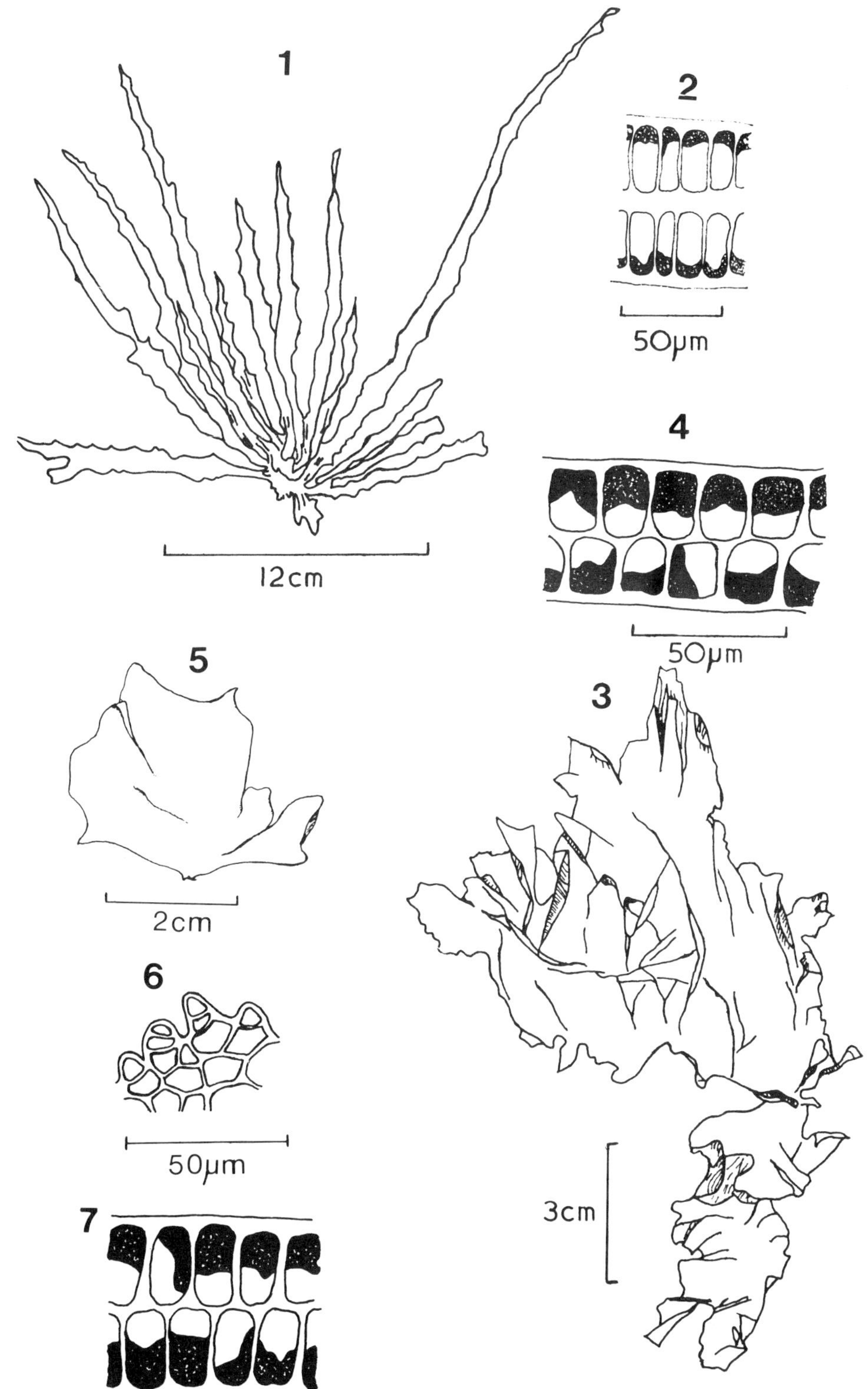
1
2
50µm
12cm
4
50µm
5
3
2cm
6
50µm
3cm
7

Occurs as solitary individuals or occasionally as low mats associated with much sand, often growing on rocks and sometimes on larger algae. Found over a very wide range of exposure to wave action in the eulittoral zone, but usually most abundant in the lower part of the barnacle subzone.

Distribution. World: as for genus. West Africa: Benin: John 1977a, John & Lawson 1972b. Côte d'Ivoire: John 1972c, 1977a. Gambia: John & Lawson 1977b. Ghana: Dickinson & Foote 1951, John 1977a, Lawson 1956, Lawson & Price 1969, Lieberman *et al.* 1979, Steentoft 1967. Liberia: John 1972c, 1977a. Príncipe: John 1977a, Lawson & Price 1969, Steentoft 1967. São Tomé: John 1977a, Lawson & Price 1969, Steentoft 1967. Sénégal (Casamance): Chevalier 1920. Sierra Leone: Aleem 1978. Togo: John 1977a, John & Lawson 1972b.

Ulva popenguinensis P. Dangeard Pl. 2, figs 5, 6.

Dangeard, 1958.

Plants to about 4 cm in height, initially entire and spathulate, later developing a stipe and becoming obovate, often lobed or bifid, with the margin having small triangular teeth or spines; frond in section about 30-35 μm thick; cells almost quadrate, from (6-)12-14 μm in diameter, with 2 to 4 pyrenoids in each cell.

For additional figures see Dangeard, 1958, p. 169, pl. 1, figs D-H.

Occurs in wave-sheltered situations in a river estuary on rocks in the lower eulittoral subzone, often associated with sand or mud.

Distribution. World: in tropical parts of the eastern Atlantic Ocean. West Africa: Gambia: John & Lawson 1977b.

Ulva rigida C. Agardh Pl. 2, fig. 7.

C. Agardh, 1823.

Plants firm in texture and to about 12 cm in height, ovate to ovate-lanceolate, sometimes orbicular and deeply lobed, with the margins undulate or folded, basally having a distinct stipe; frond in section 50-70(-110) μm thick; cells rectangular, from 13-22 μm in diameter and about 30 μm in height, with thick walls.

For additional figures see Bliding, 1968, p. 547 to 549, 551, 552, figs 6-10.

Occurs in the eulittoral zone on rocks and stranded logs and grows over a wide range of exposure to wave action, occasionally extending upwards into the littoral fringe.

Distribution. World: probably widespread from boreal-antiboreal to tropical seas. West Africa: Côte d'Ivoire: John 1977a. Ghana: Dickinson & Foote 1951, John 1977a, Lawson 1956, Lawson & Price 1969. Liberia: John 1977a. Sierra Leone: John & Lawson 1977a.

Remarks. Bliding (1968) mentions the presence of tooth-like and generally microscopic protuberances to be found along the margin of the disc. These are not obvious in the West African plants although in all other features they closely correspond to this species.

Doubtful Record

Ulva tropica Mertens

Hornemann (1819) records this species from "Danish Guinea" which is now Ghana. The attribution of this record is doubtful and will have to remain so until the material collected by P. E. Isert can be traced and examined. This has not been located in the Copenhagen Herbarium (C) and, according to the Curator (J. B. Hansen, *pers. comm.*), the possibility exists that it might have been lost when part of the Isert collection was destroyed by fire in 1807.

Order Chaetophorales · Family Chaetophoraceae

Entocladia Reinke 1879

Plants epiphytic or endophytic, consisting of procumbent and irregularly branched filaments, with or without elongate hyaline hairs; chloroplast parietal, with one or more bilenticular pyrenoids; uninucleate; reproduction by quadriflagellate zoospores and biflagellate gametes, produced by sequential cleavage of sporangial and gametangial mother cell contents.

World Distribution: cosmopolitan.

Most species belonging to this genus were transferred to *Phaeophila* when the two were merged by Nielsen (1972) and Burrows in Parke and Burrows (1976). Recently O'Kelly and Yarish (1981) have refuted this proposed merger since they were able to demonstrate that the two genera could be distinguished on the basis of their vastly different sporangial development pattern.

Key to the Species

1. Filaments from 5-12 μm in diameter; lenticular cells present . . . . *E. vagans*
1. Filaments from 3-6 μm in diameter; lenticular cells absent. . . . . *E. viridis*

Entocladia vagans (Børgesen) W.R. Taylor Pl. 3, fig. 5.

Taylor, 1960; *Endoderma vagans* Børgesen, 1920.

Plants consisting of irregularly divided filaments, from 5-12 μm in diameter; cells from 1 to 10 times longer than broad, almost cylindrical or with an elevation on the side; lenticular cells borne laterally on the filaments and containing acicular bodies.

For additional figures see Børgesen, 1920, p. 419, fig. 400 (as *Endoderma vagans*).

Occurs in or on the wall of larger algae and most readily detected towards the often colourless tips of the filaments of *Chaetomorpha antennina.*

Distribution. World: probably widespread in warm temperate and tropical seas. West Africa: Ghana: *Endoderma vagans,* John & Lawson 1972a.

Remarks. The Ghanaian plant has not been found with hyaline hairs although these have been reported on plants collected in other parts of its range (see Thivy, 1943). Until such time as the pattern of sporangial and gametangial development is determined in this species its generic identity must remain uncertain.

Entocladia viridis Reinke Pl. 3, fig. 6.

Reinke, 1879.

Plants consisting of freely branched filaments, sometimes congested and forming a pseudoparenchymatous disc; cells from 3-6 μm in diameter and 1 to 5 times longer than broad, very irregularly shaped near the centre of the disc, becoming more cylindrical towards the periphery.

For additional figures see Børgesen, 1920, p. 417, figs 398, 399 (as *Endoderma viride*).

Occurs on and in the walls of various larger algae, and like the other species, is commonly found growing on *Chaetomorpha antennina*.

Distribution. World: as for genus. West Africa: Ghana: John & Lawson 1972a. Sierra Leone: Aleem 1978.

Order Cladophorales · Family Cladophoraceae

Chaetomorpha Kützing 1845

Plants consisting of simple filaments of cylindrical to barrel-shaped cells, with walls thick and sometimes lamellated, attached usually by a disc on the elongate or obconical basal cell and also sometimes by rhizoid-like processes arising from the lower end of this cell; cells containing a reticulate and often perforated chloroplast, with numerous pyrenoids and nuclei; asexual reproduction by quadriflagellate zoospores usually developing in slightly inflated upper cells; sexual reproduction by biflagellated isogametes.

World Distribution: cosmopolitan.

Determinations within this genus must often be regarded as somewhat putative in the absence of a critical monographic treatment of this very polymorphic group. Many of the species are still little known and ill-defined with considerable overlap often found in the vegetative features normally used for specific determinations. Due to the lack of recent studies on many of the species the only illustrations are often to be found in very early publications.

Key to the Species

1. Plants consisting of straight and often stiff filaments having obvious basal attachment 2
1. Plants consisting of flexuous filaments, usually entangled and lying prostrate, often without clear basal attachment 6

2. Filaments showing a marked increase in width from base to apex, to about 1.5 mm in diameter . . . . *C. clavata*
2. Filaments of more or less equal width throughout, if showing a slight increase above then never more than 700 μm in diameter 3

3. Cells never more than 100 μm in diameter . . . . *C. nodosa*
3. Cells from 120-500 μm in diameter above the basal cell 4

4. Plants less than 1 cm in length; basal cell from 1.6-3 mm long and wall to about 56 μm thick . . . . *C. pachynema*
4. Plants always greater than 1 cm in length; basal cell either shorter or longer than above and wall never thicker than 30 μm 5

5. Basal cell usually from 4-5.6 mm long . . . . *C. antennina*
5. Basal cell less than 1.2 mm long . . . . *C. linum*, in part

6. Filaments less than 100 μm in diameter 7
6. Filaments normally greater than 100 μm in diameter 8

7. Filaments from 40-100 μm in diameter; cells from 2 to 3 times longer than broad . . . . *C. capillaris*
7. Filaments from 33-70 μm in diameter; cells from 2 to 6 times longer than broad . . . . *C. gracilis*

8. Filaments from 300-700 μm in diameter . . . . *C. crassa*
8. Filaments always less than 300 μm in diameter 9

9. Cells up to 180 μm in diameter and rarely as much as twice as long as broad . . . . *C. brachygona*
9. Cells normally greater than 180 μm in diameter and the upper cells as much as 3 times longer than broad . . . . *C. linum*, in part

Chaetomorpha antennina (Bory) Kützing Pl. 3, figs 1-4.

Kützing, 1869; *Conferva antennina* Bory, 1804

Plants forming coarse, stiff tufts, from 5-15(-40) cm in length; basal cell from 112-380(-420) μm in diameter and 4-5.6 mm in length, with the wall from 18-30 μm in thickness; cells above the basal cell from 250-350(-500) μm in diameter and 1 to 3 times longer than broad, rarely more.

For additional figures see Børgesen, 1913, p. 17, figs 4, 5; 1925, p. 38, 39 (as *C. media*).

Occurs most commonly as tufts on very wave-exposed rocks in the lower part of the upper eulittoral subzone and on occasion may form extensive mats in less exposed situations often extending into the lithothamnia subzone. In many moderately wave-sheltered localities it is found only as small tufts confined to the seaward facing vertical sides of rock ledges.

Distribution. World: in warm temperate and tropical seas. West Africa: Cameroun: Fox 1957, John 1977a, Lawson 1954b, 1955, 1966, Lawson & Price 1969, Stephenson & Stephenson 1972. Côte d'Ivoire: John 1972a, 1977a, Lawson & Price 1969. Ghana: Fox 1957, John 1977a, John & Lawson 1972a, Lawson 1953, 1954a, 1956, 1957b, 1966, Lawson & Price 1969, Stephenson & Stephenson 1972. Liberia: De May *et al.* 1977, John 1972c, 1977a. Nigeria: Fox 1957, John 1977a, Lawson 1966, Lawson & Price 1969, Steentoft Nielsen 1958, Stephenson & Stephenson 1972. São Tomé: John 1977a, Lawson & Price 1969, Steentoft 1967. Sierra Leone: John & Lawson 1977a.

Chaetomorpha brachygona Harvey

Harvey, 1858.

Plants typically forming a distinct stratum of entangled and flexuous filaments; cells from (40-)80-175(-180) μm in diameter and 40-240 μm in length, often equal in length and breadth, rarely as much as 4 times longer than broad, with the walls to about 20 μm thick.

For figures see Børgesen, 1920, p. 422, 423, figs 402-404.

The only information on the ecology of this plant in the region is that it was found growing entangled with a collection of *Siphonocladus tropicus* from the island of São Tomé.

Distribution. World: probably pantropical. West Africa: Ghana: Lawson & Price 1969. São Tomé: unpublished.

Remarks. Taylor (1960) suggested that this species might be conspecific with *Chaetomorpha linoides* Kützing.

Chaetomorpha capillaris (Kützing) Børgesen

Børgesen, 1925; *Rhizoclonium capillare* Kützing, 1847.

Plants known to form coarse and entangled mats; cells from 40-100 μm in diameter and from 2 to 3 times longer than broad, with thick walls.

For figures see Børgesen, 1925, p. 45, fig. 13 (as *C. capillare*).

There is no information on the ecology of this plant in the region but elsewhere it commonly grows entangled with other algae.

Distribution. World: widespread in warm temperate and tropical seas. West Africa: Nigeria: Lawson & Price 1969; *Lola tortuosa*, Fox 1957. Sierra Leone: *C. capillare*, Aleem 1978.

Remarks. There is some doubt regarding the record of this species from the region (see under *Rhizoclonium riparium*). Some authorities have expressed doubt as to the validity of the genus *Lola* (see Lawson and Price, 1969) and until further information is available this species is retained within the genus *Chaetomorpha*. Chapman (1961) considers the distinction between *Lola tortuosa* (possibly *C. capillaris* as sometimes applied) and *Lola gracilis* (= *C. gracilis*) to be open to question.

Chaetomorpha clavata (C. Agardh) Kützing

Kützing, 1847; *Conferva clavata* C. Agardh, 1824.

Plants forming erect tufts, often known to reach lengths of from 30-60 cm; cells towards the base of the filaments often subcylindrical to barrel-shaped, from 500-750 μm in diameter and 3 to 4 times longer than broad, with the distal cells inflated and moniliform, to about 1.5 mm in diameter and equal in length and breadth.

For figures see Chapman, 1961, p. 78, fig. 87a, b; Kützing, 1853, pl. 62, fig. 1.

There is no information regarding the ecology of this species in the region.

Distribution. World: probably pantropical. West Africa: São Tomé: Lawson & Price 1969, Steentoft 1967.

Remarks. Lawson and Price (1969) point out the need for re-assessing the limits of *Chaetomorpha antennina*, *C. clavata* and *C. robusta* in the light of comments by Papenfuss (1940b) and Taylor (1960).

?**Chaetomorpha crassa** (C. Agardh) Kützing

Kützing, 1845; *Conferva crassa* C. Agardh, 1824.

Plants free floating or else entangled with other algae; cells cylindrical or slightly inflated, from (300-)400-650(-700) μm in diameter, often just shorter than broad and only rarely to twice as long as broad.

For figures see Kützing, 1853, pl. 59, fig. 2.

Occurs entangled with other algae or on occasion free floating and outside the region reported also from brackish-water habitats.

Distribution. World: from boreal-antiboreal to tropical seas. West Africa: São Tomé: Lawson & Price 1969, Steentoft 1967.

Remarks. Some doubt attaches to this record due to the small size of the collection (Steentoft, 1967). It may be no more than an aegagropilous form of *Chaetomorpha linum* (Lawson and Price, 1969).

Chaetomorpha gracilis (Kützing) Kützing

Kützing, 1845; *Conferva gracilis* Kützing, 1843b.

Plants usually growing amongst other algae; filaments from 33-70 μm in diameter and the cells from 2 to 6 times longer than broad.

For figures see Børgesen, 1920, p. 423, fig. 405.

There is no information on the ecology of this species in the region.

Distribution. World: probably pantropical. West Africa: São Tomé: *Lola gracilis*, Lawson & Price 1969, Steentoft 1967. Sierra Leone: Aleem 1978.

Remarks. See the note under *Chaetomorpha capillaris*.

Chaetomorpha linum (O.F. Müller) Kützing — Pl. 4, figs 1-3.

Kützing, 1845; *Conferva linum* O.F. Müller, 1778.

Plants free and entangled with larger algae or epilithic and obviously basally attached, with the filaments straight or more usually flexuous; basal cell of a filament, if present, from 56-280 μm in diameter and 490-1,200 μm in length; cells in distal parts from 120-420(-560) μm in diameter, equal in length and breadth and increasing to 3 times longer than broad towards the tips of the filaments.

For additional figures see Kützing, 1853, pl. 55, fig. 3.

Occurs as plants having almost straight filaments when growing on open rocks and somewhat flexuous if through sand, most commonly on gently sloping and sheltered to moderately wave-exposed beach rocks. Found on occasion entangled with *Sargassum* plants growing in sheltered situations in the sublittoral fringe and as solitary plants on rocky platforms at about 10 m.

Plate 3: Figs 1-4. *Chaetomorpha antennina:* 1. Tufted habit of plant; 2. Lower portion of basal cell and a new erect filament developing from a creeping rhizoid; 3. Middle portion of filament showing the barrel-shaped cells; 4. Upper portion of a filament showing a cylindrical cell containing zoospores. Fig. 5. *Entocladia vagans:* Portion of plant growing over a cell of *Chaetomorpha.* Fig. 6. *Entocladia viridis:* Portion of a plant showing the irregularly shaped cells of the branched filaments. Fig. 7. *Chaetomorpha nodosa:* Lower portion of a filament showing the long basal cells. Figs 8, 9. *Urospora* sp. 8. Lower portion of a filament showing the short basal cell, cylindrical cells and somewhat inflated upper cells; 9. Lower cells containing an incomplete ring-like parietal chloroplast and pyrenoids.

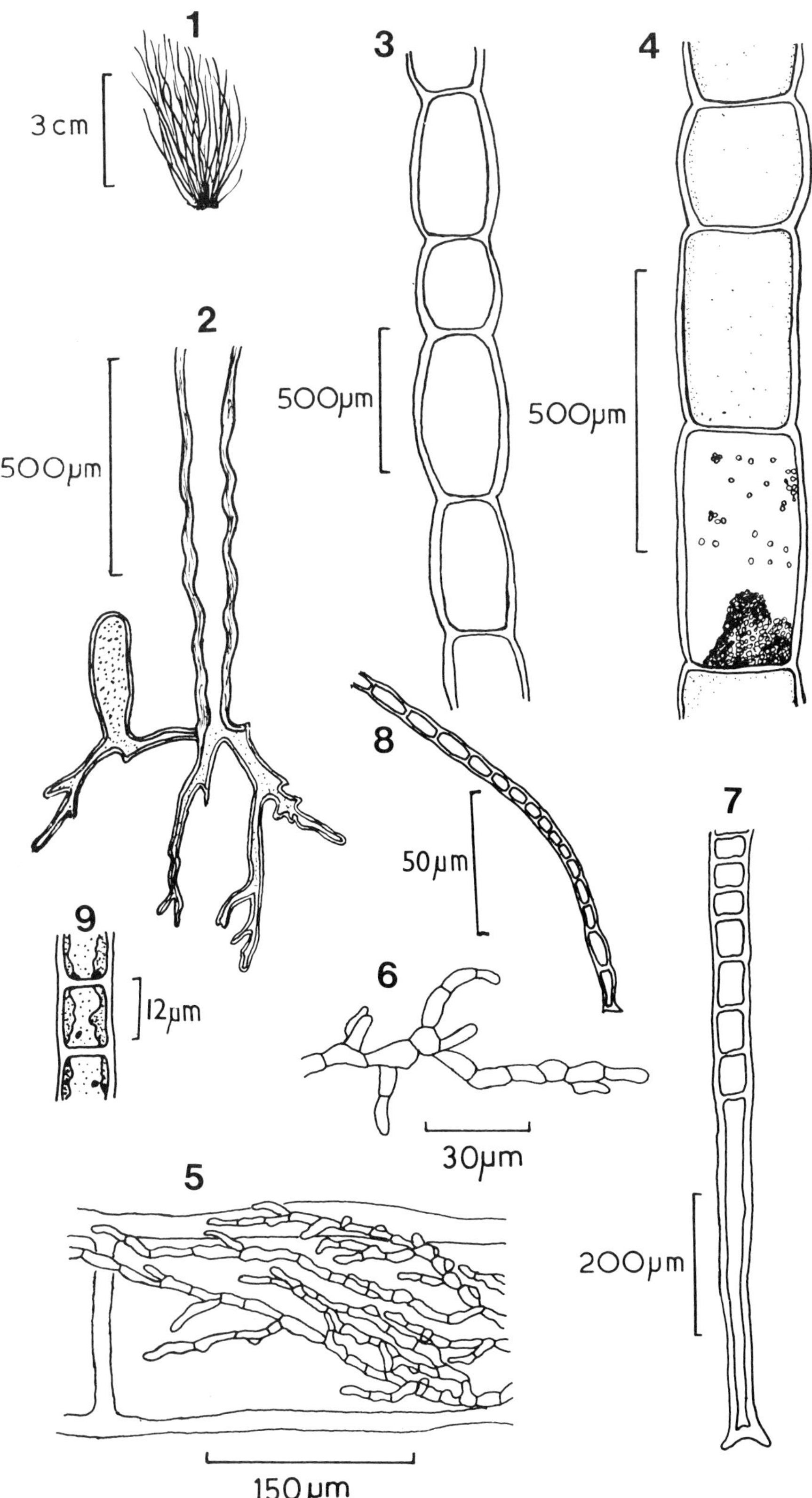
1
3 cm
2
500µm
3
500µm
4
500µm
8
50µm
7
9
12µm
6
30µm
5
200µm
150µm

Distribution. World: from boreal-antiboreal to tropical seas. West Africa: Cameroun: John 1977a, Lawson & Price 1969. Côte d'Ivoire: John 1972c, 1977a. Gabon: John & Lawson 1974a. Ghana: John 1977a; *C. aerea*, Lawson 1956, Lawson & Price 1969. Liberia: De May *et al.* 1977. Sierra Leone: Aleem 1978, John 1977a, John & Lawson 1977a. Togo: John 1977a, John & Lawson 1972b.

Remarks. According to Christensen (1957) *Chaetomorpha aerea* and *C. linum* are growth forms of the same taxon; *C. linum* has priority.

Chaetomorpha nodosa Kützing Pl. 3, fig. 7.

Kützing, 1849.

Plants forming erect tufts, to about 12 cm in height; basal cell about 40 μm in diameter near point of attachment and increasing to about 80 μm in diameter at distal end, to 450 μm in length; cells from 80-95 μm in diameter and 0.5 to 1.5 times as long as broad.

For additional figures see Kützing, 1853, pl. 52, fig. 4.

Occurs together with other species of green algae growing in the lower eulittoral subzone on moderately sheltered to moderately wave-exposed rocks and concrete structures; occasionally in tidepools in the upper subzone.

Distribution. World: in some warm temperate and tropical seas. West Africa: Cameroun: Brand 1911, John 1977a, Lawson & Price 1969. Côte d'Ivoire: John 1977a. Togo: John 1977a, John & Lawson 1972b. Sierra Leone: ?John & Lawson 1977a.

Chaetomorpha pachynema (Montagne) Montagne

Montagne, in Kützing, 1849; *Conferva pachynema* Montagne, 1839-1841.

Plants forming gregarious tufts, to about 1 cm in height; basal cell about 12 μm in diameter near point of attachment and increasing to 150-210(-350) μm at distal end, from 1.6-3 mm in length, with the wall from (20-)36-56 μm thick; cells above the basal cell from 150-350 μm in diameter and 1 to 4(-5) times longer than broad.

For figures see Børgesen, 1925, p. 42, figs 11, 12.

Occurs in moderately wave-sheltered situations in the eulittoral zone growing on lithothamnia in tidepools or in a low algal turf on open rocks.

Distribution. World: in warm temperate and tropical parts of the eastern Atlantic Ocean. West Africa: Sierra Leone: Aleem 1978, John & Lawson 1977a, Lawson 1954b, Lawson & Price 1969.

Plate 4: Figs 1-3. *Chaetomorpha linum:* 1. Basal portion of a filament showing the long basal cell with annular thickenings of the wall; 2, 3. Variations in the size and form of the cells found along the same length of filament. Fig. 4. *Cladophora ruchingeri:* Tufted habit of plant. Fig. 5. *Cladophora prolifera:* Basal portion of a plant showing a downgrowing rhizoid arising from a cell near the base of the filament.

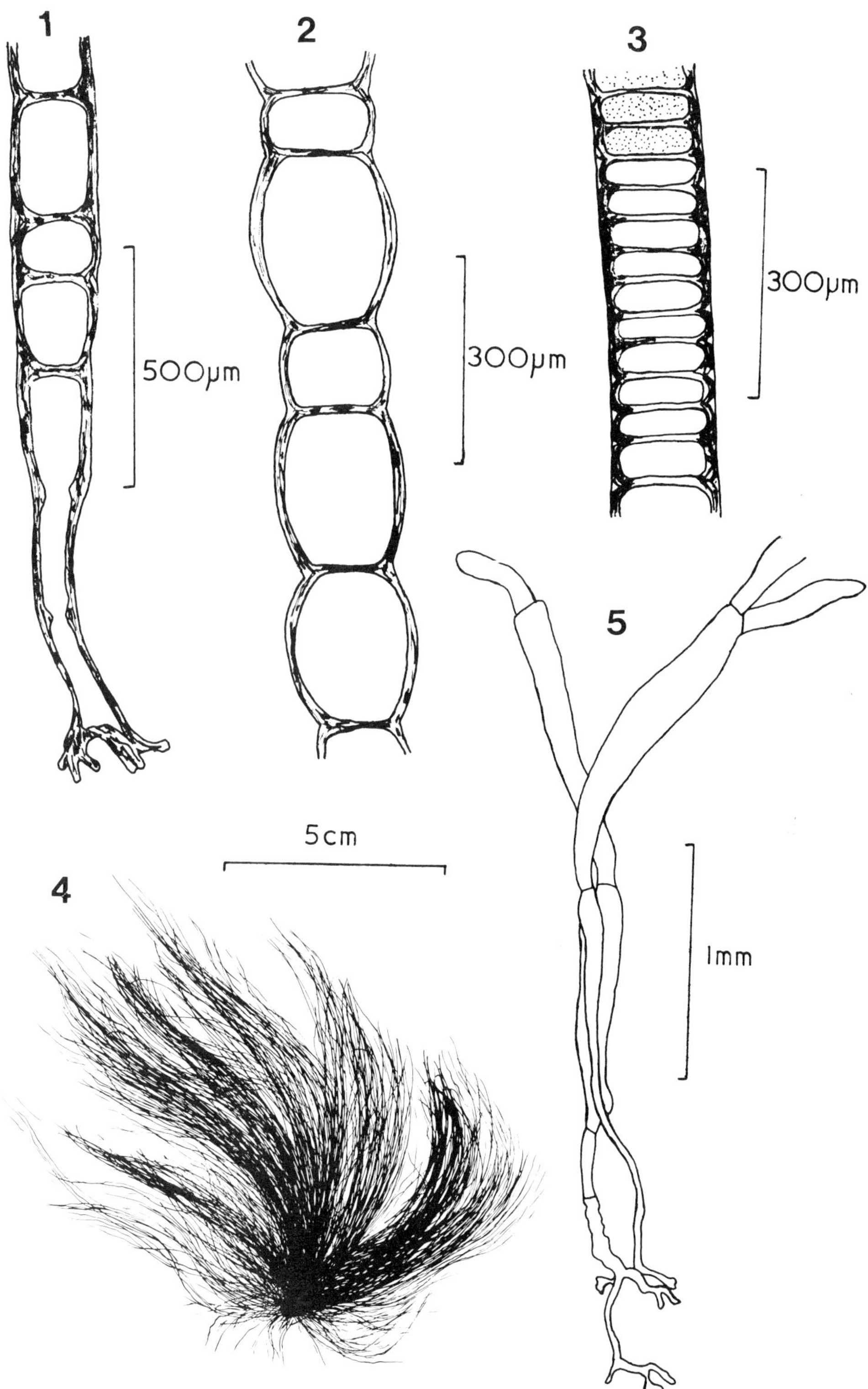
1
2
3
500µm
300µm
300µm
5
1mm
5cm
4

Cladophora Kützing 1843b

Plants consisting of sparingly to profusely divided filaments, attached by a discoid holdfast or branched rhizoids; growth from an apical cell or sometimes an intercalary; meristem; cells having numerous discoid chloroplasts or a reticulate chloroplast and multinucleate; asexual reproduction by quadriflagellate zoospores; sexual reproduction by biflagellate isogametes.

World Distribution: cosmopolitan.

This is one of the most difficult of all the genera covered by this flora and really requires global revision. The only relatively complete regional revisions are those of Hoek (1963) and Söderström (1963) for European species, and Hoek (1982) for American species. The great degree of age-dependent and environmental morphological variation exhibited by most *Cladophora* species often raises difficulties in determination. Hoek (1982) believes identification of species is best undertaken by referring to a table which permits comparison of a combination of taxonomic characters many overlapping to a lesser or greater degree. For this reason identifications arrived at using the key provided below can at best be regarded as tentative.

Key to the Species

1. Plants attached by basal rhizoids and rhizoids arising in distal parts 2
1. Plants attached only by basal rhizoids.................................... 4

2. Apical cells from 23-54 μm in diameter and ultimate branches to 55 μm in diameter .... *C. socialis*
2. Apical cells and ultimate branches always 55 μm or more in diameter 3

3. Cells of branches often more than 12 times longer than broad, with the apical cells sometimes reaching 22 times longer than broad .... *C. coelothrix*
3. Cells of branches never more than 12 times longer than broad.... *C. kamerunica*

4. Growth predominantly by intercalary cell division resulting in branchlets of varying length intercalated along the main axes 5
4. Growth predominantly or almost exclusively by division of conspicuous apical cells resulting in branchlets usually becoming progressively longer away from the tip of the main axes 6

5. Plants forming undulating, hair-like or rope-like tufts; apical cells from 24-65 μm in diameter and ultimate branches from 24-85 μm in diameter.. *C. ruchingeri*
5. Plants forming dense spongy to looser penicillate tufts; apical cells from 10-42 μm in diameter and ultimate branches from 10-53 μm in diameter............. .... *C. montagneana*

6. Growth almost exclusively by division of conspicuous apical cell resulting in regular organization.. 7
6. Growth predominantly apical, with some intercalary divisions towards base resulting in some irregular organization 8

7. Apical cells from 95-240 μm in diameter; branches usually arising at an angle of less than 45° and mostly straight .... *C. prolifera*
7. Apical cells from 25-45 μm in diameter; branches arising mostly at angles of less than 45° and often curved upwards .... *C. conferta*

8. Plants forming coarse, flexuous, entangled masses; apical cells commonly over 15 times longer than broad; estuarine species .... *C. vadorum*
8. Plants forming spongy to penicillate tufts; apical cells rarely more than 10 times longer than broad; strictly marine species, rarely estuarine (except *C. vagabunda*) 9

9. Cells of the main branches often greater than 10 times longer than broad; branches mostly markedly falcate or refracto-falcate.............. *C. dalmatica*
9. Cells of the main branches generally less than 10 times longer than broad; branches usually straight, sometimes curved or falcate..................... 10

10. Plants often with many fascicles (except if floating); maximum number of branches arising at a node is 4 (rarely 5), less in more irregularly organized plants .... *C. vagabunda*
10. Plants without fascicles; maximum number of branches at a node is 2 (rarely 3) in this normally well-organized plant .... *C. laetivirens*

Cladophora coelothrix Kützing Pl. 5, fig. 5.

Kützing, 1843b.

Plants forming cushion-like tufts or dense mats, to about 4 cm in height; branching dense and irregular; main branches from 60-275 μm in diameter and cells from 2 to 16 times as long as broad; ultimate branches from 50-200 μm in diameter, with the cells to about 19 times longer than broad; apical cells cylindrical and occasionally clavate, from 55-215 μm in diameter and from 2.5 to 22 times longer than broad; rhizoids often arising from the basal poles of cells scattered throughout the plant, simple and sometimes unicellular.

For additional figures see Hoek, 1963, pl. 5 to 8, figs 55-78; Hoek, 1982, p. 194 to 196, pls 2-4, figs 11-29.

Occurs in the upper eulittoral subzone as small compact tufts and as rather looser plants when growing in tidepools at the same level on the shore.

Distribution. World: widespread from boreal-antiboreal to tropical seas. West Africa: Cameroun: Hoek 1982; *C. camerunica,* Lawson 1955. Ghana: Hoek 1982. Liberia: De May *et al.* 1977, Hoek 1982. São Tomé: Lawson & Price 1969; *C. (Aegagropila) repens,* Grunow 1868.

Cladophora conferta Crouan frat. ex Mazé & Schramm

Mazé and Schramm, 1870-1877.

Plants forming dense tufts, rarely more than 5 cm in height, dull green; branches pseudodi- or trichotomously divided, with undulating rhizoids arising from the basal cell and the basal end of cells towards the base of the plant; main branches often indistinct, with the cells from 32-95 μm in diameter and 2 to 8 times longer than broad; cells of the ultimate branches 27-53 μm in diameter and 3.5 to 7.5 times longer than broad; apical cells cylindrical or somewhat tapering, occasionally hooked, from 25-45 μm in diameter and 2 to 9 times longer than broad.

For figures see Hoek, 1982, p. 225 to 227, pls 33-35, figs 332-354.

Growing in the eulittoral zone on moderately wave-exposed rocks mixed with *Boodlea composita.*

Distribution. World: widespread in tropical Atlantic particularly on western side. West Africa: Hoek 1982.

Remarks. According to Hoek (1982: 175) this "...is a narrow relation of *Cl. prolifera,* from which species it differs mainly by being much more delicate...".

Cladophora dalmatica Kützing

Kützing, 1843b.

Plants usually occurring as loose or spongy tufts, to about 20 cm in height; branching pseudodichotomous and the branches markedly falcate or refracto-falcate; cells of the main branches from 25-150 μm in diameter and 2.5 to 14 times as long as broad; cells of the ultimate branches from 4-42 μm in diameter and 3 to 10 times as long as broad; apical cells cylindrical or tapering, from 14-32 μm in diameter and 2 to 13 times longer than broad.

For figures see Hoek, 1963, pl. 46, 47, figs 601-635; Hoek, 1982, p. 222, pl. 30, figs 295-312.

Occurs on moderately wave-exposed rocks in the lower eulittoral subzone, often growing mixed with *Cladophora coelothrix.*

Distribution. World: in warm temperate and tropical parts of the Atlantic Ocean as well as in the Mediterranean. West Africa: Ghana: Hoek 1982. São Tomé: Lawson & Price 1969; *C. oblitterata,* Steentoft 1967.

Remarks. The confusion that exists regarding the delimitation and identification of this species is discussed in Lawson and Price (1969).

Cladophora kamerunica Brand　　Pl. 5, fig. 6.

Brand, in Pilger, 1911.

Plants forming soft spongy cushions; branches sparingly dichotomously or occasionally oppositely divided; rhizoids unicellular and unbranched, numerous, to about 20 μm in diameter, arising from the basal parts of cells scattered throughout the plant; cells from 112-120(-140) μm in diameter and 3 to 10 times as long as broad; apical cells cylindrical and blunt, about 120 μm in diameter and to 12 times as long as broad.

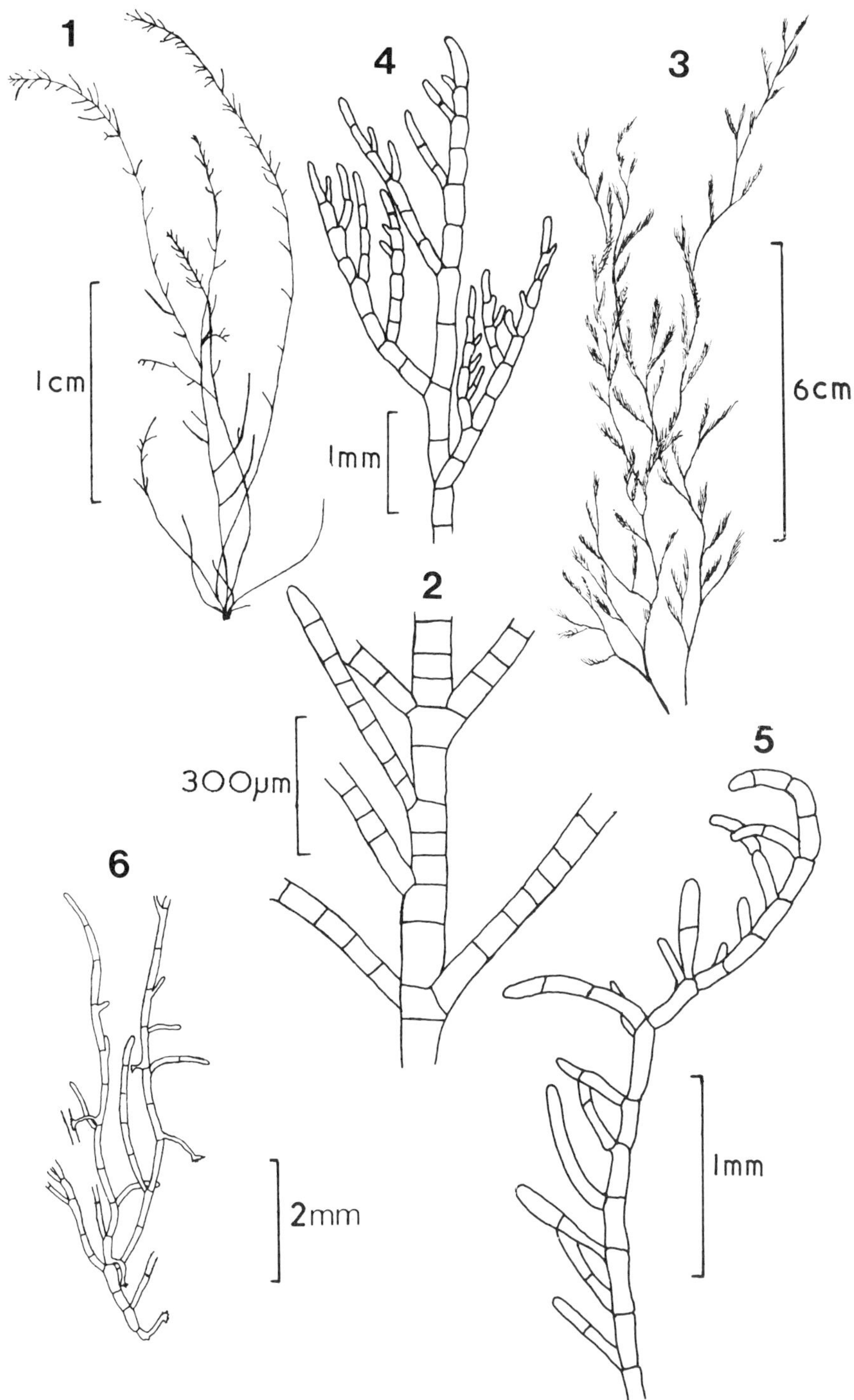

Plate 5: Figs 1, 2. *Cladophora montagneana:* 1. Habit of the looser form of the plant; 2. Upper portion of a branch with some of the branches arising opposite one another. Figs 3, 4. *Cladophora vagabunda:* 3. Portion of plant showing the alternately divided main axes; 4. Terminal region of a branch showing the often secundly arranged ultimate divisions. Fig. 5. *Cladophora coelothrix:* Upper portion of a branch showing the arrangement of the ultimate division. Fig. 6. *Cladophora kamerunica* (after Pilger, 1911): Portion of a plant showing the development of rhizoids on the branches.

For additional figures see Brand, 1911, p. 315, fig. 26.

Occurs in the upper eulittoral subzone and the littoral fringe growing over a wide range of wave-exposure, often confined to cracks and crevices in rocks or only on open rocks when shaded by overhanging trees. Sometimes on the stilt roots of *Rhizophora* fringing the muddy shores of open lagoon systems.

Distribution. World: so far only known from the tropical parts of the eastern Atlantic Ocean. West Africa: ?Bioko: unpublished. Cameroun: Brand 1911, Lawson & Price 1969, Pilger 1911, Schmidt & Gerloff 1957; *C. camerunica,* Stephenson & Stephenson 1972. Sierra Leone: John & Lawson 1977a.

Remarks. It remains to be seen to what extent this plant may be regarded as distinct from *Cladophora coelothrix.* The Camerounian plant is in general smaller though there is a certain overlap in the dimensions and both species have rhizoids arising from cells scattered throughout the plant.

Cladophora laetevirens (Dillwyn) Kützing

Kützing, 1843b; *Conferva laetevirens* Dillwyn, 1809.

Plants forming spongy tufts, about 1.5 cm in height; branching pseudodi- or trichotomous, with branches almost straight or sometimes refracto-falcate; main branches from 53-205 μm in diameter, with the cells from 3 to 11 times as long as broad; ultimate branches from 48-105 μm in diameter and the the cells 2 to 9 times as long as broad; apical cells often cylindrical, sometimes tapering or slightly clavate, from 37-150 μm in diameter and 2 to 11 times longer than broad.

For figures see Hoek, 1963, pl. 31 to 33, figs 418, 429, 433, 440; Hoek, 1982, p. 215 to 217, pls 23-25, figs 238-261.

Occurs on moderately wave-exposed rocks in the lower eulittoral subzone, often growing with other algae.

Distribution. World: in warm temperate and tropical parts of the Atlantic Ocean as well as in the Mediterranean. West Africa: Ghana: Hoek 1982. Príncipe: Lawson & Price 1969, Steentoft 1967. São Tomé: Lawson & Price 1969, Steentoft 1967.

Cladophora montagneana Kützing — Pl. 5, figs 1, 2.

Kützing, 1847.

Plants forming dense spongy tufts, to about 5 cm in height, in wave-exposed situations, and as larger, penicillate tufts, in more sheltered places; branching densely pseudodichotomous and the ultimate divisions secund or opposite; cells of the distinct main branches from 20-135 μm in diameter and 0.8 to 5 times as long as broad; cells of the ultimate branches from 10-53 μm in diameter and 1 to 6 times as long as broad; apical cells cylindrical to somewhat swollen, or distinctly tapering with an obtuse tip, from 10-42 μm in diameter and 1.5 to 10 times longer than broad.

For additional figures see Hoek, 1982, p. 208, 209, pls 16, 17, figs 145-173.

Occurs over a wide range of exposure to wave action on open rocks and in shallow sandy tidepools in the eulittoral zone, and on a few occasions to a depth of about 25 m.

Distribution. World: widespread in the tropical Atlantic and extends into the warm temperate on the western side. West Africa: Cameroun: *C. albida,* Fox 1957, John 1977a, Lawson & Price 1969; *C. tenuis,* Brand 1911. Côte d'Ivoire: *C. albida,* John 1977a. Ghana: Hoek 1982; *C. albida,* Fox 1957, John 1977a, Lawson & Price 1969. Liberia: *C. albida,* De May *et al.* 1977, John 1977a. Nigeria: Hoek 1982; *C. albida,* Fox 1957, John 1977a, Lawson & Price 1969. Sierra Leone: *C. albida,* John & Lawson 1977a. Togo: Hoek 1982; *C. albida,* John 1977a, John & Lawson 1972b.

Cladophora prolifera (Roth) Kützing Pl. 4, fig. 5.

Kützing, 1843b; *Conferva prolifera* Roth, 1797.

Plants forming coarse and dark green tufts, from 10-20 cm or more in height; branches di- or trichotomously divided and densely fasciculate towards the tip, with downwardly directed rhizoids arising from some of the cells near the base of the plant; cells of main branches from 240-650 µm in diameter and 5 to 18 times as long as broad; cells of the ultimate branches from 95-240 µm in diameter and 2 to 16 times as long as broad; apical cells blunt, from 95-240 µm in diameter and 2 to 18 times longer than broad.

For additional figures see Hoek, 1963, pl. 51, figs 677-682, pl. 52, fig. 686; Hoek, 1982, p. 224, pl. 32, figs 318-327.

Occurs as dense tufts in wave-exposed situations and as larger and laxer plants where there is less wave action such as in the sublittoral zone. Commonly growing on lithothamnia in moderately wave-exposed places and sometimes in the algal turf in moderately sheltered situations; often especially abundant in tidepools and also occasionally occurring on rocky platforms to a depth of about 10 m.

Distribution. World: widespread in warm temperate and tropical seas. West Africa: Cameroun: ?John 1977a, ?Lawson & Price 1969, ?Steentoft 1967; *C. multifida,* Brand 1911. Côte d'Ivoire: Hoek 1982, John 1972c, 1977a. Gabon: Hoek 1982, ?John 1977a, John & Lawson 1974a, ?Lawson & Price 1969, ?Steentoft 1967; *C. multifida,* De Toni 1889, Kützing 1849, 1853. Ghana: Hoek 1982, John 1977a, John *et al.* 1977, Lieberman *et al.* 1979; *C. pellucida,* Lawson & Price 1969; *C. trichotoma,* Lawson 1956, Stephenson & Stephenson 1972. Liberia: De May *et al.* 1977, Hoek 1982, John 1972c, 1977a. São Tomé: Hariot 1908, Henriques 1917, John 1977a, Lawson & Price 1969, Steentoft 1967; *C. catenata,* Hariot 1908, Henriques 1917. Sierra Leone: Aleem 1978, John & Lawson 1977a.

Cladophora ruchingeri (C. Agardh) Kützing Pl. 4, fig. 4.

Kützing, 1845; *Conferva ruchingeri* C. Agardh, 1824.

Plants forming undulating, hair or rope-like tufts, from 6-20 cm in height; branching sparingly pseudodichotomous; main branches from 50-160 μm in diameter, with the cells from 2 to 8 times as long as broad; ultimate branches from 24-85 μm in diameter, with the cells from 2 to 8 times as long as broad, having the walls from 1.5-4 μm in thickness; apical cells from 24-65 μm in diameter and 2 to 11 times longer than broad.

For additional figures see Hoek, 1963, pls 16, 17, figs 164-183; Hoek, 1982, p. 205, pl. 13, figs 120-127.

Occurs over a wide range of wave-exposure and is most common in the upper part of the lower eulittoral subzone on often sand-covered beach rocks.

Distribution. World: in warm temperate and tropical parts of the Atlantic Ocean as well as in the Mediterranean. West Africa: Ghana: Hoek 1982.

Cladophora socialis Kützing

Kützing, 1849.

Plants forming dense spongy tufts or mats, from 2-5 cm high, dull green; branching somewhat irregular, often arising at wide angles; main branches from 45-90 μm in diameter, with cells from 2 to 20 times longer than broad; ultimate branches from 22-65 μm in diameter and cells from 2 to 17 times longer than broad; apical cells cylindrical with rounded tips, from 20-54 μm in diameter and from 7 to 30 times longer than broad; rhizoids arising from the basal poles of cells throughout the plant and attached by a rather simple basal rhizoid.

For figures see Hoek, 1982, p. 197, pl. 5, figs 30-40.

Occurs as a dense carpet on the floor of tidepools in the littoral fringe.

Distribution. World: widely distributed in tropical and warm temperate seas. West Africa: Liberia: Hoek 1982.

Remarks. According to Hoek (1982: 55) it "...is narrowly related to *Cl. coelothrix,* from which it differs by being much more slender...".

Cladophora vadorum (Areschoug) Kützing

Kützing, 1849; *Conferva vadorum* Areschoug, 1843.

Plants forming coarse, flexuous tufts, up to 30 cm high, light or pale green; branching pseudodichotomous, with somewhat irregularly organized branchlets; cells of the main branches 33-120 μm in diameter and 2.5 to 10 times longer than broad; ultimate branches from 30-90 μm in diameter and cells 3 to 20 times longer than broad; apical cell from 30-60 μm in diameter and 4 to 21 times longer thanbroad.

For figures see Hoek, 1982, p. 223, pl. 31, figs 313-317.

Occurs as very entangled masses on wave-sheltered estuarine rocks in the sublittoral fringe and just above.

Distribution. World: from cold temperate to tropical parts of the Atlantic and in the Mediterranean. West Africa: Gabon: Hoek 1982; *C. laetevirens?,* John and Lawson 1974a.

Cladophora vagabunda (Linnaeus) Hoek Pl. 5, figs 3, 4.

Hoek, 1963; *Conferva vagabunda* Linnaeus, 1753.

Plants forming lax and light green tufts, from 5-15 cm in height; branches pseudodi-, tri-, or tetrachotomously divided and the ultimate branches fasiculate and pectinate; cells of the main branches from 80-250(-350) μm in diameter and to about 12 times as long as broad; cells of the ultimate branches from 18-160 μm in diameter and to 10 times longer than broad; apical cells from 17-135 μm in diameter and 1.5 to 11 times as long as broad.

For additional figures see Hoek, 1963, pl. 36, 37, figs 470-503, pl. 39, figs 505-514; Hoek, 1982, p. 218 to 221, pls 26-29, figs 264-294.

Occurs on moderately sheltered to wave-exposed rocks in the lower subzone of the eulittoral and often common in shallow tidepools; occasionally to a depth of 14 m growing on calcareous cobbles.

Distribution. World: in most warm temperate seas and probably pantropical. West Africa: Benin: Hoek 1982. Cameroun: Hoek 1982; *C. sertularina,* Brand 1911, Lawson and Price 1969. Côte d'Ivoire, Hoek 1982, John 1977a. Ghana: Hoek 1982, John 1977a; *C. fascicularis,* Fox 1957, Lawson and Price 1969. Liberia: De May *et al.* 1977, Hoek 1982. Nigeria: John 1977a; *C. fascicularis,* Fox 1957, Lawson and Price 1969, Steentoft Nielsen 1958. Sierra Leone: John and Lawson 1977a. Togo: John 1977a, John and Lawson 1977b.

Remarks. There is some confusion regarding this species which is briefly summarised by Lawson and Price (1969). According to Hoek (1982: 137) the three type specimens he examinded of *C. sertularina* (Montagne) Kützing were "...old plants of *Cl. vagabunda,* reduced to old main axes with new, robust proliferations...".

Doubtful Records

Cladophora crystallina (Roth) Kützing

Aleem (1978) reports this species from one locality on the Sierra Leone peninsula. According to Hoek (1963), based on culture experiments, this species may be conspecific with *Cladophora vagabunda* or *C. glomerata* variety *glomerata*. John and Lawson (1977a) have found *C. vagabunda* in Sierra Leone and Aleem's record may simply represent a form of this species.

Cladophora senegalensis Kützing

This species is recorded for both Gambia and Sénégal by Kützing (1849) and De Toni (1889). It has not been reported in the many recent collections from Sénégal by French algologists, nor was it re-discovered in Gambia in the recent survey by John and Lawson (1977b). We have not included this little-known species and consider it to be a doubtful record.

Rhizoclonium Kützing 1843b

Plants consisting of thick-walled filaments, simple or occasionally bearing tapering rhizoidal branches one to several cells in length; growth always intercalary; cells containing a reticulate chloroplast, numerous pyrenoids and nuclei; asexual reproduction by bi- or quadriflagellate zoospores; sexual reproduction by biflagellate isogametes.

World Distribution: cosmopolitan.

Key to the Species

1. Filaments from 10-18 μm in diameter and the cell walls from 1.5-2 μm thick . . . . *R. implexum*
1. Filaments usually greater than 28 μm in diameter and the cell wall more than 2 μm thick 2
2. Filaments from 28-40(-60) μm in diameter and the walls from 2-6 μm thick . . . . *R. riparium*
2. Filaments from 50-80(-100) μm in diameter and the walls from 4-10 μm thick . . . . *R. ambiguum*

Rhizoclonium ambiguum (Hooker f. & Harvey) Kützing Pl. 6, figs 2, 3.

Kützing, 1849; *Conferva ambigua* Hooker and Harvey, 1845.

Plants forming coarse entangled masses, cells of the filaments from 50-80(-100) μm in diameter and 1.5 to 4 times longer than broad, with the walls from 4-10 μm thick; rhizoidal branches from 1- to 3-celled and knee-like joints common along the filaments.

For additional figures see Kützing, 1853, pl. 67, fig. 2 (as *R. africanum*).

Occurs on the roots of mangroves, stranded logs, in cracks and crevices, and sometimes on rocks shaded by overhangs or trees; occasionally forming a distinct belt but more commonly associated with a mat of small red algae. Found in the upper eulittoral subzone and the littoral fringe on open parts of the coast as well as in brackish-water habitats such as estuaries and mangrove swamps.

Distribution. World: probably widespread from cold temperate to tropical seas. West Africa: Bioko: unpublished. Cameroun: *R. africanum,* John 1977a. Côte d' Ivoire: *R. africanum,* John 1972c, 1977a. Gabon: *R. africanum,* John & Lawson 1974a. Ghana: *R. africanum,* John 1977a; *R. hookeri,* ?Lawson 1960c, Lawson & Price 1969. Sierra Leone: *R. africanum,* John & Lawson 1977a.

Rhizoclonium implexum (Dillwyn) Kützing

Kützing, 1845; *Conferva implexa* Dillwyn, 1809.

Plants solitary or forming felt-like patches; cells of the filaments from 10-18 μm in diameter and from 2 to 5 times as long as broad, with the walls from 1.5-2 μm in thickness; lateral rhizoidal branches uncommon.

For figures see Abbott and Hollenberg, 1976, p. 91, fig. 45.

Occurs usually with algae such as *Murrayella periclados* and *Bostrychia* spp. on the roots of mangroves or else entangled with other algae growing in moderately wave-exposed situations on the open coast.

Distribution. World: cosmopolitan. West Africa: Benin: John & Lawson 1972b. Cameroun: Lawson & Price 1969; *R. riparium* var. *implexum,* Brand 1911. Gambia: John & Lawson 1977b. Ghana: Lawson & Price 1969. Nigeria: Fox 1957, Lawson & Price 1969, Steentoft 1967. São Tomé: Lawson & Price 1969, Steentoft 1967. Sierra Leone: Fox 1957, John & Lawson 1977a, Lawson & Price 1969, Steentoft 1967.

Plate 6: Fig. 1. *Willeella ordinata:* Portion of a plant showing the characteristic branching pattern with simple branches alternating with compound branches. Figs 2, 3. *Rhizoclonium ambiguum:* 2. Portion of a filament with a lateral unicellular rhizoid; 3. Portion of a filament showing a "knee-joint". Figs 4-6. *Acetabularia pusilla:* 4. Terminal portion of a plant with the stalk bearing a disc of fused branchlets; 5. Disc of 8 lightly connected branchlets; 6. Section of a branchlet showing the knob-like superior corona bearing hairs.

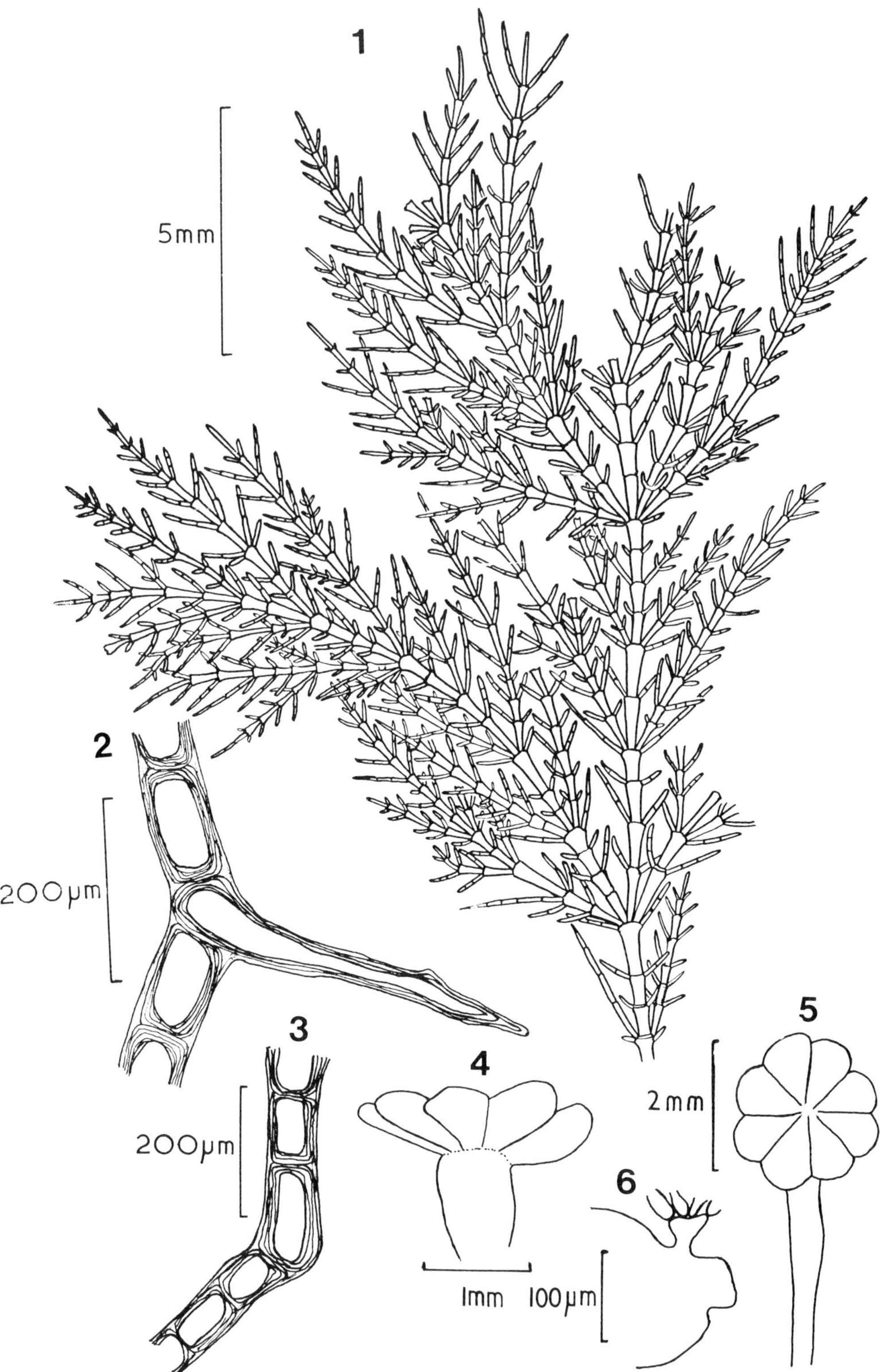
1
5mm
2
200µm
3
200µm
4
1mm
5
2mm
6
100µm

Rhizoclonium riparium (Roth) Harvey

Harvey, 1849; *Conferva riparia* Roth, 1806.

Plants often forming entangled mats; cells of the filaments from 28-40(-60) μm in diameter and from 1.5 to 4.5 times longer than broad, with the walls from 2-6 μm in thickness; rhizoidal branches very rare.

For figures see Taylor, 1962a, pl. 1, fig. 3.

Occurs most commonly with *Bostrychia* spp. growing in the littoral fringe in shaded places such as rock crevices or on rocks beneath palm trees; occasionally in the red algal turf in the lower eulittoral subzone.

Distribution. World: cosmopolitan. West Africa: Bioko: unpublished. Cameroun: John 1977a, Lawson 1955, Lawson & Price 1969, Trochain 1940, Stephenson & Stephenson 1972. Côte d'Ivoire: John 1977a. Gabon: John & Lawson 1974a. Ghana: John 1977a, Stephenson & Stephenson 1972; *R. riparium* f. *riparium* = *validum*, Lawson & Price 1969. Sierra Leone: *R. kochianum*, Aleem 1978. Togo: John 1977a, John & Lawson 1972b.

Remarks. It has been suggested by Lawson and Price (1969), based on the work of Koster (1955), that as *Rhizoclonium riparium* f. *validum* Foslie and *Lola tortuosa* (Dillwyn) V.J. Chapman are possibly conspecific, the record from Nigeria under the latter name may actually represent *R. riparium* and not *Chaetomorpha capillaris* (q.v.).

Family Anadyomenaceae

Microdictyon Decaisne 1839

Plants appearing membranaceous, consisting of closely branched monosiphonous filaments, with the branching in one plane and the branchlets fused to one another either by ring-like thickenings or special hapteral cells, giving a network having angular interstices, attached basally by simple or sparingly divided rhizoids.

World Distribution: in a number of warm temperate and tropical seas.

Microdictyon calodictyon (Montagne) Decaisne Pl. 7, fig. 2.

Decaisne, 1841; *Anadyomene calodictyon* Montagne, 1839-1841.

Plants sessile, with the net about 5 cm across; cells of the branches from 100-150 μm in diameter and to about 260 μm in length, from 1 to 3 times as long as broad; branchlets often 5 to 7 arising from a clavate cell, fusing to form an irregular network, with the interstices small as compared to the cell diameters.

For additional figures see Børgesen, 1925, p. 33 to 36, figs 5-8.

There is no information concerning the ecology of this plant in the region but in its type locality in the Canary Islands it occurs on larger algae growing in rather wave-exposed situations in the lower eulittoral subzone and below.

Distribution. World: in warm temperate and tropical parts of the eastern Atlantic Ocean. West Africa: São Tomé: Lawson & Price 1969, Steentoft 1967.

Remarks. The dried specimens from the region are black and do not adhere to paper.

Willeella* Børgesen 1930b

Plants filamentous, consisting of erect branches regularly and oppositely divided in one plane, sometimes having several branches issuing from a single cell in a fan-like manner, with a distinctive pattern of simple branches alternating with compound ones, basally attached by an intricate system of rhizoids.

World Distribution: probably pantropical.

Willeella ordinata Børgesen　　　　Pl. 6, fig. 1.

Børgesen, 1930b.

Plants forming bushy tufts, from 4-12 cm in height; branching opposite and in one plane, with sometimes several branches arising at acute angles from a somewhat swollen cell of the main axis, simple branches alternating with compound ones; basal cell from 120-170 μm in diameter and from 200-800 μm in length.

For additional figures see Børgesen, 1930b, p. 156, 157, figs 3, 4, pl. 1, fig. 1.

Occurs on rocky platforms and calcareous cobbles at depths from 8 to 14 m; never very common.

Distribution. World: as for genus. West Africa: Ghana: John *et al.* 1977, Lawson & Price 1969, Lieberman *et al.* 1979; *Cladophora ordinata,* Hoek 1982.

Remarks: Papenfuss and Egerod (1957) believe that *Willeella mexicana* Dawson and *W. ordinata* are conspecific; if this is correct, the range extends into the Pacific Ocean.

*Hoek (1982) considers *Willeela* to be a section of the broadly conceived genus *Cladophora* and earlier transferred to it *W. ordinata.*

Order Siphonocladiales · Family Dasycladaceae

Acetabularia Lamouroux 1812

Plants consisting of a creeping base bearing cylindrical or rugose stipes terminating in a disc or cup of connected or free whorled rays, often superimposed on one another; disc or cup bearing adaxially on the rays a ring of small projections forming collectively the superior corona, occasionally also with an inferior corona; superior corona often having delicate deciduous hairs or bearing their scars; rays sometimes functioning as aplanosporangia.

World Distribution: widespread in warm temperate and tropical seas.

Acetabularia pusilla (Howe) Collins Pl. 6, figs 4-6.

Collins, 1909; *Acetabulum pusillum* Howe, 1909.

Plants small and about 2-3 mm high, with the transversely rugose stipe bearing a flat disc about 2 mm in diameter consisting of about 8(7-9) lightly connected rays; disc rays obovate or somewhat cylindrical, blunt or slightly tapering at their tip, readily separated; corona superior with each projection about 40 μm in diameter and bearing 2 to 4 hairs or hair scars.

For additional figures see Taylor, 1960, p. 675, fig. 13.

Occurs on moderately wave-sheltered rocks in the sublittoral fringe and the immediate sublittoral zone on stones with other small algae.

Distribution. World: tropical parts of the Atlantic Ocean. West Africa: Liberia: De May *et al.* 1977.

Family Boodleaceae

Boodlea Murray & De Toni 1890

Plants forming small spongy cushion-like tufts or mats of freely branched filaments, with the cross walls initially absent from the base of the lateral branches, commonly having the branches attached to one another by tentacular cells developing at the branch apices.

World Distribution: in warm temperate and tropical seas.

Plate 7: Fig. 1. *Boodlea composita:* Terminal portion of a plant with a cross wall initially absent from the base of the lateral branches. Fig. 2. *Microdictyon calodictyon:* Portion of a plant showing the anastomosing net-work of branchlets. Figs 3, 4. *Trichosolen retrorsa:* 3. Plant dichotomously divided and arising from a rhizoidal plexus; 4. Gametangium arising from near the swollen base of a branchlet and backwardly directed towards the central axis. Fig. 5. *Ernodesmis verticillata:* Portion of a plant showing the successive series of whorled branches. Fig. 6. *Cladophoropsis membranacea:* Terminal portion of a plant showing the lateral branching and the absence of basal cross walls.

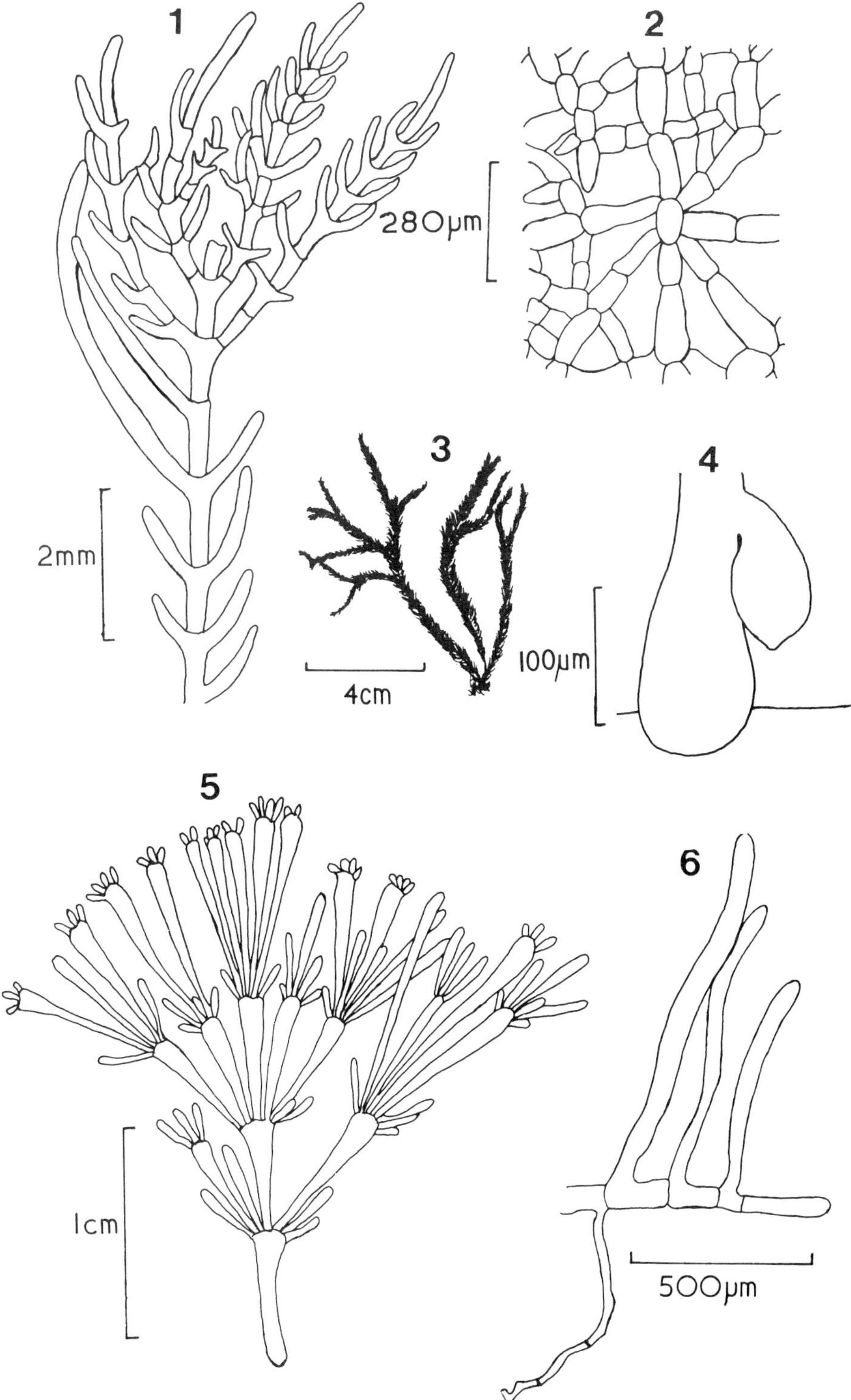
1
2
280μm
2mm
3
4
4cm
100μm
5
6
1cm
500μm

Boodlea composita (Harvey) Brand Pl. 7, fig. 1.

Brand, 1904; *Cladophora composita* Harvey, 1834.

Plants forming dense, very firm cushion-like tufts or mats, about 2 cm in height, attached by branched and septate rhizoids; branches irregularly divided and becoming somewhat whorled towards the apices, with the cells of the main axes from 200-350 μm in diameter and 0.8-2 mm in length.

For additional figures see Brand, 1904, pl. 6, figs 28-35.

Occurs as extensive spongy mats, or sometimes mixed with other low-growing algae, on moderately sheltered and almost horizontal rocky platforms in the lower eulittoral subzone while as small cushion-like tufts where there is moderate wave-exposure.

Distribution. World: pantropical. West Africa: Bioko: unpublished. Cameroun: unpublished. Ghana: Lawson 1956, 1966, Lawson & Price 1969, Stephenson & Stephenson 1972.

Doubtful Record

Boodlea struveoides Howe

The report from Sierra Leone (Aleem, 1978) is the first mention of this species from the eastern side of the Atlantic Ocean. Until verified by examination of the collection on which it is based we prefer to regard the record as doubtful.

Family Valoniaceae

Cladophoropsis Børgesen 1905

Plants filamentous, consisting of a basal system of colourless branches bearing intertwined and many times laterally divided erect branches, with the branchlets arising at the distal ends of the cells and often without a basal cross wall, attached by rhizoids; growth normally apical; chloroplasts reticulate, club-shaped or disrupted, abundant in the cells of the upper branches and the branchlets, with numerous pyrenoids.

World Distribution: widespread in warm temperate and tropical seas.

Cladophoropsis membranacea (C. Agardh) Børgesen Pl. 7, fig. 6.

Børgesen, 1925; *Conferva membranacea* C. Agardh, 1824.

Plants forming dense cushion-like tufts or small mats, to about 2 cm in height; branches of the erect system much entangled, irregularly divided below and becoming alternately divided above, from 140-220(-280) μm in diameter, with the branchlets narrow and secund towards the apex.

For additional figures see Børgesen, 1925, p. 24, fig. 1.

Occurs growing together with other small algae in cracks and crevices in rocks in the upper eulittoral subzone as well as on rocky platforms at a depth of about 10 m.

Distribution. World: as for genus. West Africa: Cameroun: unpublished. Ghana: unpublished. Sierra Leone: *Cladophora membranacea*, Aleem 1978.

Ernodesmis Børgesen 1912

Plants filamentous, consisting of successive series of whorled branches originating from a single basally annulate clavate cell, attached by irregularly divided, septate rhizoids.

World Distribution: in warm temperate and tropical seas.

Ernodesmis verticillata (Kützing) Børgesen — Pl. 7, fig. 5.

Børgesen, 1912; *Valonia verticillata* Kützing, 1849.

Plants usually forming dense and spongy hemispherical clumps, from 2-5 cm in height and sometimes to about 20 cm across; basal cell clavate, from 0.5-1.2(-2) mm in diameter and 5-12 mm in length, bearing at its apex clusters of similar but smaller branch cells which in turn bear whorls of cells terminally, repeating the pattern a number of times.

For additional figures see Børgesen, 1913, p. 67 to 69, figs 52-54.

Occurs in cracks and crevices in moderately sheltered to moderately wave-exposed rocks extending from the upper eulittoral subzone down into the sublittoral where it grows as large, loose and very light green clumps on rocky platforms down to a depth of 30 m.

Distribution. World: in warm temperate and tropical parts of the Atlantic and Indian Oceans. West Africa: Ghana: Dickinson & Foote 1951, John *et al.* 1977, Lawson 1956, Lawson & Price 1969, Lieberman *et al.* 1979, Steentoft 1967. Príncipe: Lawson & Price 1969, Steentoft 1967. São Tomé: Lawson & Price 1969, Steentoft 1967.

Siphonocladus Schmitz 1879

Plants filamentous, consisting initially of a large and often somewhat inflated cell, later becoming divided by the formation of irregular walls into cells from which arise long and radially arranged branches, attached basally by multicellular rhizoids.

World Distribution: probably widespread in warm temperate and tropical seas.

Siphonocladus tropicus J. Agardh Pl. 8, fig. 3.

J. Agardh, 1887.

Plants to about 4 cm in height; basal cell cylindrical to clavate, from 0.5-1 mm in diameter and to about 2 cm in length; branches arising from the upper part of the basal cell, slender and usually further divided into short branchlets.

For additional figures see Børgesen, 1913, p. 62 to 66, figs 44-51.

There is no information on the ecology of this plant in the region, although the specimens examined from São Tomé were mixed with *Centroceras clavulatum* and a *Chaetomorpha* suggesting that they were collected from the eulittoral zone. Found elsewhere in the Atlantic Ocean either unattached or in sheltered situations growing on larger algae; reported as having been dredged from depths to 18 m (Taylor, 1960).

Distribution. World: probably pantropical. West Africa: São Tomé: Lawson & Price 1969, Steentoft 1967.

Doubtful Record

Siphonocladus brachyartus Svedelius Pl. 10, figs 6, 7.

This plant has been tentatively identified from a collection made at the entrance to the Tabou River in Côte d'Ivoire (John, 1977a). It was growing as dense, spongy cushion-like tufts or as an extensive springy "turf" in the eulittoral zone and extended into the littoral fringe only in shaded tidepools or beneath rock overhangs.

The African plants closely resemble Svedelius's (1895) plant from the Magellan Straits in habit, branching, cell dimensions and even habitat. Nevertheless, it has been decided to regard this record as most doubtful until the South American material can be traced, since there is a wide geographical separation and climatic disparity between the two regions; the Magellan Straits are boreal-antiboreal. From the figures in Svedelius (pl. 16, figs 2, 3) the specimen was probably preserved in formalin and the curator of the Uppsala Museum (UPS) believes that this collection might now be lost.

Plate 8: Figs 1, 2. *Struvea anastomosans:* 1. Portion of a plant with the opposite lateral branches also branching in one plane; 2. Branchlets fused to another branchlet by a small hapteral cell. Fig. 3. *Siphonoclados tropicus:* Plant showing the lateral branching and the irregular walls dividing the cells. Fig. 4. *Bryopsis plumosa:* Portion of a plant with the branches having a somewhat triangular to lanceolate outline. Fig. 5. *Bryopsis pennata:* Plant with the bilaterally arranged branchlets. Figs 6, 7. *Bryopsis stenoptera:* 6. Plant having a somewhat lanceolate outline; 7. Terminal portion of a branch showing the development of the branchlets and the absence of septa. Fig. 8. *Boodleopsis pusilla:* portion of a branch showing the presence of constrictions particularly at the dichotomies.

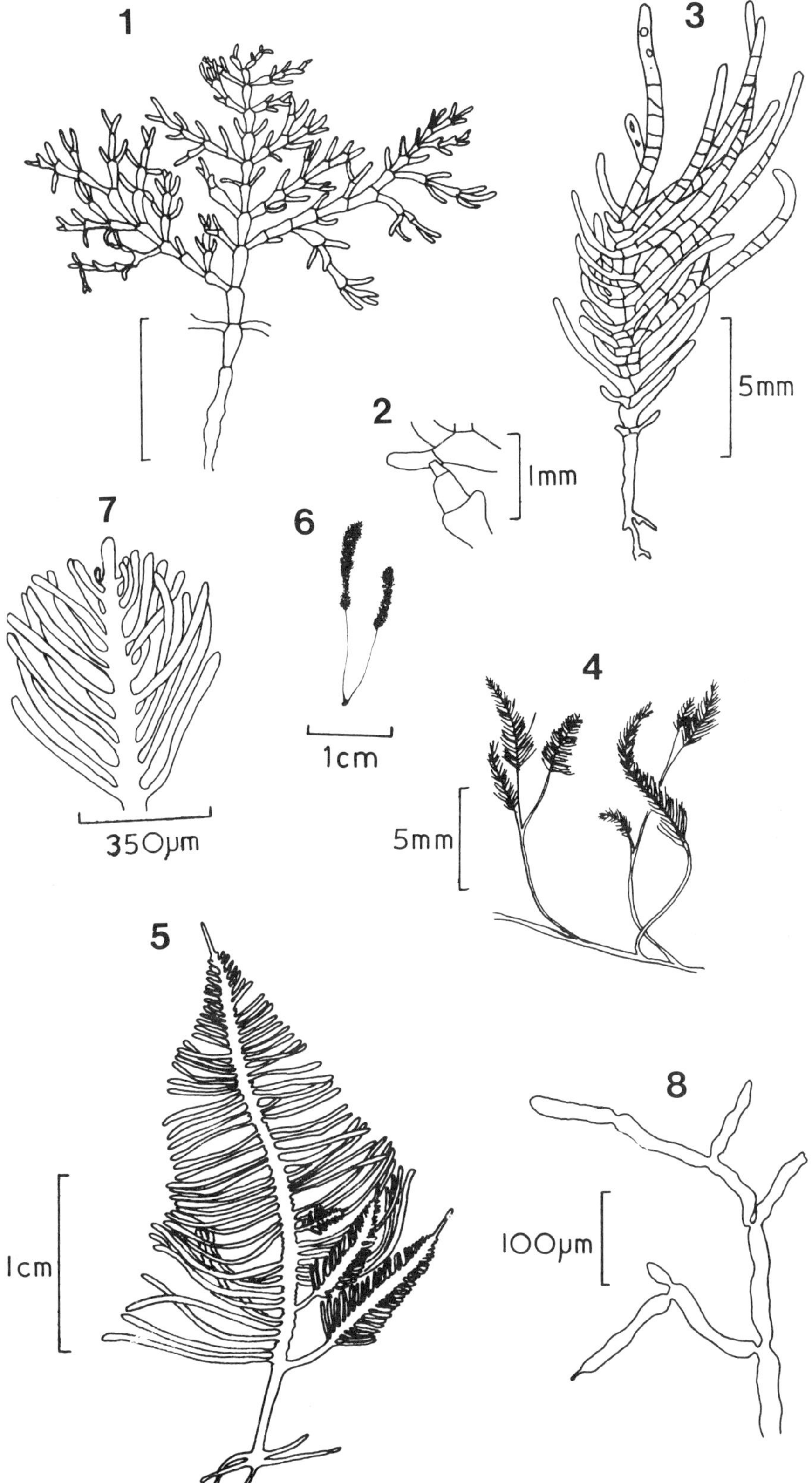
1
3
5mm
2
1mm
7
6
4
1cm
350µm
5mm
5
8
1cm
100µm

Struvea Sonder 1845

Plants filamentous, consisting of a simple or branched monosiphonous stalk, bearing from the upper end one or more branches which give rise in turn to successive series of branches and branchlets that become fused to one another by small hapteral cells and form a regular network, attached to the substratum by branched multicellular rhizoids.

World Distribution: in warm temperate and tropical seas.

Struvea anastomosans (Harvey) Piccone & Grunow Pl. 8, figs 1, 2.

Piccone and Grunow, in Piccone, 1884; *Cladophora anastomosans* Harvey, 1859.

Plants forming semi-erect tufts, from 1-3 cm in height; primary stalk cell only segmented above, from 350-700 μm in diameter, bearing distally pairs of opposite lateral branches which branch further in one plane; lateral branchlets consisting of cells from 100-170 μm in diameter, fused by hapteral cells to form a network.

For additional figures see Nizamuddin, 1969, p. 255, 257 and 259, figs 27-37, pl. 7.

Occurs growing over a wide range of wave-exposure in the lower eulittoral subzone mixed with the algal turf and found most commonly on the flat tops of rocky platforms; occasionally as small plants having a poorly developed net at depths of about 10 m.

Distribution. World: widespread in warm temperate and tropical seas. West Africa: Bioko: unpublished. Cameroun: John 1977a, Lawson 1955, Lawson & Price 1969, Steentoft 1967, Stephenson & Stephenson 1972; *S. delicatula,* Schmidt & Gerloff 1957; *S. delicatula* var. *caracasana,* Pilger 1911. Côte d'Ivoire: John 1972c, John 1977a. Gabon: John 1977a, John & Lawson 1974a; *S. delicatula,* Hariot 1896. Gambia: John & Lawson 1977b. Ghana: Dickinson & Foote 1950, John 1977a, John *et al.* 1977, Lawson 1956, Lieberman *et al.* 1979, Stephenson & Stephenson 1972. Liberia: De May *et al.* 1977. São Tomé: John 1977a, Lawson & Price 1969, Steentoft 1967; *S. delicatula,* Hariot 1908, Henriques 1917; *S. multipartita,* Pilger 1920. Sierra Leone: John & Lawson 1977a.

Remarks. Steentoft (1967) refers the São Tomé material to the variety *Struvea anastomosans* var. *anastomosans,* "which has some irregularly branched nets and shows some attachment of the end cells".

Order Codiales · Family Bryopsidaceae

Bryopsis Lamouroux 1809b

Plants often soft and flaccid, consisting of prostrate and erect aseptate, coenocytic branches attached by rhizoids; erect branches having a distinct main axis bearing pinnately or more rarely radially arranged branchlets; branchlets decreasing somewhat in length towards the apex; chloroplasts numerous, discoid, each with a conspicuous pyrenoid; reproduction by septa formation at the base of the branchlets converting them to gametangia producing biflagellate gametes.

World Distribution: from boreal-antiboreal to tropical seas.

Key to the Species

1. Plants having a somewhat triangular to lanceolate outline, with the branchlets arising bilaterally throughout .... *B. plumosa*
1. Plants usually having an oblong or linear-lanceolate outline, with the branchlets arising bi- or unilaterally... 2

2. Plants having the main axes from 240-360 μm in diameter and the branchlets from (47-)60-150 μm in diameter................................ *B. pennata*
2. Plants having the main axes from 80-260 μm in diameter and the branchlets from 30-42 μm in diameter.................................. *B. stenoptera*

Bryopsis pennata Lamouroux Pl. 8, fig. 5.

Lamouroux, 1809b.

Plants usually forming dark green and sometimes iridescent clumps or mats, from 5-10 cm in length, oblong to linear-lanceolate in outline; branches sparingly divided and from 240-360 μm in diameter, with the lateral branchlets from (47-)60-150 μm in diameter towards the base and slightly tapering towards the tip.

For additional figures see Taylor, 1960, p. 681, pl. 9, fig. 2.

var. **secunda** (Harvey) Collins & Hervey

Collins and Hervey, 1917; *B. plumosa* var. *secunda* Harvey, 1858.

Plants in which the branchlets arise for the most part in a single row and so having an asymmetric outline.

For figures see Taylor, 1928, pl. 11, figs 10-12.

Occurs over a wide range of exposure to wave action growing most commonly on the steep sides of tidepools and on open rocks in the lower eulittoral subzone; sometimes on the shells of various animals especially large barnacles and mussels. sels.

Distribution. World: apparently widespread in warm temperate and tropical seas. West Africa: Benin: John 1977a, John & Lawson 1972b. Bioko: unpublished. Cameroun: John 1977a, Lawson & Price 1969, Steentoft 1967. Côte d'Ivoire: John 1972c, 1977a. Gabon: John & Lawson 1974a. Ghana: Fox 1957, John 1977a, Lawson & Price 1969, Steentoft 1967; *B. plumosa* var. *pennata,* Lawson & Price 1969. *B. pennata* var. *secunda,* Lawson & Price 1969, Steentoft 1967; *B. plumosa* var. *harveyana,* Lawson & Price 1969. Liberia: De May *et al.* 1977, John 1972c, 1977a. Nigeria: Fox

1957, John 1977a, Lawson & Price 1969, Steentoft 1967. São Tomé: John 1977a, Lawson & Price 1969, Steentoft 1967. *B. pennata* var. *secunda*, Lawson & Price 1969, Steentoft 1967. Sierra Leone: Fox 1957, John & Lawson 1977a, Sourie 1954, Steentoft 1967. Togo: John 1977a, John & Lawson 1972b.

Remarks. Womersley and Bailey (1970) have pointed out the close similarity between *Bryopsis indica* Gepp & Gepp, a species found in the Indian and Pacific Oceans, and *B. pennata* var. *secunda* and suggest that they might be conspecific.

Bryopsis plumosa (Hudson) C. Agardh Pl. 8, fig. 4.

C. Agardh, 1821b; *Ulva plumosa* Hudson, 1778.

Plants forming erect and usually light green clumps, from 3-4 cm in height, somewhat triangular to lanceolate in outline; branches simple or sparingly divided, with the main axes about 350 μm in diameter near the base and tapering above, branchlets to about 220 μm in diameter and to about 5 mm long in lower parts.

For additional figures see Taylor, 1928, pl. 11, fig. 14.

Occurs on moderately wave-sheltered rocks in the upper part of the lower eulittoral subzone and is found occasionally growing in tidepools.

Distribution. World: as for genus. West Africa: Ghana: *B. plumosa* var. *typica*, Lawson & Price 1969. Liberia: De May *et al.* 1977. Sierra Leone: Aleem 1978, John & Lawson 1977a. Togo: John & Lawson 1972b.

Bryopsis stenoptera Pilger Pl. 8, figs 6, 7.

Pilger, 1911.

Plants loosely intertwined in the algal turf, from 2-10 cm long, lanceolate in outline; branches sparingly divided, with the main axes from 220-260(-335) μm in diameter near the base and gradually tapering above, branchlets from 30-42 μm in diameter and nearly equal in length and breadth along much of the branch axis.

For additional figures see Pilger, 1911, p. 295, figs 1, 2.

Occurs often together with *Bryopsis pennata* in the algal turf and is especially common on wave-sheltered rocks; occasionally found on calcareous cobbles to a depth of about 14 m.

Distribution. World: so far known only from the tropical parts of the eastern Atlantic Ocean. West Africa: Bioko: unpublished. Cameroun: Dawson 1962, Fox 1957, Lawson & Price 1969, Pilger 1911. Ghana: Lieberman *et al.* 1979. Nigeria: Fox 1957, Lawson & Price 1969, Steentoft Nielsen 1958.

Doubtful Records

Bryopsis balbisiana Lamouroux ex C. Agardh

There is a single collection of a *Bryopsis* from Sussex in Sierra Leone which might be referable to this species. The erect branches are sparingly divided, the main axis is very obvious, and the lateral branchlets are confined to the tips of the branches. This is only a very tentative determination and must remain so until further material is obtained from this locality as the plant may represent no more than a very depauperate form of *B. pennata*.

Bryopsis hypnoides Lamouroux

The report of this species from a number of localities on the Sierra Leone peninsula by Aleem (1978) is its first mention on the mainland coast of West Africa. The record requires verification as the most common species previously known from the peninsula is *Bryopsis pennata*.

Derbesia Solier 1846

Doubtful Record

Derbesia tenuissima (De Notier) Crouan frat.

We regard the report of a sterile collection of this plant from the Sierra Leone peninsula by Aleem (1978) as doubtful. Four other species have been reported from the West African mainland but the criteria used in separating them are not entirely satisfactory (see Lawson and Price, 1969).

Trichosolen Montagne 1860

Plants soft and straggly, consisting of erect branches arising from a rhizoidal plexus or more rarely from a creeping base; branches having a very distinctive main axis, often dichotomously divided, densely ramellate; branchlets irregularly arranged on the main axes, usually swollen at the base and having a small pore where they are joined to the axis, simple or occasionally divided; gametangia borne laterally on the branchlets, sessile or shortly stalked, variously shaped.

World Distribution: widely and discontinuously distributed in warm temperate and tropical seas.

Trichosolen retrorsa John — Pl. 7, figs 3, 4.

John, 1977b.

Plants to about 10 cm long, densely tufted, arising from a rhizoidal plexus; branches dichotomously divided, from 140-400(-440) μm in diameter, tapering towards the apex; branchlets simple, from 12-40 μm in diameter, swollen at the base and reaching 80 μm in diameter, from (0.8)1-2 mm in length and with a rounded apex; chloroplasts discoid to lenticular, 4-10 μm long, each with a single pyrenoid; gametangia single, adaxial near the base of each branchlet, backwardly directed towards the central axis, sessile or subsessile, oblongate, 50-70 μm in diameter and 100-140 μm long, with 1 or 2 terminal papillae.

For additional figures see John, 1977b, p. 408, figs 1-5.

Occurs on calcareous cobbles at a depth of about 15 m; rare and of sporadic occurrence.

Distribution. World: in tropical West Africa. West Africa: Ghana: John 1977b.

Family Caulerpaceae

Caulerpa Lamouroux 1809b

Plants aseptate and coenocytic, consisting usually of a variously elaborated erect portion arising from a creeping base of terete branches, attached by descending rhizoids; thallus strengthened by an internal system of branched rods of wall material (trabeculae).

World Distribution: widespread in warm temperate and tropical seas.

Many of the species within this genus are very polymorphic and in the past numerous varieties or forms have been recognised. A comprehensive discussion of the difficulties to be encountered when delimiting these infraspecific entities in the western tropical Atlantic is given in Taylor (1960).

Key to the Species

1. Branches of erect and creeping systems similar and the distinction between them not obvious .... *C. fastigiata*
1. Branches of erect and creeping systems readily distinguishable 2

2. Branchlets terete and terminally swollen .... *C. racemosa*
2. Branchlets flattened or slender and terete................................. 3

3. Branches of the erect system terete throughout and branchlets slender 4
3. Branches of the erect system more or less flattened......................... 5

4. Plants less than 1.5 cm in height; branchlets to about 0.5 mm in length .... *C. ambigua*
4. Plants from 2-4 cm or more in height; branchlets from 3-5 mm in length .... *C. sertularioides*

5. Branchlets arising in more than two rows .... *C. cupressoides*
5. Branchlets in two distinct rows 6

6. Branchlets having the margins distinctly dentate .... *C. scalpelliformis*
6. Branchlets having an entire margin .... *C. taxiformis*

Caulerpa ambigua Okamura Pl. 9, fig. 1.

Okamura, 1897.

Plants consisting of a slender and inconspicuous system of creeping branches, less than 0.5 mm in diameter, with the erect branches up to 1.5 cm in height; erect branches a few times irregularly divided, bearing two rows of terete branchlets; branchlets up to 0.5 mm in length, occasionally bi- or trifid and rounded at the apex.

For additional figures see Børgesen, 1911, p. 129, fig. 2 (as *C. vickersiae*).

Occurs on moderately wave-exposed rocks in the lower eulittoral subzone and sometimes growing with other algae such as *Bryopsis;* commonly found in the low mat of algae covering fish-grazed areas of rocky banks at a depth of about 10 m.

Distribution. World: apparently pantropical. West Africa: Ghana: *C. vickersiae* Dickinson & Foote 1951, Lawson 1960a, Lawson & Price 1969. Liberia: *C. vickersiae,* De May *et al.* 1977.

Caulerpa cupressoides (West ex Vahl) C. Agardh

C. Agardh, 1823; *Fucus cupressoides* West, in Vahl, 1802.

Plants consisting of a robust system of creeping branches, from 0.5-2.5 mm in diameter, bearing erect branches to about 8 cm in height; erect branches simple or more usually many times forked, often naked below and giving rise above to more than 2 rows of branchlets; branchlets terete or slightly flattened, ovoid to conical and having a mucronate tip.

For figures see Børgesen, 1913, p. 138 to 143, 145, 146, figs 108-116.

There is no information on the ecology of this plant in the region but according to Taylor (1960) it forms in the western Atlantic "large colonies in sandy shallows, attaching to shells, stones and coral fragments and anchored in the sand itself".

Distribution. World: pantropical. West Africa: Cameroun: Hieronymous 1895, Lawson & Price 1969, ?Steentoft 1967. São Tomé: Hariot 1908, Henriques 1917, Lawson 1960a, Lawson & Price 1969, Steentoft 1967; *Chauvinia cupressoides,* Henriques 1886b, 1887.

Remarks. A key to eight of the varieties and five forms of this species is given in Taylor (1960). Hieronymus (1895) implied in his publication that the plant is found not only in Cameroun but also on all of the offshore islands.

Caulerpa fastigiata Montagne Pl. 9, fig. 4.

Montagne, 1842.

Plants to about 2 cm in height, consisting of descending rhizoidal branches and erect branches, with often no marked or clear distinction between them; branching alternate, irregularly dichotomous or opposite; branches slender, to about 120 μm in diameter, often somewhat fastigiate and not bearing any distinct lateral branchlets, with the ultimate branches sometimes slightly clavate and having obtuse apices.

For additional figures see **Børgesen**, 1913, p. 119, fig. 93.

Occurs with other algae such as *Centroceras clavulatum* growing in the algal turf in sheltered and somewhat brackish-water habitats.

Distribution. World: in warm temperate and tropical seas. West Africa: Sierra Leone: unpublished.

Caulerpa racemosa (Forsskål) J. Agardh Pl. 9, fig. 5.

J. Agardh, 1872; *Fucus racemosus* Forsskål, 1775.

var. **racemosa**

Plants consisting of creeping branches, to about 2 mm in diameter, bearing erect branches from 1-7 cm high; erect branches simple or sparingly divided; branchlets radially arranged, each terete for over two-thirds of its length and then abruptly swollen terminally, with the lower branchlets often less swollen than those above.

For additional figures see **Børgesen**, 1913, p. 148 to 151, 153, 155, 156, figs 117-126 (also varieties).

Occurs as extensive grape-like masses on sheltered rocks in the lower eulittoral subzone and the sublittoral fringe, especially common in brackish-water situations; elsewhere in the Atlantic Ocean it has been reported from depths in excess of 50 m.

Distribution. World: probably pantropical. West Africa: Gabon: John & Lawson 1974a. Ghana: Lawson 1960a, 1966, Lawson & Price 1969; *C. racemosa* var. *occidentalis,* Dickinson & Foote 1951, Steentoft 1967; *C. racemosa* f. *occidentalis,* Nizamuddin 1964. Liberia: De May *et al.* 1977. São Tomé: Henriques 1917, Lawson 1960a, Lawson & Price 1969; *C. racemosa* var. *clavifera,* Hariot 1908; *C. racemosa* var. *occidentalis,* Steentoft 1967; *C. clavifera,* Hariot 1908; *Chauvinia clavifera,* Henriques 1886b, 1887. Sierra Leone: Aleem 1978, 1980b, John & Lawson 1977a,

Plate 9: Fig. 1. *Caulerpa ambigua* (after Lawson, 1960a): Portion of a plant showing the erect branches bearing terete and bilaterally arranged branchlets. Fig. 2. *Caulerpa sertularioides* (after Lawson, 1960a): Portion of a plant with the erect branches bearing incurved and mucronately tipped branchlets. Fig. 3. *Caulerpa taxifolia:* Portion of a plant with the flattened erect branches bearing oval to oblong shaped branchlets. Fig. 4. *Caulerpa fastigiata:* Portion of a plant showing the slender branches of the little differentiated erect and creeping axes. Fig. 5. *Caulerpa racemosa* variety *racemosa* (after Lawson, 1960a): Portion of a plant with the erect branches bearing terminally swollen branchlets.

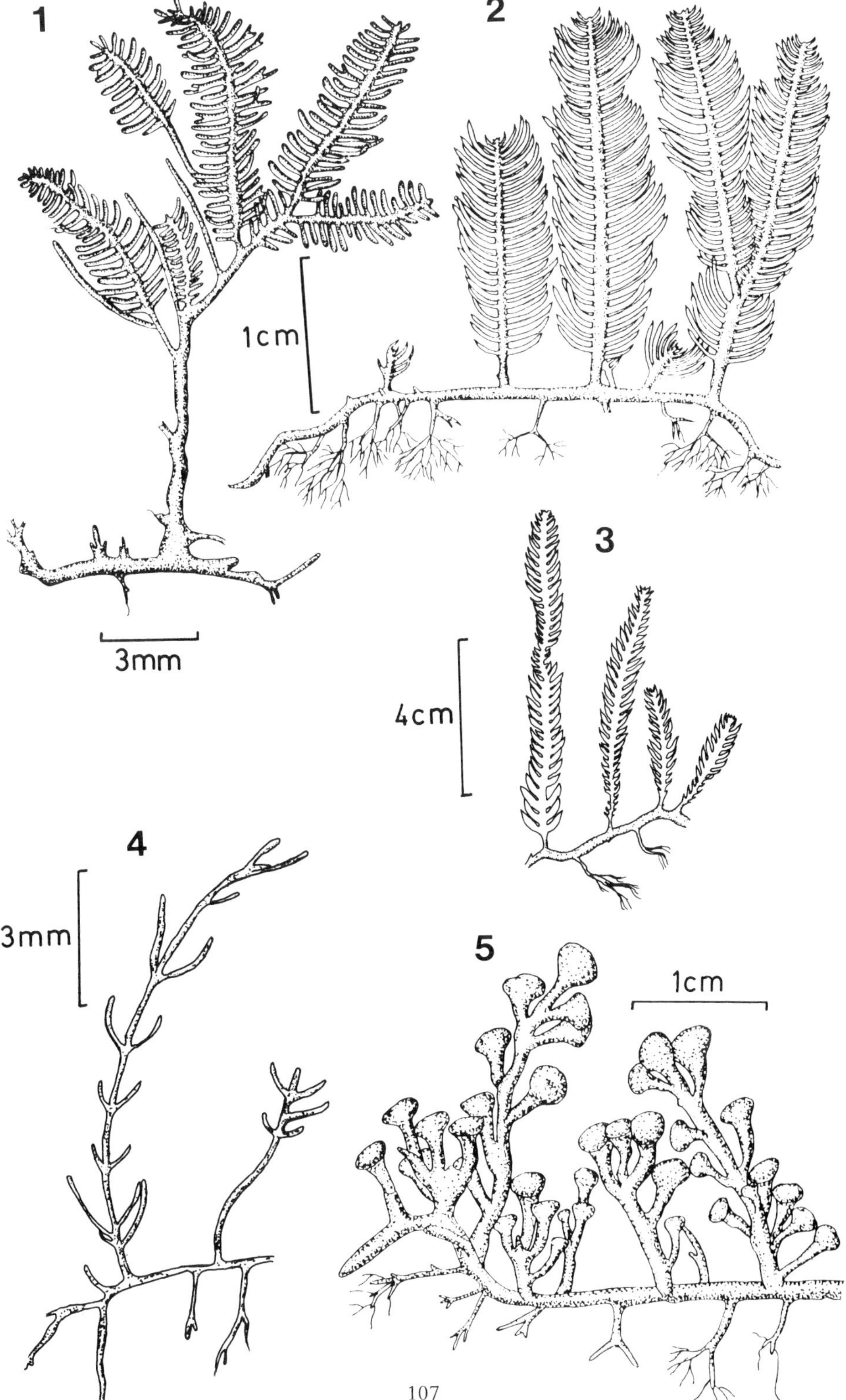
1
2
1cm
3mm
3
4cm
4
3mm
5
1cm

Lawson 1957a, 1960a, Levring 1969, Longhurst 1958, Michanek 1971, 1975; *C. racemosa* var. *clavifera?,* Lawson 1954b, Steentoft 1967.

Remarks. There is a certain divergence of opinion regarding the delimitation of the numerous varieties and forms within this species. We have accepted the view of Lawson and Price (1969) and for the sake of clarity and simplicity have adopted the scheme of Papenfuss and Egerod (1957) for the delimitation of such entities.

Caulerpa scalpelliformis (R. Brown ex Turner) C. Agardh

C. Agardh, 1823; *Fucus scalpelliformis* R. Brown, in Turner, 1811.

Plants consisting of creeping branches, to about 2 mm in diameter, bearing erect branches to about 22 cm in height, with the erect portions having a linear to lanceolate outline; erect branches simple or very occasionally divided; branchlets flattened and usually opposite, incurved upwards and with the margins dentate.

For figures see Nizamuddin, 1964, p. 215, fig. 2, p. 219, pl. 5.

Occurs on moderately wave-sheltered rocks in the lower eulittoral subzone and particularly common in tidepools; on one occasion it has been dredged to a depth of 7 m.

Distribution. World: probably pantropical. West Africa: São Tomé: Barton 1897, 1901, Carpine 1959, Hariot 1908, Henriques 1917, Lawson 1960a, Lawson & Price 1969, Steentoft 1967, Tandy 1944; *C. scalpelliformis* f. *denticulata,* Carpine 1959; *C. scalpelliformis* var. *intermedia,* Hariot 1908; *C. denticulata,* Hariot 1908, Henriques 1886b, 1887, 1917, Lawson 1960a.

Remarks. Forms and varieties in this species, as in other members of this genus, have little meaning as many of the characters used to differentiate between them are often clinal in variation. A detailed knowledge of locations and environmental conditions, and their effect on the species and its morphological development is necessary for the establishment of new forms and varieties.

Caulerpa sertularioides (Gmelin) Howe Pl. 9, fig. 2.

Howe, 1905; *Fucus sertularioides* Gmelin, 1768.

Plants consisting of slender creeping branches, to about 1 mm in diameter, bearing erect branches to about 5 cm in height; erect branches simple or more rarely divided, bearing 2 rows of fine terete branchlets; branchlets from 3-5 mm in length, sometimes slightly curved and having mucronate tips.

For additional figures see Børgesen, 1913, p. 134, fig. 106 (as *C. sertularioides* f. *farlowii*).

Occurs as small patches growing on moderately wave-sheltered rocks in the sublittoral fringe and found extending downwards to a depth of about 3 m below low water; occasionally also in tidepools in the lower eulittoral subzone and often associated with a certain amount of sand.

Distribution. World: probably pantropical. West Africa: Cameroun: Lawson 1960a, Lawson & Price 1969, Steentoft 1967; *C. plumaris*, Pilger 1911, Schmidt & Gerloff 1957. Ghana: Dickinson & Foote 1951, Lawson 1956, 1960a, Lawson & Price 1969, Nizamuddin 1964, Steentoft 1967, Stephenson & Stephenson 1972. São Tomé: Hariot 1908, Lawson 1960a, Lawson & Price 1969, Steentoft 1967; *C. plumaris*, Hariot 1908, Henriques 1886b, 1887, 1917. Sierra Leone: Aleem 1978, John & Lawson 1977a.

Caulerpa taxifolia (Vahl) C. Agardh — Pl. 9, fig. 3.

C. Agardh, 1823; *Fucus taxifolius* Vahl, 1802.

Plants consisting of a system of slender creeping branches, less than 1 mm in diameter, bearing erect branches from 2-14 cm high, with the erect portion linear to lanceolate in outline; erect branches simple or a few times forked; branchlets flattened and in 2 opposite rows, oval to oblong and narrow at the base, ascending and often somewhat curved, with the tips mucronate.

For additional figures see Børgesen, 1913, p. 130, 131, figs 102, 103 (as *C. crassifolia*), p. 132, figs 104, 105.

Occurs over a wide range of exposure to wave action growing on rocks, often sand-covered in sheltered situations, in the lower eulittoral subzone and to a depth of about 28 m where it is found as large plants creeping over a bottom of calcareous cobbles.

Distribution. World: probably pantropical. West Africa: Côte d'Ivoire: John 1977a. Gabon: John & Lawson 1974a. Ghana: John 1977a, John *et al.* 1977, 1979, Lawson & Price 1969, Nizamuddin 1964; *C. crassifolia*, Dickinson & Foote 1950, Lawson 1956, Stephenson & Stephenson 1972; *C. mexicana*, Lawson 1960a, Steentoft 1967. ?Príncipe: John 1977a, Steentoft 1967. São Tomé: Carpine 1959, Hariot 1908, Henriques 1886b, 1887, 1917, John 1977a, Steentoft 1967; *C. taxifolia* f. *asplenioides*, Trochain 1940; *C. crassifolia*, Carpine 1959; *C. crassifolia* var. *crassifolia*, De Toni 1889, Henriques 1885, 1886a; *C. crassifolia* var. *mexicana*, Askenasy 1896; *C. mexicana*, Lawson 1960a; *taxifolia* var. *crassifolia*, Henriques 1885, 1886a. Sénégal (Casamance): *C. pinnata*, Chevalier 1920.

Family Codiaceae

Boodleopsis A. & E. S. Gepp 1911

Plants consisting of siphonaceous filaments, much intertwined and di- or trichotomously or irregularly divided, in upper parts smaller and thinner walled than below and with distinctive constrictions at the origins of the branches, attached by rhizoids.

World Distribution: probably widespread in warm temperate and tropical seas.

Boodleopsis pusilla (Collins) W. R. Taylor *et. al.* Pl. 8, fig. 7.

Taylor *et al.*, 1953; *Dichotomosiphon pusillus* Collins, 1909.

Plants of densely intertwined filaments, about 1 cm in height; filaments di- or trichotomously or somewhat irregularly divided; filaments in lower parts up to 80 μm in diameter, often colourless and with an irregular outline, sometimes forming rhizoids; filaments in upper parts from (10-)22-40 μm in diameter, usually markedly constricted at intervals of (150-)200-300 μm especially at points of branching.

For additional figures see Taylor *et al.*, 1953, pl. 1, figs 1-12, pl. 2, figs 1-18, pl. 3, figs 1-10.

Occurs as a bright green turf mixed with other algae growing in the lower eulittoral subzone over estuarine rocks.

Distribution. World: in warm temperate and tropical parts of the Atlantic Ocean. West Africa: Bioko: unpublished. Sierra Leone: John & Lawson 1977a, Lawson 1957a, Lawson & Price 1969; *Dichotomosiphon tuberosus*, Lawson 1954b, Longhurst 1958.

Remarks. Taylor (1960) suggests that this plant might be the protonemal stage of a large and more elaborate member of this family. *Boodleopsis pusilla* might be distinguished from species of *Vaucheria*, which often occur in similar situations though usually in somewhat shaded places at a higher level on the shore, by having constrictions at intervals along the branches.

Codium Stackhouse 1787

Plants consisting of terete or flattened branches, irregularly or regularly dichotomously divided, attached by a matted rhizoidal holdfast; branches having a filamentous medulla and these filaments terminating in cylindrical to turbinate swollen utricles, radially arranged and forming the cortex; utricles often bearing simple hairs or hair scars at the distal end and containing discoid chloroplasts; reproduction by biflagellate gametes produced in fusiform to cylindrical gametangia, borne laterally on the utricles and separated by annular thickenings of the wall.

World Distribution: probably cosmopolitan.

Key to the Species

1. Branches divaricate, irregularly dichotomous and appearing cervicorn . *C. taylorii*
1. Branches usually regularly dichotomous at angles of less than 45° 2

2. Plants usually greater than 15 cm in length and branches from 0.6-6 cm in breadth; utricles greater than 500 μm in length . . . . *C. decorticatum*
2. Plants less than 15 cm in length and branches from 2-3 cm in breadth; utricles always less than 500 μm in length. . . . . *C. guineënse*

Codium decorticatum (Woodward) Howe Pl. 10, fig. 5.

Howe, 1911; *Ulva decorticata* Woodward, 1797.

Plants consisting of one to several branches arising from a crustose base, up to 1 m or more in length; branches regularly dichotomously divided, terete or occasionally compressed, from 0.6-2.5 cm in breadth, with the dichotomies often markedly flattened, expanded, cuneate and reaching 6 cm or more in breadth; utricles cylindrical or clavate, from (115-)220-500(-850) μm in diameter and usually from about 1,000-2,000 μm in length, with the distal wall to about 8 μm thick and hairs or hair scars abundant when present; gametangia often several per utricle, borne more than 400 μm below the distal end, stalked, lanceolate to ovoid, usually from 70-120 μm in diameter and to 400 μm in length.

For additional figures see Silva, 1960, pl. 114, pl. 120c, pl. 121.

There is no information on the ecology of this plant in the region.

Distribution. World: widespread in warm temperate and tropical parts of the Atlantic Ocean. West Africa: Sénégal (Casamance): *C. elongatum,* Chevalier 1920. Sierra Leone: *C. elongatum,* Aleem 1978.

Codium guineënse Silva ex Lawson & John nov. sp. Pl. 9, figs 3, 4.

Plantae erectae fruticosae, ca. 5 cm altae, interdum usque ad. 12 cm; rami plus minusve regulariter dichotomi, teretes, in furcis interdum leviter complanati, ca. 1-3 mm diametro; utriculi clavati, ca. (56-)70-102(-154) μm diametro, ca. 210-350 (-420) μm alti, apice late rotundato pariete inspissato lenticulari, 12-18 μm crassitie, pili aut pilorum cicatrices probabiliter nulli; gametangia saepe 1 per utriculum, ca. 160-220 μm infra apicem portata, sessilia vel breviter pedicellata, fusiformia, ca. (56-)60-84 μm diametro et ca. (112-)140-190 μm alta.

Plants erect and bushy, about 5 cm in height, occasionally reaching 12 cm; branches more or less regularly dichotomous, terete, sometimes slightly compressed at the forkings, from 1-3 mm in diameter; utricles clavate, from (56-)70-102(-154) μm in diameter and from 210-350(-420) μm in height, with the apices broadly rounded and the distal wall having a lens-shaped thickening from 12-18 μm in thickness, generally without hairs or hair scars; gametangia often 1 per utricle, borne about 160-220 μm from the distal end, sessile or shortly stalked, spindle-shaped, from (56-)60-84 μm in diameter and about (112-)140-190 μm in length.

Holotype: A230, Prampram, Ghana, 1 November 1951. Deposited in the British Museum (Natural History), London (BM).

Occurs over a wide range of wave-exposure in the lower eulittoral subzone and often common in tidepools; larger and looser plants have been dredged to a depth of 55 m.

Distribution. World: in warm temperate and tropical parts of the eastern Atlantic Ocean. West Africa: Côte d'Ivoire: John 1972c, 1977a. Ghana: John *et al.* 1977, Lawson 1956, Lawson & Price 1969, Lieberman *et al.* 1979; *Codium tomentosum,* Hornemann 1819, Lawson & Price 1969.

Remarks. For nearly a quarter of a century this name has been used for *Codium* plants collected along the West African coast. It was first mentioned from Sénégal by Dangeard (1955), later from Ghana by Lawson (1956) and from South West Africa by Silva (1960), and more recently from Angola and the Canaries (Lawson and Price, 1969). The West African material was first appreciated as being a new species by Dr. P. Silva who gave it the name of *Codium guineënse* but never apparently published a description of it. We have retained his name in validating the taxon. According to Silva (1960) its nearest relative is *Codium isthmocladum* Vickers which is known only from tropical and subtropical parts of the western Atlantic. The plant Aleem (1978) briefly describes collected from fishermen's nets on the Sierra Leone peninsula is probably referrable to this *Codium* species.

Codium taylorii Silva Pl. 9, figs 1, 2.

Silva, 1960.

Plants often semi-prostrate and occasionally to about 10 cm in height; branches divaricate, irregularly dichotomously divided and appearing cervicorn, terete but compressed at the forkings, from 2-9 mm in width and 3-4 mm in thickness, often having markedly obtuse tips; utricles cylindrical or clavate, from 56-250 μm in diameter and from 370-1,180 μm in length, with the distal wall 10 μm or more in thickness and hairs or hair scars usually abundant; gametangia usually 2 per utricle, arising more than 250 μm below the distal end, shortly stalked, cylindrical to ellipsoidal, from 50-85 μm in diameter and to about 360 μm in length.

For additional figures see Silva, 1960, pl. 112, figs 118, 119, 120a, b.

Occurs as coarse and dark green plants growing on rocks in the lower eulittoral zone over a wide range of exposure to wave action.

Distribution. World: in warm temperate and tropical parts of the Atlantic Ocean. West Africa: Ghana: Chapman 1961, Hoppe 1969, Lawson 1956, Lawson & Price 1969, Silva 1960.

Remarks. This species is usually, but incorrectly, spelled "*taylori*".

Plate 10: Figs 1, 2. *Codium taylorii.* 1. Plant showing the somewhat compressed, irregularly dichotomously divided and widely divergent branches; 2. Club-shaped utricle. Figs 3, 4. *Codium guineënse:* 3. Plant showing the terete branches which are slightly compressed at the dichotomies; 4. Utricle having a distal thickening and bearing a lateral gametangium. Fig. 5. *Codium decorticatum:* Plant with the branches compressed at the dichotomies. Figs 6, 7. *?Siphonocladus brachyartus:* 6. Portion of a plant showing the branching pattern; 7. Portion of a branch bearing a rhizoid and showing details of the reticulate chloroplast in one of the cells.

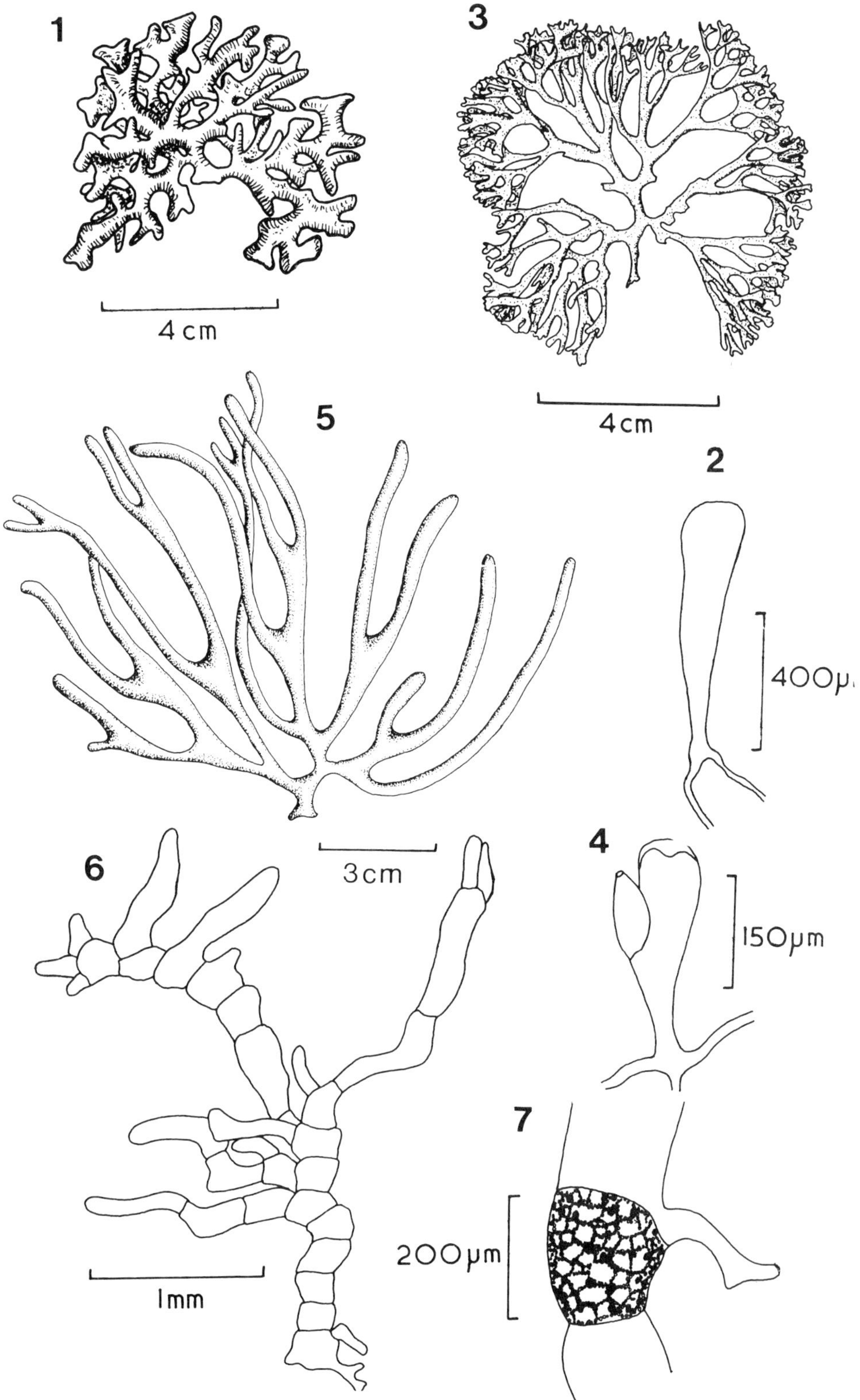
1
4 cm
3
4cm
5
2
400μ
6
3cm
4
150μm
7
200μm
1mm

Doubtful Record

Codium tomentosum Stackhouse

This species has been reported from Ghana ("Guinea Coast") by Hornemann (1819) São Tomé by Barton (1897, 1901), from São Tomé and Príncipe by Steentoft (1967), and from Gabon by Hariot (1896). The Hornemann plant has been examined and found to be referable to *Codium guineënse;* none of the Welwitsch specimens on which the offshore islands records are based have been located by Steentoft (1967). She comments that "it is very unlikely that it is *Codium tomentosum*". The Gabon material has not been seen but, as has been found to be the case elsewhere for the West African coast, records attributed to this plant have often proved to be mis-identifications. It seems best to regard the presence of this plant in the region as uncertain.

DIVISION PHAEOPHYTA · CLASS PHAEOPHYCEAE

Order Ectocarpales* · Family Ectocarpaceae

Bachelotia Kuckuck 1929

Plants filamentous, simple or sparingly and irregularly divided erect branches arising from a loosely branched prostrate system; growth always intercalary; chloroplasts numerous, rod-shaped to discoid, forming one or two stellate clusters in mature cells; unilocular sporangia seriate and intercalary; plurilocular sporangia intercalary or lateral and shortly stalked or sessile.

World Distribution: widespread in warm temperate and tropical seas.

Bachelotia antillarum (Grunow) Gerloff Pl. 11, fig. 7.

Gerloff, 1959; *Ectocarpus antillarum* Grunow, 1868.

Plants forming tufts or extensive entangled mats, from 1-4 cm in height; branches sparsely and irregularly divided, often bearing many spur-like lateral branchlets; cells to about 40 μm in diameter and from shorter than broad to more than 3 times as long as broad; chloroplasts arranged in 1 or 2 stellate groupings in each mature cell, often irregular in immature or degenerating cells; unilocular sporangia intercalary and usually barrel-shaped; plurilocular sporangia sessile or stalked, rarely intercalary, either cylindrical and tapering above, from 25-35 μm in diameter and 80-115 μm in length, with or without a terminal hair, or less commonly globose, from 22-28 μm in diameter and 27-37 μm in length.

For additional figures see Børgesen, 1920, p. 431, 432, figs 408, 409 (as *Pylaiella fulvescens*).

Commonly found as extensive mats on gently sloping beach rocks and is then often associated with a good deal of sand, sometimes it is the dominant alga in shallow tidepools in the upper eulittoral subzone. On open rocks this plant is found only in sheltered to moderately wave-exposed situations in the lower part of the eulittoral zone. Fox (1957) reports it from Nigeria growing on pebbles in a freshwater stream.

Distribution. World: as for genus. West Africa: Bioko: unpublished. Côte d'Ivoire: John 1972c, Price *et al.* 1978. Gabon: John & Lawson 1974a, Price *et al.* 1978. Gambia:John & Lawson 1977b, Price *et al.* 1978. Ghana: John 1977a, Price *et al.* 1978; *B. fulvescens,* Fox 1957, Lawson 1966; *Pylaiella fulvescens,* Lawson 1956, Stephenson & Stephenson 1972. Liberia: De May *et al.* 1977, John 1977a, Price *et al.* 1978.

*On the acceptance and validity of extant generic concepts in this order, particularly regarding *Ectocarpus, Giffordia* and *Feldmannia,* see the summary in Price *et al.* (1978).

Nigeria: John 1977a, Price 1973, Price *et al.* 1978; *B. fulvescens,* Fox 1957, Lawson 1966, Steentoft Nielsen 1958. São Tomé: John 1977a, Price *et al.* 1978, Steentoft 1967. Sierra Leone: John & Lawson 1977a, Price *et al.* 1978; *B. fulvescens,* Fox 1957.

Remarks. The study by Price (1973) has cast doubt upon the validity of many of the morphological characteristics used for distinguishing between this genus and *Pilayella.*

Ectocarpus Lyngbye 1819

Plants filamentous, consisting of simple or variously divided erect branches arising from a prostrate, rhizoidal or penetrating system of branches; filaments very occasionally having rhizoidal cortication near the base; growth zone intercalary and indistinct; chloroplasts one to several per cell, parietal, discoid, band-shaped or sometimes forked; sporangia (unilocular, plurilocular) and gametangia (plurilocular) scattered and sessile or shortly stalked.

World Distribution: cosmopolitan.

Ectocarpus, like many other allied genera in the Ectocarpales, is very plastic in form and many of the morphological characters used in their separation are known to be subject to environmental modification. Ravanko (1970) has on the basis of her culture studies concluded that many genera in this order (including *Giffordia* and *Feldmannia*) are no more than developmental stages of others, or that their mature form reflects no more than a response to the environment of another genus. Recently Clayton (1974) has summarised the current taxonomic and biological situation in the Ectocarpales and defended the integrity of such genera as *Ectocarpus, Giffordia* and *Feldmannia,* though G. Russell (*pers. comm.*) is proposing to combine all three under *Ectocarpus.*

Key to the Species

1. Plants forming spongy tufts, from 2-5 cm in height; branches bearing hook-shaped lateral branches . . . . *E. breviarticulatus*
1. Plants usually forming low felty patches, often less than 2 cm in height; branches never hook-shaped . . . . *E. rhodochortonoides*

Ectocarpus breviarticulatus J. Agardh Pl. 11, figs 3, 4.

J. Agardh, 1848a.

Plants consisting of filaments twisted to form rope-like tufts, from 2-5 cm or more in height; branches irregularly divided and bearing many short hook-like lateral branches which are occasionally transformed into rhizoids; cells about 27 μm in diameter and from 1 to 2 times longer than broad; plurilocular reproductive organs borne on 1- or 2-celled stalks and at right angles to the branches, spherical to ovoid, from about 38-42(-57) μm in diameter and 46-50(-62) μm in length.

For additional figures see Børgesen, 1914, p. 173, fig. 136.

Occurs most commonly in moderately to very wave-exposed situations growing on rock surfaces, mussels or larger algae in the upper eulittoral subzone and just above where there is considerable wave splash.

Distribution. World: probably widespread in warm temperate and tropical seas. West Africa: Benin: John 1977a, John & Lawson 1972b, Price *et al.* 1978. Côte d'Ivoire: John 1977a, Price *et al.* 1978. Ghana: Fox 1957, John 1977a, John & Pople 1973, Lawson 1956, Price *et al.* 1978, Stephenson & Stephenson 1972. Liberia: De May *et al.* 1977, John 1972c, 1976, Price *et al.* 1978. Nigeria: Fox 1957, Price *et al.* 1978, Steentoft Nielsen 1958. Togo: John 1977a, John & Lawson 1972b, Price *et al.* 1978.

Remarks. The exact taxonomic position of this species is somewhat problematical. It has numerous small discoid chloroplasts and so is not strictly an *Ectocarpus* in the sense of Hamel (1931-1939) but it also lacks the well-defined growth regions found in *Giffordia* and *Feldmannia.*

Ectocarpus rhodochortonoides Børgesen Pl. 11, fig. 5.

Børgesen, 1914.

Plants most commonly forming low felty patches over the surface of larger algae, from 1-2 cm in height; branches somewhat irregularly and sparingly divided, tapering gradually towards the apices; cells from 7-15 μm in diameter and 2 to 8 times longer than broad; plurilocular reproductive organs sessile or occasionally borne on a 1-celled stalk, sometimes terminal on a branch, ovoid to truncate-cylindrical to clavate, to about 15 μm in diameter and reaching to 43 μm in length.

For additional figures see Børgesen, 1926, p. 9, 10, 12, 13, figs 3-6; Jaasund, 1969, p. 271, fig. 7.

Occurs most often on larger algae such as *Gracilaria dentata* growing on moderately wave-exposed rocks in the lower eulittoral subzone.

Distribution. World: widespread in tropical seas. West Africa: Côte d'Ivoire: John 1977a, Price *et al.* 1978. Ghana: John 1977a, John & Lawson 1972a, Price *et al.* 1978.

Remarks. There seems to be a case for putting this species into the genus *Giffordia* as it has rod-shaped or plate-like chloroplasts similar to those found in *Giffordia rallsiae* (see Jaasund, 1969; Ravanko, 1970).

Feldmannia Hamel 1931-1939

Plants filamentous, sparingly to densely irregularly divided, with the branches generally arising beneath the growth zones; growth zone intercalary and distinct; chloroplasts numerous, discoid to rod-shaped; sporangia (unilocular, plurilocular) and gametangia (plurilocular) frequently stalked, occasionally sessile, developing below the growth zones.

World Distribution: widespread from boreal-antiboreal to tropical seas.

Refer to remarks under *Ectocarpus* and the footnote to the order Ectocarpales.

Key to the Species

1. Plants epiphytic and minute, growing to a height of 3 mm in the cryptostomata of *Sargassum* . . . . *F. elachistaeformis*
1. Plants epiphytic or epilithic, usually greater than 2 cm in height . . . . *F. indica*

Feldmannia elachistaeformis (Heydrich) Pham-Hoàng Pl. 11, fig. 6.

Pham-Hoàng, 1969; *Ectocarpus elachistaeformis* Heydrich, 1892.

Plants ramifying throughout the cryptostomata of *Sargassum*, often less than 3 mm in height; branches simple or sparingly divided from near the base, from 16-22 μm in diameter below and tapering gradually above to 12-13 μm, often terminating in a hair-like tip; cells of the branches shorter than broad near the base and increasing above to several times longer than broad; unilocular sporangia borne on 2-celled stalks, ovoid, from 30-40 μm in diameter and 40-60 μm in length; plurilocular reproductive organs sessile or shortly stalked, ovate to spindle-shaped, from 20-28 μm in diameter and 60-90(-135) μm in length.

For additional figures see Børgesen, 1914, p. 175, fig. 137 (as *Ectocarpus elachistaeformis*); 1920, p. 435, fig. 411 (as *E. elachistaeformis*).

Found so far only on drifting specimens of *Sargassum filipendula* which have probably become dislodged from the large beds of this plant growing on rocky platforms at about 10 m.

Distribution. World: probably widespread in warm temperate and tropical seas. West Africa: Ghana: *Ectocarpus elachistaeformis*, John & Lawson 1972a, Price *et al.* 1978.

Plate 11: Fig. 1. *Giffordia mitchelliae:* Portion of a plant showing plurilocular reproductive organs borne towards the base of tapering lateral branches. Fig. 2. *Giffordia rallsiae:* Portion of a branch bearing ovate to cylindrical plurilocular reproductive organs. Figs 3, 4. *Ectocarpus breviarticulatus:* 3. Dense and spongy appearance of plant; 4. Portion of a filament bearing the distinctively hooked lateral branches. Fig. 5. *Ectocarpus rhodochortonoides:* terminal portion of a plant with the plurilocular reproductive organs arising laterally and terminally. Fig. 6. *Feldmannia elachistaeformis:* Basal portion of a plant showing the plurilocular reproductive organs towards the base of the erect filaments. Fig. 7. *Bachelotia antillarum:* Portion of a filament showing the cells containing 1 or 2 stellate groups of chloroplasts. Fig. 8. *Hecatonema* sp..: Portion of a plant showing the plurilocular gametangia and erect filament arising from the two-layered creeping base.

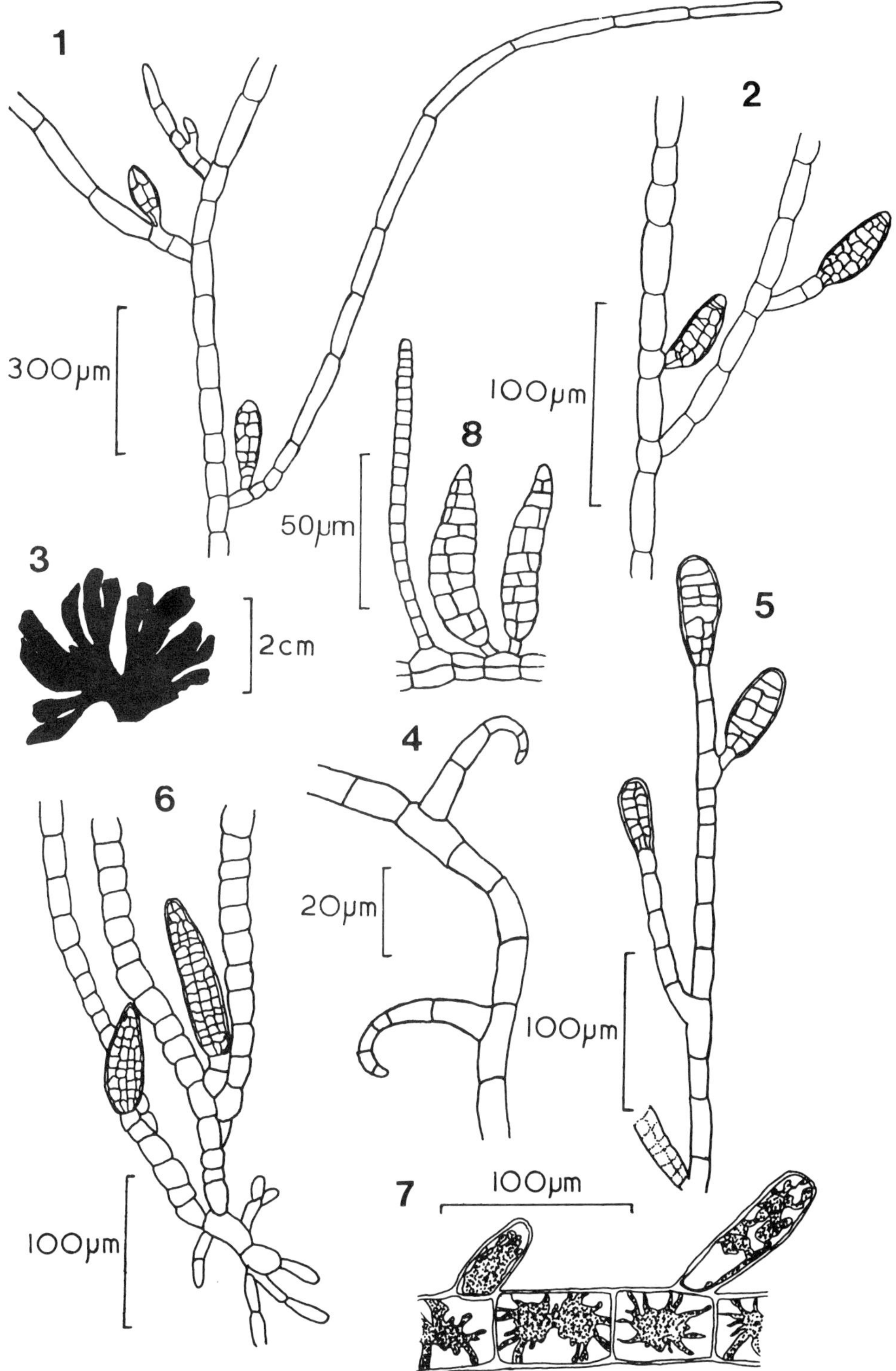
1
300µm
2
100µm
8
50µm
3
2cm
5
4
6
20µm
100µm
7
100µm
100µm

Feldmannia indica (Sonder) Womersley & Bailey

Womersley and Bailey, 1970; *Ectocarpus indicus* Sonder, in Zollinger, 1854.

Plants forming soft, light brown tufts or mats, from 2-5 cm in height, with the erect filaments arising from an entangled and somewhat decumbent base; branches sparingly irregularly to alternately divided, occasionally terminating in hair-like tips; cells of the main branches from 20-35(-45) μm in diameter and to 3 times longer than broad, often somewhat barrel-shaped; cells of the tapering lateral branches almost cylindrical, from 5-12 μm in diameter, sometimes as much as 4 to 5 times longer than broad; plurilocular reproductive organs sessile or occasionally on a 1- or 2-celled stalk, elongate and subcylindrical or a little clavate in shape, with an uneven outline and a blunt tip, from (15-)28-32 μm in diameter and (80-)140-160 μm in length.

For figures see Clayton, 1974, p. 762, fig. 10; Børgesen, 1941, p. 18, 21, figs 6, 7 (as *Ectocarpus indicus)*; Jaasund, 1969, p. 265, figs 1a-e (as *Giffordia duchassaingiana*).

Occurs in moderately to very wave-exposed places as an epilith or epiphyte in the eulittoral zone, occasionally associated with a good deal of sand when forming extensive mats on very sheltered beach rocks.

Distribution. World: widespread in warm temperate and tropical seas. West Africa: Cameroun: Price *et al.* 1978; *Ectocarpus indicus,* Pilger 1911, Schmidt & Gerloff 1957. Gabon: Price *et al.* 1978; *Giffordia indica,* John & Lawson 1974a. Gambia: John & Lawson 1977b. Nigeria: Price *et al.* 1978; *Ectocarpus indicus,* Fox 1957.

Remarks. Some doubt attaches to Pilger's (Pilger, 1911) report of this plant from the Cameroun as Fox (1957) believes that his description more closely corresponds to *Giffordia mitchelliae*.

Giffordia Batters 1893

Plants filamentous, densely irregularly branched and sometimes distally alternate, attached basally by rhizoids; growth zone intercalary, diffuse or distinct; chloroplasts numerous, parietal, discoid or occasionally as short bands; unilocular sporangia usually sessile, often serially arranged; plurilocular sporangia or gametangia of two or three types having different sized loculi, commonly asymmetrical in shape.

World Distribution: cosmopolitan.

See the remarks under *Ectocarpus* and the footnote to the order Ectocarpales.

Key to the Species

1. Plants commonly forming low felty patches less than 5 mm in height; plurilocular reproductive organs with an acute to tapering apex .... *G. rallsiae*
1. Plants forming bushy tufts or mats greater than 1 cm in height; plurilocular reproductive organs having an obtuse apex .... *G. mitchelliae*

Giffordia mitchelliae (Harvey) Hamel Pl. 11, fig. 1.

Hamel, 1931-1939; *Ectocarpus mitchelliae* Harvey, 1852.

Plants forming soft light brown to yellowish-brown tufts or mats, from 1-5 cm in height, arising from a creeping system of filaments; branches many times alternately divided and tapering to hair-like tips; cells of the main branches cylindrical, to about 36 μm in diameter and 1 to 3 times as long as broad; cells near the base of the lateral branches from 10-20 μm in diameter and gradually becoming narrower towards the branch tips, from 1 to 7 times longer than broad; plurilocular reproductive organs arising usually in series from the lower cells of the lateral branches, sessile or rarely stalked, elongate and cylindrical to ellipsoidal, with base and apex obtuse, and with an irregular outline, from 15-24 μm in diameter, usually reaching a length of 55 μm and only rarely to 150 μm.

For additional figures see Børgesen, 1941, p. 10 to 12, 14, 15, figs 1-5; Abbott & Hollenberg, 1976, p. 144, fig. 104; Jaasund, 1969, p. 266, fig. 2.

Occurs on moderately to very wave-exposed rocks in the eulittoral zone as tufts or as a mat associated with a good deal of sand; rarely epiphytic. Found on occasion in the sublittoral zone growing on calcareous cobbles at depths down to 10 m.

Distribution. World: widespread in warm temperate and tropical seas. West Africa: Benin: John & Lawson 1972a, 1972b, Price *et al.* 1978. Cameroun: ?Fox 1957, John & Lawson 1972a, Price *et al.* 1978. Gambia: John & Lawson 1977b, Price *et al.* 1978. Ghana: John & Lawson 1972a, John *et al.* 1977, Lieberman *et al.* 1979, Price *et al.* 1978. Liberia: De May *et al.* 1977, Price *et al.* 1978. Nigeria: Fox 1957, John & Lawson 1972a, Price *et al.* 1978. São Tomé: Price *et al.* 1978, Steentoft 1967. Sénégal (Casamance): *Ectocarpus virescens*, Chevalier 1920, Price *et al.* 1978. Sierra Leone: Aleem 1978. Togo: John & Lawson 1972a, 1972b, Price *et al.* 1978.

Remarks. We have followed Earle (1969) in considering this species and *Ectocarpus virescens* to be conspecific but see detailed comments in Price *et al.* (1978). See the note under *Feldmannia indica* regarding Fox's (Fox, 1957) record from Cameroun.

Giffordia rallsiae (Vickers) W. R. Taylor Pl. 11, fig. 2.

Taylor, 1960; *Ectocarpus rallsiae* Vickers, 1905.

Plants commonly forming a felty covering over larger algae, rarely epilithic, usually less than 5 mm in height; branches sparingly and irregularly divided, with the erect branches arising from a well-developed prostrate system of filaments; cells of the main branches often barrel-shaped, from 10-28 μm in diameter and 1 to 6 times longer than broad, with the lateral branches narrower and usually terminating in a hair-like tip; plurilocular reproductive organs often lying obliquely to the branch, sessile or with a 1- or 2-celled stalk, ovate to cylindrical, acute to tapering at the apex, to about 33 μm in diameter and reaching a length of about 88 μm.

For additional figures see Børgesen, 1926, p. 24, fig. 11; Jaasund, 1969, p. 269, fig. 5.

Occurs most frequently on larger algae particularly *Chnoospora minima* plants growing on moderately sheltered to very wave-exposed rocks; rarely epilithic. On occasion found on calcareous cobbles and rocky banks at depths from 10 to 15 m.

Distribution. World: widespread in warm temperate and tropical seas. West Africa: Benin: John 1977a, John & Lawson 1972b, Price *et al.* 1978. Ghana: John1977a, John & Lawson 1972a, Lieberman *et al.* 1979, Price *et al.* 1978. Liberia: De May *et al.* 1977, John 1977a, Price *et al.* 1978. Togo: John 1977a, John & Lawson 1972b, Price *et al.* 1978.

Remarks. There is some doubt as to whether this species more closely accords to the currently accepted features of *Feldmannia* or *Giffordia* (Earle, 1969).

Doubtful Record

Giffordia sandriana (Zanardini) Hamel

The only report of the species from the West African coast is its mention by Chevalier (1920) under *Ectocarpus sandrianus* in a list of algae from the Casamance region of Sénégal. It would seem at present best to consider this record as most doubtful until further collections are made in the region and pending examination of the original specimen on which the record is based.

Spongonema Kützing 1849

Doubtful Record

Spongonema tomentosum (Hudson) Kützing

This plant has been reported by Hornemann (1819) as *Fucus tomentosus* Hudson from "Danish Guinea", the coastal part of present-day Ghana. Hornemann's early report is considered with much doubt as it is the only mention of this species from one of the phycologically better known countries bordering the Gulf of Guinea region of West Africa. It is not usually considered to be a tropical species being known off the West African coast from the Cape Verde Islands (see Price *et al.*, 1978) which are bathed by the cold Canary Current.

Family Ralfsiaceae

Basispora John & Lawson 1974b

Plants crustose, thallus consisting of one to several layers of prostrate filaments, giving rise to simple and narrowly clavate erect filaments; chloroplasts discoid and several per cell ; unilocular sporangia arising from the base of the erect vegetative filaments, terminal on a distinct stalk several cells in length; plurilocular sporangia unknown.

World Distribution: possibly widespread in warm temperate and tropical seas.

Basispora africana John & Lawson Pl. 12, figs 1-3.

John and Lawson, 1974b.

Plants forming disc or ring-like black crusts, often coalescing; thallus consisting of 3 to several layers of prostrate filaments bearing erect assurgent filaments; erect filaments simple, free and closely packed, distinctly clavate towards the upper end; cells of the erect filaments subglobose above, from 8-11 μm in diameter, to cylindrical below, from 2.5-5 μm in diameter and 2.4 to 5 times longer than broad; unilocular sporangia oblong, from 15-33(-45) μm in diameter and 70-106(-125) μm in length, with the stalks from (4-)6 to 10(-15) cells in length and from 9.5-12 μm in diameter.

For additional figures see John and Lawson, 1974b, p. 287, fig. 1, p. 288, figs 2-4.

Occurs on rocky outcrops and stable boulders in the lower part of the upper eulittoral subzone where wave action is moderate and the barnacles are in low abundance.

Distribution. World: in warm temperate and tropical parts of the eastern Atlantic Ocean. West Africa: Cameroun: John & Lawson 1974b, Price *et al.* 1978; *?Mesospora,* John 1972c, Lawson 1955. Côte d'Ivoire: John 1977a, John & Lawson 1974b, Price *et al.* 1978. Gambia: John & Lawson 1977b. Ghana: John & Lawson 1974b, Price *et al.* 1978; *?Mesospora,* John 1972c, Lawson 1956; *?Hapalospongidium spongiosum,* ?Lawson 1966. Liberia: De May *et al.* 1977, John 1977a, John & Lawson 1974b, Price *et al.* 1978. Sierra Leone: John & Lawson 1974b, 1977a, Price *et al.* 1978; *?Mesospora,* John 1972c, Lawson 1957a; *?Hapalospongidium spongiosum,* Lawson 1966.

Nemoderma Schousboe ex Bornet 1892

Doubtful Record

Nemoderma tingitana Schousboe ex Bornet

Reported as black glossy crusts on the Sierra Leone peninsula (Aleem, 1978). This is likely to be a misdetermination for *Basispora africana* which is common along the shores of the peninsula (John & Lawson, 1977a). The material on which the record is based requires re-examination in order to clarify this point.

Ralfsia Berkeley 1843

Plants forming leathery crusts of indefinite extent, sometimes the crusts overlapping and forming several layers; thallus consisting of erect and laterally united filaments, often bearing terminally groups of inconspicuous hairs, mostly without rhizoids on the lower surface; chloroplast single, parietal; unilocular sporangia arising laterally at the base of free paraphysal filaments; plurilocular sporangia terminal on the erect filaments.

World Distribution: cosmopolitan.

Ralfsia expansa (J. Agardh) J. Agardh Pl. 12, figs 6, 7.

J. Agardh, 1848b; *?Myrionema expansum* J. Agardh, 1848a.

Plants forming dark brown to almost black crusts, with the surface initially smooth and later becoming irregularly wrinkled, sometimes having concentric zones present; thallus consisting of filaments curving upwards from the base and sometimes downwards from a median layer; unilocular sporangia stalked, oval to pyriform, about 30 μm in diameter and 75-120 μm in length; paraphyses clavate and from 5 to 20 cells long; plurilocular sporangia cylindrical and partly biseriate, from 15 to 20 cells in length.

For additional figures see Børgesen, 1914, p. 190, 191, figs 146-148; Weber-van Bosse, 1913, p. 147, fig. 45.

Occurs as often continuous and extensive crusts growing over a wide range of wave-exposure in the lower eulittoral subzone and over more limited areas in tidepools at higher levels.

Distribution. World: probably widespread in warm temperate and tropical seas. West Africa: Benin: John 1977a, John & Lawson 1972b, Price *et al.* 1978. Bioko: unpublished. Cameroun: John 1977a, Lawson 1966, Price *et al.* 1978. Côte d'Ivoire: John 1972c, 1977a, Price *et al.* 1978. Gabon: John & Lawson 1974a, Price *et al.* 1978. Ghana: John 1977a, Lawson 1956, 1966, Price *et al.* 1978, Stephenson & Stephenson 1972. Guinée: Lawson 1966, Price *et al.* 1978, Sourie 1954. Liberia: De May *et al.* 1977, John 1972c, 1977a, Price *et al.* 1978. Sierra Leone: John & Lawson 1977a, Price *et al.* 1978. Togo: John 1977a, John & Lawson 1972b, Price *et al.* 1978.

Remarks. Culture studies suggest that in some cases members of this genus are involved in the life histories of various members of the family Scytosiphonaceae (see Nakamura, 1965; Tatewaki, 1966; Wynne, 1969; among others).

Doubtful Record

Ralfsia macrocarpa J. Feldmann

Aleem (1978) reports "*Ralfsia macrophysa* (in Newton, 1930)" from Cape Club rocks on the Sierra Leone peninsula. We can trace no such species of *Ralfsia* and believe this may be an orthographic error for *Ralfsia macrocarpa* described by J. Feldmann (1931: 211) from Algeria. This species has larger paraphyses and sporangia than the species previously reported from the West African coast (*R. expansa*) and the colder water species (*R. verrucosa*) known from as far south as the Canary

Islands. The Sierra Leone report is considered as doubtful until the material on which it is based is examined and the determination verified.

Family Myrionemataceae

Hecatonema Sauvageau 1898

Plants minute, crustose or cushion-like, consisting of radially arranged creeping filaments, sometimes dividing to form two cell layers which give rise to simple or divided erect filaments; hairs borne terminally on the erect filaments or from the creeping filaments; chloroplasts numerous, disc-like or as short bands, each with a pyrenoid; unilocular reproductive structures unknown; plurilocular gametangia arising on the creeping or erect filaments.

World Distribution: widespread from boreal-antiboreal to tropical seas.

Undetermined Species

Hecatonema sp. Pl. 11, fig. 8.

Plants forming epiphytic discs, with the basal layer of filaments often 2 cells thick and giving rise to erect assimilatory filaments; erect filaments from 10-12 μm in diameter and up to 175 μm in height, with the cells to about 1.5 times longer than broad and decreasing above to shorter than broad; chloroplasts numerous and discoid; gametangia on a 1- to 6-celled stalk, from (10-)15-20 μm in diameter and up to 52 μm in length, cylindrical to somewhat spindle-shaped, with the loculi from 4-6 μm in diameter.

Found on only one occasion growing over the surface of *Gracilaria dentata* in the lower eulittoral subzone.

Distribution. West Africa: Gambia: John & Lawson 1977b.

Remarks. This plant is similar to the little-known *Hecatonema floridana* (W.R. Taylor) W.R. Taylor described by Taylor (1928, p. 109, pl. 14, fig. 19) as *Phycocelis floridana* growing on *Zonaria* in Florida although its gametangia are usually shorter.

Family Chordariaceae

Levringia Kylin 1940

Plants soft, gelatinous and lubricous; branches composed of a filamentous medulla giving rise to a cortex of assimilatory filaments having a trichothallic zone of growth; chloroplasts small and discoid; plurilocular sporangia arising on short stalks from the base of the assimilatory filaments.

World Distribution: in warm temperate and tropical parts of the Atlantic Ocean.

The taxonomic and nomenclatural confusion in this genus is discussed at some length by Price *et al.* (1978).

Levringia brasiliensis (Montagne) Joly Pl. 12, figs 4, 5.

Joly, 1953; *Mesogloea brasiliensis* Montagne, 1843.

Plants solitary, from 5-12 cm in length, greenish-brown; branches sparingly and irregularly divided towards the base, from 1.5-2.5 mm in diameter; cortex consisting of densely branched filaments below and simple above, with the cells from 13-17 μm in diameter and to about 33 μm in length; plurilocular sporangia arising terminally on short stalks, elongate and pod-shaped, from 13-16 μm in diameter and to about 85 μm in length.

For additional figures see Ganesan, 1968, p. 131, 132; Gayral, 1960, p. 52, fig. 1.

Occurs on moderately to very wave-exposed rocks in the upper eulittoral subzone, occasionally fringing large eulittoral tidepools; rare and of erratic appearance in the region but may be locally abundant.

Distribution. World: as for genus. West Africa: Ghana: John & Lawson 1972a, Price *et al.* 1978.

Family Chnoosporaceae

Chnoospora J. Agardh 1848a

Plants erect and bushy, with the branches subterete and gradually tapering towards the tip, arising from a large discoid holdfast; branches consisting of large and elongate medullary cells, with an inner cortex of less elongate cells and an outer cortex of smaller assimilatory cells; unilocular sporangia densely packed around groups of hairs and forming more or less continuous sori.

World Distribution: widespread in warm temperate and tropical seas.

Plate 12: Figs 1-3. *Basispora africana:* 1. General appearance of the crust; 2. Section showing the large unilocular sporangia occurring amongst the erect filaments; 3. Unilocular sporangium borne on a stalk arising from near the base of an erect filament. Figs 4, 5. *Levringia brasiliensis:* 4. Plant showing the sparingly irregularly divided and tomentose branches; 5. Structure of cortex with the plurilocular sporangia arising towards the base of the assimilatory filaments. Figs 6, 7. *Ralfsia expansa:* 6. Crust in section showing the fused filaments curving upwards and downwards from a median layer; 7. Upper portion of the crust showing rows of cells.

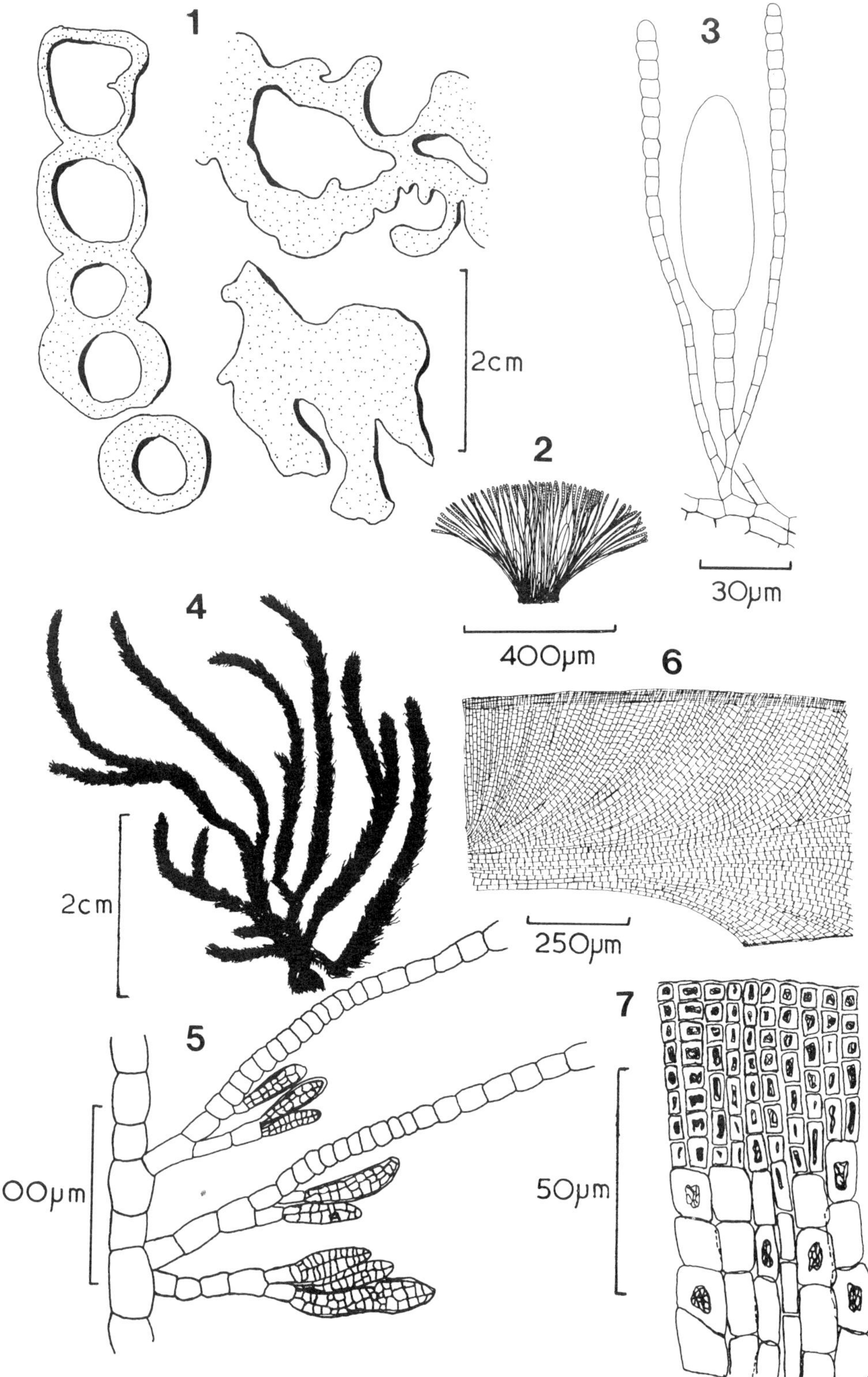
1
2cm
2
400µm
3
30µm
4
2cm
5
100µm
6
250µm
7
50µm

Chnoospora minima (Hering) Papenfuss Pl. 13, fig. 5.

Papenfuss, 1956; *Fucus minima* Hering, 1842.

Plants forming coarse clumps or more extensive patches, from 10-15 cm or more in height, olive brown; branches di- or polydichotomously divided and loosely or closely fastigiate, subterete, occasionally distinctly flattened, about 2 mm in breadth and often somewhat wider at the forks than elsewhere, tapering to a rounded apex.

For additional figures see Levring, 1938, p. 22, fig. 10, pl. 4, fig. 12 (as *C. pacifica*).

Occurs on moderately to very wave-exposed rocks and most common in the lower part of the upper eulittoral subzone or the lower lithothamnia subzone.

Distribution. World: as for genus. West Africa: Benin: John 1977a, John & Lawson 1972b, Price *et al.* 1978. Côte d'Ivoire: John 1972c, 1977a, Price *et al.* 1978. Ghana: John 1977a, John & Lawson 1972a, 1977a, John & Pople 1973, Lawson 1966, Price *et al.* 1978, Steentoft 1967, Townsend & Lawson 1969; *C. atlantica*, Lawson 1966. Liberia: De May *et al.* 1977, John 1972c, 1977a, Price *et al.* 1978. Nigeria: Fox 1957, John 1977a, Lawson 1966, Price *et al.* 1978, Steentoft 1967. São Tomé: John 1977a, Price *et al.* 1978, Steentoft 1967. Sierra Leone: John & Lawson 1977a. Togo: John 1977a, John & Lawson 1972b, Price *et al.* 1978.

Family Punctariaceae

Hydroclathrus Bory 1825

Plants initially spherical and hollow, later becoming flattened and perforated by numerous pores or apertures, whose margins are often inrolled and with the free edges sometimes fused together by the development of rhizoid-like projections from the surface cells; thallus consisting of an outer cortex of small assimilatory cells and an inner medulla of large colourless cells; hairs arising in groups confined to depressions in the thallus surface; plurilocular sporangia forming a continuous surface layer, normally confined to young plants.

World Distribution: in warm temperate and tropical seas.

Hydroclathrus clathratus (C. Agardh) Howe Pl. 13, fig. 6.

Howe, 1920; *Encoelium clathratum* C. Agardh, 1823.

Plants subspherical or more usually forming irregular and elongated masses; perforations very numerous, circular at first but later often becoming elongated and a few centimetres in length, with the strands bordering the apertures inrolled and from 1-3 mm across.

For additional figures see Taylor, 1928, pl. 15, fig. 19, pl. 19, fig. 1.

Occurs as one of the most common algae locally on moderately wave-sheltered rocks in the lower part of the lithothamnia subzone; rare in the region.

Distribution. World: as for genus. West Africa: Bioko: Schmidt & Gerloff 1957. Gabon: John & Lawson 1974a. Sierra Leone: Aleem 1978.

Rosenvingea Børgesen 1914

Plants consisting of terete or compressed branches, tubular above and solid below, alternately, dichotomously or pseudodichotomously divided, occasionally proliferous, attached by a basal disc; thallus consisting of a cavity sometimes partly filled with branched filaments, an inner cortical layer of colourless cells and an outer layer of small assimilatory cells; plurilocular sporangia clavate or cylindrical and in irregular sori, paraphyses absent.

World Distribution: widespread in warm temperate and tropical seas.

Rosenvingea intricata (J. Agardh) Børgesen. Pl.13, fig. 7.

Børgesen, 1914; *Asperococcus intricatus* J. Agardh, 1848a.

Plants forming intricate and procumbent clumps, to about 2 cm in height and from 5-10 cm in diameter, olive-brown, soft in texture; branches tubular and often somewhat flattened, to about 5(-8) mm in breadth, irregularly to dichotomously divided, often fused to one another, contorted, narrowing to attenuate or obtuse tips.

For additional figures see Earle, 1969, p. 204, 206, 208, figs 108-112.

Occurs on rocks in the immediate sublittoral and found on cobbles at depths down to 15 m; rare and of sporadic occurrence in the region.

Distribution. World: as for genus. West Africa: Cameroun: Price *et al.* 1978; *Asperococcus intricatus,* Pilger 1911. Ghana: John *et al.* 1977, Lieberman *et al.* 1979, Price *et al.* 1978.

Family Scytosiphonaceae

Colpomenia* (Endlicher) Derbès & Solier 1851

Plants more or less globose, solid at first and later becoming hollow, more irregularly shaped and thin-walled, with multicellular hairs scattered over the surface; thallus composed of one or more outer layers of assimilatory cells and three or more rows of large inner colourless cells; plurilocular sporangia initially in sori and later spreading over the surface, paraphyses present.

World Distribution: widespread from boreal-antiboreal to tropical seas.

*It has not been possible to identify a small and sterile *Colpomenia* plant found in the drift on the coast of Gambia (John and Lawson, 1977b).

Key to the Species

1. Plants having the punctate or punctiform sori covered with a distinct cuticle; medulla 4 to 6 cell layers thick . . . . *C. sinuosa*
1. Plants having extensive and confluent sori without a cuticle being present; medulla 3 to 4 cell layers thick . . . . *C. peregrina*

Colpomenia peregrina (Sauvageau) Hamel

Hamel, 1931-1939; *Colpomenia sinuosa* var. *peregrina* Sauvageau, 1927.

Plants forming smooth-surfaced and unfolded bladders, thin-walled and delicate, usually olive-brown; medulla of 3 to 4 layers of cells; sori confluent as raised patches over the surface and without a covering cuticle.

For figures see Sauvageau, 1927, p. 331, 333, 336, 339 to 341, 344, 347, figs 1-8 (as *C. sinuosa* var. *peregrina*).

Occurs in the lower eulittoral subzone growing in greatest abundance on moderately wave-sheltered rocks.

Distribution. World: as for genus but more common in colder waters. West Africa: Liberia: De May *et al.* 1977, Price *et al.* 1978.

Remarks. Clayton (1976) has studied morphological and anatomical variation in Australian *Colpomenia*. She has shown that most of the characters used commonly to distinguish subgeneric taxa in this genus are very variable and do not provide a basis for taxonomic separation. The shape of the sorus, presence of a cuticle over the plurilocular sporangia, and to a lesser extent the thickness of the medulla, appear to be the only reliable characters for separating *C. peregrina* and *C. sinuosa*.

Colpomenia sinuosa (Roth) Derbès & Solier. Pl. 14, figs 1, 2.

Derbès and Solier, 1851; *Ulva sinuosa* Roth, 1806.

Plants solitary or more usually aggregated together, initially globose and becoming somewhat flattened, folded, wrinkled and lobed, from 1-5 cm or more across, yellowish-brown to olive-brown; medulla of 4 to 6 layers of cells; sori punctate or punctiform and covered at least initially with a cuticle.

Plate 13: Fig. 1. *Sphacelaria rigidula:* Portion of a plant bearing propagulae in various stages of development. Figs 2, 3. *Sphacelaria brachygonia* (after John, 1972b): 2. Portion of a filament with a developing propagulum and the thick-walled stalks of former generations of propagulae; 3. Mature propagulum. Fig. 4. *Sphacelaria tribuloides* (after John and Lawson, 1972a): Portion of a filament bearing a broadly triangular propagulum with short biradiate arms. Fig. 5. *Chnoospora minima:* Plant showing the dichotomously divided branches arising from a discoid holdfast. Fig. 6. *Hydroclathrus clathratus:* Young plant with the convoluted and irregular thallus perforated by many apertures. Fig. 7. *Rosenvingea intricata:* Plant showing the contorted and somewhat flattened tubular branches.

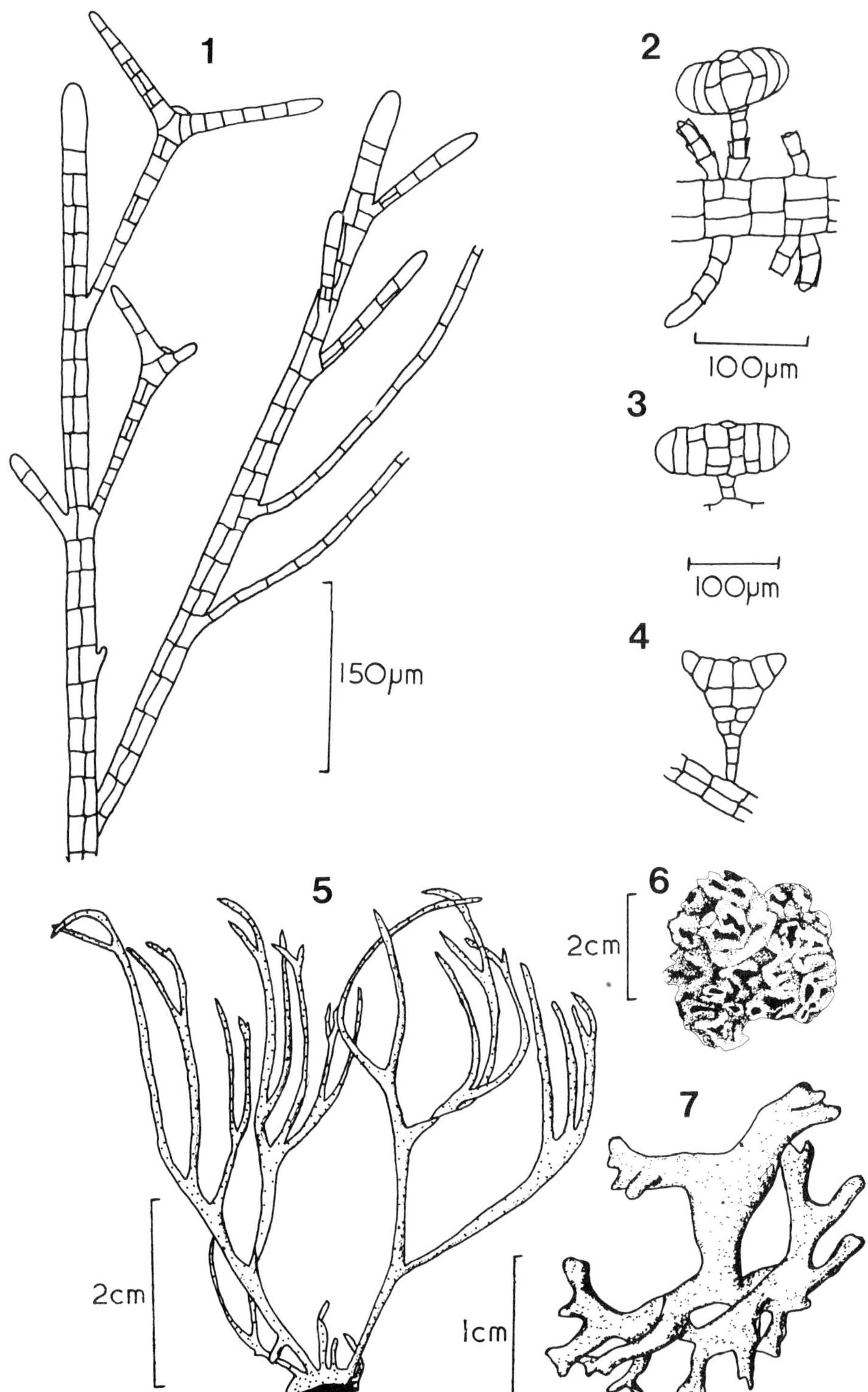
1
2
100µm
3
100µm
4
150µm
5
6
2cm
7
2cm
1cm

For additional figures see Earle, 1969, p. 201, figs 100, 101.

Occurs in the eulittoral zone and in the sublittoral fringe as large and well-developed plants only on wave-sheltered rocks or larger algae, occasionally found as small individuals on wave-exposed rocks.

Distribution. World: as for genus but most commonly found in warmer waters. West Africa: Bioko: unpublished. Côte d'Ivoire: John 1972c, 1977a, Price *et al.* 1978. Gabon: John & Lawson 1974a, Price *et al.* 1978. Ghana: Dickinson & Foote 1951, John 1977a, Lawson 1956, Price *et al.* 1978, Steentoft 1967. São Tomé: Carpine 1959, John 1977a, Price *et al.* 1978, Steentoft 1967. Sierra Leone: Aleem 1978. Togo: John 1977a, John & Lawson 1972b, Price *et al.* 1978.

Remarks. See the comments under the previous species.

Order Cutleriales · Family Cutleriaceae

Cutleria Greville 1830

Doubtful Record

Cutleria multifida (J. E. Smith) Greville.

This plant has been reported by Henriques (1917) from São Tomé and in earlier publications (Hariot 1908; Henriques, 1886b, 1887) as *Aglaozonia reptans.* Carpine (1959) also mentions a species of *Aglaozonia* dredged from 3 to 12 m at the island of Príncipe. *Aglaozonia parunda* (Greville) Zanardini (=*A. reptans*) is now known to be the diploid phase of *Cutleria multifida.* According to Steentoft (1967) the São Tomé material is in such a poor state of preservation as not to permit examination. This species is not reported elsewhere in the literature from further south than Morocco; at the present time we therefore prefer to regard it as an uncertain record for the islands.

Order Laminariales · Family Laminariaceae

Macrocystis C. Agardh 1821a

Doubtful Record

Macrocystis pyrifera (Linnaeus) C. Agardh.

The type locality cited by Linnaeus (1771, p. 311) as "oceano aethiopico" is believed to be somewhere off the West African coast below the "bulge of Africa". According to Womersley (1954) and Papenfuss (1940b), the type material is insufficient to identify as a presently recognised species, since holdfast characters are necessary for separating the species. Hooker (1845-1847) mentions that the Agulhas Current, which sweeps around the southern tip of Africa, "swarms" with *Macrocystis* and

Papenfuss (1940b) suggests that the plants of Linnaeus may have arisen from South Africa. See also the remarks in section "An Historical Review".

Saccorhiza de la Pylaie 1830

Doubtful Record

Saccorhiza polyschides (Lighfoot) Batters

The reasons for considering the report of this plant from present day Ghana by Palisot de Beauvois (1805) as uncertain are given on page 9.

Order Sphacelariales · Family Sphacelariaceae

Sphacelaria Lyngbye 1819

Plants erect and tufted, with the erect branches irregularly or alternately divided, arising from a crustose base or from rhizoidal filaments; apical cell prominent and dividing transversely to give segments which further divide longitudinally to form tiers of cells; sheathed hairs often present; propagulae characteristic in form; sporangia and gametangia similar or dissimilar in form, shortly stalked.

World Distribution: cosmopolitan.

Key to the Species

1. Propagulae consisting of slender arms many times longer than broad .... *S. rigidula*
1. Propagulae stout and with the arms absent or little-developed 2
2. Propagulae ovoid to elliptical .... *S. brachygonia*
2. Propagulae triangular or shield-shaped .... *S. tribuloides*

Sphacelaria brachygonia Montagne. Pl. 13, figs 2, 3.

Montagne, 1843.

Plants forming dense hemispherical tufts or small mats, from 1.5-2.5 cm in height; branches sparingly irregularly divided, from 50-70 μm in diameter; segments less than half as long as broad to equal in length and breadth; propagulae borne on 2- or 3-celled stalks, ovoid to elliptical, with the long axis about 150 μm and the short axis up to 70 μm.

For additional figures see Prud'homme van Reine, 1982, p. 189, 190, figs 457-472.

Occurs on rocks in the lower eulittoral subzone in moderately wave-exposed situations, often associated with a certain amount of sand especially when growing in shallow tidepools.

Distribution. World: in warm temperate and tropical seas. West Africa: Ghana: John 1972b, John & de Graft-Johnson 1977, Price *et al.* 1978, Prud'homme van Reine 1982; *S. elliptica,* Dickinson 1952, Lawson 1956, Stephenson & Stephenson 1972.

Sphacelaria rigidula Kützing Pl. 13, fig.1.

Kützing, 1843.

Plants forming loose tufts or compact cushions, to about 2 cm in height; branches sparingly and irregularly divided, from 16-26 μm in diameter; segments half to twice as long as broad; propagulae biradiate, with the two arms cylindrical or slightly tapering towards the tips, having the stalk and the arms of nearly equal length when mature.

For additional figures see Prud'homme van Reine, 1982, p. 204, 205, 207, 208, figs 508-555.

Occurs in the eulittoral zone on very sheltered to moderately wave-exposed rocks and down to a depth of about 10 m; occasionally found as an epiphyte on drift plants of *Sargassum.*

Distribution. World: as for genus. West Africa: Benin: *S. furcigera,* John 1977a, John & Lawson 1972b, Price *et al.* 1978. Cameroun: *S. furcigera,* John 1977a, John & Lawson 1972a, Price *et al.* 1978. Côte d'Ivoire: *S. furcigera,* John 1977a, Price *et al.* 1978. Gabon: *S. furcigera* John & Lawson 1974 a, Price *et al.* 1978. Gambia: *S. furcigera* John & Lawson 1977b, Price *et al.* 1978. Ghana: *S. furcigera* John 1972b, 1977a, John & Lawson 1972a, John *et al.* 1977, John & Pople 1973, Lieberman *et al.* 1979, Price *et al.* 1978. Liberia: *S. furcigera,* John 1977a, Price *et al.* 1978; *S. fusca,* De May *et al.* 1977. Nigeria: *S. furcigera,* Lawson 1980. Sierra Leone: *S. furcigera,* Aleem 1978, John & Lawson 1977a, Price *et al.* 1978. Togo: *S. furcigera,* John & Lawson 1972b, Price *et al.* 1978.

Sphacelaria tribuloides Meneghini Pl. 13, fig. 4.

Meneghini, 1840.

Plants forming subglobose or cushion-like tufts, about 1 cm in height; branches irregularly divided and erect ones arising from a well-developed system of prostrate branches, from 20-40(-65) μm in diameter; segments about equal in length and breadth; propagulae borne on 1- or 2-celled stalks, broadly triangular and with the short biradiate arms having a span of to about 120 μm.

For additional figures see Prud'homme van Reine, 1982, p. 180, 181, 183, figs 422-453.

Occurs on shells or rocks in moderately wave-exposed situations in the upper eulittoral subzone.

Distribution. World: widespread in warm temperate and tropical seas. West Africa: Côte d'Ivoire: John 1972c, Price *et al.* 1978. Ghana: John 1972b, 1977a, John & Lawson 1972a, Price *et al.* 1978; *S. hancockii,* John 1972b, John & Lawson 1972a,

John & Pople 1973, Price *et al.* 1978. Liberia: De May *et al.* 1977, John 1972c, 1977a, Price *et al.* 1978. Togo: John 1977a, John & Lawson 1972b, Price *et al.* 1978.

Remarks. Plants bearing only immature propagulae have often been misidentified as *Sphacelaria hancockii* since the normally distinctive short biradiate arms are undeveloped.

Halopteris Kützing 1843b

Plants usually forming erect tufts, with the branches pinnately divided and occurring across the wall between two adjacent segments; apical cell prominent and giving rise by transverse and longitudinal divisions to segments, corticated in older parts; propagulae absent; unilocular sporangia shortly stalked, borne in axils of the lateral branches and branchlets.

World Distribution: cosmopolitan.

Halopteris scoparia (Linnaeus) Sauvageau

Sauvageau, 1903; *Conferva scoparia* Linnaeus, 1753.

Plants usually known to form erect tufts, attached to the substratum by a fibrous disc; branching irregularly alternate and distichous; main branches having a small celled cortex and the lateral branches and branchlets ecorticate throughout; sporangia arising on 1- or 2-celled stalks, ovoid in shape.

For figures see Newton, 1931, p. 196, fig. 124 (as *Stypocaulon scoparium*).

This rather distinctive plant is known in the region only from a single specimen collected at Lagos, Nigeria "in the sterile condition on Victoria Beach rocks, where it was constantly wave washed" according to Fox (1957).

Distribution. World: widespread from boreal-antiboreal to tropical seas, but rarer in the latter. West Africa: Nigeria: Fox 1957, Lawson 1980, Price *et al.* 1978.

Remarks. This plant is common in warm and cold temperate waters around Europe and the Nigerian record is the only one for the whole of tropical West Africa.

Cladostephus C. Agardh 1817

Rejected Record

Cladostephus spongiosus (Hudson) C. Agardh forma **verticillatus** (Lightfoot) Prud' homme van Reine

Sampaio (1962b), when referring to the occurrence of the blue-green alga *Microcoleus lyngbyaceus* (as *Hydrocoleum lyngbyaceum*) on São Tomé and Príncipe, mentions that in other areas it often grows on this species of *Cladostephus*. In mentioning that Sampaio recorded this brown alga from the two islands Steentoft (1967) has apparently made a mistranslation.

Dictyopteris Lamouroux 1809b

Plants consisting of erect or semi-erect branches, flattened and more or less dichotomously divided, with a conspicuous midrib and sometimes also having smaller lateral veins, attached by a well-developed rhizoidal holdfast; branches terminating in a row of apical cells; sporangia usually arising in sublinear or spherical sori on both sides of the midrib; antheridia in scattered and sunken sori; oogonia solitary and scattered over the surface of the thallus.

World Distribution: widespread in warm temperate and tropical seas.

Dictyopteris delicatula Lamouroux — Pl. 15, fig. 3.

Lamouroux, 1809b.

Plants forming loose clumps or low mats of indefinite extent, from 2-8 cm in height, attached to the substratum and often one branch to another by multicellular rhizoids arising from the conspicuous midrib; branches irregularly to dichotomously divided, from 1-4 mm in width.

For additional figures see Børgesen, 1914, p. 216, figs 166, 167.

Occurs as low mats on moderately sheltered to moderately wave-exposed rocks in the sublittoral fringe and when growing on beach rocks it is often associated with a good deal of sand. Larger and looser clumps are found in sheltered tidepools and at depths down to 10 m.

Distribution. World: as for genus. West Africa: Côte d'Ivoire: John 1972c, 1977a, Price *et al.* 1978. Gabon: John & Lawson 1974a, Price *et al.* 1978. Gambia: John & Lawson 1977b, Price *et al.* 1978. Ghana: Dickinson & Foote 1951, Fox 1957, Gauld & Buchanan 1959, Hartog 1959, John 1977a, John & Pople 1973, John *et al.* 1977, 1980, Lawson 1953, 1956, 1957b, 1966, Lieberman *et al.* 1979, Nizamuddin & Saifullah 1967, Price *et al.* 1978, Stephenson & Stephenson 1972. Liberia: De May *et al.* 1977, John 1972c, 1977a, Price *et al.* 1978. Nigeria: Fox 1957, John 1977a, Lawson 1966, Price *et al.* 1978, Stephenson & Stephenson 1972. Sierra Leone: John & Lawson 1977a, Price *et al.* 1978.

Dictyota Lamouroux 1809b

Plants consisting of erect and flattened branches, regularly or irregularly divided, or more rarely divisions alternate and pinnate, attached by a fibrous holdfast; branches terminating in a conspicuous lens-shaped apical cell; thallus having a single layered medulla of large colourless cells bounded by a layer of smaller assimilatory cells in longitudinal rows; sporangia single or in sori, each producing 4 aplanospores, sometimes associated with hairs and with a cup-shaped structure formed from the stalk cells and a ring of enlarged cells; gametangia in conspicuous spherical or ellipsoidal sori.

World Distribution: cosmopolitan.

Jaasund (1970b) discusses six of the species considered below and the confusion that has existed regarding their specific delimitations.

Key to the Species

1. Branching always distinctly alternate .... *D. mertensii*
1. Branching dichotomous or subdichotomous 2

2. Branches having a dentate or crenulate margin 3
2. Branches having a smooth and entire margin 4

3. Plants very bushy, densely branched and the margins obviously crenulate .... *D. crenulata*
3. Plants lax and loosely branched, with the margins bearing many small teeth arising at irregular intervals .... *D. ciliolata*

4. Branches cervicorn, with the reduced fork either spur-like (f. *pseudobartayresiana*) or sometimes curved and backwardly (f. *curvula*) .... *D. cervicornis* .... *D. cervicornis*
4. Branches otherwise 5

5. Branches narrowing abruptly towards the apex .... *D. divaricata*
5. Branching showing no sudden change in width towards the apex 6

6. Branches usually rounded or truncate at the tips; surface outgrowths absent; sporangia not surrounded at the base by a ring of enlarged cells ... *D. dichotoma*
6. Branches often having acute tips; surface outgrowths spiny or tongue-shaped; sporangia having a ring of enlarged cells at the base forming a cup-shaped structure .... *D. bartayresii*

Dictyota bartayresii Lamouroux — Pl. 14, figs 3, 4.

Lamouroux, 1809b.

Plants either large, to about 20 cm in height, with the branches from 2-3 mm in width, or smaller (‹3 cm high), with the branches from 2-5 mm in width, forming entangled clumps or mats; branching always dichotomous and the upper divergences greater than the lower, with the angle of divergence from (30-)45-90°; young branches having broadly rounded tips and older ones often with characteristically acute and tapering tips.

For additional figures see Rodriques, 1960, p. 586, fig. 1a; Jaasund, 1970b, p. 72, fig. 1d, p. 73, fig. 2c, p. 74, fig. 3c.

Occurs commonly as the smaller form on wave-exposed rocky ledges and as the larger, less entangled form usually in more wave-sheltered situations. Found in the lower eulittoral subzone especially in tidepools, and occasionally on cobbles and rocky platforms at depths down to 10 m.

Distribution. World: widespread in warm temperate and tropical seas. West Africa: Cameroun: John 1977a, Price *et al.* 1978; *D. bartayresiana,* Pilger 1911, Rodrigues 1960, Steentoft 1967. Côte d'Ivoire: Egerod 1974, Price *et al.* 1978. Ghana: Chapman 1963, Cribb 1954, John 1977a, John *et al.* 1977, Lieberman *et al.* 1979, Price *et al.* 1978; *D. bartayresiana,* Dickinson 1952, Lawson 1956, Rodrigues 1960, Steentoft 1967. Liberia: De May *et al.* 1977, John 1977a, Price *et al.* 1978. Príncipe: John 1977a, Price *et al.* 1978; *D. bartayresiana,* Rodrigues 1960, Steentoft 1967. São Tomé: John 1977a, Price *et al.* 1978; *D. bartayresiana,* Carpine 1959, Hariot 1908 (pro parte), Henriques 1917 (p.p.), Rodrigues 1960, Steentoft 1967; *D. ciliata,* Hariot 1908 (p.p.), Henriques 1886b (p.p.), 1887 (p.p.). Sierra Leone: Aleem 1978.

Dictyota cervicornis Kützing

Kützing, 1859.

f. pseudobartayresii W. R. Taylor

Taylor, 1928.

Plants usually bushy, from 10-25 cm in height; branching subdichotomous and cervicorn to irregular, with the reduced fork of a branch often short and spur-like; branches from 2-4 mm in width below and not narrowing markedly above, with the branch tips often acute; surface bearing small and flattened tongue-shaped outgrowths; sporangia enclosed by a ring of elongated cells.

For figures see Taylor, 1928, pl. 16, fig. 15.

f. curvula W.R. Taylor Pl. 15, fig. 4.

Taylor, 1969.

Plants similar to the form described above but smaller, less than 10 cm in height; branches narrower and the angle of divergence less than in other forms, with one member of each ultimate dichotomy simple or less divided than the other and sometimes strongly backwardly curved.

For additional figures see Taylor, 1969, p. 159, figs 17, 18.

Occurs in the lower part of the lithothamnia subzone where wave-action is moderate and also found growing on rocky platforms, calcareous cobbles, pebbles and shell fragments to a depth of about 12 m; occasionally in shallow tidepools in the eulittoral zone.

Plate 14: Figs 1, 2. *Colpomenia sinuosa:* 1. Young, globose to irregularly lobed plants growing on *Padina;* 2. Section of the wall showing the plurilocular sporangia associated with a tuft of hairs. Figs 3, 4. *Dictyota bartayresii:* 3. Large and lax form of the plant having long narrow branches; 4. Small compact form with the branches wider, more closely divided and tips characteristically acute. Fig. 5. *Dictyota mertensii:* Plant having a well-defined and alternately branched main axis.

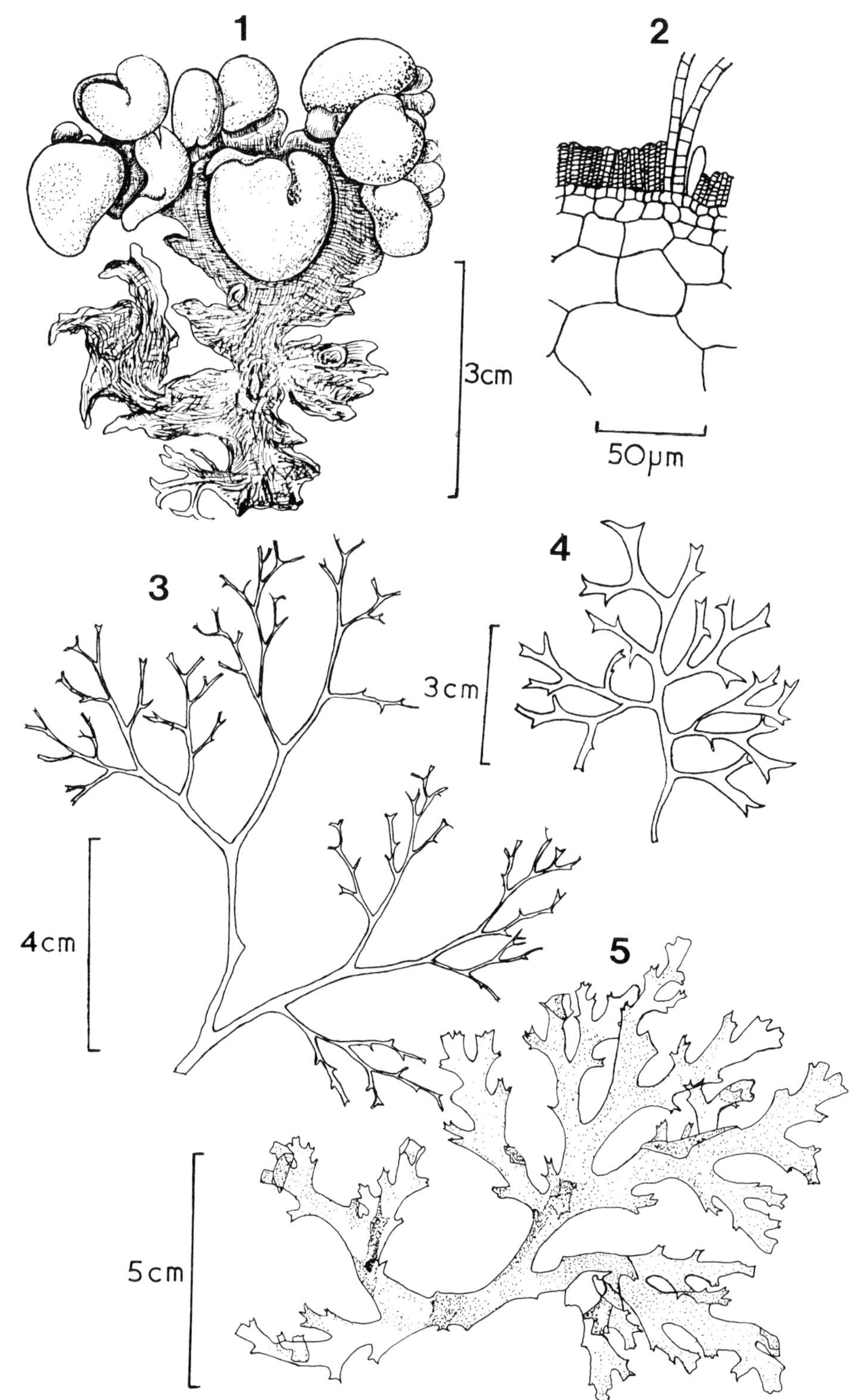
1
2
3cm
50µm
3
4
3cm
4cm
5
5cm

Distribution. World: widespread in warm temperate and tropical seas. West Africa: Bioko: *D. cervicornis* f. *curvula*, unpublished. Côte d'Ivoire: *D. cervicornis* f. *curvula*, John 1977a, Price *et al.* 1978. Gabon: *D. cervicornis* f. *curvula*, John & Lawson 1974a, Price *et al.* 1978. *D. cervicornis* f. *pseudobartayresii*, John & Lawson 1974a, Price *et al.* 1978. Gambia: *D. cervicornis* f. *curvula*, John & Lawson 1977b, Price *et al.* 1978. *D. cervicornis* f. *pseudobartayresii*, John & Lawson 1977b, Price *et al.* 1978. Ghana: John *et al.* 1977, Lieberman *et al.* 1979; *D. cervicornis* f. *curvula*, John 1977a, Price *et al.* 1978. Liberia: *D. cervicornis* f. *curvula*, De May *et al.* 1977, Price *et al.* 1978. São Tomé: *D. cervicornis* f. *curvula*, John 1977a, Price *et al.* 1978. *D. cervicornis* f. *pseudobartayresii*, Price *et al.* 1978, Rodrigues 1960, Steentoft 1967; *D. bartayresiana*, Hariot 1908 (pro parte), Henriques 1917 (p.p.); *D. ciliata*, Hariot 1908 (p.p.), Henriques 1886b (p.p.), 1887 (p.p.). Sierra Leone: Aleem 1978. *D. cervicornis* f. *pseudobartayresii*, John & Lawson 1977a, Price *et al.* 1978.

Remarks. The form referred to as *Dictyota cervicornis* f. *curvula* might equally well be identified as *D. pardalis* Kützing. We have accorded the Gulf of Guinea plants form status accepting the view of Taylor (1969) that this variant appears to be of no major taxonomic importance since the extent of the development of recurved branches is extremely variable. Further comments regarding this species and other problematical taxa in this genus are to be found in Price *et al.* (1978).

Dictyota ciliolata Sonder ex Kützing Pl. 15, figs 1, 2.

Sonder, in Kützing, 1859.

Plants usually solitary, from 10-20 cm in height; branching regularly and loosely dichotomous, with divergences not wide and sometimes spirally twisted, proliferations often present; branches from 4-12 mm in width, bearing along the margin small acute and upwardly directed teeth, with subacute tips; sporangia without an enclosing ring of elongated vegetative cells.

For additional figures see Rodrigues, 1960, p. 586, fig. 16, pl. 2; Jaasund, 1970b, p. 73, fig. 2a, p. 74, figs g-h.

Occurs as the most common member of the genus along the Gulf of Guinea being most commonly found in sheltered tidepools in the lower eulittoral subzone and on rocky surfaces to depths of about 10 m.

Distribution. World: probably widespread in warm temperate and tropical seas. West Africa: Cameroun: Price *et al.* 1978. Côte d'Ivoire: John 1972c, 1977a, Price *et al.* 1978. Gabon: John & Lawson 1974a, Price *et al.* 1978. Ghana: John 1977a, John *et al.* 1977, Lieberman *et. al.* 1979, Price *et al.* 1978; *D. ciliata*, Dickinson 1952, Fox 1957, Lawson 1956, 1957b, 1966, Rodrigues 1960, Steentoft 1967. Liberia: De May *et al.* 1977, Price *et al.* 1978. Nigeria: John 1977a, Price *et al.* 1978; *D. ciliata*, Fox 1957, Rodrigues 1960, Steentoft 1967. São Tomé: John 1977a, Price *et al.* 1978; *D. ciliata*, Fox 1957, Hariot 1908 (pro parte), Henriques 1886b (p.p.) 1887 (p.p.), 1917 (p.p.), Rodrigues 1960 (p.p.), Steentoft 1967. Sierra Leone: John & Lawson 1977a, Price *et al.* 1978.

Remarks. The plant is readily determined when the small marginal teeth are well-developed but individuals may be found having the teeth almost entirely absent.

Dictyota crenulata J. Agardh

J. Agardh, 1848b.

Plants bushy, to about 10 cm in height; branching dense, regularly dichotomous, with angles of divergence often wide, occasionally proliferous; branches linear, to 7 mm in width, with margins dentate and crenulate especially at the rounded tips.

For figures see Taylor, 1945, p. 337, pl. 10, fig. 1.

Occurs on the lower part of the eulittoral zone and is "not infrequent" on the island of Príncipe according to the inscription on a Welwitsch specimen (No. 258).

Distribution. World: probably widespread in warm temperate and tropical seas. West Africa: Príncipe: Price *et al.* 1978, Steentoft 1967; *D. dichotoma*, Barton 1897, 1901. São Tomé: ?Price *et al.* 1978, Steentoft 1967; *D. ciliata*, Hariot 1908 (pro parte), Henriques 1917 (p.p.), Rodriguez 1960 (p.p.).

Remarks. The possibility exists that this species is conspecific with *Dictyota jamaicensis* described by Taylor (1960) from the western side of the Atlantic Ocean.

Dictyota dichotoma (Hudson) Lamouroux

Lamouroux, 1809b; *Ulva dichotoma* Hudson, 1762.

Plants erect and bushy, to about 20 cm in height; branching regularly dichotomous, with the angles of divergence from 15-45(-90)°, occasionally proliferous from near the base; branches from 2-15 mm in width and decreasing but little in width from the base to the apex, sometimes twisted, with the tips blunt to somewhat tapering; sporangia not enclosed by a ring of vegetative cells.

For figures see Earle, 1969, p. 157, 158, figs 50, 51.

Occurs in the lower eulittoral subzone on moderately sheltered to moderately wave-exposed rocks and occasionally found to a depth of about 12 m.

Distribution. World: as for genus. West Africa: Côte d'Ivoire: John 1977a, Price *et al.* 1978. Gabon: Hariot 1896, ?John & Lawson 1974a, ?Price *et al.* 1978. Gambia: John & Lawson 1977b, Price *et al.* 1978. Ghana: ?John 1977a, John *et al.* 1977, Lieberman *et al.* 1979, Price *et al.* 1978. Liberia: ?De May *et al.* 1977, ?Price *et al.* 1978. Príncipe: ?Carpine 1959, John 1977a, Price *et al.* 1978, Steentoft 1967. São Tomé: Carpine 1959, Hariot 1908, ?Henriques 1885, 1886a, John 1977a, Price *et al.* 1978, Steentoft 1967. Sierra Leone: ?John & Lawson 1977a, Price *et al.* 1978. Togo: ?John 1977a, ?John & Lawson 1972b, Pilger 1911, Price *et al.* 1978.

Remarks. This is an extremely variable "species" and many of its features overlap with those of other members of the genus making a definitive identification difficult especially since various interpretations have been put as to the limits of the taxon. The specimens from Togo and Ghana, which we have tentatively attributed to this species, possess an entire margin and have none of the branching characteristics of other species reported from the Gulf of Guinea. The only somewhat unusual feature of many of the collections from this region is the presence of a bluish iridescence in the form of a margin or as transverse banding which is soon lost after gathering.

Dictyota divaricata Lamouroux Pl. 15, fig. 5.

Lamouroux, 1809b.

Plants forming entangled clumps, to about 10 cm in height; branching dichotomous and the angle of divergence from 100-130° in the upper parts, somewhat less below, with many adventitious branches often arising along the margins; branches commonly 1-3 mm in width below and abruptly narrowing above to less than 0.5 mm, with the tips blunt or rounded; sporangia without a ring of elongated vegetative cells.

For additional figures see Earle, 1969, p. 152, fig. 47, p. 164, fig. 58; Jassund, 1970b, p. 72, fig. 1f, p. 73, fig. 2e, p. 74, fig. 3e.

Occurs usually along with other species of *Dictyota* on rocky platforms at depths down to about 10 m; found also in the sublittoral fringe in wave-sheltered situations.

Distribution. World: probably widely distributed in warm temperate and tropical seas. West Africa: Gabon: John & Lawson 1974a, Price *et al.* 1978. Gambia: John & Lawson 1977b, Price *et al.* 1978. Ghana: John *et al.* 1977, Lieberman et al. 1979, Price *et al.* 1978. Sierra Leone: Aleem 1978.

Remarks. The plants found at shallow depths often have a marked bluish iridescence but this phenomenon does not seem to be exhibited by those at about 10 m.

Dictyota mertensii (Martius) Kützing Pl. 14, fig. 5.

Kützing, 1859; *Ulva mertensii* Martius, 1828.

Plants bushy, from 15-20 cm in height; branching alternate and pinnate, with the main axes well-defined; branches from 4-7 mm in width and the lateral branches commonly bearing spur-like branchlets to 2 mm in length, with rounded or acute apices.

For additional figures see Taylor, 1928, pl. 16, figs 4, 5 (as *D. dentata*).

Occurs on rocks and shell fragments in shallow water in places relatively sheltered from wave action.

Distribution. World: in warm temperate and tropical parts of the Atlantic Ocean. West Africa: Gabon: John & Lawson 1974a, Price *et al.* 1978. São Tomé: Agardh 1894, De Toni 1895, Hariot 1908, Price *et al.* 1978; *D. martensii,* Henriques 1917; *D. dentata,* Hariot 1908, Henriques 1886b; 1887, 1817, Rodrigues 1960, Steentoft 1967.

Plate 15: Figs 1, 2. *Dictyota ciliolata:* 1. Plant dichotomously divided and with small dentations along the margins of the branches; 2. Section showing the large medullary cells bounded by a layer of assimilatory cortical cells. Fig. 3. *Dictyopteris delicatula:* Plant subdichotomously divided and with a prominent midrib. Fig. 4. *Dictyota cervicornis* forma *curvula:* Plant having one member of a dichotomy sometimes strongly backwardly curved. Fig. 5. *Dictyota divaricata:* Plant having many adventitious proliferations arising along the margins of the dichotomously divided branches.

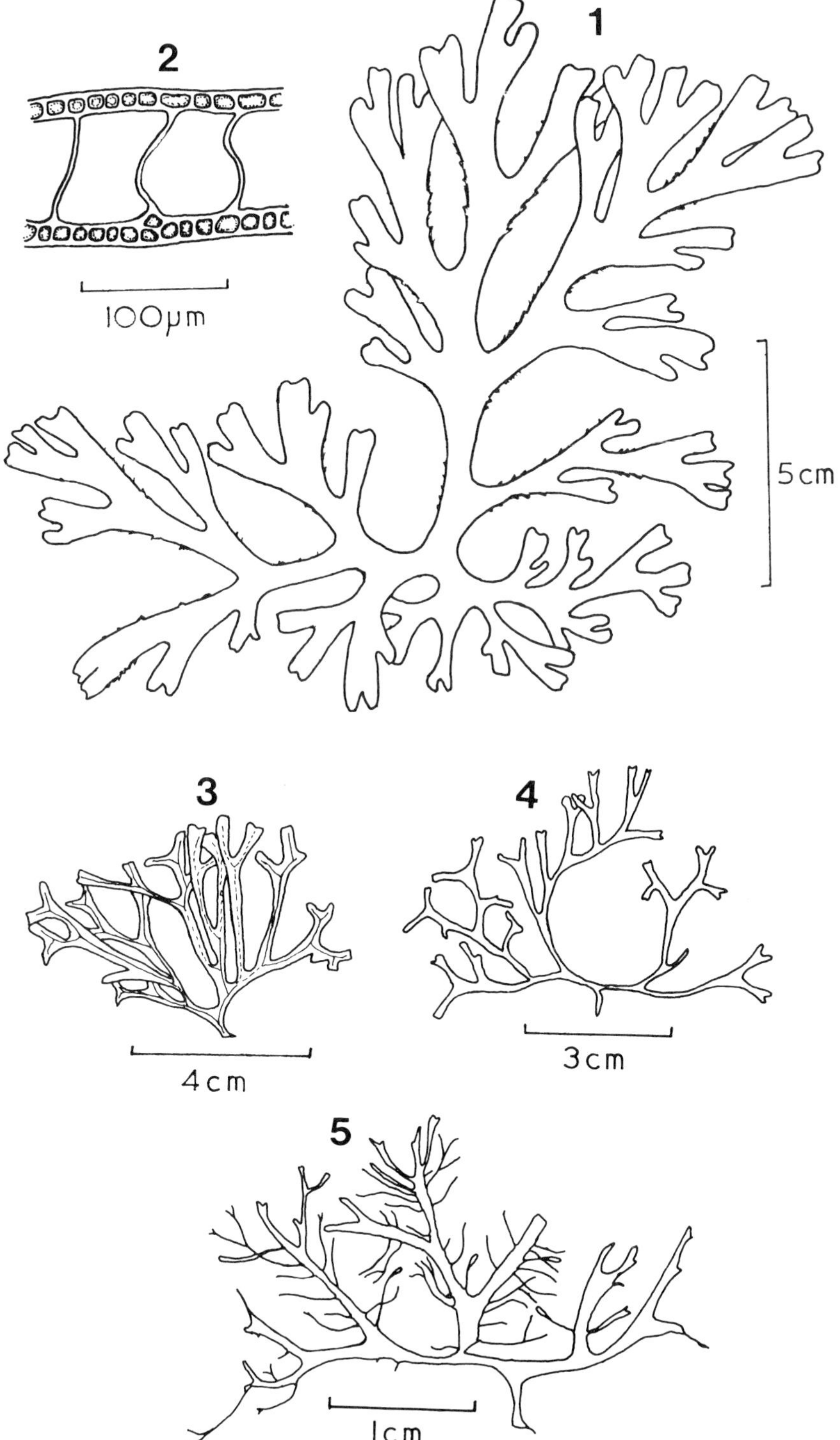
1
2
100µm
5cm
3
4cm
4
3cm
5
1cm

Remarks. *Dictyota dentata* (Gmelin) Lamouroux is a superfluous name for *Fucus atomarius* Gmelin, itself the basionym of *Taonia atomaria* (Woodward) J. Agardh. The earliest legitimate name to apply to alternately branched *Dictyota* species is *D. mertensii.* The taxonomic and nomenclatural reasons for taking this name are discussed in Price *et al.* (1978).

Doubtful Records

Dictyota indica Sonder

The only West African record of this plant is the mention of it growing at the bases of the stilt roots of *Rhizophora* at one locality on the Sierra Leone peninsula (Aleem, 1978). We regard this as a doubtful record of a plant usually considered to be a western Atlantic species.

Dictyota volubilis Kützing *sensu* Vickers

This plant is usually considered to be a western Atlantic species but is reported from two localities on the Sierra Leone peninsula by Aleem (1978). The characteristic twisting of the blades of this species appears to be only an extreme condition of a feature commonly found in many *Dictyota* spp. (see Earle, 1969). We consider this a doubtful West African record requiring further verification.

Dilophus J. Agardh 1881

Doubtful Record

Dilophus guineensis (Kützing) J. Agardh

This plant has been reported as having been collected by Ehrenberg from the island of St. Thomas (Kützing, 1843b, 1849; Murray, 1888). Murray states that the type locality is São Tomé in the Gulf of Guinea. This seems most improbable since Ehrenberg is known to have never visited the region but did most of his plant collecting in the West Indies between 1828-1829 (see Steentoft, 1967). We attach much doubt to the report by Aleem (1978) of this species from Sierra Leone.

Lobophora J. Agardh 1894

Plants erect to prostrate, consisting of flattened fronds, initially deltoid to suborbicular but later becoming cleft and intricately lobed; growth from a marginal row of cells; thallus in transverse section composed of a single-layered medulla of tall cells and a cortex of one to several layers of small cells arranged in tiers; reproductive organs in definite sori on one or both surfaces of the frond; sporangia producing four or eight spores and lacking paraphyses.

World Distribution: widespread in warm temperate and tropical seas.

Lobophora variegata (Lamouroux) Womersley Pl. 16, figs 1-3.

Womersley, 1967; *Dictyota variegata* Lamouroux, 1809b.

Plants prostrate or occasionally erect, with the whole of the undersurface of the prostrate form and only the base of the erect form attached to the substratum by a felty mass of moniliform rhizoids; fronds usually from 3-8 cm in width, consisting of little divided and somewhat overlapping segments, deltoid to suborbicular, often bearing a number of rounded lobes.

For additional figures see Papenfuss, 1943, p. 464, figs 1-8, p. 466, figs 9-14 (as *Pocockiella variegata*).

The erect form is usually best developed at depths greater than 22 m whilst semi-erect individuals are to be found in shallower water and occasionally in the eulittoral zone provided wave action is minimal. The closely adherent prostrate form is that most commonly found, often growing over lithothamnia in moderately wave-exposed situations where such crustose algae are dominant.

Distribution. World: as for genus. West Africa: Bioko: unpublished. Cameroun: John 1977a, Price *et al.* 1978; *Pocockiella variegata*, Lawson 1955. Gabon: Price *et al.* 1978; *Padina lobata*, Jardin 1851, 1891; *Pocockiella variegata*, John & Lawson 1974a; *?Stypopodium fissum*, Kützing 1859. Ghana: John 1977a, John *et al.* 1977, Lieberman *et al.* 1979, Price *et al.* 1978; *Pocockiella variegata*, Dickinson & Foote 1950, Lawson 1956, 1966, Rodrigues 1960, Stephenson & Stephenson 1972. Liberia: De May *et al.* 1977, John 1972c, 1977a, Price *et al.* 1978. Sierra Leone: John & Lawson 1977a, Price *et al.* 1978; *P. variegata*, Aleem 1978, Lawson 1954b, 1957a, Rodrigues 1960.

Remarks. Secondary records from Príncipe should be discounted as they are based on the misidentification of material from that island by Rodriques (1960). The prostrate form of this species may superficially resemble *Ralfsia* from which it can be readily distinguished by its somewhat lighter colour and the darkly pigmented marginal row of meristematic cells.

Padina Adanson 1763

Plants consisting of one or more flattened fan-shaped fronds, simple or divided into segments, bearing on the surface concentric rows of hairs, with or without calcification, attached by a compact rhizoidal holdfast; growth by a row of cells along the inrolled margin; thallus from two to several cells in thickness; tetrasporangia scattered or in sori, forming bands between the hair rows; oogonial sori having an indusium.

World Distribution: widespread in warm temperate and tropical seas, rare in boreal-antiboreal seas.

A curious growth phase may sometimes be found in this genus which consists of an entangled mat of semi-prostrate branches and these may or may not bear the more typical fan-shaped segments. The branches are lobed or ribbon-like, from 1-2 mm in width, often once or twice pinnate and have a distinct apical cell. The branches are often monostromatic at the margin and up to four cells thick elsewhere. When the fan-shaped segments are absent then this growth phase bears a close resemblance to *Dilophus pinnatus* described by Dawson (1950b), but when present we have often been able to identify the plant as *Padina durvillei*

Key to the Species

1. Plants completely prostrate and attached to the substratum by rhizoids arising from the undersurface .... *P. mexicana*
1. Plants erect or semi-erect and attached only at the base..................... 2

2. Fronds having the surface distinctly corrugated and uncalcified .... *P. tetrastromatica*
2. Fronds having the surface more or less flat and with variable amounts of calcification 3

3. Fronds from 6 to 8 cells thick, at least towards the base..................... 4
3. Fronds never more than 4 cells thick throughout........................... 5

4. Fronds 6 to 8 cells thick near the base and decreasing above to 2 cells towards the margin .... *P. vickersiae*
4. Fronds 6 to 8 cells thick throughout .... *P. durvillei*

5. Plants lightly calcified on the lower surface only; fronds 3 cells thick throughout. .... *P. boryana*
5. Plants moderately to strongly calcified on both surfaces; fronds from 2 to 4 cells thick .... *P. australis*

Padina australis Hauck Pl. 16, fig. 9.

Hauck, 1887.

Plate 16: Figs 1-3. *Lobophora variegata:* 1. Erect form of the plant; 2. Prostrate form of the plant showing the overlapping and lobed segments; 3. Section showing the rectangular medullary cells and the tiers of cortical cells. Figs 4, 5. *Padina tetrastromatica:* 4. Plant with narrow segments and having conspicuous rows of surface hairs; 5. Section showing the 4 layers of cells. Figs 6, 7. *Padina durvillei:* 6. Plant with the segments having a fan-shaped outline, concentric rows of surface hairs and an inrolled distal margin; 7. Section of an older portion of the frond showing several layers of cells and developing reproductive organs. Fig. 8a-c. *Padina mexicana* (after Dickinson and Foote, 1951): a. Section showing several layers of cells and an antheridial sorus; b. Oogonia developing from the superficial layers of cells; c. Tetrasporangia developing on the upper surface and rhizoids on the lower surface. Fig. 9. *Padina australis:* Plant showing the fan-shaped outline to the frond and the inconspicuous hair zones.

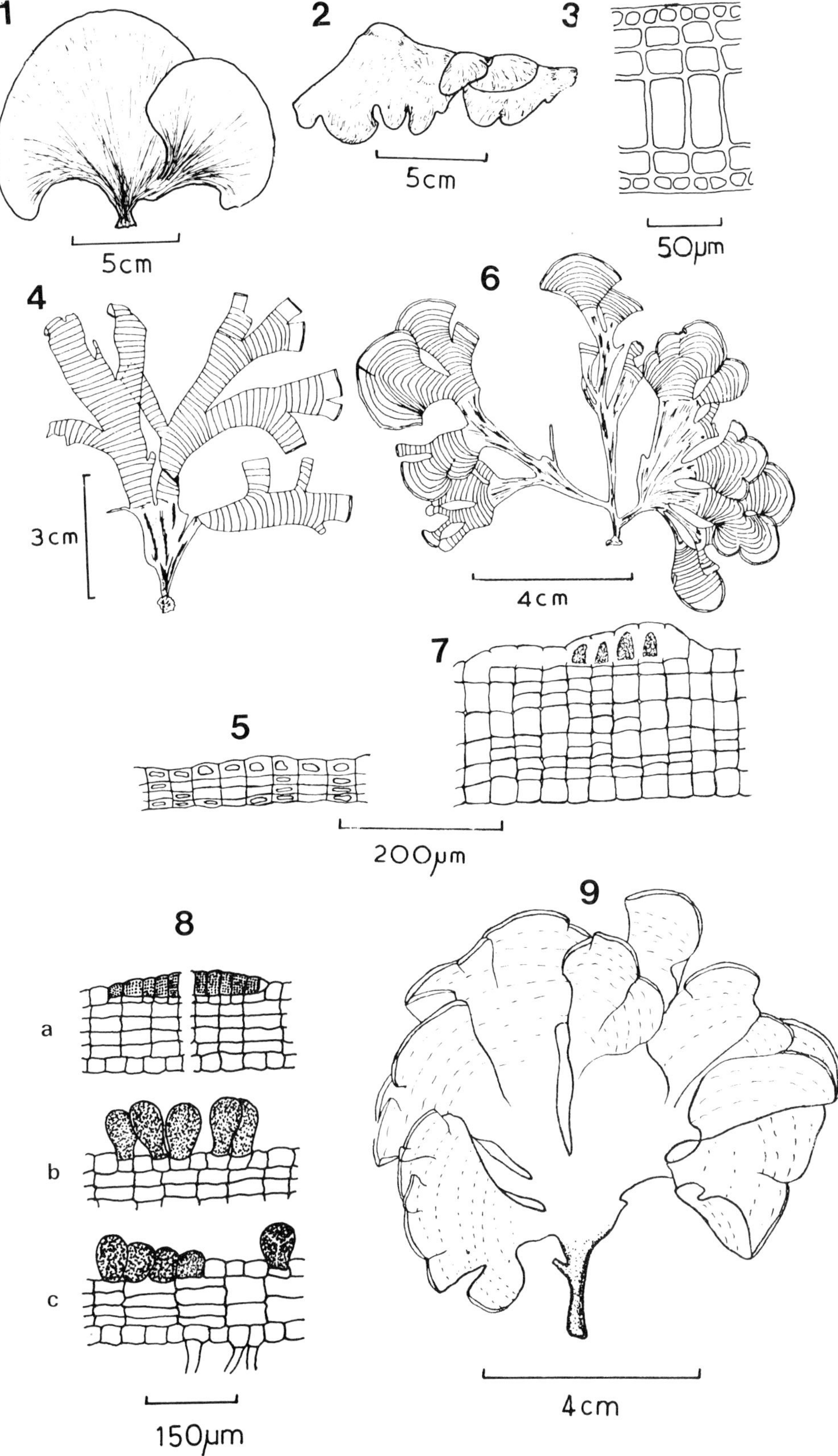
1
5cm
2
5cm
3
50µm
4
3cm
6
4cm
7
5
200µm
8
a
b
c
150µm
9
4cm

Plants forming clumps, from 2-4 cm in height, usually divided into strap-like segments, with the surface moderately calcified and the hair zones about 2 to 4 mm apart; thallus only 2 cells thick at the margin, increasing to 3 cells below and finally to 4 cells at the base; sporangial sori arising midway between the concentric rows of hairs, without an indusium; antheridia and oogonia arising in a similar position.

For additional figures see **Børgesen**, 1914, p. 203, 204 (as *P. gymnospora*).

Occurs on moderately sheltered to moderately wave-exposed rocks in the algal turf in the lower eulittoral subzone and the sublittoral fringe; sometimes in sheltered muddy situations.

Distribution. World: probably in warm temperate and tropical seas. West Africa: **Cameroun: John 1977a, Price *et al.* 1978; *P. gymnospora*, Pilger 1911. Côte d'Ivoire:** ?John 1977a, Price *et al.* 1978. Gabon: John 1977a, John & Lawson 1974a, Price *et al.* 1978.

Remarks. The concentric zones of hairs are often very indistinct in this species.

Padina boryana Thivy

Thivy, in W.R. Taylor, 1966.

Plants erect and to about 5 cm in height; fronds usually fan-shaped, with rounded **lobes, lightly calcified only on the lower surface, having hair zones about 2 mm** apart; thallus 3- or rarely 2-layered; sori arising just above each hair zone and absent just below, without an indusium.

For figures see Durairatnam, 1961, p. 131, pl. 7, figs 1-3, p. 167, pl. 25 (as *P. commersonii*).

There is no information on the ecology of this plant in the region but elsewhere it is known from wave-sheltered situations, often growing in the algal turf in the lower eulittoral subzone and the sublittoral fringe.

Distribution. World: probably pantropical. West Africa: São Tomé: Price *et al.* 1978; *P. commersonii*, Steentoft 1967.

Padina durvillei Bory Pl. 16, figs 6, 7.

Bory, 1827-29.

Plants forming small or occasionally extensive patches, to about 20 cm in height, somewhat leathery, entire or sometimes lobed and kidney-shaped; surface of the fronds lightly calcified and the hair zones to about 3 mm apart and 1 mm in width; thallus from 6 to 8 cells in thickness throughout; oogonial sori in two concentric rows between each consecutive zone of hairs.

For additional figures see Bory, 1827-29, pl. 21, figs 1a - c.

Occurs most commonly on sand-covered rocks in sheltered to moderately wave-exposed situations in the lower eulittoral subzone and especially abundant in tide-pools.

Distribution. World: in warm temperate and tropical seas. West Africa: Ghana: Price *et al.* 1978; *P. durvilliae.* Dickinson & Foote 1951. Liberia: Price *et al.* 1978; *P. durvilliae.* De May *et al.* 1977. Sierra Leone: Price *et al.* 1978; *P. durvilliae.* John & Lawson 1977a. Togo: Price *et al.* 1978; *P. durvilliae.* ?John & Lawson 1972b.

Remarks. The Togolese and some of the Ghanaian plants are unusual in that the thallus is sometimes just 2 to 3 cells in thickness towards the margin although in all other respects they possess features characteristic of this species.

Padina mexicana Dawson Pl. 16, figs 8a - c.

Dawson, 1944.

Plants prostrate and attached by rhizoids arising in indistinct rows over much of the undersurface; fronds fan-shaped, bearing hairs in rows 3-3.5 mm apart on the upper surface, calcification only between the hair rows; thallus only 2 cells thick near the margin, from 4 to 7 cells in thickness in the lower and older parts; sporangial sori on the upper surface only, discontinuous or scattered in groups between the hair zones, with an indusium; antheridia and oogonia forming 2 broad rows between each of the zones of hairs.

For additional figures see Dawson, 1944, p. 403, pl. 52, fig. 2.

Of rare occurrence in the region having been found only in the western region of Ghana growing on rocks and over articulated corallines in moderately wave-exposed situations in the eulittoral zone.

Distribution. World: in tropical parts of the Pacific and Atlantic Oceans. West Africa: Ghana: Dickinson & Foote 1951, Price *et al.* 1978.

Padina tetrastromatica Hauck Pl. 16, figs 4, 5.

Hauck, 1887.

Plants solitary or as patches of indefinite extent, to about 20 cm in height; fronds often repeatedly divided into delicate and narrow segments, uncalcified and often distinctly corrugated, with the zones of hairs from 1-1.5 mm apart; thallus 4 cells thick throughout; sori developing close to and on either side of each hair zone, without an indusium.

For additional figures see Gaillard, 1967, p. 448, 450, 452, 455, 457, 462, figs 1-6.

Occurs on moderately wave-exposed rocks as small and solitary individuals whilst in more sheltered situations it often grows as large plants covering many square metres on open rocks and rocks in tidepools in the lower subzone of the eulittoral; occasionally on cobbles at a depth of about 10 m below low water.

Distribution. World: probably pantropical. West Africa: Côte d'Ivoire: John 1972c, 1977a, Price *et al.* 1978. Gambia: John & Lawson 1977b, Price *et al.* 1978. Ghana: Dickinson & Foote 1951, Gaillard 1967, John 1977a, John *et al.* 1977, Lawson 1956, 1957b, 1966, Lieberman *et al.* 1979, Price *et al.* 1978, Stephenson & Stephenson

1972. Liberia: De May *et al.* 1977, Price *et al.* 1978. Sierra Leone: John & Lawson 1977a. Togo: John 1977a, John & Lawson 1972b, Price *et al.* 1978.

Padina vickersiae Hoyt in Howe Pl. 17, fig. 5.

Hoyt, in Howe, 1920.

Plants erect and forming variously sized clumps, from 6-15 cm in height; fronds fan-shaped and usually divided, occasionally proliferous, with calcification light or absent, having hair zones about 4 mm apart and from 2-5 mm in width; thallus only 2 cells thick at the margin, increasing to 4 cells below and from 6 to 8 cells in thickness at the base; sori arising midway between the hair zones, with an indusium.

For additional figures see Børgesen, 1914, p. 205 to 297, figs 157-207 (as *Padina variegata*).

Occurs on shells, fragments of lithothamnia, as well as on moderately sheltered to moderately wave-exposed rocks in the lower eulittoral subzone; dredged to a depth of 14 m in other parts of the Atlantic Ocean.

Distribution. World: in warm temperate and tropical parts of the Atlantic Ocean. West Africa: Cameroun: John 1977a, Price *et al.* 1978, Rodrigues 1960, Steentoft 1967. Côte d'Ivoire: John 1972c, 1977a, Price *et al.* 1978. Gabon: John & Lawson 1974a, Price *et al.* 1978. Gambia: John & Lawson 1977b, Price *et al.* 1978. Liberia: De May *et al.* 1977, John 1972c, 1977a, Price *et al.* 1978. Príncipe: John 1977a, Price *et al.* 1978, Rodrigues 1960, Steentoft 1967. São Tomé: John 1977a, Price *et al.* 1978, Rodrigues 1960, Steentoft 1967; *P. pavonia*, Henriques 1917; *P. variegata*, Hariot 1908; *Pocockiella variegata*, Rodrigues 1960; *Zonaria pavonia*, Henriques 1886b, 1887; *Z. variegata*, Henriques 1917. Sierra Leone: Aleem 1978, John & Lawson 1977a.

Doubtful Records

Padina jamaicensis (Collins) Papenfuss

This is reported under the name of *Padina sanctae-crucis* Børgesen by Aleem (1978) from shallow water on the Sierra Leone peninsula. We regard this to be a doubtful record of a species only previously reported from the western side of the Atlantic Ocean.

Padina pavonia (Linnaeus) Lamouroux

All the reports of this plant from the Gulf of Guinea have been discounted when the material has been re-examined (see Steentoft, 1967). For this reason we regard the reports of this species from Gabon by Hariot (1896) and from the Casamance region of Sénégal by Chevalier (1920) to be uncertain until such time as they can be verified by examination of the material on which they are based.

Spatoglossum Kützing 1843b

Plants erect, flattened, subpalmately to pinnately divided into somewhat irregular lobes, shortly stipitate and attached by a fibrous holdfast; growth by meristematic cells at the tips of the lobes; thallus composed of a small-celled cortex and a several layered medulla; tetrasporangia scattered over the surface of the fronds; antheridia arising in inconspicuous sori; oogonia scattered or occasionally in small groups.

World Distribution: widespread in warm temperate and tropical seas.

Key to the Species

1. Plants having an irregular and dentate margin .... *S. schroederi*
1. Plants having an entire and undulating margin .... *S. solierii*

Spatoglossum schroederi (C.Agardh) Kützing Pl. 17, fig. 4.

Kützing, 1859; *Zonaria schroederi* C. Agardh, 1824.

Plants bushy and having a ragged appearance, to about 30 cm in height, dichotomously or occasionally subpalmately divided; segments elongate and linear, about 2.5 cm in width and to about 13 cm in length, often bearing narrow sublinear proliferations, with the margin irregular and obviously dentate.

For additional figures see Taylor, 1960, p. 720, fig. 5.

Occurs in tidepools in the lower part of the eulittoral zone though the largest and most developed plants are to be found growing on rocky platforms and calcareous cobbles at depths down to 28 m.

Distribution. World: as for genus. West Africa: Cameroun: Price *et al.* 1978. Gabon: John & Lawson 1974a, Price *et al.* 1978. Gambia: John & Lawson 1977b, Price *et al.* 1978. Ghana: Dickinson & Foote 1950, John *et al.* 1977, 1980, Lieberman *et al.* 1979, Price *et al.* 1978. Sierra Leone: Aleem 1978, John & Lawson 1977a, Price *et al.* 1978.

Remarks. The distinction between this species and *Spatoglossum solierii* is open to doubt as even within a single population there may be considerable variation between plants as to the degree of dentition of the margin. Other characters for separating the two species as given by De Toni (1895), based largely on surface features, appear to be so variable and ill-defined as to be of little use in distinguishing them. The description of this plant by Martius (1833) is antedated by one of C. Agardh (1824). For the typification of the species see Price *et al.* (1978).

Spatoglossum solierii (Chauvin ex Montagne) Kützing

Kützing, 1843b; *Dictyota solierii* Chauvin, in Montagne, 1836.

Plants forming clumps, from 10-40 cm in height, dichotomously or pseudodichotomously divided; segments cuneate or oblong, basal segments from 6-10 mm in width, with the margin entire, undulate and sometimes proliferous.

For figures see Gayral, 1958, p. 235, pl. 36, p. 236, fig. 34b.

There is no information on the ecology of this plant in the Gulf of Guinea though elsewhere it is usually reported from the lower part of the eulittoral zone.

Distribution. World: as for genus. West Africa: Gabon: Hariot 1896, Price *et al.* 1978, ?Trochain 1940.

Remarks. There is some confusion regarding the earliest description and attribution of this species. The reasons for accepting the description of Montagne (1836) as that first valid are given in Price *et al.* (1978).

Stypopodium Kützing 1843b

Plants usually erect, fan- or strap-shaped and divided into narrow spathulate or cuneate segments, with concentric rows of hairs on the surface, attached by a large rhizoidal holdfast; growth by a marginal row of meristematic cells; thallus composed of a medulla of large irregularly disposed cells and a cortex of several layers of small cells; tetrasporangia in sori and with paraphyses.

World Distribution: probably widespread in warm temperate and tropical seas.

Stypopodium zonale (Lamouroux) Papenfuss Pl. 17, fig. 3.

Papenfuss, 1940a; *Fucus zonalis* Lamouroux, 1805.

Plants bushy, to about 15 cm or more in height, irregularly divided into cuneate or fan- or strap-shaped segments; segments from 1-1.5 cm in width, with the concentric rows of hairs arising irregularly on the surface and from 3-10 mm apart; tetrasporangia in irregular sori and developing near the hair rows.

For additional figures see Taylor, 1960, p. 719, pl. 28, fig. 1.

Known in tropical West Africa from a single specimen collected on moderately wave-exposed rocks in the eulittoral zone; dredged from depths of up to 55 m in other parts of the Atlantic Ocean.

Distribution. World: as for genus. West Africa: Ghana: Dickinson & Foote 1950, Price *et al.* 1978.

Plate 17: Fig. 1. *Sargassum cymosum:* Portion of a plant showing the terete main axes bearing the smooth-margined foliar appendages. Fig. 2. *Sargassum filipendula:* Portion of a plant showing the air bladders and the receptacles borne at the base of the serrated-margined foliar appendages. Fig. 3. *Stypopodium zonale:* Plant irregularly divided into segments without an inrolled margin and with concentric rows of surface hairs. Fig. 4. *Spatoglossum schroederi.* Plant having a somewhat ragged appearance with the margin irregular and dentate. Fig. 5. *Padina vickersiae* (after Rodrigues, 1960): Plant having a fan-shaped outline, inrolled margin and rows of sori between the concentric zones of surface hairs.

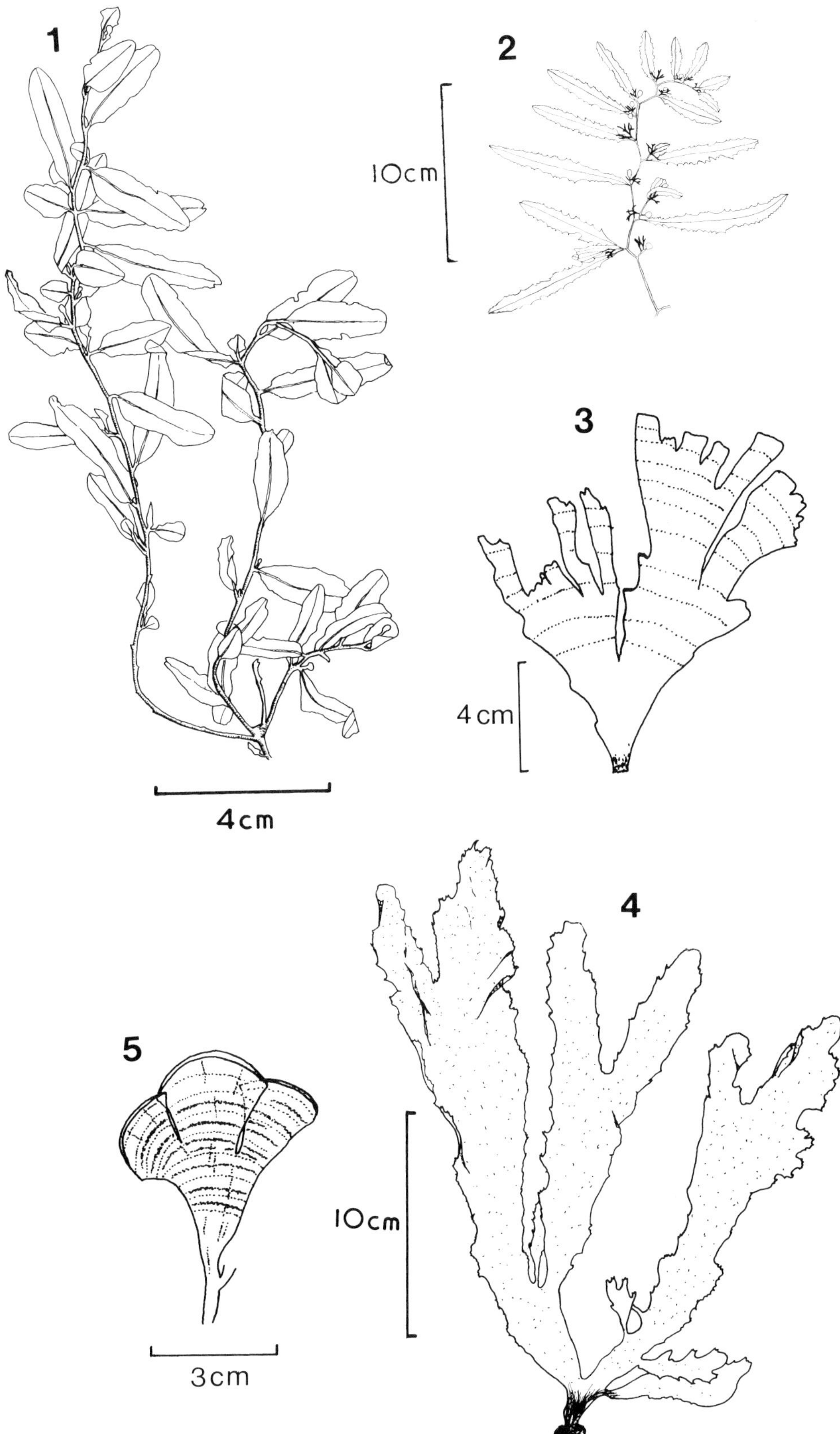
1
2
10cm
3
4cm
4cm
4
5
10cm
3cm

Zonaria C. Agardh 1817

Doubtful Record

Zonaria tournefortii (Lamouroux) Montagne

Reported from a single locality on the Sierra Leone peninsula (Aleem, 1978). This is the first mention of this species from the mainland of West Africa and we consider the record as doubtful until it can be further verified.

Order Fucales* · Family Cystoseiraceae

Bifurcaria Stackhouse 1809

Doubtful Record

Bifurcaria Stackhouse 1809

Sourie (1954) mentions this species (as *Bifurcaria tuberculata*) along with *Caulacanthus ustulatus, Centroceras clavulatum, Cystoseira foeniculacea* (as *C. abrotanifolia*), *Gigartina acicularis, Laurencia pinnatifida* and *Lithophyllum tortuosum*, as some of the more common littoral algae occurring on the Mauritanian side of the Cap Blanc peninsula. He states that these species exist "... au moins assez générale sur le littoral ouest-africain, du Maroc à la Guinée française.". This contention appears to be true for only some of the species mentioned but not for *Bifurcaria bifurcata, Cystoseira foeniculacea* and *Lithophyllum tortuosum* which have yet to be reported from further south than Mauritanie or the northern part of Sénégal.

Cystoseira C. Agardh 1821a

Plants erect, solitary or clumped, attached by a conical disc; main axis terete, smooth or covered by tubercles, simple or radially or bilaterally branched usually in a pinnate manner, bearing short spine-like or filiform and dichotomous appendages; air bladders often in the ultimate branches; receptacles generally terminal on the ultimate branch divisions and usually distal to an air bladder, often clustered, simple or branched, cylindrical or lanceolate, verrucose, frequently covered by spine-like appendages.

World Distribution: widespread from warm temperate to tropical parts of the Atlantic Ocean, also common in the Mediterranean but rare in the Pacific Ocean.

**Ascophyllum nodosum* Stackhouse (Fucaceae) has been reported by John (1974) as drift plants found just outside the general area of the Gulf of Guinea (00° 47'S, 14° 10'W - 01° 21'S, 13° 32'W).

Cystoseira nodicaulis (Withering) M. Roberts.

Roberts, 1967; *Fucus nodicaulis* Withering, 1796.

Plants bushy, usually to about 25 cm in length, with the branches alternately pinnate or dichotomously divided, beset with small, acute, spine-like appendages; branches filiform with the swollen bases of the lateral branches often persisting after the loss of the lateral; air bladders elliptical to oblong, borne on the branches often below an axil and 2 or 3 together on the ultimate branch division; receptacles simple or bifurcate, cylindrical, obtuse or accuminate, irregularly nodulose, covered by spine-like appendages.

For figures see Roberts, 1977, p. 177 to 180, 182 to 189, 191 to 193, 195, 197, figs 1-23.

There is no information on the ecology of this species in the region.

Distribution. World: from warm temperate to tropical parts of the Atlantic Ocean as well as in the Mediterranean. West Africa: Sénégal (Casamance): Price *et al.* 1978; *C. granulatum*, Chevalier 1920.

Doubtful Record

Cystoseira foeniculacea (Linnaeus) Greville

See the remarks made under *Bifurcaria bufurcata* regarding the doubt that attaches to the record of this plant from Guinée (Sourie, 1954).

Family Sargassaceae

Sargassum C. Agardh 1821a

Plants erect and usually basally attached, rarely free-floating; main axis alternately divided, terete, angular or somewhat flattened, bearing simple or forked foliar appendages; foliar appendages variously shaped, with an entire, serrate or dentate margin; air bladders when present stalked, spherical, ellipsoidal, obovate or somewhat flattened; receptacles usually axillary in position, generally terete and more rarely angular or somewhat flattened, simple or branched.

World Distribution: widespread in warm temperate and tropical seas.

This genus cannot be treated in anything like a satisfactory manner until there is an up-to-date revision of the group as a whole. Many of the characters believed in earlier monographs to have taxonomic significance are now known to undergo marked alteration depending on local environmental differences. Taylor (1960) has summarised the present state of our knowledge regarding this especially problematic genus, and emphasises the need for a thorough knowledge of the plants in any one area to be first undertaken so as "to determine the variation shown with regard to habitat and age" before trying to identify any of the local entities. In the Gulf of

Guinea we make no attempt to recognise any of the many forms which have been described from other regions, such as the numerous subsidiary taxa given in Grunow's Monograph on *Sargassum* (Grunow, 1915, 1916), until more is known regarding their delimitation.

Key to the Species

1. Foliar appendages with an entire or slightly crenulate margin....... *S. cymosum*
1. Foliar appendages usually with a distinctly serrate margin.................. 2

2. Plants rather lax and pale yellowish-brown; foliar appendages thin and delicate, up to 1(-1.5) cm in width .... *S. filipendula*
2. Plants bushy and olive brown; foliar appendages thick and somewhat leathery, upper ones usually less than 5 mm in width......................... *S. vulgare*

Sargassum cymosum C. Agardh — Pl. 17, fig. 1.

C. Agardh, 1821a.

Plants to about 50 cm in length, with the branches sparingly divided and bearing the foliar appendages on short lateral branches; main axis generally from 1-2 mm in diameter; foliar appendages elliptical to lanceolate or linear, usually from 4-8(-10) mm in width and up to 4 cm in length, with the margin entire or only slightly crenulate; cryptostomata small and scattered over the surface of the foliar appendages; vesicles usually few in number, borne on a stalk about equal in length to their diameter, spherical or occasionally apiculate; receptacles filiform and dichotomously divided, often somewhat lax.

For additional figures see Taylor, 1960, p. 739, pl. 38, fig. 4.

Occurs in the sublittoral fringe and immediate sublittoral on moderately sheltered to moderately wave-exposed rocks.

Distribution. World: in warm temperate and tropical parts of the Atlantic Ocean. West Africa: Gabon: De Toni 1895, Hariot 1896, John & Lawson 1974a, Price *et al.* 1978. Sierra Leone: Price *et al.* 1978; *S. cymosum* f. *latifolium*, C. Agardh 1821a.

Sargassum filipendula C. Agardh — Pl. 17, fig. 2.

C. Agardh, 1824.

Plants usually rather lax, to about 50(-100) cm in length, with the branches sparingly alternately divided and bearing spirally disposed foliar appendages; main branches about 1 mm in diameter; foliar appendages yellowish-brown, thin and delicate, linear to lanceolate, up to 1(-1.5) cm in width and from 3-8 cm in length, with the margin distinctly serrate; cryptostomata conspicuous and scattered or in rows along the midrib; vesicles axillary, spherical, occasionally apiculate, up to 5 mm in diameter; receptacles axillary on the branches, simple or forked and in the form of a raceme.

For additional figures see Taylor, 1960, p. 737, fig. 3, p. 743, fig. 2.

Occurs as extensive beds, sometimes covering several square metres, growing on sand-covered rocks at about 10 to 15 m and at shallower depths only in relatively wave-sheltered situations. Large plants may be found floating in the drift or lying unattached on the bottom near the beds of attached individuals. These floating plants are particularly abundant towards the end of the dry season when storms are frequent which cause them to be cast-up on the shore in large numbers.

Distribution. World: in tropical parts of the Atlantic Ocean. West Africa: Gabon: John & Lawson 1974a, Price *et al.* 1978. Ghana: John & Lawson 1972a, John & Pople 1973, John *et al.* 1977, 1980, Lieberman *et al.* 1979, Price *et al.* 1978.

Sargassum vulgare C. Agardh Pl. 18, figs 1-3.

C. Agardh, 1821a.

Plants bushy, to about 50 cm in length, with alternately divided branches bearing at short intervals spirally disposed foliar appendages; main branches about 2 mm in diameter; foliar appendages olive-green, usually thick and leathery, lanceolate to ovate or oblong, about 7 mm in width and to 10 cm in length towards the base of plant, those above from 1-5 mm wide and to 5 cm long, with acute serrations along margins of the upper appendages and those below occasionally having an undulate margin; cryptostomata indistinct and irregularly scattered on the surface of the appendages; vesicles variable in number, to about 8 mm in diameter; receptacles clustered, with the main axis irregularly divided and forming a fasciculate cyme.

For additional figures see Taylor, 1960, p. 739, fig. 1, p. 743, fig. 5.

var. **foliosissimum** (Lamouroux) J. Agardh

J. Agardh, 1889; *Fucus foliosissimus* Lamouroux, 1813.

Differs from the typical form in that the foliar appendages are often closer together, broader and shorter in length, from 0.5-2.5 cm long, often more obtuse, and with undulate margins; receptacles usually shorter than in the typical form.

For figures see Vickers and Shaw, 1908 (Part 2), pl. 2.

Occurs in greatest abundance in the sublittoral fringe where in many localities it is the dominant alga. This plant extends only into the immediate sublittoral giving way in many wave-exposed situations to lithothamnia-covered rocks and large populations of sea urchins. It may be particularly common in large and deep tidepools in the lower eulittoral subzone but absent where there are large numbers of sea urchins. The variety has only been found growing in moderately wave-sheltered situations although it is often reported solely as drift plants.

Distribution. World: in warm temperate and tropical seas. West Africa: Benin: *S. vulgare* var. *foliosissimum*, John & Lawson 1972b, Price *et al.* 1978, Steentoft 1967. Cameroun: Hoppe 1969, Lawson 1955, Price *et al.* 1978, Schmidt & Gerloff 1957,

?Steentoft 1967, Stephenson & Stephenson 1972. *S. vulgare* var. *foliosissimum*, Pilger 1911, Price *et al.* 1978, Steentoft 1967. Côte d'Ivoire: John 1972c, 1977a, Price *et al.* 1978. Gabon: John & Lawson 1974a, Price *et al.* 1978, ?Steentoft 1967; *?S. cheirifolium*, Kützing 1849, Steentoft 1967; *S. cymosum* var. *esperi* f. *hapalophylla*, Grunow 1916; *?S. tenue*, Kützing 1849, 1861, ?Srinivasan 1967; *S. tenue* var. *gabonensis*, Grunow 1915; *S. tenue* var. *gabonensis* f. *intermedia*, Grunow 1915. Gambia: John & Lawson 1977b, Price *et al.* 1978. Ghana: Gauld & Buchanan 1959, Hartog 1959, John & Lawson 1972a, John *et al.* 1977, John & Pople 1973, Lawson 1954a, 1956, 1957b, 1959, Michanek 1971, 1975, Price *et al.* 1978, Steentoft 1967, Stephenson & Stephenson 1972. Liberia: De May *et al.* 1977, John 1972c, 1977a, Price *et al.* 1978. Nigeria: *S. vulgare* var. *foliosissimum*, Price *et al.* 1978, ?Steentoft 1967. Príncipe: *S. vulgare* var. *foliosissimum*, Price *et al.* 1978, Steentoft 1967. São Tomé: Hariot 1908, Henriques 1917, Price *et al.* 1978, Rodrigues 1960, Steentoft 1967, Tandy 1944; *S. boryanum* (pro parte), Henriques 1886b, 1887; *?S. cheirifolium*, Kützing 1861. *S. vulgare* var. *foliosissimum*, Hariot 1908, Price *et al.* 1978, Rodrigues 1960, Steentoft 1967; *S. vulgare* var. *foliosissimum* f. *pteropus*, Rodrigues 1960; *Marginaria boryana*, Henriques 1917. Sénégal (Casamance): Chevalier 1920, Price *et al.* 1978. Sierra Leone: Aleem 1978, John & Lawson 1977a, ?Lawson 1957a, ?Levring 1969, Longhurst 1958, Michanek 1971, 1975, Price *et al.* 1978, ?Steentoft 1967, Stephenson & Stephenson 1972; *S. cymosum* f. *latifolium*, C. Agardh 1821a. Togo: *S. vulgare* var. *foliosissimum*, John & Lawson 1972b, Price *et al.* 1978.

Remarks. *Sargassum tenue* Kützing and *S. cheirifolium* Kützing have been placed by De Toni (1895) in synonymy with *S. cymosum*, but we have followed Price *et al.* (1978) in provisionally placing them under *S. vulgare* since both species have obvious serrations along the margins of the foliar appendages. There is some uncertainty as to whether the plants described as *Sargassum tenue* and *S. cheirifolium* were collected from the mouth of the Gabon River or in present-day Guinée. The former location seems the more likely since the coast of Guinée consists largely of mangrove swamps. A more detailed account of some of the problems outlined above is to be found in Price *et al.* (1978).

Doubtful Records

Sargassum hystrix J. Agardh

This is a widely distributed species in the western Atlantic but the reports from Sierra Leone (Aleem, 1978) and Sénégal (Bodard and Mollion, 1974) are the only two for the West African coast. We consider both records as doubtful.

Sargassum natans (Linnaeus) Gaillon

Reported from the drift at Gabon as *Sargassum bacciferum* by Jardin (1851) and from "Danish Guinea" (present-day Ghana) as "*F.* [*Fucus*] sp. e cohorte *F. natantis* Lin." by Hornemann (1819). The source of the material on which these records are based must remain uncertain as it is an entirely pelagic plant. See Price *et al.* (1978) for further discussion.

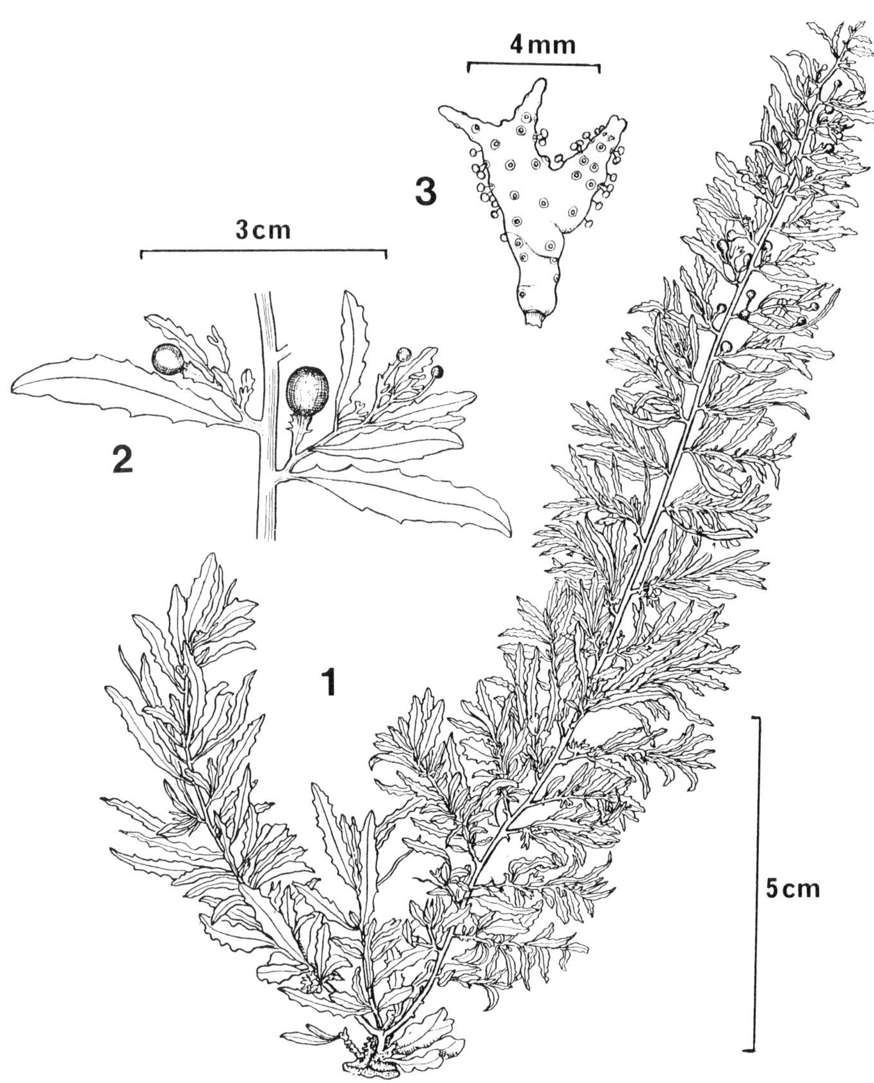

Plate 18: Figs 1-3. *Sargassum vulgare:* 1. Plant showing the alternately divided branches bearing serrated foliar appendages; 2. Portion of a branch showing air bladders and the acute serrations along the margin of the foliar appendages; 3. An irregularly divided receptacle with oogonia hanging on threads protruding through the pores of the conceptacles.

DIVISION RHODOPHYTA · CLASS BANGIOPHYCEAE

Order Porphyridiales · Family Goniotrichaceae

Goniotrichum Kützing 1843b

Plants filamentous, consisting of uniseriate or occasionally multiseriate branches, irregularly or pseudodichotomously divided; cells ovoid to cylindrical, with very thick gelatinous walls; chloroplast single, stellate, with or without a central pyrenoid; asexual reproduction by naked monospores.

World Distribution: cosmopolitan.

Goniotrichum alsidii (Zanardini) Howe Pl. 19, fig. 7.

Howe, 1914; *Bangia alsidii* Zanardini, 1839.

Plants solitary and epiphytic, to about 1 mm or more in height, rose-red; branches mostly uniseriate, rarely multiseriate, sparingly and regularly pseudodichotomously divided, from 10-20 μm in diameter; cells cylindrical to somewhat elliptical, from 6-10 μm in diameter and half to twice as long as broad.

For additional figures see Abbott and Hollenberg, 1976, p. 281, fig. 222; Taylor, 1962a, pl. 423, p. 28, figs 1-4.

Commonly found as an epiphyte on larger algae growing over a wide range of wave-exposure in the eulittoral zone and down to a depth of 25 m.

Distribution. World: as for genus. West Africa: Gabon: John & Lawson 1974a, John *et al.* 1979. Gambia: John & Lawson 1977b, John *et al.* 1979. Ghana: John & Lawson 1972a, John *et al.* 1977, 1979, Lieberman *et al.* 1979. Nigeria: John & Lawson 1972a, John *et al.* 1979; *G. elegans,* Fox 1957. Sierra Leone: John & Lawson 1977a, John *et al.* 1979; *G. elegans,* Aleem 1978. Togo: John & Lawson 1972b, John *et al.* 1979.

Chroodactylon Hansgirg 1885

Doubtful Record

Chroodactylon ornatum (C. Agardh) Basson

We consider the report of this species (as *Asterocytis ramosa*) growing epiphytically on *Dictyota* and *Gelidiella* in two localities on the Sierra Leone peninsula by Aleem (1978) as doubtful. This is the only mention of this species from the African mainland though previously it has been reported on a number of occasions from the Canary Islands.

Erythrocladia Rosenvinge 1909

Plants microscopic, forming monostromatic discs of irregularly alternate or more often pseudodichotomously divided filaments radiating from a common centre; filaments either free from one another or compacted laterally and pseudoparenchymatous, with the cells at the margin of the disc usually deeply cleft; chloroplast single, plate-like, containing a pyrenoid; asexual reproduction by monosporangia formed in the older parts by the oblique division of a vegetative cell.

World distribution: cosmopolitan.

Erythrocladia irregularis Rosenvinge — Pl. 19, figs 1, 2.

Rosenvinge, 1909.

Plants forming minute discs on larger algae, to about 120 μm in diameter or more; cells of the filaments oblong or irregular in shape near the center of the disc, usually cylindrical to oblong and longer towards the margin, with the marginal cells subentire or bifurcate; monosporangia subspherical and produced by division of the cells in the central part of the disc.

For additional figures see Abbott and Hollenberg, 1976, p. 285, fig. 227 (as *E. subintegra*); Børgesen, 1915, p. 8, figs 3, 4 (as *E. subintegra*).

Occurs on larger algae growing over a wide range of wave-exposure in the eulittoral zone. Common on *Chaetomorpha antennina* and readily detectable on this plant where the cells at the distal end of the filaments are often colourless having lost their contents.

Distribution. World: as for genus. West Africa: Cameroun: John *et al.* 1979. Côte d'Ivoire: John 1977a, John *et al.* 1979. Gambia: John & Lawson 1977b, John *et al.* 1979. Ghana: John 1977a, John *et al.* 1979; *E. subintegra*, John & Lawson 1972a. Liberia: De May *et al.* 1977, John 1977a, John *et al.* 1979. Nigeria: John 1977a, John *et al.* 1979; *E. subintegra*, Fox 1957, John & Lawson 1972a. Sierra Leone: John & Lawson 1977a, John *et al.* 1979; *E. subintegra*, Aleem 1978.

Remarks. Heerebout (1968) has concluded from his culture studies on the Erythropeltidaceae that there is considerable morphological variability in the genus *Erythrocladia* with *E. subintegra* and *E. irregularis* probably conspecific.

Erythrotrichia Areschoug 1850

Plants erect, filamentous and uniseriate or occasionally multiseriate, sometimes strap-shaped and monostromatic, attached by a single basal cell, rhizoidal filaments or a group of cells forming a multicellular disc; chloroplast usually single and stellate, with a single central pyrenoid; reproduction by monosporangia produced in the upper vegetative cells by oblique divisions; sexual reproduction reported but much doubt surrounding the events and structures observed.

World Distribution: cosmopolitan.

Erythrotrichia carnea (Dillwyn) J. Agardh Pl. 19, figs 3, 4.

J. Agardh, 1882-1883; *Conferva carnea* Dillwyn, 1809.

Plants usually solitary and epiphytic, to about 1 mm in height, attached by a basal cell bearing short lobes; filaments simple and uniseriate throughout, to about 10 μm in diameter at the base, widening above to 16 μm and again tapering towards the apex; cells from half to twice as long as broad; monosporangia arising in somewhat swollen cells, from 13-15 μm in diameter.

For additional figures see Abbott and Hollenberg, 1976, p. 287, fig. 228; Taylor, 1937, p. 423, figs 13-15.

Occurs on larger algae growing over a wide range of wave-exposure in the eulittoral zone and down to a depth of 10 m.

Distribution. World: as for genus. West Africa: Benin: John 1977a, John & Lawson 1972b, John *et al.* 1979. Gabon: John & Lawson 1974a, John *et al.* 1979. Gambia: John & Lawson 1977b. Ghana: John 1977a, John & Lawson 1972a, John *et al.* 1979, Lieberman *et al.* 1979. Liberia: De May *et al.* 1977, John 1977a, John *et al.* 1979. Sierra Leone: John & Lawson 1977a, John *et al.* 1979; *E. investiens.* Aleem 1978. Togo: John 1977a, John & Lawson 1972b, John *et al.* 1979.

Family Bangiaceae

Bangia Lyngbye 1819

Macroscopic stage and microscopic stage in the life history, with the former only initially simple and uniseriate but soon becoming multiseriate above and tubular or constricted at intervals, attached by the lowermost cell and rhizoidal extensions of the lower cells; chloroplast single, stellate, with a central pyrenoid; asexual reproduction claimed to take place in the genus.

World Distribution: cosmopolitan.

Bangia atropurpurea (Roth) C. Agardh Pl. 19, figs 5, 6.

C. Agardh, 1824; *Conferva atropurpurea* Roth, 1906.

Plants entangled below and forming tufts or extensive mats, from 1-6 cm in height, dark purplish to almost black; filaments lubricous, uniseriate towards the base and multiseriate above, from 20-200 μm in width; cells in more or less transverse rows, irregular to rectangular, from 15-30 μm in size.

For additional figures see Abbott and Hollenberg, 1976, p. 295, fig. 237 (as *B. fusco-purpurea*); Newton, 1931, p. 239, fig. 145 (as *B. fusco-purpurea*).

Occurs on sand-covered rocks in moderately sheltered to moderately wave-exposed situations in the upper eulittoral subzone; commonly to be found on rocks in estuaries and lagoon entrances.

Distribution. World: as for genus. West Africa: Benin: John *et al.* 1979; *B. fuscopurpurea*, John & Lawson 1972b. Gabon: John & Lawson 1974a, John *et al.* 1979. Ghana: John *et al.* 1979; *B. fuscopurpurea*, John & Lawson 1972a, Nigeria: unpublished.

Remarks. This species and *Bangia fuscopurpurea* were until recently considered to be separate species. Culture work by Geesink (1973) has established that they are conspecific, with *B. atropurpurea* the first validly published name.

Porphyra C. Agardh 1824

Doubtful Record

Porphyra ledermannii Pilger

Pilger (1911) provides the following diagnosis of the sterile plant collected by C. Ledermann from Cameroun (p. 298): "Thallus monostromaticus, valde tenuis, 5-6 cm circ. longus, stipite brevi affixus, ambitu ovatus vel magis rotundatus et irregulariter lobatus.". This description would appear to be too vague and insufficient to warrant considering the Cameroun plant as a distinct new species at the present time.

Plate 19: Figs 1, 2. *Erythrocladia irregularis:* 1. Portion of a crust in surface view showing the subentire or bifurcate marginal cells; 2. Portion of a crust showing the monosporangia formed by the oblique division of some inner cells. Figs 3, 4. *Erythrotrichia carnea:* 3. Upper portion of a filament showing monosporangia formed by division of a vegetative cell and a lateral branch formed by the *in situ* development of a spore; 4. Lowermost portion of a filament showing the attachment lobes growing from the basal cell. Figs 5, 6. *Bangia atropurpurea:* 5. Upper portion of a plant with the cells arranged in irregular transverse rows; 6. Lower portion showing the monosiphonous base. Fig. 7. *Goniotrichum alsidii:* Portion of a plant showing the irregularly or pseudodichotomously divided filaments.

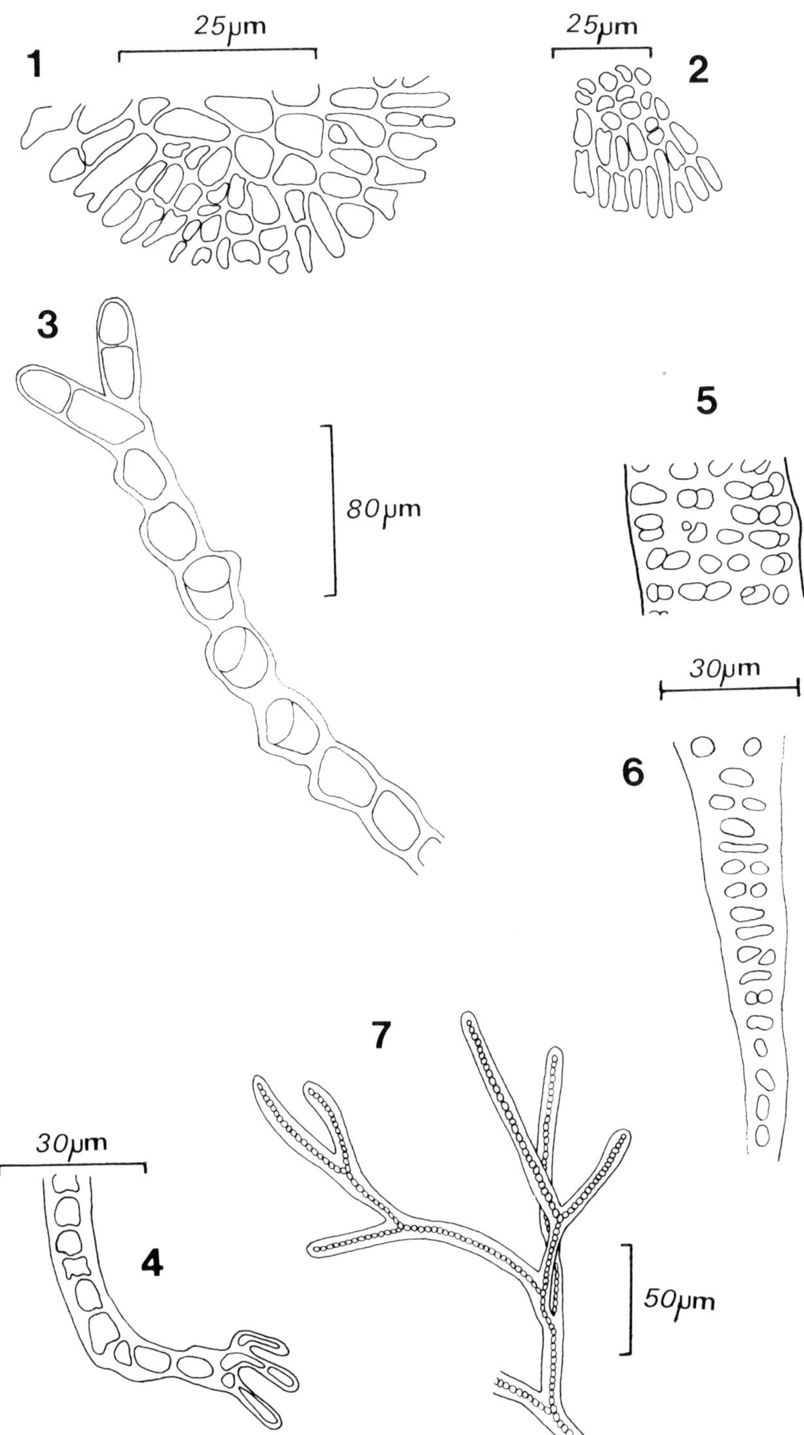
1
25µm
2
25µm
3
80µm
5
30µm
6
7
30µm
4
50µm

CLASS FLORIDEOPHYCEAE

Order Nemaliales · Family Acrochaetiaceae

Audouinella Bory 1823

Plants filamentous, consisting of simple or divided branches arising from a basal cell or a prostate system of filaments which may sometimes be pseudoparenchymatous; chloroplast variously shaped, one to several per cell; asexual reproduction by monosporangia, bisporangia, polysporangia and/or tetrasporangia; spermatangia single or more usually clustered, terminal or lateral on the filaments; carpogonia solitary or more rarely in small groups, sessile, intercalary or borne on one- or two-celled stalks.

World Distribution: cosmopolitan.

Woelkerling (1971), who has reviewed the morphology and taxonomy of the southern Australian representatives of the "*Acrochaetium-Rhodochorton*" complex, considers that *Audouinella* and *Acrochaetium* are not distinct enough to be kept as separate genera. He transferred many species of *Acrochaetium* to *Audouinella* whilst asexual taxa of uncertain systematic status were placed in the "form genus" *Colaconema.* Recently both the asexual and sexual species of acrochaetoid algae that occur in the British Isles have all been placed under *Audouinella* (Dixon and Irvine, 1977b; Parke and Dixon, 1976). The genus *Chantransia* is no longer accepted (Papenfuss, 1947) and many of its members have now been referred to the synonymy of *Audouinella.*

Key to the Species

1. Plants attached by a system of prostrate filaments, often pseudoparenchymatous 2
1. Plants attached by a single basal cell or more rarely also having one to several accessory cells 4

2. Monosporangia forming clusters on branched stalks .... *A. daviesii*
2. Monosporangia sessile or borne on 1-celled stalks 3

3. Monosporangia from 10-14 μm in length and arising in distinct secund series .... *A. seriata*
3. Monosporangia from 16-18 μm in length and scattered or in secund series .... *A. saviana*

4. Basal cell and the cells above similar in size and shape; filaments terminating in a hair .... *A. microscopica*
4. Basal cell usually dissimilar to the cells above; filaments without a terminal hair (non *A. parvula*) 5

5. Basal cell often pyriform or fiddle-shaped, usually bearing accessory cells or filaments. . . . . *A. dasyae*
5. Basal cell quadrate or sometimes rectangular, without accessory cells or filaments 6

6. Plants usually greater than 500 μm in height; monosporangia on a small lateral branch or on a 1-celled stalk . . . . *A. hallandica*
6. Plants usually less than 200 μm in height; monosporangia sessile . . . . . *A. parvula*

Audouinella dasyae (Collins) Woelkerling — Pl. 20, fig. 4.

Woelkerling, 1973a; *Acrochaetium dasyae* Collins, 1906.

Plants epiphytic, with the erect filaments arising from a swollen basal cell; basal cell often pyriform or fiddle-shaped, about 20 μm in diameter and to about 30 μm in length, usually bearing accessory filaments; erect filaments irregularly branched, with the cells from 6-8 μm in diameter and to 2 to 4 times as long as broad.

For additional figures see Woelkerling, 1973a, p. 598, 599, figs 10-31.

Occurs on larger algae such as *Sargassum* or *Dictyota* growing in moderately wave-sheltered tidepools and also found to a depth of about 10 m.

Distribution. World: as for genus. West Africa: Bioko: unpublished. Gambia: John & Lawson 1977b. Ghana: John 1977a. Lieberman *et al.* 1979. Liberia: De May *et al.* 1977, John 1977a.

Audouinella daviesii (Dillwyn) Woelkerling — Pl. 20, fig. 5.

Woelkerling, 1971; *Conferva daviesii* Dillwyn, 1809.

Plants tufted and usually epiphytic, with the erect filaments arising from an irregularly divided and entangled mass of prostrate filaments that often appear pseudoparenchymatous; erect filaments irregularly divided and sometimes tapering towards their apices; cells of the erect filaments from 6-11 μm in diameter and 3 to 5 times as long as broad; monosporangia ovoid, to about 10 μm in diameter and from 9-14 μm in length, forming clusters on branched stalks.

For additional figures see Woelkerling, 1971, p. 71, fig. 7, p. 86, fig. 22.

Occurs on *Chaetomorpha antennina* growing on moderately wave-exposed rocks in the eulittoral zone.

Distribution. World: as for genus. West Africa: Côte d'Ivoire: John 1977a. Gabon: John & Lawson 1974a. Liberia: De May *et al.* 1977, John 1977a. Sierra Leone: *Acrochaetium daviesi,* Aleem 1978.

Audouinella hallandica (Kylin) Woelkerling — Pl. 20, figs 1, 2.

Woelkerling, 1973b; *Chantransia hallandica* Kylin, 1906.

Plants usually epibiotic, with the erect filaments usually from 0.5-1.0 mm in height, arising from a thick-walled basal cell; basal cell usually quadrate or somewhat rectangular, from 10-20 μm in diameter; erect filaments irregularly branched, occasionally tapering towards their tips; cells of the erect filaments from 5-7 μm in diameter, sometimes decreasing above to about 3.5 μm in diameter, to about 6 times longer than broad towards tips but much less in the lower parts; monosporangia ovoid, about 10 μm in diameter and to about 16 μm in length, arising on a small side branch or sessile or on 1-celled stalks on the main filaments, occasionally in series.

For additional figures see Woelkerling, 1973b, p. 83, figs 1-4.

Occurs as a fine felty layer on the surface of larger algae and polyzoans growing over a wide range of wave-exposure both in the eulittoral zone and in shallow water to a depth of about 2 m.

Distribution. World: widespread in warm temperate and tropical seas. West Africa: Togo: *Acrochaetium* sp., John & Lawson 1972b. Sierra Leone: *Acrochaetium sargassi*, Aleem 1978.

Audouinella microscopica (Nägeli) Woelkerling. Pl. 20, fig. 7.

Woelkerling, 1971; *Callithamnion microscopicum* Nägeli, in Kützing, 1849.

Plants usually epiphytic, with the erect filaments less than 100 μm in height, arising from a little modified basal cell; basal cell somewhat similar to the cells above; erect filaments irregularly or sometimes secundly branched, often having a long terminal or pseudoterminal hair; cells of the filaments from 6-10 μm in diameter and 3 to 5 times longer than broad; monosporangia about 5 μm in diameter and up to 10 μm in length, usually solitary and borne laterally or terminally on the filaments.

For additional figures see Woelkerling, 1971, p. 74, fig. 10, p. 87, fig. 23a.

Occurs on plants of *Chaetomorpha antennina* growing on moderately wave-exposed rocks in the eulittoral zone.

Distribution. World: as for genus. West Africa: Ghana: unpublished. Nigeria: Lawson 1980.

Plate 20: Figs 1, 2. *Audouinella hallandica:* 1. Upper portion of a filament bearing ovoid monosporangia; 2. Basal portion showing the thick-walled attachment cell. Fig. 3. *Audouinella saviana:* Portion of a filament with the stalked monosporangia arising in series. Fig. 4. *Audouinella dasyae:* Basal portion of a filament showing the lozenge (a) or fiddle-shaped (b) basal cell. Fig. 5. *Audouinella daviesii:* Portion of a filament showing the monosporangia arising in clusters on branched stalks. Fig. 6. *Audouinella parvula:* Plant with sparingly divided filaments arising from a moderately thick-walled basal cell. Fig. 7. *Audouinella microscopica:* Plant showing the little modified basal cell and a terminal hair.

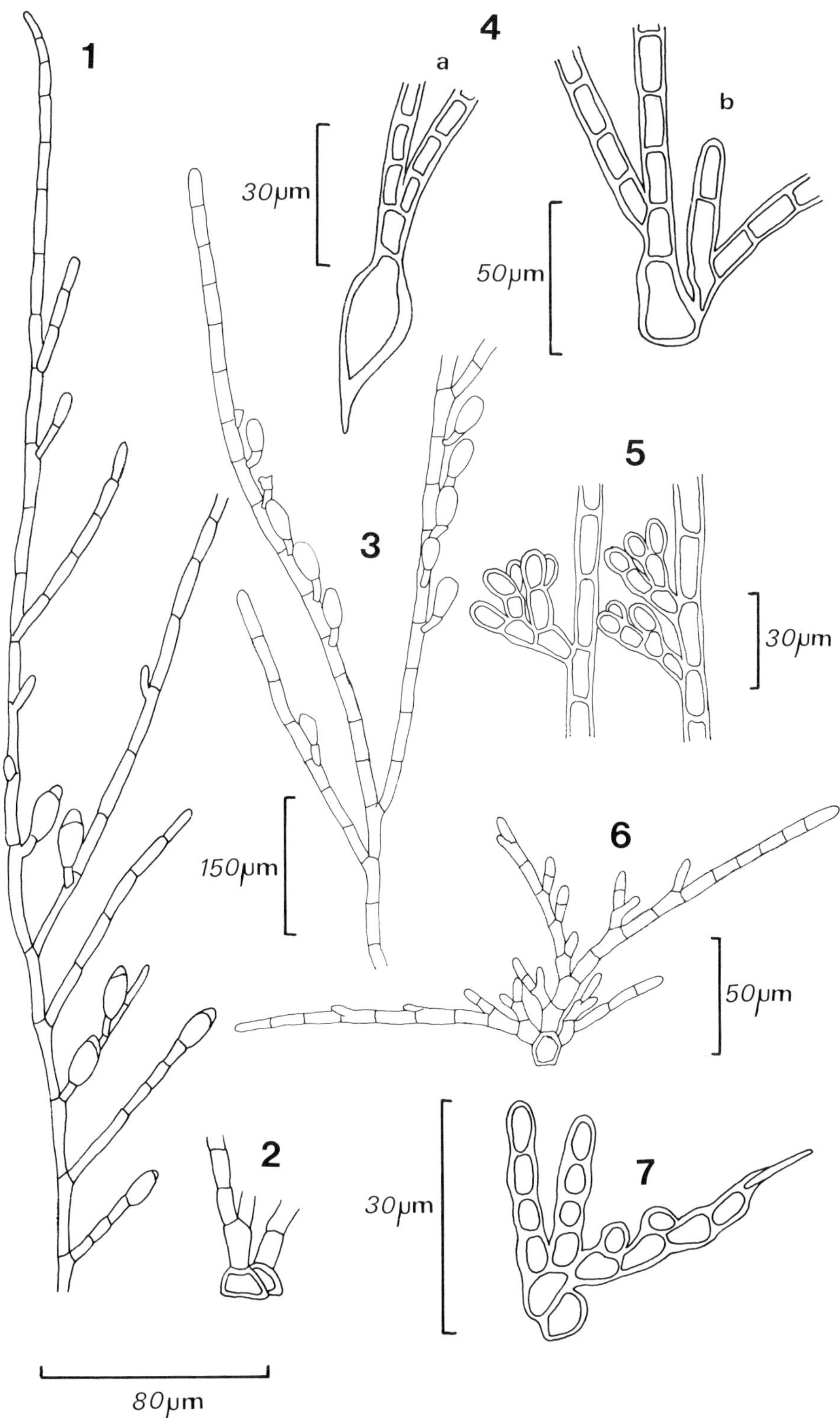
1
2
3
4
a
b
5
6
7
30µm
50µm
30µm
150µm
50µm
30µm
80µm

Audouinella parvula (Kylin) Dixon Pl. 20, fig. 6.

Dixon, in Parke and Dixon, 1976; *Chantransia parvula* Kylin, 1906

Plants epiphytic, with the erect filaments to about 150 μm in height, arising from a relatively distinct basal cell; basal cell usually quadrate and about 14 μm in diameter, with the wall moderately thick; erect filaments sparingly irregularly branched below and somewhat secund above, sometimes bearing unicellular hairs; cells of the filaments from 5-7 μm in diameter and 1 to 3(-4) times longer than broad; monosporangia oblong, about 5-7 μm in diameter and from 7-9 μm in length, sessile and occasionally arising in series.

For additional figures see Kylin, 1944, p. 13, 14, figs 4, 5 (as *Kylinia parvula*).

Found as a felty fringe to the frond of *Halymenia duchassaingii* growing in shallow water in moderately wave-sheltered situations; uncommon.

Distribution. World: from boreal-antiboreal to tropical parts of the Atlantic. West Africa: Cameroun: unpublished.

Remarks. This species is possibly no more than a small form of *Audouinella hallandica* (Woelkerling, *pers. comm.*).

Audouinella saviana (Meneghini) Woelkerling Pl. 20, fig. 3.

Woelkerling, 1973a; *Callithamnion savianum* Meneghini, 1840.

Plants usually epiphytic, with the erect filaments arising from a basal layer of short, simple or branched filaments, or from a small and irregularly shaped disc; erect filaments many times irregularly branched, sometimes with branching unilateral and secund, often tapering towards their tips; cells of the erect filaments from 5-7 (-9) μm in diameter and 3 to 5(-8) times longer than broad; monosporangia ovoid, from 7-10 μm in diameter and from 16-18 μm in length, sessile or on a 1-celled stalk, scattered or in secund series along the side branches.

For additional figures see Woelkerling, 1973a, p. 602, figs 56-60; Pilger, 1911, p. 299, figs 8, 9 (as *Chantransia mollis*).

Occurs as a felty layer on the surface of *Gracilaria* and *Sargassum* plants growing in the immediate sublittoral and sublittoral fringe in moderately wave-sheltered situations.

Distribution. World: as for genus. West Africa: Cameroun: *Acrochaetium molle*, Hamel 1928, Papenfuss 1945; *Chantransia mollis*, De Toni 1924, Pilger 1911. Gabon: John & Lawson 1974a.

Remarks. *Acrochaetium molle* was described from Cameroun by Pilger (1911) as *Chantransia mollis* and later transferred by Hamel (1928) to the genus *Acrochaetium*. There is no mention in Pilger's original description of the nature of the basal parts of his plant which provide characters that are now considered important in separating species of acrochaetoid algae. An examination of one of the syntype collections (No. 254) has revealed that the erect filaments arise from a well-developed filamentous basal layer. The nature of the basal region together with the arrangement and dimensions of the monosporangia suggest that Pilger's plant is identical to *Audouinella saviana.* Pilger (1911, p. 299) had recognised its close relation to this species stating "Die neue Art ist verwandt mit *Callithamnion (Chantransia) byssaceum* Kütz. (nach De Toni = *Chantransia saviana* [Menegh] Ardiss., einer mir nicht bekannten Art), ..". See remarks under *A. seriata* regarding the close relation between the two species.

Audouinella seriata (Børgesen) Garbary

Garbary, 1979; *Acrochaetium seriatum* Børgesen, 1915.

Plants epiphytic, with the erect filaments arising severally from a prostrate system of irregularly shaped filaments forming a pseudoparenchymatous disc; erect filaments irregularly branched below and somewhat secund in the upper parts, with the apical cell blunt; cells of the erect filaments about 9 μm in diameter towards base and twice as long as broad, becoming narrower above, from 7-8 μm in diameter and 5 times longer than broad; monosporangia ovoid, from 7-9 μm in diameter and from 10-14 μm in length, usually with a 1-celled stalk and rarely sessile, arising in long and distinctive secund series.

For figures see Børgesen, 1915, p. 32 to 43, figs 25-28 (as *Acrochaetium seriatum*).

Occurs as a felty layer on the surface of *Sargassum* plants growing in shallow water in moderately wave-sheltered localities where the water is somewhat brackish.

Distribution. World: in warm temperate and tropical parts of the Atlantic Ocean. West Africa: Gabon: *?Acrochaetium seriatum,* John & Lawson 1974a. Gambia: John & Lawson 1977b. Nigeria: Lawson 1980.

Remarks. The Gabon material differs in a number of minor respects from the type description and illustrations of Børgesen (1915) and for this reason the determination is regarded with some doubt. This species and *A. saviana* show a close resemblance in having a similar basal system and often stalked monosporangia which arise in series (rare in the latter), although the dimensions of the sporangia are somewhat less in *A. saviana.*

Doubtful Records

Audouinella codicola (Børgesen) Garbary

Garbary, 1979; *Acrochaetium codicola* Børgesen, 1927.

In marking this new combination Garbary followed Dixon (Dixon, in Parke and Dixon, 1976) who has placed several genera of acrochaetoid algae occurring in the British Isles into the genus *Audouinella.* He argues that there seems no logical reason for delineating genera in this closely interrelated group in which the circumscription of species is often confused.

This plant is reported growing on *Codium decorticatum* (as *C. elongatum*) at Aberdeen on the Sierra Leone peninsula (Aleem, 1978). We consider as doubtful the report of this little-known species which until now has only been found on the Canaries.

Audouinella endozoica (Darbishire) Dixon

A small plant found in Ghana growing on the utricles of *Codium **guineënse*** has been tentatively referred to this species by Woelkerling (*pers. comm.*). The determination is regarded with some doubt as all the Ghanaian plants were sterile and the chloroplast structure has not been determined.

Audouinella hypneae (Børgesen) nov. comb.

Chantransia hypneae Børgesen, 1909: 2.

See remarks under *Audouinella codicola.*

Reported growing epiphytically on *Hypnea* spp. at two localities on the Sierra Leone peninsula (Aleem, 1978). The record is considered doubtful since this small epiphyte has only previously been reported from the western side of the Atlantic Ocean.

Family Gelidiaceae

Gelidiella J. Feldmann & Hamel 1934

Plants usually coarse and bushy, with the stiff branches terete or slightly compressed, alternately, pinnately or secundly divided; growth from a distinct apical cell; branches consisting of a medulla of elongate and thick-walled cells, and a cortex of a few layers of short cells; tetrasporangia either in regular divergent rows in the swollen tips of the branches or in stichidial branchlets, sporangia variously divided.

World Distribution: in warm temperate and tropical seas.

Gelidiella acerosa (Forsskål) J. Feldmann & Hamel Pl. 21, fig. 2.

J. Feldmann and Hamel, 1934; *Fucus acerosus* Forsskål, 1775.

Plate 21: Fig. 1. *Gelidium crinale:* Portion of a plant showing the terete or slightly flattened branches irregularly or pinnately divided in the upper parts. Fig. 2. *Gelidiella acerosa:* Portion of a plant showing the erect branches bearing subopposite branchlets. Fig. 3. *Gelidiopsis variabilis:* Plant with terete branches irregularly or suboppositely divided. Figs 4, 5. *Gelidium corneum:* 4. Portion of a plant showing the distinctly flattened and pinnately divided branches; 5. Tip of a branch showing the single apical cell and its initials.

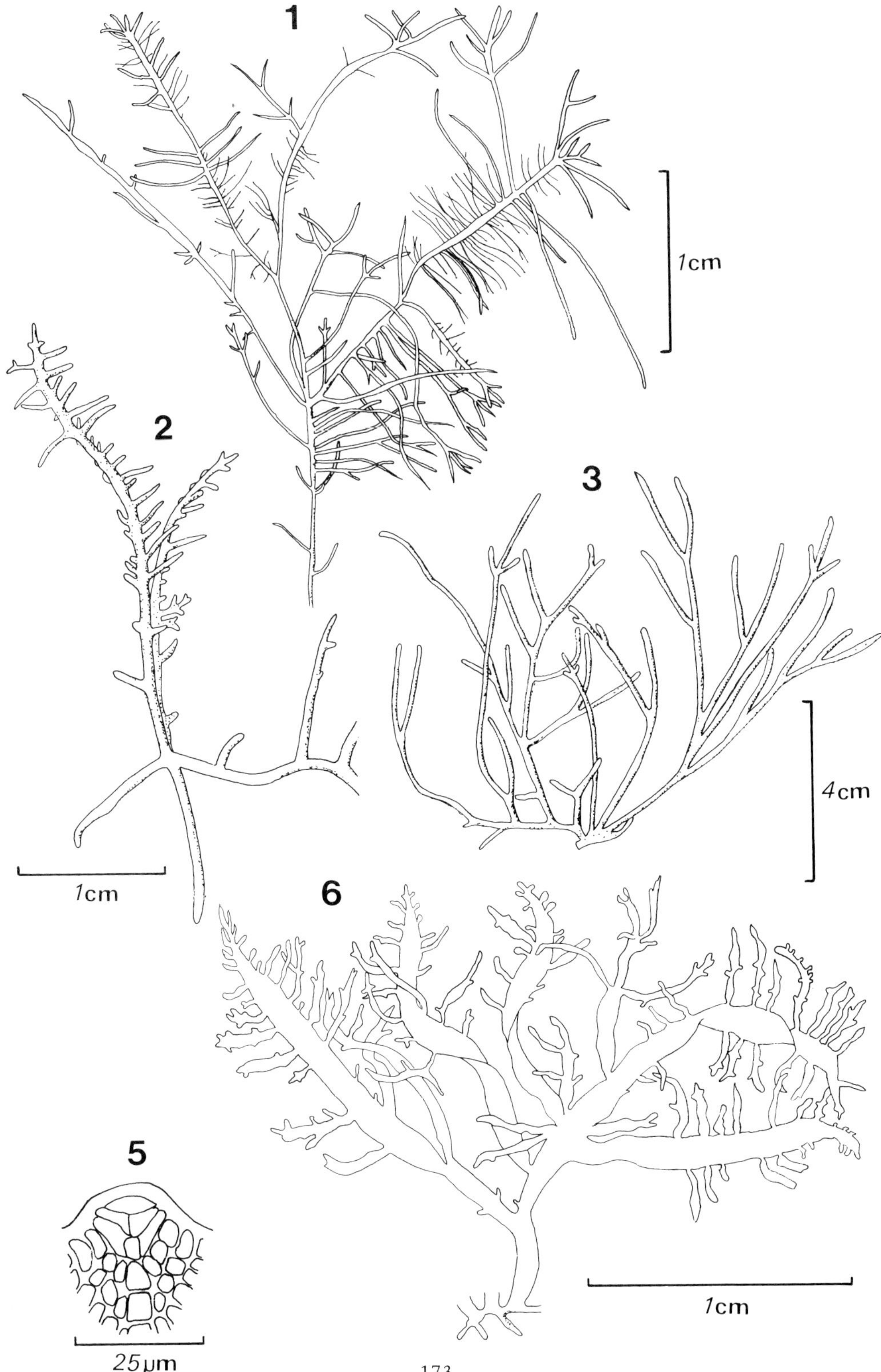
1
1cm
2
1cm
3
4cm
6
1cm
5
25μm

Plants forming loose clumps or mats, from 2-4 cm in height, reddish-purple but often bleached to a yellowish-green, with semi-erect or arcuate branches arising from a basal system of prostrate branches; branches terete to slightly compressed above, from 0.5-0.8(-1.0) mm in width, irregularly alternate below, often pinnate and subopposite or even pseudodichotomously divided in upper parts; tetrasporangia borne towards the tips of somewhat swollen branchlets.

For additional figures see Rao, 1970, p. 65, 66, figs 1, 2, pl. 1.

Occurs in the algal turf on moderately wave-sheltered rocks in the lower eulittoral subzone and in tidepools at higher levels on the shore; occasionally found growing on shell fragments in shallow water.

Distribution. World: as for genus. West Africa: Gabon: John & Lawson 1974a. São Tomé: Steentoft 1967; *Gelidium claviferum*, Henriques 1886b, 1887, 1917. Sierra Leone: Aleem 1978.

Gelidiopsis Schmitz 1895

Plants usually stiff and erect, consisting of irregularly or dichotomously divided branches; growth from an apical meristem; branches pseudoparenchymatous, with the medulla composed of thick-walled elongate cells and the cortex of cells gradually becoming smaller towards the surface; tetrasporangia embedded in the cortex of the ultimate branches, tetrahedrally divided; cystocarps prominent, solitary or grouped towards the tips of the branches.

World Distribution: widespread in warm temperate and tropical seas.

Key to the Species

1. Branches occasionally compressed and palmately divided .... *G. planicaulis*
1. Branches terete throughout and sometimes having a trifid appearance .... *G. variabilis*

Gelidiopsis planicaulis (W.R. Taylor) W.R. Taylor Pl. 22, fig. 6.

Taylor, 1960; *Wurdemannia miniata* var. *planicaulis* Taylor, 1943.

Plants forming coarse and loose clumps, from 4-8 cm in height, purplish-brown; branching irregularly alternate or opposite and distichous, sometimes appearing palmate; branches terete at the base and towards the tip, compressed near the middle where they attain a width of about 1 mm.

Occurs in the algal turf on moderately wave-exposed to somewhat sheltered rocks in the lower eulittoral subzone.

Distribution. World: in tropical parts of the Atlantic Ocean. West Africa: ?Bioko: unpublished. Cameroun: unpublished. Gabon: John & Lawson 1974a. Liberia: De May *et al.* 1977. Sierra Leone: John & Lawson 1977a.

Remarks. This species is very variable in form particularly as regard to the extent to which the branches are flattened. According to Taylor (*pers. comm.*) it corresponds closely to material from the western side of the Atlantic although the flattening in the West African plants appears to be more localised (see John and Lawson, 1974a).

Gelidiopsis variabilis (J. Agardh) Schmitz Pl. 21, fig. 3.

Schmitz, 1895; *Gelidium variabile* J. Agardh, 1852.

Plants usually forming coarse clumps, from 5-40 cm in height, purplish-brown and occasionally becoming bleached towards the tips of the branches; branching sparingly irregular or sometimes subopposite, with the ultimate divisions commonly subopposite and giving a trifid appearance to the branch tips; branches terete or subterete in the upper parts, from 100-500 μm in width, sometimes clothed in the younger parts with unicellular hairs.

For additional figures see Kützing, 1869, pl. 23, figs c, d (as *Gelidium variabile*).

Occurs as small clumps on wave-exposed rocks and as more extensive patches together with other algae in more sheltered situations, extending from the lower eulittoral zone to a depth of about 10 m. Large and very well-developed plants are commonly to be found in tidepools and on sheltered rocks in brackish-water habitats such as river estuaries.

Distribution. World: as for genus. West Africa: Bioko: unpublished. Cameroun: Dangeard 1952, John 1977a, Pilger 1911, Schmidt & Gerloff 1957, Steentoft 1967. Côte d'Ivoire: John 1977a. Gambia: John & Lawson 1977b. Ghana: John 1977a, John *et al.* 1977, 1980, John & Pople 1973, Lawson 1956, Lieberman *et al.* 1979, Steentoft 1967, Stephenson & Stephenson 1972. Guinée: ?Sourie 1954, ?Steentoft 1967. Liberia: De May *et al.* 1977. São Tomé: Carpine 1959, John 1977a, Steentoft 1967; *Gelidiopsis intricata*, Tandy 1944. Sierra Leone: John & Lawson 1977a.

Remarks. Steentoft (1967) has expressed doubt as to whether this species and *Gelidiopsis intricata* (Kützing) Vickers should be regarded as separate species. She states: "Larger tuft-forming plants seem to be placed in the former species [*G. variabilis*], low growing turf-forming ones in the latter.".

Gelidium Lamouroux 1813

Plants consisting of terete or flattened branches, pinnately divided at least in the upper parts; growth by an apical cell; branches consisting of a filamentous medulla, with a cortex of cells decreasing in size towards the surface and thick-walled filaments (rhizines) occurring particularly among the inner cortical cells; tetrasporangia scattered or in sori, often immersed in swollen portions near the tips of the branches, sporangia tetrapartitely, tetrahedrally or irregularly divided; spermatangia often forming extensive superficial patches; cystocarps bilocular and opening by a pore on each side of a branch.

World Distribution: cosmopolitan.

Although a number of species are keyed out below it is far from clear as to what extent they may or may not be regarded as distinct species. In fact, it is widely recognised that this is one of the most polymorphic genera of red algae. A world revision of the genus is necessary and some progress has been made in this direction through the studies of Dixon (1958, 1966) on the European representatives. He recognises two broad categories into which most of the West African plants may be placed and has indicated that three of the species given below, namely *G. crinale*, *G. elminense* and *G. pusillum*, are probably all part of the same complex. In the first volume of the seaweed flora of the British Isles, Dixon and Irvine (1977a) place *G. crinale*, *G. pusillum* and *G. pulchellum* (*sensu* Feldmann and Hamel, 1936) under *G. pusillum*.

Key to the Species

1. Plants having a well-developed and conspicuous system of prostrate branches ... . *G. pusillum*
1. Plants without a prostrate system of branches or when present then little-developed as compared to the erect system 2

2. Branches filiform and terete throughout .... *G. crinale*
2. Branches subterete or distinctly flattened, at least in the upper parts.......... 3

3. Branches terete to subterete, less than 500 μm in width .... *G. elminense*
3. Branches usually markedly flattened throughout, often greater than 1 mm in width 4

4. Tetrasporangia borne in small marginal appendages .... *G. arbuscula*
4. Tetrasporangia borne in ordinary vegetative branches .... *G. corneum*

Gelidium arbuscula (Montagne) Børgesen

Børgesen, 1927; *Gelidium corneum* var. *nereideum* Montagne, 1839-1841.

Plants usually known to form dense clumps, to about 15 cm in height; branches repeatedly pinnate, opposite or subopposite, flattened, usually up to 1 mm in width, with the branchlets often spine-like and the branches having obtuse or emarginate tips; tetrasporangia borne in marginal leaf-like appendages, short-stalked and heart-shaped to oval, about 70 μm in width and to about 1 mm in length.

For figures see Børgesen, 1927, p. 86 to 88, figs 45-47.

There is no information on the ecology of this plant in the region but further north it often forms a belt or band in the lower eulittoral subzone.

Distribution. World: in warm temperate and tropical parts of the eastern Atlantic Ocean. West Africa: Guinée: ?Sourie 1954.

Remarks. This plant has been reported with some doubt from Guinée by Sourie (1954) although it is well-known from nearby Sénégal. Bory is often cited as the authority for this species although there is no evidence that he ever provided a description of it. Børgesen (1927), who has examined Bory's material, believed it to be the same plant described by Montagne (1839-1841) as *Gelidium corneum* var. *nereideum*. Therefore, Børgesen appears to be the authority for making the combination *Gelidium arbuscula* based on this variety of *G. corneum*.

Gelidium corneum *sensu* Børgesen Pl. 21, figs 4, 5.

Børgesen, 1916.

Plants forming bushy and dark reddish-brown clumps, to about 4 cm in height; branches bi- or tripinnately divided, often arising irregularly to opposite along the margin, flattened throughout, to about 1.5 mm in width, with the apices slightly tapering to rounded or emarginate; tetrasporangia usually in broad, rounded and somewhat spathulate ultimate branches.

For additional figures see Børgesen, 1916, p. 115, fig. 124.

Occurs in the lower eulittoral subzone in the algal turf and often especially abundant on sand-covered rocks in shallow tidepools.

Distribution. World: widespread in boreal-antiboreal to tropical parts of the Atlantic Ocean. West Africa: Gabon: John & Lawson 1974a. Gambia: John & Lawson 1977b. Ghana: John 1977a. Liberia: De May *et al.* 1977, John 1977a; *G. arbuscula*, John 1972a. Nigeria: John 1977a; *G. pusillum*, Fox 1957. Togo: John 1977a; *G. arbuscula*, John & Lawson 1972b.

Remarks. According to Dixon (1967) this name is illegitimate since the description of *Fucus corneus* by Hudson (1762) is of material now known to be referable to *G. sesquipedale* (Clemente) Thuret. Nigerian plants previously reported as *Gelidium pusillum* (Fox, 1957) have on subsequent re-examination been found to be more closely related to this "species".

Gelidium crinale (Turner) Desmazières Pl. 21, fig. 1.

Desmazières, 1827; *Fucus crinalis* Turner, 1819.

Plants forming wiry clumps, from 2-5 cm in height, often yellowish-brown, with the erect branches arising from a little-developed system of spreading prostrate branches; branching irregularly alternate below and irregular or pinnate above; branches terete or slightly flattened above, to about 250 μm in width below and tapering upwards towards the apices; tetrasporangia in flattened and spathulate branchlets.

For additional figures see J. Feldmann and Hamel, 1936, p. 218, 241.

Occurs together with other algae growing over a wide range of wave-exposure usually on sand-covered beach rocks in the lower eulittoral subzone.

Distribution. World: as for genus. West Africa: Gabon: John & Lawson 1974a. Gambia: John & Lawson 1977b. Ghana: Dickinson 1952, Lawson 1956, Steentoft 1967. Liberia: De May *et al.* 1977. São Tomé: Hariot 1908, Henriques 1917, Steentoft 1967. Sierra Leone: Aleem 1978, John & Lawson 1977a.

Gelidium elminense Dickinson Pl. 22, fig. 1.

Dickinson, 1952.

Plants forming loose clumps, to about 6 cm in height; branching sparingly pinnate in upper parts, often giving branches a trifid appearance, with the lower third of the plant usually unbranched; branches terete and slightly flattened in upper parts, to about 300 μm in width, with the branchlets attenuate towards base and apex; tetrasporangia in irregular sori near apices of the branchlets.

For additional figures see Dickinson, 1952, p. 42, fig. 2a-d.

Occurs on rocks in the lower eulittoral subzone and is rare in the region.

Distribution. World: in tropical parts of the eastern Atlantic Ocean. West Africa: Ghana: Dickinson 1952, John 1977a. Côte d'Ivoire: John 1977a.

Remarks. This species, originally described from plants collected at Elmina and nearby Iture in Ghana, falls within the *Gelidium crinale* complex as defined by Dixon (1958, 1966). In the British Museum (Natural History) there are two specimens on a sheet (no. 35, Elmina, 25. 4. 45) that compare almost exactly with two of those illustrated by Dickinson (1952, p. 47, fig. 2a, b). These specimens may be regarded as syntypes in the absence of a designated holotype.

Gelidium pusillum (Stackhouse) Le Jolis Pl. 22, fig. 2.

Le Jolis, 1863; *Fucus pusillus* Stackhouse, 1795.

Plants growing as ring-like felty patches or occasionally together with other algae, less than 1 cm in height, with the erect branches arising from a well-developed prostrate system of branches; branching opposite or subopposite and pinnate; branches terete to flattened and from 50-210(-400) μm in width.

For additional figures see J. Feldmann and Hamel, 1936, p. 237.

var. **pulvinatum** (C. Agardh) J. Feldmann

Feldmann, in Hamel, 1927; *Sphaerococcus corneus* f. *pulvinatus* C. Agardh, 1823.

Plants larger than the typical form, up to 2 cm in height, forming dense intricate cushions or mats; erect branches often well-developed and foliose, usually flattened, to about 1-1.5 mm in width.

Plate 22: Fig. 1. *Gelidium elminense:* Plant showing the sparingly divided pinnate branches. Fig. 2. *Gelidium pusillum:* Portion of a plant with the erect branches arising from a creeping base and tetrasporangia immersed in the swollen tip of an erect branch. Fig. 3. *Pterocladia capillacea:* Portion of a cystocarpic plant showing the flattened pinnate branches. Figs 4, 5. *Wurdemannia miniata:* 4. Plant growing as a dense cushion on shell fragments; 5. Portion of a plant showing the terete and irregularly divided branches. Fig. 6. *Gelidiopsis planicaulis:* Portion of a plant showing the irregularly or occasionally suboppositely divided branches sometimes appearing palmate.

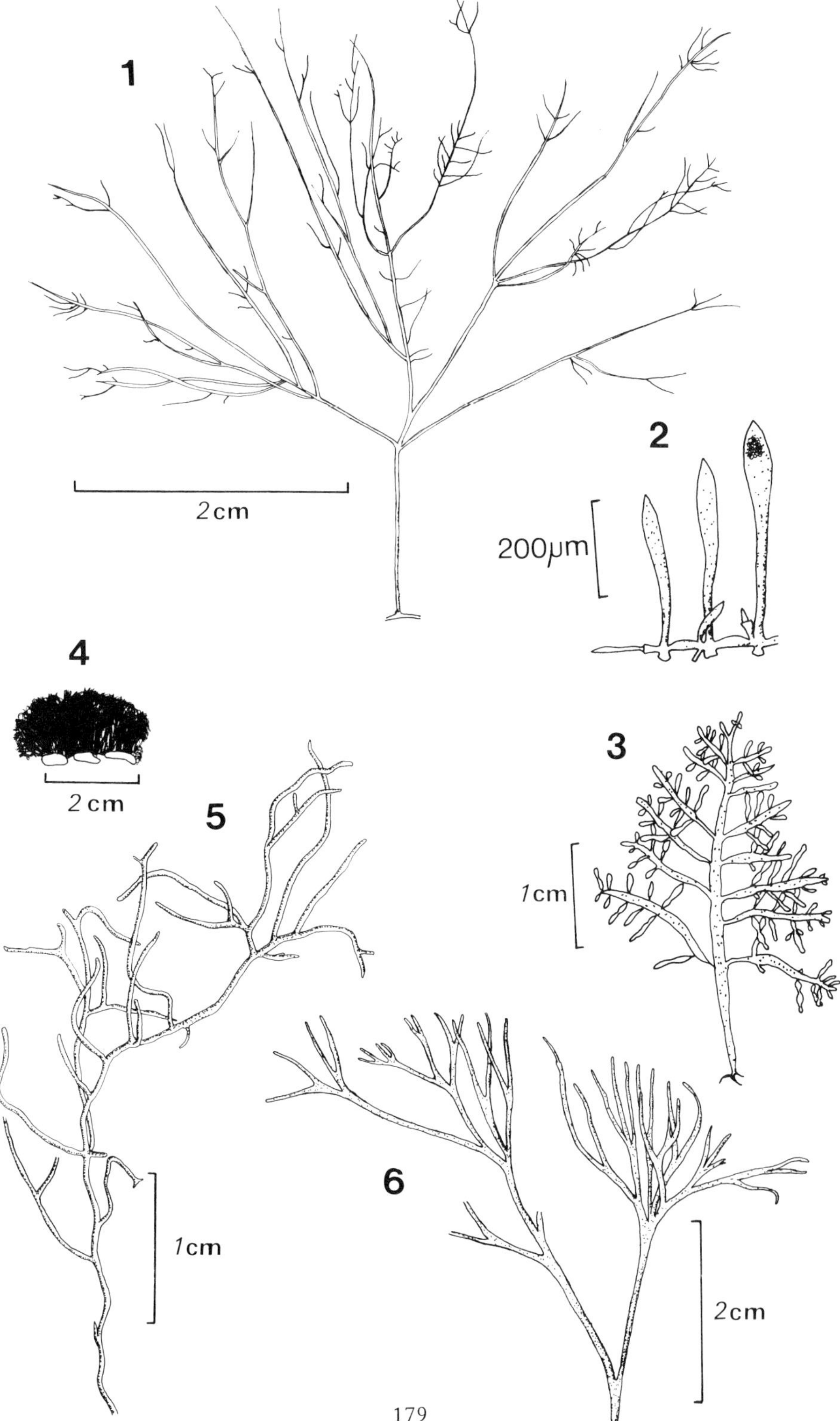
1
2cm
2
200µm
4
2cm
3
1cm
5
1cm
6
2cm

For figures see Feldmann and Hamel, 1936, p. 113, fig. 19c; Børgesen, 1943, p. 6, fig. 1.

Occurs over a wide range of exposure to wave action though the variety *pulvinatum* is usually most common on moderately wave-sheltered rocks and in more exposed situations it is often confined to the shells of *Patella*. Found over the lower eulittoral subzone down to a depth of about 1 m, sometimes associated with a good deal of sand.

Distribution. World: as for genus. West Africa: Benin: John 1977a, John & Lawson 1972b. Bioko: unpublished. Cameroun: Hoppe 1969, John 1977a, Lawson 1955, Pilger 1911, Richardson 1972, Schmidt & Gerloff 1957, Stephenson & Stephenson 1972. Gabon: *G. pusillum* var. *pulvinatum*, John & Lawson 1974a. Gambia: *G. pusillum* var. *pulvinatum*, John & Lawson 1977b. Ghana: Fox 1957, John 1977a, John & Pople 1973. Liberia: De May *et al.* 1977, John 1972c, 1977a; *G. pusillum* var. *pulvinatum*, De May *et al.* 1977, John 1972c, 1977a. São Tomé: John 1977a; *G. pusillum* var. *pulvinatum*; as *G. reptans*, Carpine 1959, Steentoft 1967. Sierra Leone: Aleem 1978, Fox 1957, John & Lawson 1977a, Lawson 1954b, 1957a, Levring 1969, Longhurst 1958; *G. pusillum* var. *pulvinatum*, John & Lawson 1977a. Togo: John 1977a, John & Lawson 1972b.

Remarks. *Gelidium reptans* (Suhr) Kylin has been reported from São Tomé but we have followed Børgesen (1943) in considering it to be no more than a variety of *G. pusillum*.

Pterocladia J. Agardh 1852

Plants usually erect, consisting of terete or flattened branches, laterally and pinnately divided, with each branch ending in a distinct apical cell; branches having a medulla of large filaments mixed with small, thick-walled refractive filaments (rhizines), and a compact small-celled cortex; tetrasporangia usually arranged in parallel rows and often in sori, formed initially towards the apices of the branchlets, sporangia tetrapartite; cystocarps somewhat elevated, with a single pore opening on one side of the branch.

World Distribution: from boreal-antiboreal to tropical seas.

Pterocladia capillacea (S. G. Gmelin) Bornet & Thuret Pl. 22, fig. 3.

Bornet and Thuret, 1876; *Fucus capillaceus* Gmelin, 1768.

Plants erect and bushy, to about 5 cm in height, reddish-brown; branching simple or 3 or 4 times pinnately divided; branches terete below and compressed above, from 1-2 mm in width in widest parts, with the ultimate branches often attenuate towards the base.

For additional figures see Dixon, 1966, p. 55, fig. 3; Dixon and Irvine, 1977a, p. 136, fig. 50; Feldmann and Hamel, 1936, p. 91, fig. 3, p. 95, fig. 6, p. 131, fig. 30, pl. 6, fig. 1.

Occurs usually as clumps or occasionally as loose mats on moderately sheltered to moderately wave-exposed rocks from the lower part of the upper eulittoral subzone to a depth of about 10 m.

Distribution. World: as for genus. West Africa: Cameroun: Hoppe 1969, John 1977a, Lawson 1966; *P. pinnata*, Lawson 1955, Stephenson & Stephenson 1972. Côte d'Ivoire: John 1977a. Gabon: John & Lawson 1974a. Gambia: John & Lawson 1977b. Ghana: Dickinson 1952, John 1977a, Lawson 1966. Liberia: De May *et al.* 1977, John 1977a. Sierra Leone: John & Lawson 1977a.

Remarks. The reasons for adopting this name rather than *P. pinnata* (Hudson) Papenfuss are to be found in Dixon (1960).

Suhria J. Agardh ex Endlicher 1843

Doubtful Record

Suhria vittata (Linnaeus) J. Agardh

This is usually regarded as a cold water species and so the record from "Danish Guinea" (present-day Ghana) by Hornemann (1819) as "*F.* [*Fucus*] *vittatus* L." is somewhat surprising. The attribution can only be established by referring to the original Isert collection on which the record is based, though much of this collection was lost in a fire in 1807.

Family Wurdemanniaceae

Wurdemannia Harvey 1853

Plants consisting of a prostrate and erect system of entangled, wiry and irregularly divided branches with multiaxial apices, attached by discoid holdfasts; branches having a medulla of elongate cells becoming shorter and thicker walled towards periphery, and a single layer of rectangular cortical cells; tetrasporangia in outer tissues of the inflated branch tips, sporangia zonately divided.

World Distribution : probably widespread in warm temperate and tropical seas.

Wurdemannia miniata (Duby) J. Feldmann & Hamel Pl. 22, figs 4, 5.

J. Feldmann and Hamel, 1934; *Gigartina miniata* Duby, 1830.

Plants forming clumps or cushions, to about 3 cm in height, dull red or somewhat bleached, with intricate branches attached to one another and to substratum by small holdfasts; branching very irregular on all sides; branches terete or slightly compressed, about 150 μm in width, with ultimate branches often spine-like and having acute tips.

For additional figures see J. Feldmann and Hamel, 1934, p. 545 to 547, figs 9-11.

Occurs on rocks in the upper eulittoral subzone and extends down into the sublittoral fringe, usually to be found in somewhat wave-sheltered situations and often associated with a certain amount of sand.

Distribution. World: as for genus. West Africa: Cameroun: unpublished. São Tomé: Steentoft 1967; *Caulacanthus notulatus*, Henriques 1917; *C. ustulatus*, Hariot 1908, Henriques 1886b, 1887.

Family Helminthocladiaceae

Liagora Lamouroux 1812

Plants often bushy, either soft and somewhat gelatinous or firm and with a calcified cortex; branches having a medulla of longitudinal filaments giving rise to a cortex of fascicles of dichotomously branched assimilatory filaments, with the cells ovate to spherical and the subterminal cell usually larger than the terminal cell; spermatangia lateral or terminal on the assimilatory filaments; gonimoblast relatively large, usually surrounded by branched, sterile filaments.

World Distribution: widespread from boreal-antiboreal to tropical seas.

Liagora farinosa Børgesen. Pl. 23, figs 4, 5.

Børgesen, 1915.

Plants forming loose clumps, about 10 cm in height; branching irregularly dichotomous, with long intervals between the forkings, without lateral proliferations; branches terete, moderately calcified except towards the branch tips; assimilatory filaments extending outside the calcification and little divided; cells of assimilatory filaments nearly cylindrical, slightly swollen, from 12-20 μm in diameter and 1.5 to 3 times as long as broad.

For additional figures see Børgesen, 1927, p. 60, 61, figs 32, 33.

There is no information available on the ecology of this plant in the region but elsewhere in the Atlantic Ocean it is found in the eulittoral zone and sometimes in the immediate sublittoral.

Plate 23: Fig. 1. *Galaxaura oblongata:* Plant showing the terete and dichotomously divided branches. Fig. 2. *Galaxaura marginata:* Portion of a plant showing the prominent margin of the flattened and dichotomously divided branches. Fig. 3. *Galaxaura rugosa:* Portion of a plant showing the annulate constrictions along the branches. Figs 4, 5. *Liagora farinosa:* 4. Portion of a plant showing the irregularly dichotomous branching; 5. Terminal portion of an assimilatory filament. Fig. 6. *Nemalion inamoena* (after Børgesen, 1942): Outer cortical filament showing the moniliform rows of cells terminating in a pyriform cell.

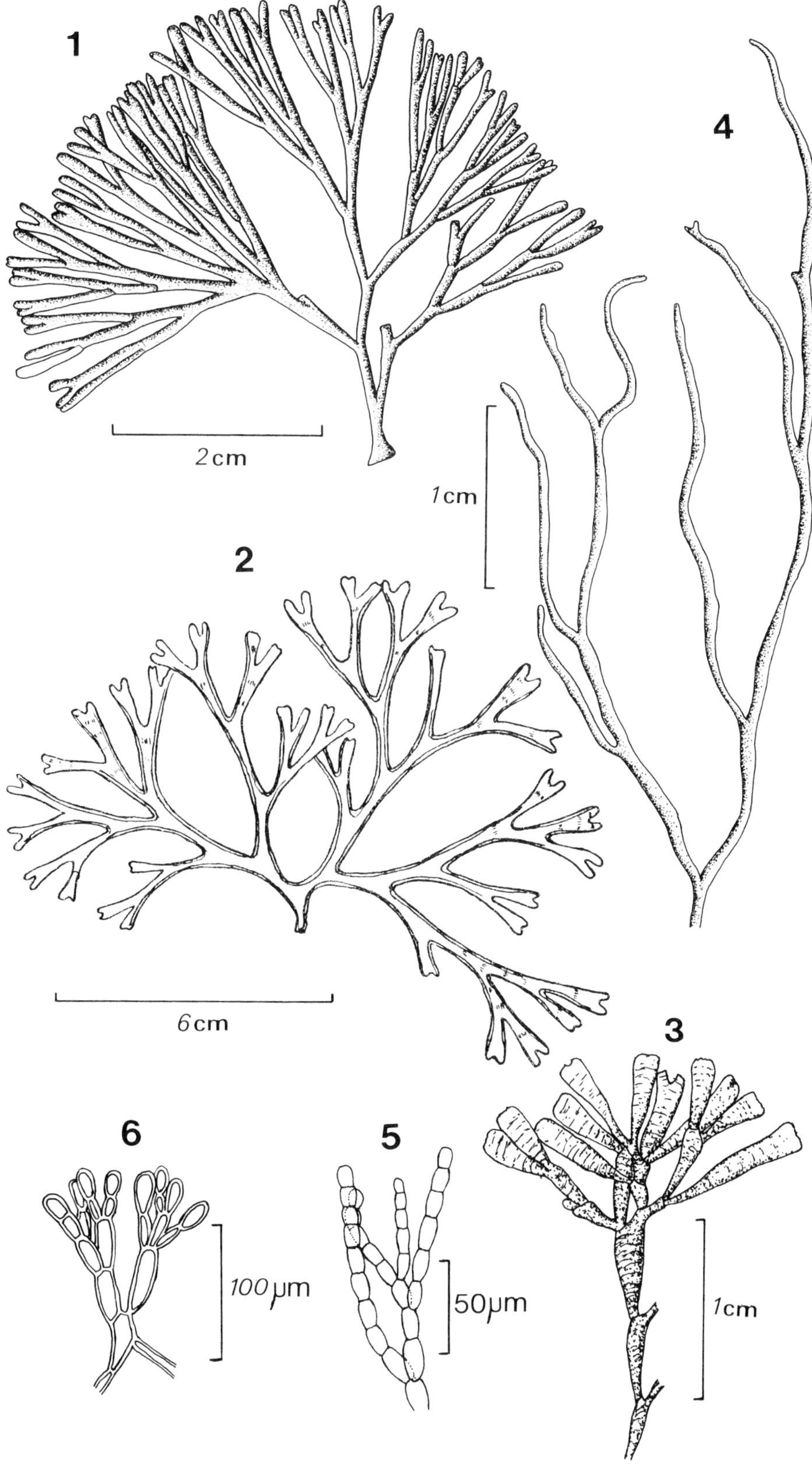
1
2cm
4
1cm
2
6cm
3
6
5
100µm
50µm
1cm

Distribution. World: widespread in warm temperate and tropical seas. West Africa: ?Príncipe: *L. megagyna*, Steentoft 1967. ?São Tomé: *L. megagyna*, Steentoft 1967.

Remarks. Steentoft (1967) lists this plant both from São Tomé and Príncipe, the latter record being based on information accompanying the specimen collected by F. Newton. It appears that Newton never actually visited Príncipe (Steentoft, *pers. comm.*) and thus the mention of it from this island is possibly a mistake on the label. We believe, after careful examination of Newton's collection, that a better disposition of this plant might be *Liagora farinosa* rather than *L. megagyna* as she reports it.

Nemalion Duby 1830

Gametangial phase usually soft and gelatinous, consisting of terete or subterete branches, simple or irregularly divided; branches having a medulla of interwoven and anastomosing filaments, and a cortex of corymbose fascicles of assimilatory filaments; cystocarps having dense clusters of gonimoblasts and not surrounded by special filaments. Sporangial plants microscopic, consisting of a creeping and erect system of uniseriate filaments, with tetrapartitely divided tetrasporangia.

Distribution. World: widespread from boreal-antiboreal to tropical seas.

?**Nemalion amoenum** (Pilger) Børgesen — Pl. 23, fig. 6.

Børgesen, 1942; *Dermonema amoenum* Pilger, 1911.

Plants forming bushy clumps, from 2-3 cm high; branches di- or trichotomously divided, terete and from 4-6 mm in diameter; outer cortex consisting of branched assimilatory filaments of moniliform rows of cells terminating in a large pyriform cell from 40-50 μm in diameter; spermatangia arising from second or third cells from the tip of the branched assimilatory filaments. Sporangial plants unknown.

For additional figures see Børgesen, 1942, p. 26.

Occurs on rocks protruding through sand presumably in the eulittoral zone.

Distribution. World: in tropical parts of the eastern Atlantic Ocean. West Africa: Cameroun: Børgesen 1942, Kylin 1956, Papenfuss 1968b; *Dermonema amoenum*, De Toni 1924, Papenfuss 1968b, Pilger 1911.

Remarks. The exact taxonomic position of Pilger's plant from Cameroun is somewhat problematical. Børgesen (1942), after examining the type, made the following comment: "The final discussion as to real place can of course not be taken before its female organs are found; nevertheless I think at present its right place is in the genus *Nemalion*.".

Family Chaetangiaceae

Galaxaura Lamouroux 1812

Plants weakly to strongly calcified, smooth or pilose, with dichotomously divided and occasionally segmented branches; surface of branches bearing long assimilatory filaments or consisting of a layer of cells which sometimes carry one-celled sub-spinulose extensions or filaments of two or three cells, occasionally having unicellular hairs; thallus in section composed of branched medullary filaments giving rise to a pseudoparenchymatous cortex of large cells; tetrasporangia borne on one- to few-celled stalks or at the base of specialised filaments; spermatangia arising in conceptacles; cystocarps immersed and surround by a pericarp of sterile filaments.

World Distribution: widespread in warm temperate and tropical seas.

There is usually a difference in the structure of the cortex of sexual and asexual plants belonging to the same species. This has caused considerable confusion in the past and has often led to the two phases in the life history being considered as distinct species. This genus has been partly reviewed by Papenfuss and Chiang (1969) who have considerably reduced the number of entities formerly recognised.

Key to the Species

1. Branches clearly flattened, with irregular and not always obvious articulations. . 2
1. Branches terete or occasionally subterete, with articulations usually obvious. . . . 3

2. Branches from 2-3 mm in width and the margin prominent especially when dry. . . . . *G. marginata*
2. Branches from 1-2 mm wide and always flat . . . . *G. tenera*

3. Surface of branches pilose, with a dense covering of assimilatory filaments. 4
3. Surface of branches glabrous or only partly pilose, with assimilatory filaments discontinuous when present 5

4. Section of branches showing no clear differentiation between medulla and cortex; assimilatory filaments borne on little-developed supporting cells and basal cells absent. . . . . *G. filamentosa*
4. Section of branches having a clear separation between the medulla and cortex, with supporting and basal cells well-developed. . . . . *G. subfruticulosa*

5. Branches partly pilose and the assimilatory filaments frequently in whorls; calcification found throughout . . . . *G. elongata*
5. Branches glabrous and often having annulate constrictions on the surface; calcification confined to the cortex 6

6. Branch segments usually greater than 2 mm in diameter, occasionally to about 4 mm . . . . *G. obtusata*
6. Branch segments rarely more than 2 mm in diameter. 7

7. Segments usually less than 10 mm in length, with the surface rugose or having conspicuous annulate constrictions; inner cortical cells more than 30 μm in diameter and outermost cells from 18-32 μm in diameter .... *G. rugosa*
7. **Segments usually greater than 10 mm in length and on occasion to 20 mm, with the surface not rugose or constricted; inner cortical cells usually less than 30 μm indiameter and outermost cells from 13-15 μm in diameter**........ *G. oblongata*

Galaxaura elongata J. Agardh Pl. 24, fig. 3.

J. Agardh, 1876.

Plants forming bushy clumps, to about 10 cm in height; branches repeatedly dichotomously divided and occasionally prolifercus, terete, up to 2 mm in diameter below and sometimes tapering above to about 1 mm, with segments from 3-20 mm in length; surface of the branches pilose, with the assimilatory filaments long and whorled at the nodes of the younger branch segments but shorter elsewhere; asexual plants having a 2- or 3-layered cortex of nearly barrel-shaped cells; sexual plants having a 3- or 4-layered cortex, with the cells of the inner layer often fused and lobed; asexual and sexual plants both with the inner cortical cells from 22-40 μm in diameter, intermediate layer or layers from 8-30 μm in diameter, and outer cells either cuneate, from 10-15 μm in height and 18-25 μm in diameter, or spherical and somewhat flattened or elevated.

For additional figures see Papenfuss and Chiang, 1969, p. 311, fig. 5a, p. 312, fig. 6.

Occurs in the immediate sublittoral together with other species of *Galaxaura* in moderately wave-sheltered situations.

Distribution. World: widespread in warm temperate seas and pantropical. West Africa: Gabon: John & Lawson 1974a. São Tomé: *G. rugosa* (pro parte), Hariot 1908, Henriques 1917; *G. squalida*, Steentoft 1967.

Galaxaura filamentosa Chou

Chou, 1945.

Plants bushy and to about 5 cm in height; branches di- or trichotomously divided, terete, from 0.5-1.5 mm in diameter, with segments from 2-10 mm in length; surface of the branches pilose being covered by a dense layer of long assimilatory filaments; asexual plants not having the tissues differentiated into a medulla and a cortex, with assimilatory filaments lacking basal cells and borne on little-developed supporting cells; cells of assimilatory filaments from 16-18 μm in diameter in lower parts and increasing above to 20-22 μm in diameter, from 1.5 to 3 times as long as broad; sexual phase not known.

For figures see Chou, 1945, pl. 1, figs 1-6, pl. 6, fig. 1.

There is no information on the ecology of this species in the region other than that it occurs in the sublittoral zone.

Distribution. World: in tropical parts of the Atlantic and Pacific Oceans. West Africa: São Tomé: Steentoft 1967; *G. lapidescens*, Henriques 1886b, 1887, 1917.

Galaxaura marginata (Ellis & Solander) Lamouroux Pl. 23, fig. 2; Pl. 24, fig. 5.

Lamouroux, 1816; *Corallina marginata* Ellis and Solander, 1786.

Plants forming clumps, from 6-13 cm in height; branches regularly dichotomously divided, terete towards the base and becoming flattened above, from 2-3 mm in width and about 1 mm in thickness; surface of the branches glabrous and becoming transversly banded above, having conspicuous and prominently thickened margins in dried specimens; asexual plants having the outermost cells small and depressed, each bearing stalk cells divided towards the apices and having 2 oval to ovoid thick-walled and often apiculate assimilatory cells; sexual plants with the outermost cells bearing oblong-ovate to subcylindrical and sometimes apiculate assimilatory cells; asexual and sexual plants both having a cortex of 2 to 3 layers of polygonal cells.

For additional figures see Børgesen, 1916, p. 110 to 112, figs 118-123 (as *G. occidentalis*).

Occurs as conspicuous light red plants usually on sand-covered rocks in sheltered to somewhat wave-exposed situations growing in the lower part of the eulittoral zone and to a depth of about 10 m, also commonly found in tidepools.

Distribution. World: in warm temperate and tropical parts of the Atlantic Ocean. West Africa: Cameroun: De Toni 1924, John 1977a; *Brachycladia marginata*, Pilger 1911; *B. marginata* f. *linearis*, Pilger 1911. Côte d'Ivoire: John 1977a. Gabon: John 1977a, John & Lawson 1974a; *Brachycladia australis*, Hariot 1896. Ghana: Dickinson & Foote 1950, John 1977a, John *et al.* 1977, Lawson 1956, Steentoft 1967. Liberia: De May *et al.* 1977. Príncipe: John 1977a, Steentoft 1967. São Tomé: Henriques 1886b, 1887, 1917, John 1977a, Steentoft 1967, Tandy 1944; *Brachytrichia marginata*, Hariot 1908.

Remarks. The plants become a dull greyish-green on drying and the margins of the branches become particularly prominent giving them a channelled appearance.

Galaxaura oblongata (Ellis & Solander) Lamouroux Pl. 23, fig. 1; pl. 24, fig. 2.

Lamouroux, 1816; *Corallina oblongata* Ellis and Solander, 1786.

Plants forming bushy clumps, from 3-10 cm in height; branches repeatedly di- or trichotomously divided, from 1-2 mm in diameter and with the segments from 4-28 mm in length; surface of the branches glabrous, with annulate constrictions on the upper segments; asexual plants not known with certainty; sexual plants having a 3- or 4-layered cortex consisting of an inner layer of polygonal cells, to about 30 μm in diameter, an intermediate layer or layers of smaller cells, and an outer layer of flattened cells of from 13-15 μm in diameter; calcification confined to the cortex.

For additional figures see Børgesen, 1926, p. 73 to 75, figs 39-41.

Occurs usually together with *G. marginata* in the region but is less common.

Distribution. World: in warm temperate and tropical seas. West Africa: Côte d'Ivoire: John 1977a. Gabon: John 1977a, John & Lawson 1974a; *G. fragilis*, Hariot 1896. Ghana: Dickinson & Foote 1950, John 1977a, John *et al.* 1977, Lawson 1956, Steentoft 1967. Liberia: De May *et al.* 1977. Príncipe: John 1977a, Steentoft 1976; *G. cylindrica*, Hariot 1908, Henriques 1917. São Tomé: John 1977a, Steentoft 1967; *G. cylindrica*, Hariot 1908, Henriques 1886b, 1887, 1917, Steentoft 1967.

Galaxaura obtusata (Ellis & Solander) Lamouroux Pl. 24, figs 1, 4.

Lamouroux, 1816; *Corallina obtusata* Ellis and Solander, 1786.

Plants forming bushy clumps, to about 10 cm in height; branches more or less regularly dichotomously divided, terete, from 1-4.5 mm in diameter, with the segments up to 2.5(-3.5) cm in length; surface of the branches glabrous; asexual plants having a cortex of a layer of large cells bearing stalk cells which distally support 1 to 2 depressed ovoid cells; sexual plants 3-layered, with a compact outermost layer of cells, an intermediate layer of swollen columnar cells, and an inner layer of large rectangular cells.

For additional figures see Papenfuss and Chiang, 1969, p. 306, 307, figs 1, 2.

Occurs as small patches growing on moderately wave-sheltered rocks in the lower eulittoral subzone though elsewhere in the Atlantic Ocean it has been reported dredged from a depth as great as 53 m (Taylor, 1960).

Distribution. World: as for genus. West Africa: Gabon: Hariot 1896, John & Lawson 1974a. Sierra Leone: Aleem 1978.

Galaxaura rugosa (Ellis & Solander) Lamouroux Pl. 23, fig. 3.

Lamouroux, 1816; *Corallina rugosa* Ellis and Solander, 1786.

Plants forming dense, bushy clumps, to about 7 cm in height; branches regularly dichotomously divided, terete and up to 2 mm in diameter, with the segments from 2-10 mm in length; surface of the branches usually glabrous and only rarely having assimilatory filaments, with annulate constrictions or rugose; asexual plants uncertain; sexual plants having a 3- or 4-layered cortex, with the inner cells ovoid and large, irregularly lobed and from 30-40 μm in diameter, intermediate cells smaller and ovoid, and outer cells somewhat flattened and from 18-32 μm in diameter; calcification of the cortex light and often almost absent from the branch tips.

Plate 24: Fig. 1. *Galaxaura obtusata:* Portion of a plant showing the di- or trichotomous branching and the often strongly constricted nodes. Fig. 2. *Galaxaura oblongata:* Cortex of a sexual plant showing the outer layer of assimilatory cells and the three layers of colourless inner cells. Fig. 3. *Galaxaura elongata:* Cortex of a sexual plant showing the two inner layers of often fused cells. Fig. 4. *Galaxaura obtusata:* Cortex of a sexual plant showing the inner layers of large rectangular cells. Fig. 5. *Galaxaura marginata:* Cortex of a sexual plant showing the often apiculate outermost layer of cells and fused inner, colourless cells.

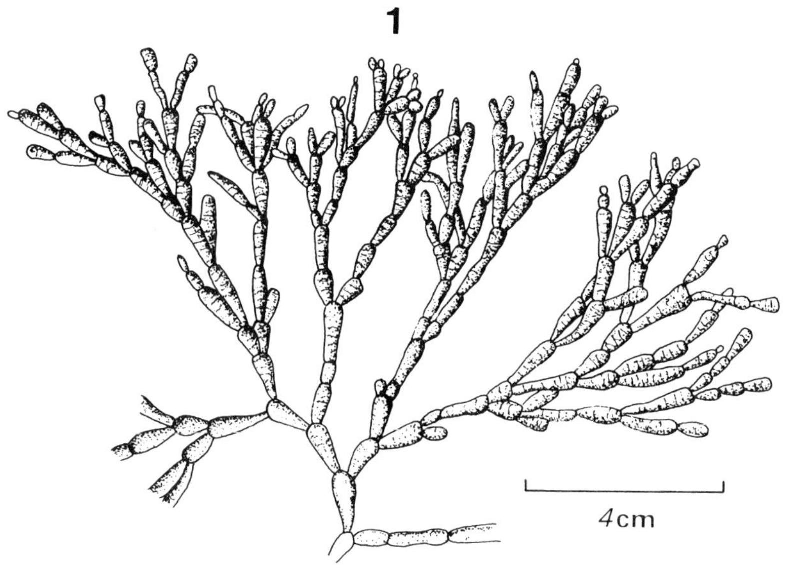
1
4cm

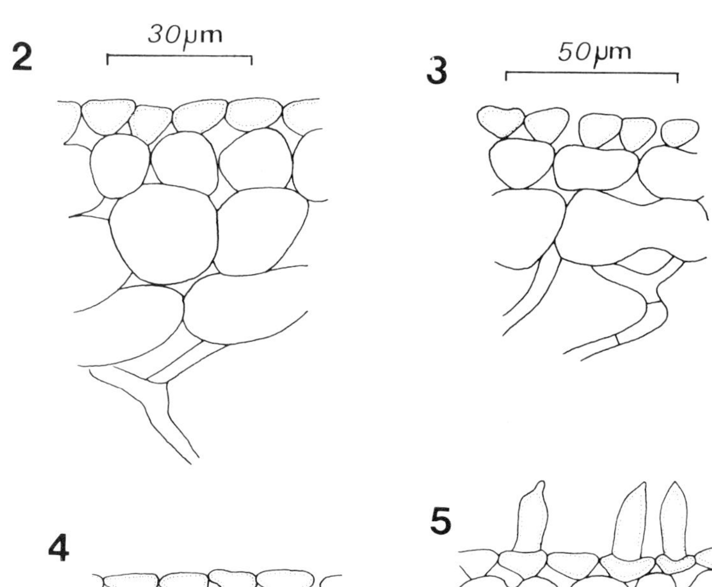
2
30µm
3
50µm

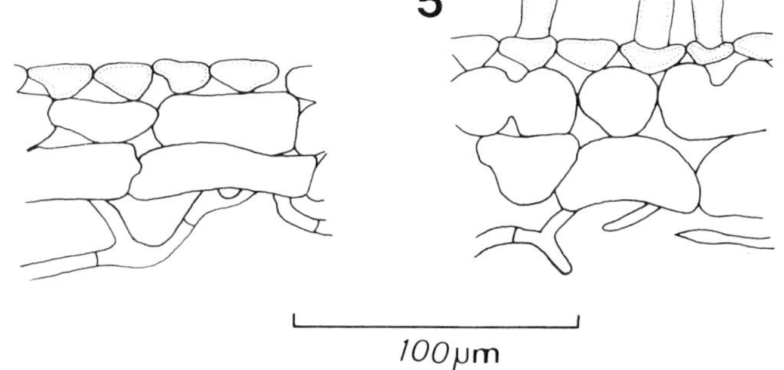
4
5
100µm

For additional figures see Chou, 1947, pl. 4, figs 12, 13, pl. 10, fig. 2.

The only information on the ecology of this plant in the region is that it occurs on rocks presumably in the eulittoral zone although elsewhere in the Atlantic Ocean it has been dredged to a depth of 18 m.

Distribution. World: as for genus. West Africa: Cameroun: John 1977a, Pilger 1911, Steentoft 1967. Côte d'Ivoire: John 1972c, 1977a. São Tomé: Hariot 1908 (pro parte), Henriques 1886b, 1917 (p.p.), John 1977a, Steentoft 1967.

Galaxaura subfruticulosa Chou

Chou, 1945.

Plants forming clumps, to about 6 cm or more in height; branches irregularly dichotomously divided, terete and from 1-1.5 mm in diameter, with the segments from 1-4 mm in length below and increasing to 10(-12) mm above; surface on the branches pilose and with the assimilatory filaments sometimes whorled near the branch apices and at the base of proliferations; asexual plants with a cortex composed of supporting cells and globose to ovoid or pyriform basal cells from 28-40 μm in diameter and from 45-65 μm in length, bearing short or long assimilatory filaments; short assimilatory filaments scattered amongst the longer ones, usually 2- or 3-celled, with a subglobose terminal cell from 20-34 μm in diameter; long assimilatory filaments having the basal parts from 14-16 μm in diameter and cells from 2 to 3 times as long as broad; sexual phase unknown.

For figures see Chou, 1945, pl. 2, fig. 6, pl. 8, fig. 2.

The only information on the ecology of this plant in the Gulf of Guinea is that it was growing in the sublittoral zone.

Distribution. World: in tropical parts of the Atlantic and Pacific Oceans. West Africa: São Tomé: Steentoft 1967; *G. rugosa* (pro parte), Hariot 1908, Henriques 1886b, 1887, 1917.

Galaxaura tenera Kjellman.

Kjellman, 1900.

Plants forming clumps, to about 12 cm or more in height; branches more or less regularly dichotomously divided, compressed, from 1-2 mm in width; asexual plants composed of a 1- or 2-layered cortex of large cells bearing cylindrical and simple or furcate cells giving rise to up to 2 stalk cells terminating in ovoid or clavate assimilatory cells; sexual plants consisting of a cortex of 3 layers, with the outer cells depressed-obovoid, the intermediate layer of large and subglobose or 2- or 3-lobed cells, and the inner layer of large cells.

For figures see Papenfuss and Chiang, 1969, p. 308, fig. 3a, p. 309, fig. 4.

There is no information on the ecology of this plant in the region other than it was collected from near the mouth of the Gabon River.

Distribution. World: as for genus. West Africa: Gabon: Papenfuss & Chiang 1969; *G. ventricosa*, Chou 1947, De Toni 1924, Kjellman 1900, Steentoft 1967, Womersley & Bailey 1970.

Remarks. Papenfuss and Chiang (1969) make the following comment "*Galaxaura tenera* shows several features of agreement with the older *G. marginata* ... and it is not inconceivable that these taxa will be found to be conspecific.".

Pseudogloiophloea Levring 1956

Plants usually bushy, consisting of repeatedly dichotomously divided, terete or somewhat flattened branches; branches having a thin medulla of longitudinal filaments giving rise to a cortex of large hyaline utricles interspersed with rows of small coloured cells; monosporangia formed on the microscopic, filamentous initial stages of development of sexual and asexual plants; spermatangia continuous over large areas of the surface of the branches; gonimoblasts embedded in the inner cortex, with the pericarp of a few layers of filament or cells.

World Distribution: probably widespread in tropical seas.

Pseudogloiophloea verae (Dickinson) Papenfuss Pl. 25, fig. 1; pl. 26, fig. 6.

Papenfuss, 1968a; *Gloiophloea verae* Dickinson, 1951.

Plants very bushy, to about 20 cm or more in height, arising from a small holdfast about 6 mm in diameter; branches dichotomously divided and upper divisions at intervals of from 2-4(-5) cm, terete and varying in diameter from 1 mm towards base to about 3 mm above; spermatangia unknown; cystocarps flask-shaped, from 130-150 μm in diameter and from 200-230 μm in length.

For additional figures see Dickinson, 1951, p. 297, pl. 4 (as *Gloiophloea verae*).

Occurs on moderately sheltered to moderately wave-exposed rocks in the lower eulittoral subzone and sublittoral fringe, often found growing on sand-covered rocks in tidepools.

Distribution. World: in the tropical parts of the eastern Atlantic Ocean. West Africa: Ghana: Ganesan 1974, Papenfuss 1968a; *Gloiophloea verae*, Dickinson 1951, Lawson 1956.

Scinaia* Bivona 1822

Plants usually solitary and bushy, consisting of firm but gelatinous branches, terete or slightly flattened and repeatedly dichotomously divided; medulla composed of

**Scinaia forcellata* Bivona has been reported both from the Congo (Hariot, 1895, 1896) and the island of Pagalu (Pilger, 1920).

loose, subdichotomously divided filaments and a two-layered cortex; cortex having a distinct inner layer of intertwined filaments and an outer layer of two or three layers of assimilatory cells terminating in swollen utricles; monosporangia scattered between the cortical cells; spermatangia arising in continuous surface sori; gonimoblasts embedded in the cortex and surrounded by a dense filamentous pericarp, with a single pore. Microscopic filamentous phase bearing tetrasporangia known only in two species.

World Distribution: widespread from boreal-antiboreal to tropical seas.

Key to the Species

1 Plants consisting of slightly flattened branches. *S. cottonii*
1. Plants having branches terete throughout. 2

2. Branches constricted at intervals into obovate or oblong segments . . *S. hormoides*
2. Branches not constricted at intervals. . . . . *S. johnstoniae*

Scinaia cottonii Setchell Pl. 26, figs 3, 4.

Setchell, 1914.

Plants usually bushy, to about 10 cm in height, rose red; branches dichotomously divided about 9 to 10 times at intervals of from 5-15 mm, slightly flattened, from 3-7 mm in width above and somewhat narrower towards the base; utricles quadrate to rectangular, from (8-)10-16(-20) μm in diameter and 18-24 μm in height; cystocarps globular, from (150-)170-220 μm in diameter, tending to be marginal.

For additional figures see Setchell, 1914, p. 143, pl. 11, fig. 24.

Occurs together with other species of *Scinaia* growing on rocky platforms and calcareous cobbles at a depth of about 10 m and rarely in shallower water.

Distribution. World: in warm temperate and tropical parts of the Atlantic and Pacific Oceans. West Africa: Ghana: unpublished.

Remarks. The Gulf of Guinea plants are referred to this species in the Complanatae group of the genus although the number of dichotomies is more than usually reported and the cystocarps are not always marginally arranged.

Plate 25: Fig. 1. *Pseudogloiophloea verae:* Plant terete and dichotomously branched. Fig. 2. *Scinaia hormoides:* Plant dichotomously divided and the branches constricted at intervals into obovate or oblong segments.

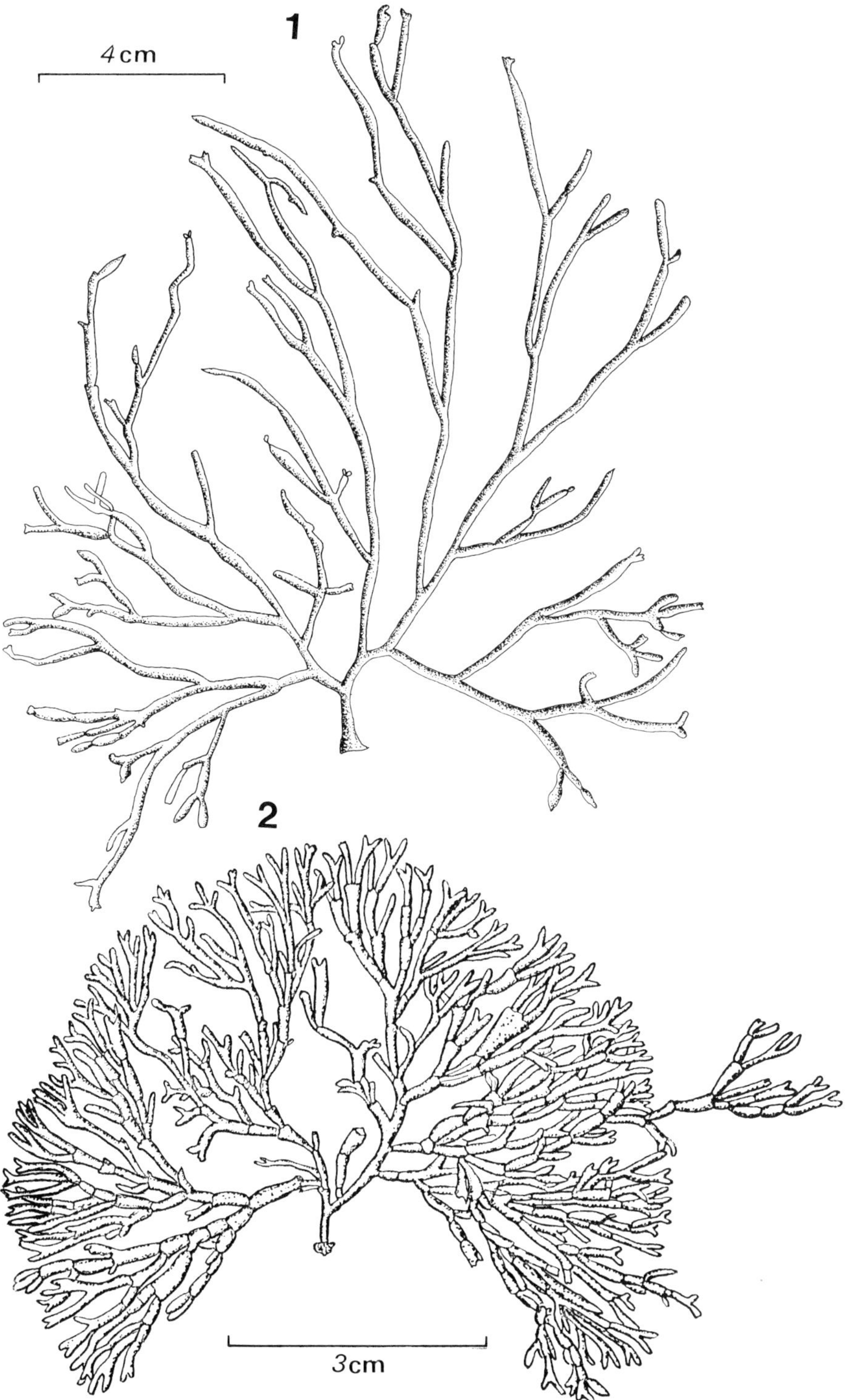
1
4cm
2
3cm

Scinaia hormoides Setchell Pl. 25, fig. 2; pl. 26, fig. 5.

Setchell, 1914.

Plants bushy and to about 8 cm in height, reddish-brown; branching dichotomous, from 7 to 8 times forked at intervals of up to 5 mm near the base and increasing to 20 mm above; branches terete, from 1.5-2.5 mm in diameter, deeply constricted into obovate or oblong segments, from 5-15 mm in length; utricles nearly quadrate or slightly rectangular, from 10-18(-20) μm in diameter and 20-24(-28) μm in height, with the outer wall a little convex; cystocarps globular and to about 210 μm in diameter.

For additional figures see Setchell, 1914, p. 145, pl. 12, figs 33-35, p. 147, pl. 13, figs 36, 37.

Occurs as solitary individuals growing on rocky platforms and calcareous cobbles at a depth of about 10 m.

Distribution. World: widespread in tropical seas. West Africa: Ghana: unpublished.

?**Scinaia johnstoniae** Setchell. Pl. 26, figs 1, 2.

Setchell, 1914.

Plants bushy, from 4-14 cm in height, rose-red to reddish-brown or purplish-brown; branching dichotomous, with up to 12(-14) divisions arising at intervals of from 2-10 mm; branches terete, from 0.5-3 mm in diameter, with acute apices; utricles quadrate or slightly rectangular, from 9(10-)13-20(-26) μm in diameter and 19-25 (-30) μm in height; cystocarps globular, from (140-)170-220 μm in diameter and to about 250 μm in height.

For additional figures see Abbott and Hollenberg, 1976, p. 332, fig. 274; Setchell, 1914, p. 143, pl. 11, figs 14, 15.

Occurs at a depth of about 10 m growing on rocky platforms and especially abundant on calcareous cobbles.

Distribution. World: in warm temperate and tropical parts of the Atlantic and Pacific Oceans. West Africa: Gabon: ?John & Lawson 1974a. Ghana: unpublished.

Remarks. The Ghanaian plants come close to this species although there are certain minor differences in the dimensions of the branches, texture of the plant, and the extent to which the axial filament is visible, which cast some doubt onto this determination. It is possible that the Ghanaian plants may represent more than one entity and this might only be resolved by the acquisition of further collections. Dawson (1952) has speculated that the Pacific *Scinaia minima* Dawson might be no more than a deep water form of *S. johnstoniae*.

Plate 26: Figs 1, 2. *Scinaia johnstoniae:* 1. Plant showing the terete and dichotomously divided branches; 2. Structure of the cortex with the medullary filaments terminating in large quadrate utricles. Figs 3, 4. *Scinaia cottonii:* 3. Plant showing the slightly flattened and dichotomously divided branches; 4. Structure of the cortex with the large quadrate to rectangular utricles. Fig. 5. *Scinaia hormoides:* Structure of cortex with the outer wall of the utricles a little convex. Fig. 6. *Pseudogloiophloea verae:* Structure of the cortex showing smaller assimilatory cells interspersed between the large hyaline utricles.

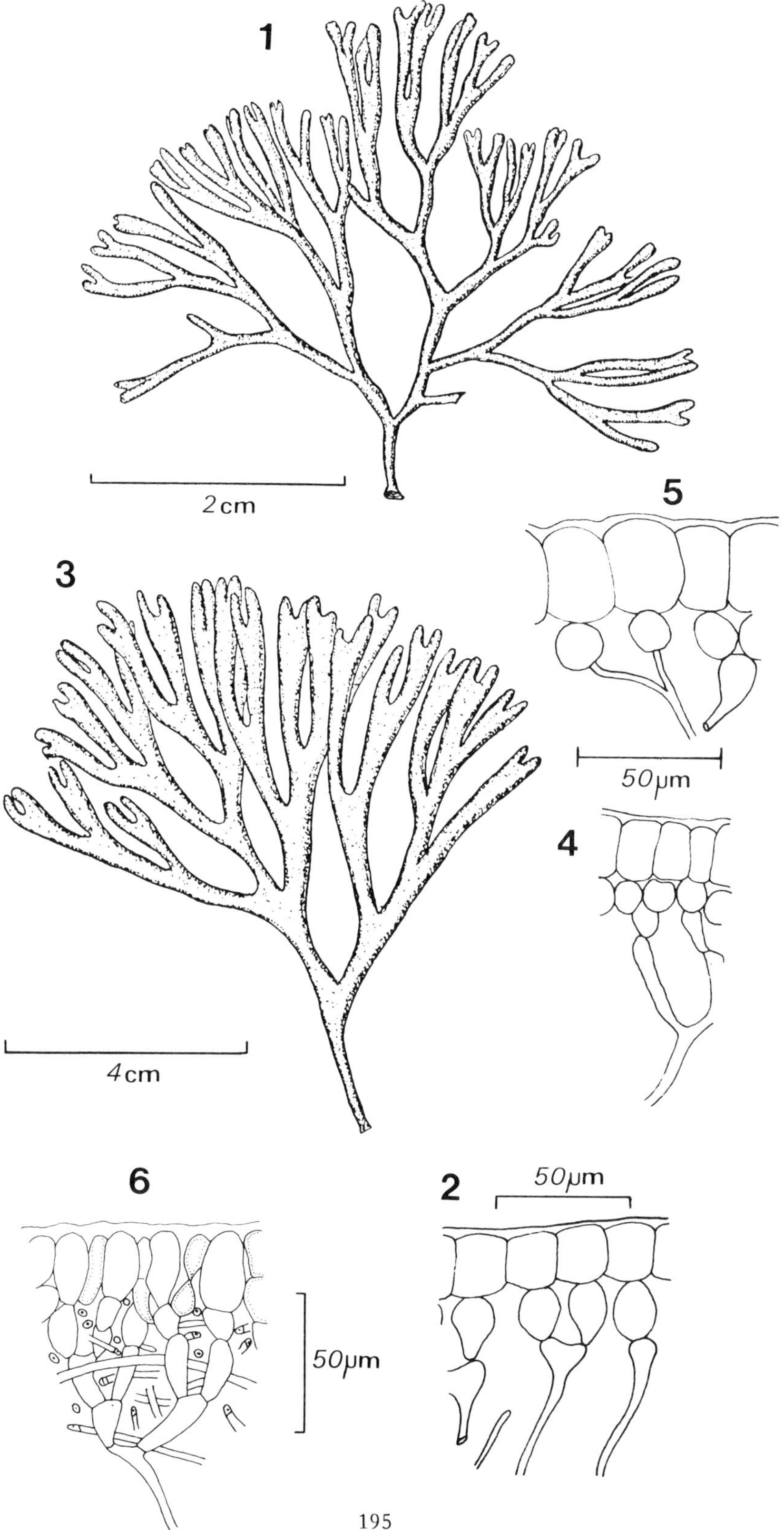
1
2 cm
5
50 µm
3
4
4 cm
6
50 µm
2
50 µm

Family Bonnemaisoniaceae

Asparagopsis* Montagne 1841

Sexual plants usually soft and plumose, consisting of alternately divided branches bearing numerous short branchlets; branches in section often subtubular and having a distinct axial filament running through the cavity, with the cortex parenchymatous and the largest cells innermost; branchlets having three pericentral cells and ecorticate; spermatangia usually covering modified branchlets; cystocarps stalked, large and subspherical. Asexual plants filamentous, composed of alternately divided branches, spreading and intertwined, attached by multicellular holdfasts; branches having three pericentral cells, ecorticate; tetrasporangia formed from the pericentral cells, tetrahedrally divided.

World Distribution: widespread in warm temperate and tropical seas.

It has been demonstrated by Magne (1964) and Chihara (1962) that the basic life history of *Asparagopsis* and the closely related *Bonnemaisonia* consists of a sequence of haploid sexual, diploid carposporic and diploid tetrasporic phases. The gametophytic and sporophytic plants are heteromorphic and the asexual phase of *Asparagopsis* was formerly considered as a separate genus *Falkenbergia.*

Asparagopsis taxiformis (Delile) Trevisan Pl. 27, figs 1-3.

Trevisan, 1845; *Fucus taxiformis* Delile, 1813.

Sexual plants purplish-brown to purplish-pink, from 7-20 cm in length, with the erect branches arising from a creeping system that may occasionally become secondarily erect; branches of the erect system sparingly alternately divided below and much divided above, bearing at narrow angles, crowded and pinnately divided branchlets. Asexual plants (= *Falkenbergia hillebrandii*) forming soft, pink and almost spherical tufts, from 2-3 cm in diameter, usually attached to larger algae by branched holdfasts; branch segments from 20-45 μm in diameter and 0.5 to 1.5 times as long as broad; a conspicuous darkly coloured body in each cell; apical cells usually prominent; tetrasporangia to about 50 μm in diameter, arising in discontinuous series.

*The asexual plants found in Gambia by John and Lawson (1977b) could not be identified to species.

Platè 27: Figs 1-3. *Asparagopsis taxiformis:* 1. Portion of a sexual plant showing an erect axis bearing crowded and pinnately divided branchlets; 2. Portion towards the tip of a branch of an asexual plant (=*Falkenbergia hillebrandii*) containing a developing tetrasporangium; 3. Section of an asexual plant showing the axial cell surrounded by three pericentral cells. Fig. 4. *Catenella impudica:* Portion of a plant showing the branches constricted into elliptical to oblong segments. Figs 5, 6. *Solieria tenera:* Variations in the form of this irregularly to sparingly alternately divided plant.

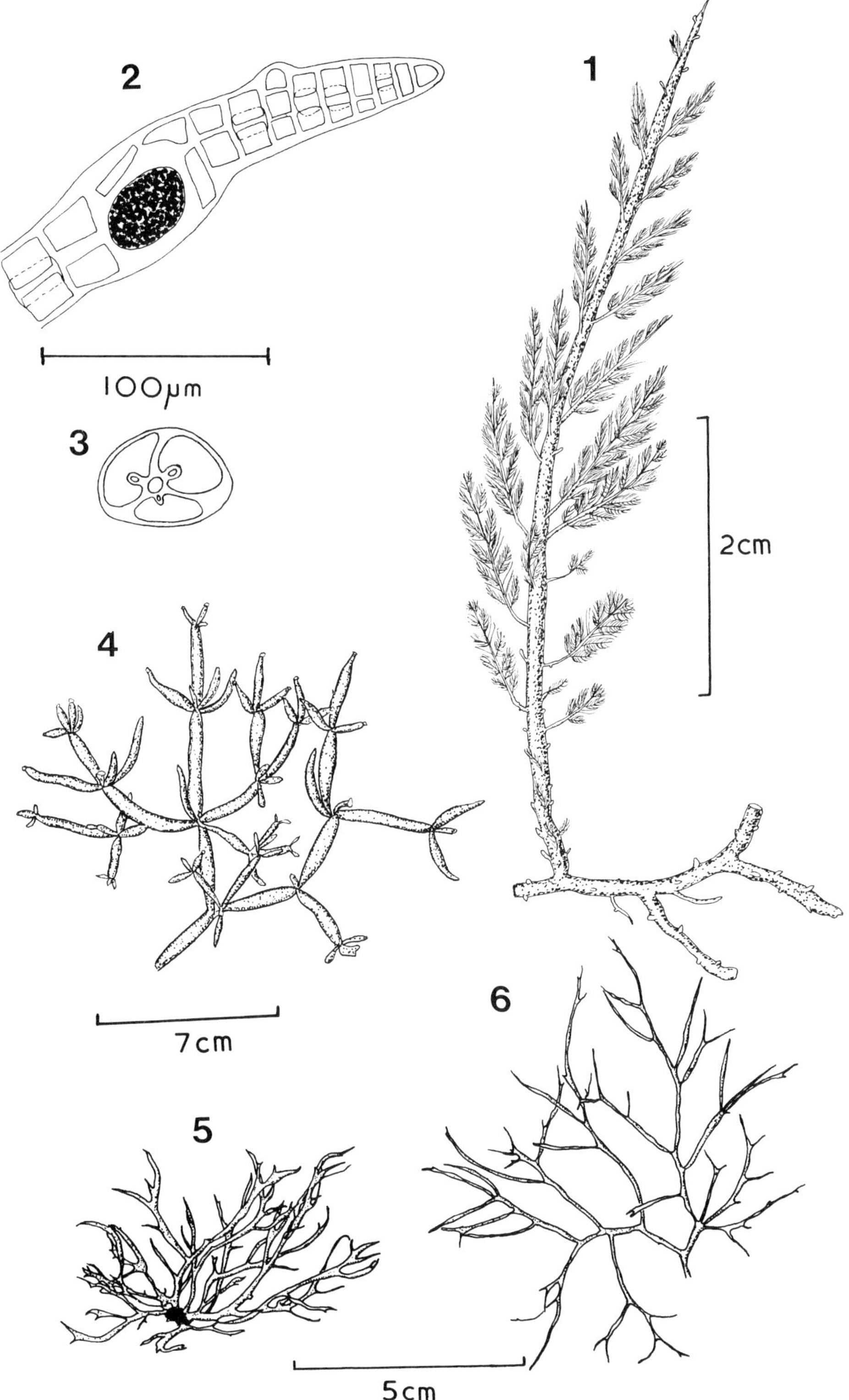
2
1
100μm
3
2cm
4
7cm
6
5
5cm

For additional figures see Børgesen, 1919, p. 352 to 355, figs 347-351 (sexual phase), p. 332, 333, figs 332, 333 (asexual phase).

Occurs most commonly on the sides of steep gullies or beneath rock ledges, usually in moderately sheltered to moderately wave-exposed situations. The sexual plants grow with other algae or occasionally as a distinct belt from the lower eulittoral subzone down to a depth of about 15 m where the largest and best developed individuals are to be found. Often more common is the minute and usually epiphytic asexual phase which occurs in the eulittoral and occasionally down to a depth of about 10 m.

Distribution. World: in warm temperate and tropical parts of the Atlantic and Pacific Oceans as well as in the Mediterranean. West Africa: Ghana: Dickinson & Foote 1950, John *et al.* 1977, Lawson 1980, Lieberman *et al.* 1979; *Falkenbergia hillebrandii*, Dickinson & Foote 1950, Lawson 1956, Lieberman *et al.* 1979. Liberia: De May *et al.* 1977. Nigeria: Lawson 1980.

Remarks. It is not clear to what extent *Asparagopsis armata* and *A. taxiformis* may be regarded as distinct species. Dixon (1964) has pointed out that the only apparent difference between the two (asexual phases) is that *A. taxiformis* possesses spines on the branches. Similarly J. Feldmann and G. Feldmann (1942) have commented on there appearing to be no morphological difference between the asexual phases of the two species.

Order Gigartinales ·Family Nemastomataceae*

Nemastoma J. Agardh 1842

Plants soft and gelatinous, varying from irregularly spherical to flattened, through to filiform and irregularly pinnate; thallus composed of loosely aggregated dichotomous medullary and cortical filaments, with subcortical gland cells present; gonimoblasts borne towards the surface of the thallus on auxiliary cells; tetrasporangia tetrapartitely divided.

World Distribution: in warm temperate and tropical seas.

Nemastoma confusum Kraft & John — Pl. 28, figs 1-3.

Kraft and John, 1976.

Plants solitary, from 3-10 cm in length, 2-4 cm wide and 1-3 mm thick, bearing on the surface varying numbers of simple to multifid, blunt proliferations, with the margins smooth, variously lobed or proliferous; thallus composed of lax medullary filaments and a cortex of submoniliform, dichotomous filaments from 5 to 10 cell layers deep; gland cells borne in the subsurface cortical layers, subspherical, to about 35 μm in diameter; gonimoblast filaments protected by a rudimentary involucre formed of filaments arising directly from the auxiliary cells.

For additional figures see Kraft and John, 1976, p. 333, figs 1, 4, 5, p. 334, figs 6-9.

*For discussion concerning the correct family name see Silva (1980, p. 83).

Occurs on rocky ledges and gullies at a depth of about 10 m.

Distribution. World: tropical West Africa. West Africa: Ghana: John *et al.* 1977, Kraft & John 1976, Lieberman *et al.* 1979.

Platoma Schmitz 1894

Doubtful Record

Platoma cyclocolpa (Montagne) Schmitz

A single drift specimen is reported by Aleem (1978) from the Sierra Leone peninsula. We consider this to be a doubtful record of a plant not previously reported from the mainland of West Africa.

Predaea De Toni fil. 1936

Plants soft and gelatinous, irregularly spherical to flattened and often bearing warty processes; thallus consisting of a loose medulla of sparingly divided filaments and a cortex of dichotomous filaments, with gland cells known only in one species; gonimoblasts developing laterally or towards the surface of the thallus either on auxiliary cells or on connecting filaments, with the cells adjacent to the auxiliary cells usually covered by clusters of small nutritive cells; tetrasporangia unknown.

World Distribution: in warm temperate and tropical seas.

Key to the Species

1. Plants flattened and with a smooth surface; gland cells present in cortex; carposporophyte forming a single ovoid mass .... *P. masonii*
1. Plants normally irregularly branched and lobed, tubular to broadly flattened; gland cells absent; carposporophyte almost pyriform and consisting of two or more distinct lobes .... *P. feldmannii*

Predaea feldmannii Børgesen Pl. 28, figs 4-6.

Børgesen, 1950.

Plants solitary, from 2-11 cm in length, irregularly branched and lobed, cylindrical to broadly flattened, attached by a fleshy disc-like holdfast; cortical cells elongate, cylindrical, from 4-6 μm in diameter and 3 to 7 times longer than broad; gland cells absent; gonimoblasts developing laterally on the connecting filaments and mature carposporophyte almost pyriform consisting of 2 or more distinct lobes.

For additional figures see Børgesen, 1950, p. 3, 4, 6, figs 1-3; Kraft and John, 1976, p. 333, figs 2, 3, p. 338 to 340, 342, figs 10-31.

Occurs as an uncommon plant growing on low rocky platforms or in the carpet of algae covering areas of cobbles at depths ranging from 10 to 27 m.

Distribution. World: in warm temperate and tropical seas. West Africa: Ghana: John *et al.* 1977, Kraft & John 1976, Lieberman *et al.* 1979.

Remarks. Information and discussion on the Gulf of Guinea material and its relationship to other closely related taxa are to be found in Kraft and John (1976).

Predaea masonii (Setchell & Gardner) De Toni Pl. 28, fig. 7.

De Toni, 1936; *Clarionea masonii* Setchell and Gardner, 1930.

Plants solitary, to about 6 cm in length, flattened, irregularly ovate and smooth-surfaced, with an indistinct stipe and attached by a fleshy holdfast; thallus having a cortex of moniliform, dichotomous filaments; gland cells numerous; gonimoblasts arising on the connecting filaments close to the point of attachment with the auxiliary cell; carposporangial masses single and irregularly lobed.

For additional figures see Schneider and Searles, 1975, p. 97, pl. 1, figs 1, 2.

Found on only a single occasion at a depth of about 25 m growing on calcareous cobbles.

Distribution. World: in warm temperate and tropical parts of the Atlantic Ocean. West Africa: Ghana: John *et al.* 1977, Kraft & John 1976.

Family Caulacanthaceae

Catenella Greville 1830

Plants usually creeping or assurgent, with the terete or compressed branches constricted at intervals into segments, attached to the substrate and one branch to another by holdfasts; branches consisting of a medulla of slender anastomosing filaments and a cortex of radial rows of cells decreasing in size towards the surface; tetrasporangia in specialised terminal segments, zonately divided; spermatangia immersed in the cortex of swollen segments; cystocarps protruding and somewhat elongated, arising on short segments, with a distinct pore.

World Distribution: widespread from boreal-antiboreal to tropical seas.

Plate 28: Figs 1-3. ***Nemastoma confusum:*** 1. Plant flattened with the margin variously lobed or proliferous; 2. A particularly proliferous portion of a plant showing the simple to multifid, blunt proliferations arising from the flat surface (after Kraft and John, 1976); 3. Cortical filaments showing a gland cell (gl) and an auxiliary cell (aux) bearing a developing gonimoblast initial (g. i) (after Kraft and John, 1976). Figs 4-6. ***Predaea feldmannii*** (after Kraft and John, 1976): 4. Plant showing the irregularly tubular lobes or branches; 5. Cortical filaments bearing a carpogonial branch (c. b) on a supporting cell (su); 6. The carpogonial branch(c. b), supporting cell acting as an auxiliary cell (aux), and the development of nutritive cell filaments (n. f). Fig. 7. ***Predaea masonii:*** Plant flattened, irregularly ovate, and smooth-surfaced.

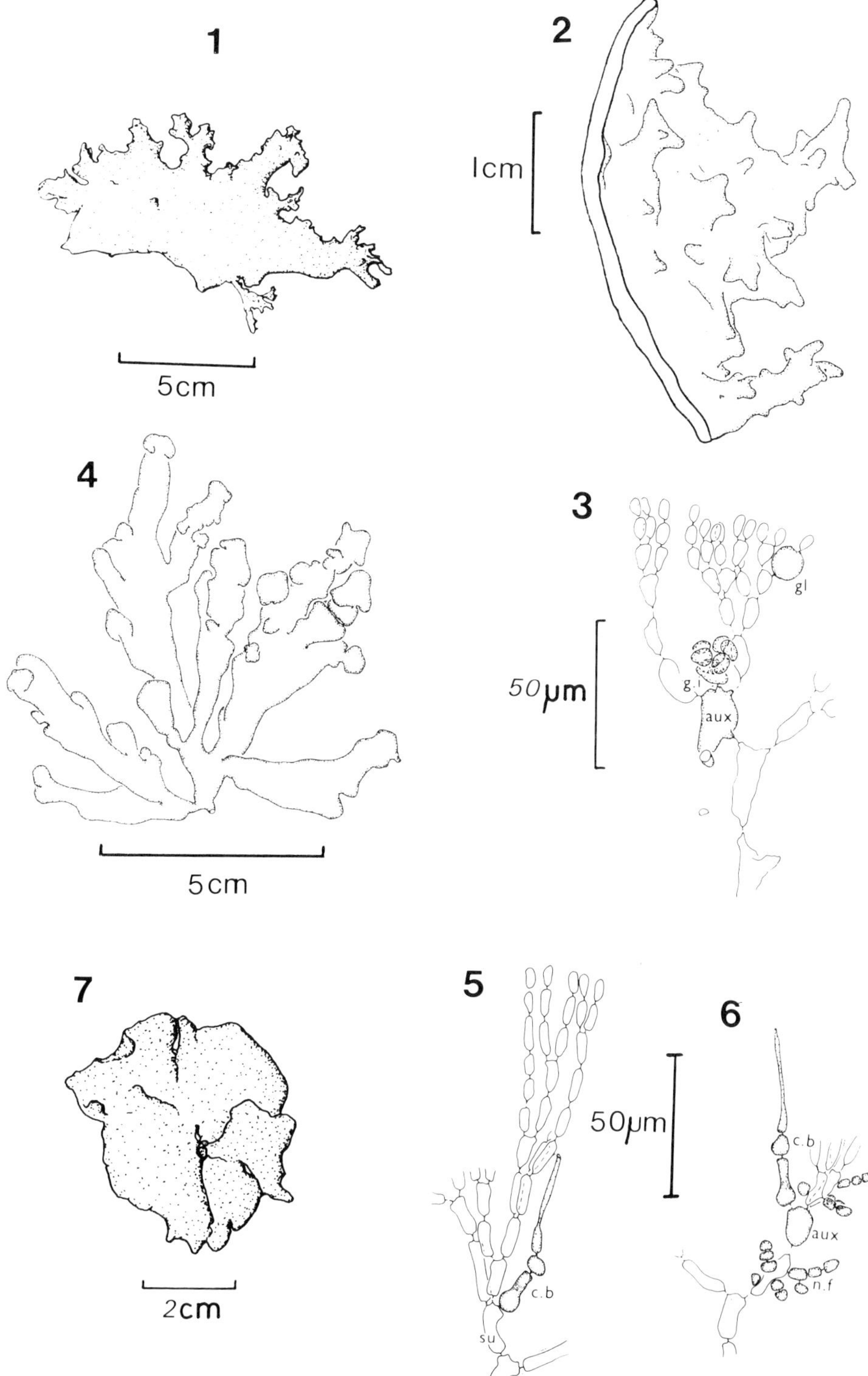
1
5cm
2
1cm
4
5cm
3
gl
50μm
g.i
aux
7
2cm
5
c.b
su
6
50μm
c.b
aux
n.f

Key to Species

1. Plants attached by flattened haptera formed terminally on attenuated branch segments. . . . . *C. impudica*
1. Plants attached by stalked haptera arising from the points of branching. . . . . *C. caespitosa*

Catenella caespitosa (Withering) L. Irvine

Irvine, in Parke and Dixon, 1976; *Ulva caespitosa* Withering, 1776.

Plants creeping or forming small cushion-like clumps, up to 2 cm in height, usually brownish-purple, with the holdfasts arising from the point of branching as short stalks terminating in a broad disc; branches irregularly divided, with segments terete to compressed, narrowly lanceolate, to about 500 μm in width, from 3 to 5 times longer than broad.

For figures see Newton, 1931, p. 420, fig. 251a-g (as *C. repens*); Dixon and Irvine, 1977a, p. 191, fig. 67.

Occurs in the littoral fringe growing on the stilt roots of *Rhizophora* and may be found along river banks often many miles from the sea.

Distribution. World: widespread from boreal-antiboreal to tropical seas. West Africa: Liberia: *C. repens,* De May *et al.* 1977. Nigeria: unpublished.

Catenella impudica (Montagne) J. Agardh Pl. 27, fig. 4.

J. Agardh, 1852; *Lomentaria impudica* Montagne, 1840.

Plants forming low mats or mixed in a turf of other small algae, to about 4 cm in height, purplish-black, with the holdfasts formed terminally on attenuated branch segments; branches di- or trichotomously divided, with the apices acute; segments terete and in older parts compressed, elliptical to oblong, to about 2 mm in width and up to twice as long as broad or occasionally longer; tetrasporangia forming a transverse band across the centre of oblong segments.

For additional figures see Kützing, 1865, pl. 92, figs a-c.

Occurs in brackish-water habitats such as on estuarine rocks or on the roots of mangroves, usually growing in the littoral fringe.

Distribution. World: as for genus. West Africa: Cameroun: John 1977a, Post 1936, 1957b. Côte d'Ivoire: De May *et al.* 1977, John 1972c, 1977a. Guinée: Post 1955a, 1955b, 1957b, 1959, Schnell 1950.

Caulacanthus Kützing 1843b

Plants usually erect, composed of terete, densely and variously divided branches; growing from a distinct apical cell; branches consisting of a central axial filament and surrounded by a loose layer of cells, becoming denser towards the surface and forming a compact cortex; tetrasporangia scattered towards the tips of the branches, sporangia zonately divided.

World Distribution: probably widespread from boreal-antiboreal to tropical seas.

Caulacanthus ustulatus (Turner) Kützing — Pl. 31, fig. 4.

Kützing, 1843b; *Fucus acicularis* var. *ustulatus* Turner, 1809.

Plants forming cushion-like tufts or mats, from 2-5 cm in height, reddish or greenish-brown, consisting of an intricately intertwined system of erect and prostrate branches; branches terete, about 500 μm in diameter, irregularly divided, with the ultimate divisions often spine-like; thallus in section having an outermost cortex of small cells elongated at right angles to the surface.

For additional figures see Gayral, 1958, p. 375, 377, fig. 56c, pl. 98.

Occurs in the algal turf in the lower eulittoral subzone and at higher levels on the shore in cracks and crevices especially in wave-exposed situations. In other parts of West Africa it may sometimes form a well-developed belt or band in the upper eulittoral subzone (see Lawson *et al.,* 1975).

Distribution. World: widespread in warm temperate and tropical seas. West Africa: Gambia: John & Lawson 1977b. Sénégal (Casamance): Sourie 1954. Sierra Leone: Aleem 1978.

Family Solieriaceae

Anatheca Schmitz 1896

Plants consisting of flattened and foliaceous fronds, irregularly or somewhat dichotomously divided or lobed; thallus having a narrow medulla of loose filaments and a thick and pseudoparenchymatous cortex in which the cells become progressively smaller towards the surface; tetrasporangia scattered in the outer cortex, zonately divided; cystocarps borne in surface papillae and having a well-developed pore.

World Distribution: in warm temperate and tropical seas.

Anatheca montagnei Schmitz — Pl. 29, fig. 1.

Schmitz, in Schmitz and Hauptfleisch, 1896.

Plants solitary, flat, widening to as much as 8 cm and to about 24 cm or more in length, fleshy in the fresh state and membranaceous on drying, arising from a short stipitate base; frond irregularly divided in the lower part and appearing dichotomous above, often proliferous from the margin, with the surface usually covered by papillae; inner cortical cells colourless, from 110-200 μm in diameter, with the cells decreasing in size towards the surface.

For additional figures see Bodard, 1966b, p. 868 to 870, 872, 874 to 883, 885 to 889, figs 1-21, pls 1-7.

Occurs on calcareous cobbles at depths ranging from 10 to 27 m and has been found just to the north of the region in Sénégal growing in the eulittoral zone (Bodard, 1966b).

Distribution. World: as for genus. West Africa: Ghana: John & Lawson 1972a, John *et al.* 1977, Lieberman *et al.* 1979. Sénégal (Casamance): *A. dentata*, Chevalier 1920.

Solieria J. Agardh 1842

Plants consisting of firm, terete and tapering branches, variously divided; branches having a filamentous medulla, with an inner cortex of large cells developing down-growing filaments and an outer cortex of smaller assimilatory cells; tetrasporangia scattered and embedded in the outer cortex, zonately divided; cystocarps large and slightly projecting, containing in the centre a large fusion cell, with a distinct pore.

World Distribution: probably widespread in warm temperate and tropical seas.

Solieria tenera (J. Agardh) Wynne & W. R. Taylor Pl. 27, figs 5, 6.

Wynne and Taylor, 1973; *Gigartina tenera* J. Agardh, 1841.

Plants usually solitary, to about 20 cm in height, reddish-pink and fleshy, attached by a loose and fibrous holdfast; branching irregular to sparingly alternate, with the branches bearing a number of soft lateral branchlets; branches terete, from 1-3 mm in diameter; branchlets constricted at the base, somewhat inflated above and with a long acuminate apex; medulla consisting of thick-walled and continuous filaments; tetrasporangia from 16-20 μm in diameter and 28-34 μm in length.

For additional figures see Taylor and Rhyne, 1970, p. 11, 12, figs 1, 2 (as *Agardhiella tenera*); Wynne and Taylor, 1973, p. 96, 97, figs 1-6.

Occurs on moderately sheltered to moderately wave-exposed rocks in the lower part of the eulittoral and commonly found in tidepools. Loosely branched plants grow in abundance at depths ranging from 8 to 30 m particularly on calcareous cobbles.

Distribution. World: in warm temperate and tropical parts of the Atlantic Ocean. West Africa: Gabon: John & Lawson 1974a. Ghana: Gabrielson & Hommersand 1982; John *et al.* 1977, 1980, Lieberman *et al.* 1979, Wynne & Taylor 1973; *Agardhiella tenera,* Dickinson & Foote 1950.

Remarks. Cystocarpic plants have yet to be found in West Africa but other features are sufficiently close to assign the asexual plants to this species (see Wynne and Taylor, 1973).

Family Rhodophyllidaceae

Rhodophyllis Kützing 1847

Plants flattened, somewhat fleshy to membranaceous, with the branches many times dichotomously or laterally divided or occasionally lobed; fronds having a weakly developed medulla of slender branched filaments associated with a few large cells and a thin cortex of one or two layers of large cells; tetrasporangia immersed in the branches or confined to marginal proliferations, zonately divided; cystocarps mostly marginal and with no pore through the thick-walled pericarp formed from the cortex.

World Distribution: widespread from boreal-antiboreal to tropical seas.

Rhodophyllis gracilarioides Howe & W.R. Taylor Pl. 29, fig. 2.

Howe and Taylor, 1931.

Plants erect and bushy, from 2-14 cm in height, reddish-purple, arising from a short stipe; fronds palmately divided from near the base and irregularly divided above, with the lateral branches subpinnate or subbipinnate; branches to about 5 mm wide in deep water plants and narrower in bushy plants normally found at shallow depths, with the ultimate divisions subterete and deltoid or narrowly lanceolate, terminally acuminate; surface of the branches often having a net-like appearance in older parts; cystocarps spherical, to about 500 μm in diameter, arising marginally on the ultimate branches.

For additional figures see Howe and Taylor, 1931, p. 13, fig. 5, pl. 2, figs 4, 5.

Occurs as sparsely branched plants up to 8 cm in height at depths greater than 20 m whilst between 10 and 20 m it grows both larger and bushier.

Distribution. World: in tropical parts of the Atlantic Ocean. West Africa: Ghana: John & Lawson 1972a, John *et al.* 1977, Lieberman *et al.* 1979.

Family Hypneaceae

Hypnea Lamouroux 1813

Plants usually bushy, consisting of slender and terete branches, many times irregularly divided and often virgate, bearing short, acute and often subulate branchlets; branches in section having an ill-defined axial cell and a pseudoparenchymatous medulla grading into a cortex of large cells becoming smaller towards the surface;

tetrasporangia immersed and scattered in modified lateral branchlets, zonately divided; cystocarps borne on the ultimate branches, prominent and nearly spherical, without a pore or indistinct.

World Distribution: widespread in warm temperate and tropical seas.

This is an extremely polymorphic genus; its many variants and their relation with environmental conditions are little understood. As with certain other genera mentioned previously the confused taxonomic situation can only be resolved by a monographic treatment on a global scale. Bodard (1968) has recently reviewed the West African species of *Hypnea* but this has contributed little to resolve the problems associated with the genus in the region.

Key to the Species

1. Plants having sparingly divided main axes, beset with short branchlets becoming gradually shorter and spine-like towards the tips of the branches .... *H. flagelliformis*
1. Plants usually bushy and of another form 2

2. Plants erect and bushy, usually 8 cm or more in height 3
2. Plants cushion-like or as low mats, usually less than 6 cm in height 4

3. Branches often naked towards their tips, with the ends thickened and hook-shaped .... *H. musciformis*
3. Branches usually completely covered by short branchlets, with the tips of the branches straight .... *H. valentiae*

4. Branches anastomosing or attached to one another by discoid holdfasts 5
4. Branches entangled and usually free from one another 6

5. Ultimate branches and branchlets having obtuse apices .... *H. cenomyce*
5. Ultimate branches and branchlets acute and spine-like .... *H. spinella*

6. Branches occasionally flattened and branchlets sometimes spathulate .... *H. arbuscula*
6. Branches terete throughout and often branchlets having a cervicorn appearance . ... *H. cervicornis*

Hypnea arbuscula P. Dangeard Pl. 29, fig. 6.

Dangeard, 1952.

Plants forming loose cushion-like tufts, from 2-4 cm in height, reddish-brown; branches dichotomously divided, terete and sometimes flattened, from 300-500 μm in width, bearing irregularly arranged simple or bifurcate branchlets, with the ultimate divisions tapering and acute or occasionally flattened and spathulate; tetrasporangia arising in inflated branchlets, often appearing siliquose.

For additional figures see Dangeard, 1952, p. 321, pl. 19, figs i-l.

Occurs in the algal turf growing on moderately wave-exposed rocks in the lower eulittoral subzone.

Distribution. World: in warm temperate and tropical parts of West Africa. West Africa: Gambia: John & Lawson 1977b.

Remarks. This plant bears a close resemblance to *Gigartina acicularis* and may only be separated from it with certainty by an anatomical examination revealing its uniaxial structure as opposed to the multiaxial axis of *Gigartina*.

Hypnea cenomyce J. Agardh

J. Agardh, 1852.

Plants forming small cushion-like clumps or balls; branches from 100-150 μm in diameter and irregularly divided, often intricate and anastomosing or attached to one another by discoid holdfasts, with the apices obtuse.

For figures see Dangeard, 1952, p. 321, pl. 19, fig. e.

There is no information on the ecology of this plant in the region although in Sénégal it grows in the eulittoral zone.

Distribution. World: in warm temperate and tropical parts of West Africa, and in Australasia. West Africa: Sierra Leone: Bodard 1968, John & Lawson 1977a.

Hypnea cervicornis J. Agardh Pl. 30, figs 3, 4.

J. Agardh, 1852.

Plants forming loosely entangled clumps or mats, from 2-6 cm in height, reddish-purple or occasionally bleached to a greenish-brown; branches from 400-500 μm in diameter, irregularly and subdichotomously divided, usually divaricate, with the ultimate divisions attenuate, curved and cervicorn; tetrasporangia and cystocarps arising in crowded lateral branchlets which are linear to ovate and often forked.

For additional figures see Bodard, 1968, figs 18-22.

Occurs as dense entangled clumps in the algal turf on moderately wave-sheltered rocks in the lower eulittoral subzone; looser and larger plants found in tidepools and down to a depth of about 10 m.

Distribution. World: widespread in warm temperate and tropical seas. West Africa: Gabon: John & Lawson 1974a. Gambia: John & Lawson 1977b. Ghana: John 1977a, John *et al.* 1977, 1980, Lawson 1953, 1956, Lieberman *et al.* 1979, Stephenson & Stephenson 1972. Liberia: De May *et al.* 1977, John 1977a. Príncipe: Carpine 1959, John 1977a, Steentoft 1967. São Tomé: John 1977a, Steentoft 1967; *H. musciformis*, Hariot 1908, Henriques 1886a, 1887, 1917. Sénégal (Casamance): Bodard 1966a. Sierra Leone: Aleem 1978, John & Lawson 1977a.

Remarks. There is much confusion regarding the distinction between *Hypnea cervicornis* and *H. spinella* taxonomically, nomenclaturally, and biologically.

Doubt has been expressed (see De Toni, 1900; Tanaka, 1941) as to whether the illustration of *H. spinella* by Kützing (1868, pl. 26, figs d-f) should be more correctly attributed to *H. cervicornis*. This question can only be finally resolved by referring back to the original material on which Kützing based his illustration.

Hypnea flagelliformis Greville ex J. Agardh Pl. 29, figs 3-5.

Greville, in J. Agardh, 1852.

Plants often solitary or occasionally as loose clumps, from 8-16 cm in length; branches to about 1 mm in diameter, irregularly divided a few times towards the base, gradually attenuate towards the tips; branchlets almost absent from the uppermost parts of the branches, long and sometimes divided when arising near the base of the branches, becoming shorter, simple and spine-like towards the branch tips; tetrasporangia arising near the base of the lower branchlets.

For additional figures see Dangeard, 1952, p. 321, pl. 19, figs a, b.

Occurs on moderately wave-exposed rocks in the lower eulittoral subzone and becomes particularly abundant where there is current or wave surge.

Distribution. World: probably widespread in tropical seas. West Africa: Ghana: John & Asare 1975.

Remarks. The small creeping *Hypnea* reported from Sierra Leone by John and Lawson (1977a) has short erect axes with a distinctly pyramidal outline and might be just a form of this species.

Hypnea musciformis (Wulfen) Lamouroux Pl. 30, fig. 1.

Lamouroux, 1913; *Fucus musciformis* Wulfen, 1789.

Plants forming loose bushy clumps or occasionally as dense cushion-like patches, from 8-16 cm in length, dull reddish-purple or often bleached to a greenish colour; branches terete, to about 1 mm in diameter, few to many times irregularly divided, bearing below alternately arranged and upwardly curving branchlets, with the branch tips often swollen and hook-shaped; branchlets short, awl-shaped, sometimes in secund series towards the ends of the branches; tetrasporic plants furry in appearance due to the presence of large number of branchlets bearing tetrasporangia; cystocarps about 650 μm in diameter, arising on spine-like or awl-shaped branchlets.

Plate 29: Fig. 1. ***Anatheca montagnei:*** Plant showing the flattened and irregularly to dichotomously divided frond arising from a short stipitate base. Fig. 2. ***Rhodophyllis gracilarioides:*** Portion of a plant showing the subterete and pinnately divided branches bearing marginal cystocarps. Figs 3-5. ***Hypnea flagelliformis:*** 3. Plant showing the sparingly divided main axes beset with short branchlets; 4. Terminal portion of a branch bearing spine-like branchlets; 5. Lower portion of a branch with the branchlets subdivided. Fig. 6. ***Hypnea arbuscula:*** Portion of plant showing the simple or bifurcate and often acute branchlets arising on the branches.

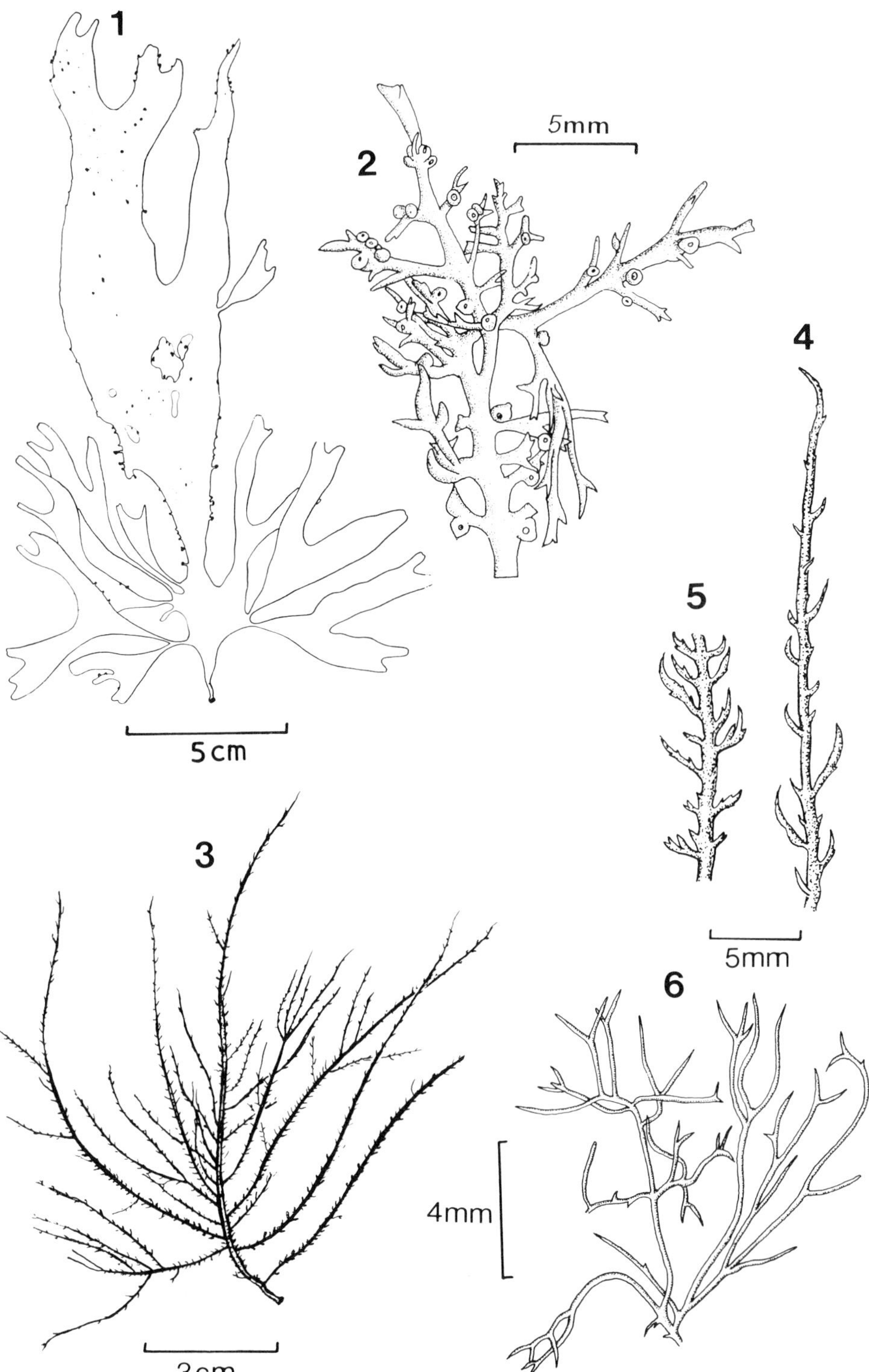
1
2
5mm
4
5
5cm
3
5mm
6
4mm
3cm

For additional figures see Taylor, 1960, p. 809, pl. 73, fig. 1.

Occurs in moderately wave-sheltered situations as large, well-developed plants growing on rocks in the lower eulittoral subzone and entangled with other algae; found in wave-exposed places are small and densely clumped plants.

Distribution. World: as for genus. West Africa: Cameroun: Fox 1957, Hoppe 1969, John 1977a, Lawson 1955, Pilger 1911, Schmidt & Gerloff 1957, Steentoft 1967, Stephenson & Stephenson 1972. Côte d'Ivoire: John 1972c, 1977a. Gabon: John & Lawson 1974a. Gambia: John & Lawson 1977b. Ghana: Fox 1957, John 1977a, John & Asare 1975, Lawson 1953, 1956, 1957b, 1966, Michanek 1971, 1975, Steentoft 1967, Stephenson & Stephenson 1972. Liberia: De May *et al.* 1977, John 1972c, 1977a. Nigeria: Fox 1957, John 1977a, Steentoft 1967. Príncipe: Carpine 1959, John 1977a, Steentoft 1967. São Tomé: Fox 1957, John 1977a, Steentoft 1967. Togo: John 1977a, John & Lawson 1972b. Sénégal (Casamance): Chevalier 1920. Sierra Leone: Aleem 1978, Fox 1957, John & Lawson 1977a, Lawson 1954b, Steentoft 1967.

Remarks. This is a very polymorphic plant with its most typical growth form usually found where there is very little wave action. The branches terminate in the characteristically thickened, hook-shaped tips in the typical growth form but where there is moderate to strong wave action the branch tips may be straight and unthickened.

Hypnea spinella (C. Agardh) Kützing Pl. 30, fig. 2.

Kützing, 1849; *Sphaerococcus spinellus* C. Agardh, 1823.

Plants forming coarse, dense clumps or mats, sometimes reaching 2 cm in height; branches to about 250 μm in diameter, irregularly divided at close intervals, intricate, obviously anastomosing or attached to one another and to the substratum by discoid holdfasts, with the ultimate branches very short and spine-like; tetrasporangial sori on the ultimate branches and branchlets, covering the surface in the form of annular bands or occasionally as a cap.

For additional figures see Børgesen, 1920, p. 384, fig. 369.

Occurs together with other small algae on moderately wave-exposed to somewhat sheltered rocks in the lower eulittoral subzone.

Distribution. World: as for genus. West Africa: Gabon: ?John & Lawson 1974a. Gambia: John & Lawson 1977b. Ghana: unpublished. São Tomé: Hariot 1908, Henriques 1886b, 1887, 1917, Steentoft 1967. Sierra Leone: Aleem 1978, John & Lawson 1977a, Lawson 1954b, ?1957a, Longhurst 1958, Steentoft 1967.

Plate 30: Fig. 1. *Hypnea musciformis:* Portion of a plant showing the characteristicically thickened hook-like tips to some of the branches. Fig. 2. *Hypnea spinella:* Portion of a tetrasporic plant with some of the branches anastomosing or attached to one another by holdfasts. Figs 3, 4. *Hypnea cervicornis:* 3. Portion of a plant showing the curved and cervicorn ultimate divisions of the branches; 4. Portion of a branch with the swollen lateral branchlets containing tetrasporangia. Figs 5, 6. *Hypnea valentiae:* 5. Bushy habit of plant; 6. Terminal portion of a branch showing the often basally swollen lateral branchlets.

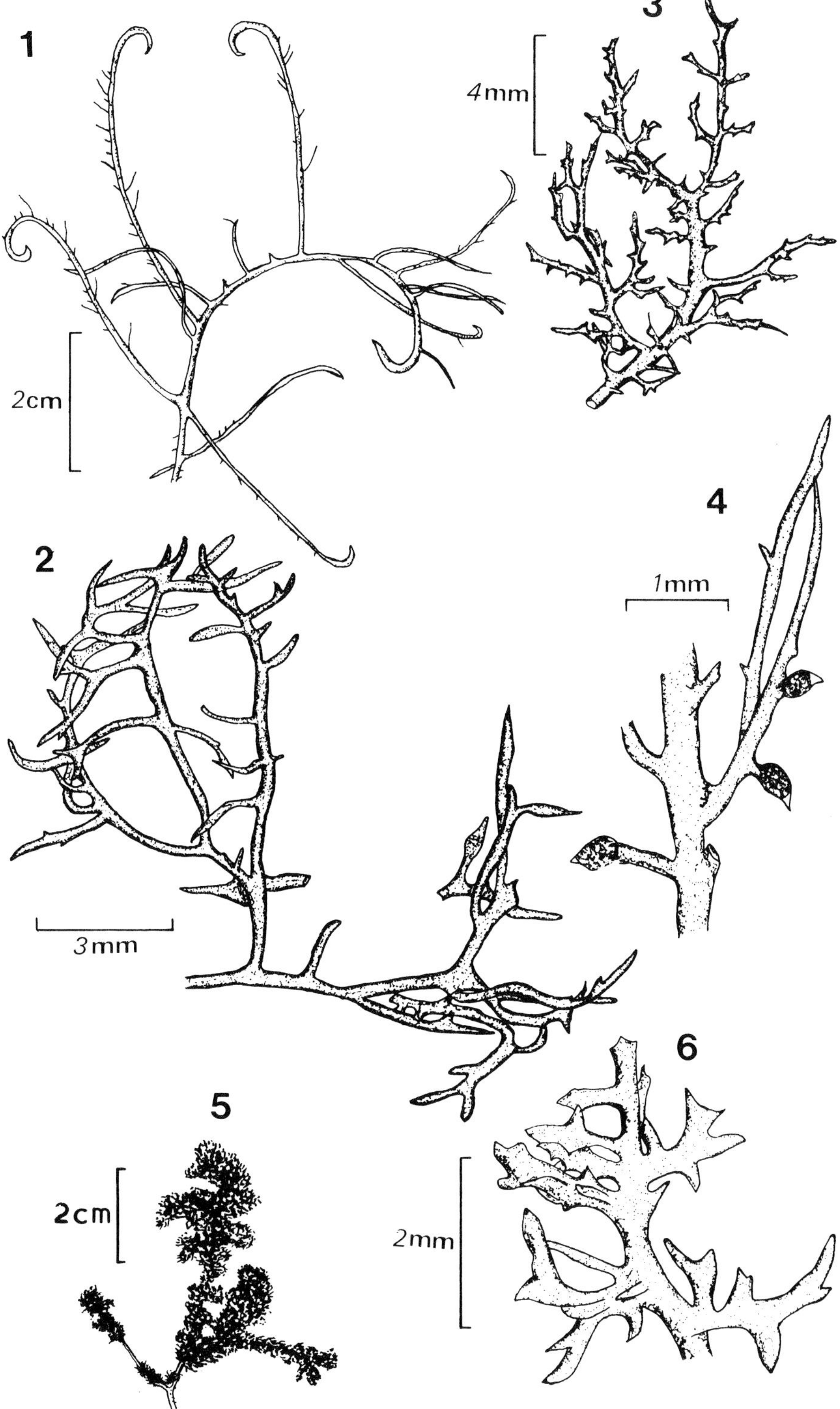
1
2cm
3
4mm
2
3mm
4
1mm
5
2cm
6
2mm

Remarks. See the remarks under *Hypnea cervicornis.*

Hypnea valentiae (Turner) Montagne Pl. 30, figs 5, 6.

Montagne, 1839-1841; *Fucus valentiae* Turner, 1809.

Plants forming dense, bushy clumps, to about 10 cm in height, dark purplish-red; branches up to 1 mm in diameter, alternately divided and densely clothed with regularly divided branchlets; branchlets sometimes basally swollen and somewhat stellate, with the ultimate divisions often divaricate and the apices acute.

For additional figures see Børgesen, 1943, p. 57, fig. 27 (as *H. charoides*); Nasr, 1947, p. 113, pl. 13, fig. 1.

Occurs on sand-covered rocks in tidepools in the lower eulittoral subzone where there is very little wave action.

Distribution. World: widespread in warm temperate and tropical seas. West Africa: Ghana unpublished. Guinée: *H. cornuta,* Agardh 1852, Børgesen 1920, De Toni 1900.

Remarks. Børgesen (1934, 1943) has transferred a number of species formerly in Section Spinuligera of this genus into the synonymy of *Hypnea valentiae* believing them to be no more than growth forms related to different environmental conditions.

Rejected Record

Hypnea unilateralis P. Dangeard

Bodard (1968) reports this species from "Golfe de Guinée" but gives no reason to substantiate this distribution record. He doubts the validity of this taxon believing it to be very close to the little-known plant described by Kützing (1849) as *Hypnea setacea.* It appears that Bodard believed that the type came from the Gulf of Guinea region of West Africa but Kützing (1849) cites "Cayenne" in French Guiana (now Guyana) as the type locality.

Family Plocamiaceae

Plocamium Lamouroux 1813

Plants consisting of subterete to flattened and sometimes almost membranaceous branches, distichous and pinnately divided, with small groups of branchlets arising in a regular arrangement on the lateral branches; branches having a large-celled medulla and a cortex of a number of layers of polygonal cells; tetrasporangia embedded in small modified branchlets, zonately divided; cystocarps arising marginally on the ultimate branches or on modified branchlets, without a pore.

World Distribution: probably cosmopolitan.

Key to the Species

1. Secund lateral branchlets mostly in pairs and a simple spine-like branchlet arising almost opposite each lateral branch .... *P. telfairiae*
1. Secund lateral branchlets usually in groups of more than two and without a branchlet opposite each lateral branch .... *P. cartilagineum*

Plocamium cartilagineum (Linnaeus) Dixon

Dixon, 1967; *Fucus cartilagineus* Linnaeus, 1753.

Plants bushy, often from 5-15 cm in length, clear red; branching distichous, alternate or subdichotomous, with the lateral branches a number of times pinnately divided; lateral branches usually having groups of one to several secund branchlets, with the lowermost branchlet often simple and the upper ones divided once or twice, again the lowermost subdivision rarely crenulate and the upper subdivisions bearing 3 or 4 subulate teeth on their adaxial side; tetrasporangia in lanceolate or dichotomous lateral branchlets, scattered along the margins of the upper branches.

For figures see Newton, 1931, p. 444, fig. 265 (as *P. coccineum*).

There is no information on the ecology of this plant in the region.

Distribution. World: from cold temperate to tropical parts of the eastern Atlantic Ocean as well as the Mediterranean. West Africa: Sénégal (Casamance): *P. vulgare*, Chevalier 1920.

Plocamium telfairiae Harvey ex Kützing Pl. 31, figs 5, 6.

Harvey, in Kützing, 1849.

Plants erect and bushy, from 2-10 cm in height, deep red to reddish-purple; branching distichous and alternate or subdichotomous, with the lateral branches two to three times pinnately divided and arising nearly opposite a simple spine-like branchlet; lateral branches having a long curved spine subtending 2 branchlets which bear on their incurved margin 2 to 3 subdivisions having subulate teeth on their adaxial side; tetrasporangia in short and dichotomously divided lateral branchlets, scattered along the margins of the older lateral branches.

For additional figures see Simons, 1964, p. 190, fig. 8.

Occurs often in some abundance as large and well-developed plants on rocky platforms and calcareous cobbles at depths from 8 to 28 m. Smaller and more stunted individuals are found in shallower water and on rare occasions it grows in the sublittoral fringe.

Distribution. World: widespread in many warm temperate and tropical seas. West Africa: Ghana: John *et al.* 1977, Lieberman *et al.* 1979.

Family Gracilariaceae

Gracilaria Greville 1830

Plants consisting of terete, compressed or flattened branches, fleshy to cartilaginous, dichotomously or irregularly divided and occasionally proliferous, attached by a discoid holdfast; branches consisting of a medulla or large parenchymatous cells and a narrow cortex of cells becoming progressively smaller outwards; tetrasporangia scattered and embedded in the outer cortex, tetrapartitely divided; spermatangia in small pit-like depressions scattered over the surface; cystocarps prominent and spherical to conical, with the gonimoblast variously shaped in section and often connected to the thick pericarp by filaments, possessing an apical pore.

World Distribution: cosmopolitan.

This genus is notoriously polymorphic and many investigators have stressed the need for obtaining as large amount of material as possible before attempting to make specific determinations. Even so the separating of species based solely on vegetative characters is unsatisfactory and many of the names in the literature will undoubtedly eventually prove to be quite superfluous. Bodard (1966a, 1967) has attempted a revision of the West African species based largely on cystocarp characters but there is still much confusion regarding the delimitation of species found in the region.

Key to the Species

1. Plants with the branches terete and filiform 2
1. Plants with the branches largely subterete or compressed 3

2. Branches and cystocarps greater than 1.5 mm in diameter; pericarp cells frequently anastomosing .... *G. camerunensis*
2. Branches and cystocarps usually less than 1.5 mm in diameter; pericarp cells not usually anastomosing .... *G. verrucosa*

3. Branches subterete and usually less than 2 mm in width, occurring close together in upper parts thus giving the plant a fan-like or corymbose appearance. *G. ferox*
3. Branches subterete to compressed and usually 2 mm or more in width, with plants not having a fan-like or corymbose appearance 4

Plate 31: Figs 1, 2. *Gracilaria dentata:* Variations in habit of this repeatedly dichotomously or occasionally pinnately divided plant. Fig. 3a-d. Cystocarp structure of *Gracilaria* spp. (after Bodard, 1967): a. *Gracilaria disputabilis;* b. *Gracilaria foliifera;* c. *Gracilaria dentata;* d. *Gracilaria camerunensis.* Fig. 4. *Caulacanthus ustulatus:* Portion of a plant showing the irregularly divided branches and spine-like ultimate divisions. Figs 5, 6. *Plocamium telfairiae:* 5. Portion of a plant showing the lateral branchlets often arising almost opposite a simple spine; 6. Portion of a lateral branch showing the curved branchlets and the marginal teeth.

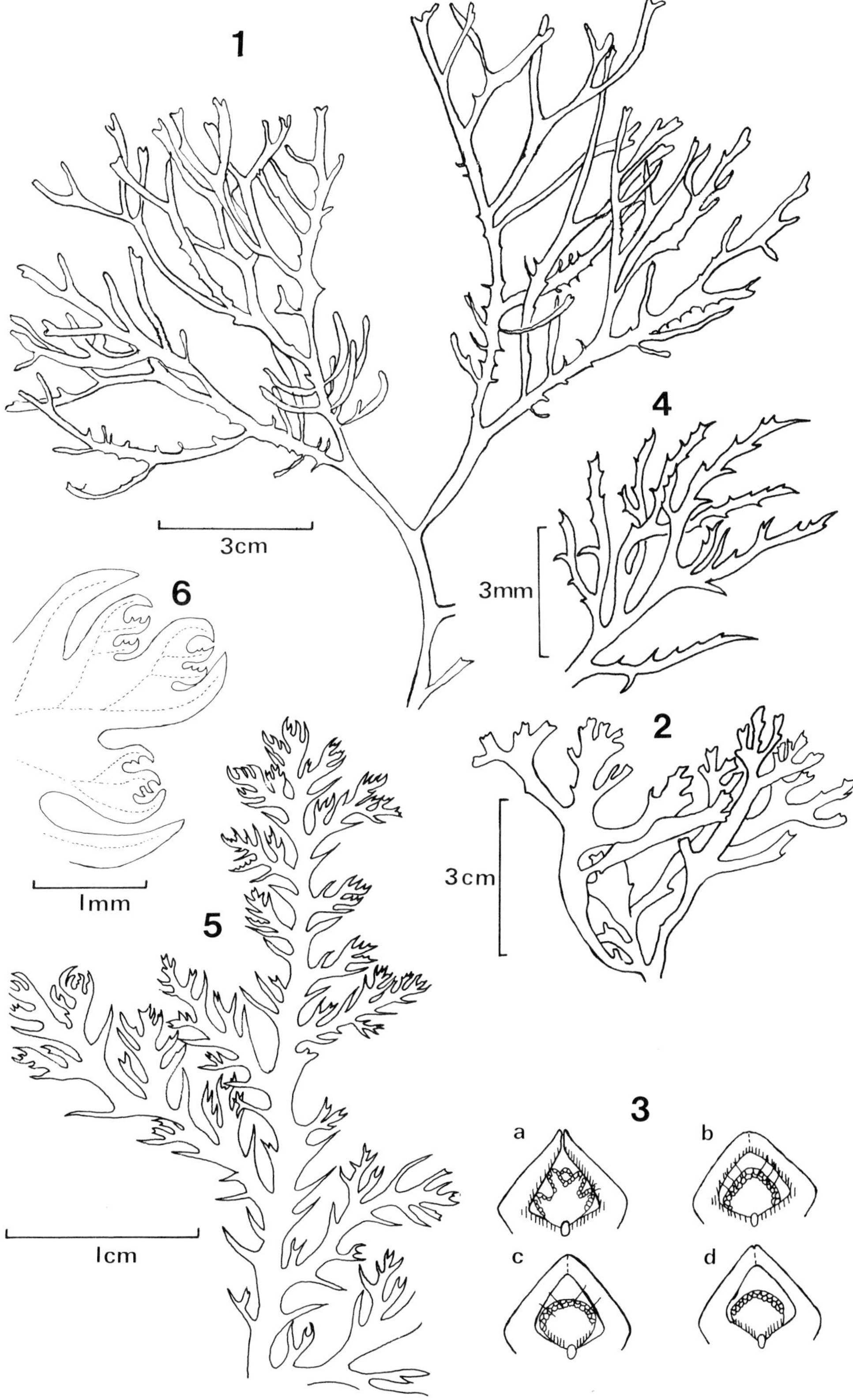

1
3cm
4
3mm
6
2
3cm
1mm
5
3
a
b
c
d
1cm

4. Plants fleshy, but membranaceous when dry; branches from 1-4 cm in width . . . . *G. foliifera*
4. Plants usually thick and fleshy, but not becoming membranaceous on drying; branches usually less than 1 cm wide 5

5. Plants bright red and not changing colour on drying, adhering well to herbarium paper; branches with few or no lateral proliferations . . . . *G. henriquesiana*
5. Plants dark purplish-red when not bleached, turning brownish on drying, not readily adhering to herbarium paper; branches commonly having many marginal proliferations 6
6. Cystocarps having the regularly shaped gonimoblast connected to the pericarp by a number of filaments . . . . *G. dentata*
6. Cystocarps having the irregularly shaped gonimoblast attached to the pericarp by its lobes . . . . *G. disputabilis*

Gracilaria camerunensis Pilger — Pl. 31, fig. 3d.

Pilger, 1911.

Plants soft and limp, to about 50 cm in length, with the branches much entangled and divided towards the base; upper branches only sparingly divided in the lower parts and simple or rarely divided above, only proliferous at damaged tips, filiform throughout and to about 2(-3) mm in diameter, tapering somewhat towards both the base and the apex; cystocarps in groups on the upper branches, usually from 1.5-3 mm in diameter, with the cells of the pericarp mostly anastomosing and not connected to the gonimoblast by filaments.

For additional figures see Bodard, 1967, p. 873 to 877, figs 1-7, pl. 1, figs 1-3 (as *Gracilariopsis camerunensis*).

The only information on the ecology of this species in the region is that it was found growing through beach sand presumably in the eulittoral zone.

Distribution. World: in warm temperate and tropical parts of the eastern Atlantic Ocean. West Africa: Cameroun: De Toni 1924, Pilger 1911, Schmidt & Gerloff 1957; *Gracilariopsis camerunensis*, Bodard 1966a, 1967.

Gracilaria dentata J. Agardh — Pl. 31, figs 1, 2, 3c.

J. Agardh, 1852.

Plants usually coarse and erect, up to 20 cm in height, reddish-purple, sometimes brownish, greyish or bleached green; branches repeatedly dichotomously divided or on occasion pinnate, subterete to markedly compressed, to about 5 mm in width, elongate and cuneate or linear in shape, with the tips often furcate and usually having many marginal proliferations; spermatangia arising in globular depressions having 3 to 6 cavities; cystocarps prominent, to about 1.5 mm in diameter, with the gonimoblast regular in shape and attached to the pericarp by a number of connecting filaments.

For additional figures see Ohmi, 1968, pl. 1a-c, pl. 2, figs 1-6 (as *G. henriquesiana*).

Occurs on moderately sheltered to moderately wave-exposed rocks as clumps or forming extensive patches of small individuals growing on the leeward side of rocks where wave action is less severe. This is one of the most common algae in the lower eulittoral subzone with the largest and bushiest individuals occurring in sheltered tidepools.

Distribution. World: probably in most warm temperate and tropical seas. West Africa: Cameroun: Bodard 1966a, Fox 1957, John 1977a, Pilger 1911, Steentoft 1967. Côte d'Ivoire: Bodard 1966a, John 1972c, 1977a. Gabon: John & Lawson 1974a. Gambia: John & Lawson 1977b, Steentoft 1967. Ghana: Fox 1957, John 1977a, John & Asare 1975, John & Lawson 1972a, John & Pople 1973, Steentoft 1967; *G. henriquesiana*, Dickinson & Foote 1950, Lawson 1953, 1956, 1957b, 1966, Levring 1969, Ohmi 1968, Stephenson & Stephenson 1972. Liberia: De May *et al.* 1977, John 1972c, 1977a. Nigeria: Fox 1957, John 1977a, Steentoft 1967. São Tomé: Carpine 1959, Fox 1957, John 1977a, Steentoft 1967. Sénégal (Casamance): Bodard 1966a.

Remarks. The distinction between this species and *Gracilaria henriquesiana* appears to be based solely on vegetative features (see Steentoft, 1967) and it is possible that they are no more than forms of the same plant. This species does not adhere well to herbarium paper or retain its natural colour on drying. In the absence of the original material involved it is impossible to attribute the record of *Fucus aeruginosus* Turner from "Danish Guinea" (now Ghana) by Hornemann (1819) to any species of *Gracilaria*. The original material might have involved *Gracilaria dentata* and/or *G. foliifera* and even a third species, *G. corticata* (see Price *et al.*, 1983).

Gracilaria disputabilis (Bodard) Bodard Pl. 31, fig. 3a.

Bodard, 1967; *?Gracilariopsis disputabilis* Bodard, 1966a.

Plants forming loose clumps, to about 20 cm in height, reddish-purple or on occasion bleached to a greenish colour; branches dichotomously divided and flattened and usually linear, with the tips blunt or acute and commonly with marginal proliferations; cystocarps having an irregularly shaped gonimoblast attached directly to the pericarp by lobes.

For additional figures see Bodard, 1966a, 54, pl. 8, figs a-h (as *Gracilariopsis disputabilis*).

Occurs in much the same situations as *G. dentata* but the two species are impossible to separate on the shore.

Distribution. World: in warm temperate and tropical parts of the eastern Atlantic Ocean. West Africa: Côte d'Ivoire: John 1977a; *?Gracilariopsis disputabilis*, Bodard 1966a; *Gracilariopsis tridactylites*, Bodard 1966a. Ghana: John 1977a. Sénégal (Casamance): *?Gracilariopsis disputabilis*, Bodard 1966a; *Gracilariopsis tridactylites*, Bodard 1966a.

Gracilaria ferox J. Agardh Pl. 32, figs 2, 3.

J. Agardh, 1852.

Plants erect and bushy, from 4-6 cm in height, reddish-brown; branches arising at close intervals in upper parts and giving plant a fan-like or corymbose appearance, terete to subterete, to about 2 mm in width, with the apices obtuse to acute.

Occurs on moderately wave-exposed rocks in the lower part of the lithothamnia subzone of the eulittoral zone and in the beds of *Sargassum* forming the sublittoral fringe.

Distribution. World: in warm temperate and tropical seas. West Africa: ?Bioko: unpublished. Cameroun: Hoppe 1969, Lawson 1955, Stephenson & Stephenson 1972. Ghana: Cordeiro-Marino 1978.

?**Gracilaria foliifera** (Forsskål) Børgesen Pl. 31, fig. 3b; Pl. 32, fig. 4.

Børgesen, 1932; *Fucus foliifer* Forsskål, 1775.

Plants usually solitary, from 8-20 cm in height, often bright red; branches several times dichotomously or occasionally alternately divided, flattened and fleshy, becoming membranaceous on drying, from 2-4 cm wide in upper parts, with the apices rounded and having proliferations from the margin; cystocarps strongly projecting and subspherical, to about 1 mm in diameter, having a very distinct pore, with the gonimoblast triangular in shape.

For additional figures see Bodard, 1966a, p. 52, pl. 6 figs a-j.

Occurs only in deep water in the region on rocky platforms and calcareous cobbles at depths ranging from 10 to 28 m; growing on the dead shells of *Turritella* at still greater depths.

Distribution. World: probably widespread in warm temperate and tropical seas. West Africa: Ghana: John *et al.* 1977, Lieberman *et al.* 1979.

Remarks. Bodard (pers. comm.) believes that the West African plants referred to this species closely resemble the Canary Island plant identified as *Gracilaria lacinulata* by Børgesen (1929). Bodard (1966a) suggests that most of the Sénégal plants formerly called *G. dentata* should more correctly be referred to the European *G. foliifera*, and that some of the African and Canary plants which had been given this latter name may possibly be a new, and as yet, undescribed species.

Gracilaria henriquesiana Hariot

Hariot, 1908.

Plants usually erect, to about 15 cm in height, bright red; branches irregularly dichotomously divided and often pseudosecund, somewhat fastigate, subterete to compressed, linear to cuneate, with the apices acute to obtuse and having many small teeth often around the tip; cystocarps to about 1 mm in diameter, often confined to the margin.

For figures see Steentoft, 1967, pl. 1, 2.

There is no information on the ecology of this species in the region.

Distribution. World: in the tropical parts of the eastern Atlantic Ocean. West Africa: São Tomé: De Toni 1924, Hariot 1908, Steentoft 1967; *G. henriquesii,* Henriques 1917; *G. wrightii,* Henriques 1886b, 1887, 1917.

Remarks. See the note under *Gracilaria dentata.*

Gracilaria verrucosa (Hudson) Papenfuss Pl. 32, fig. 1.

Papenfuss, 1950; *Fucus verrucosus* Hudson, 1762.

Plants forming loose, low and entangled clumps, extensive mats or occasionally solitary, to about 30 cm in length, yellowish- to purplish-brown, usually attached by numerous rhizoidal filaments ending in discoid holdfasts; branches sparingly divided and only occasionally opposite or dichotomous, rarely simple, terete and filiform, from 0.5-2 mm in diameter, with the branches always attenuate towards the apex and sometimes at the base, sometimes bearing short spine-like lateral branches; cystocarps ovoid to hemispherical, to about 1.5 mm in diameter, often numerous and very prominent.

For additional figures see Newton, 1931, p. 430, fig. 258 (as *G. confervoides*).

Occurs growing through sand on the floor of sheltered to moderately wave-exposed tidepools in the lower part of the eulittoral zone; solitary plants are to be found on calcareous cobbles at depths down to 25 m.

Distribution. World: widespread from boreal-antiboreal to tropical seas. West Africa: Cameroun: John 1977a, Steentoft 1967; *Gracilaria confervoides,* Goor 1923, Pilger 1911, Schmidt & Gerloff 1957. Côte d'Ivoire: John 1972c, 1977a. Ghana: John 1977a, John *et al.* 1977, Lieberman *et al.* 1979, Steentoft 1967; *G. confervoides,* Dickinson & Foote 1950, Lawson 1956. Liberia: De May *et al.* 1977. São Tomé: John 1977a, Steentoft 1967; *G. confervoides,* Carpine 1959. Sénégal (Casamance): *G. confervoides,* Chevalier 1920. Sierra Leone: John & Lawson 1977a; *G. confervoides,* Aleem 1978. Togo: John 1977a, John & Lawson 1972b.

Family Phyllophoraceae

Gymnogongrus Martius 1833

Plants usually erect and bushy, consisting of terete to flattened branches, repeatedly dichotomously divided and occasionally with lateral proliferations; branches composed of a medulla of large angular cells and a cortex of sparingly divided anticlinal rows of smaller cells; tetrasporangial phase when present having swollen nemathecia, or as pustular tetrasporophytes arising from a gonimoblast, sporangia irregularly zonate; cystocarps immersed, globose, without a pore.

World Distribution: cosmopolitan.

This genus cannot be treated in anything like a satisfactory manner not only in the region but in most parts of the world. Many plants which cannot readily be referred to a species may be no more than unusual growth forms reflecting the great degree of morphological plasticity in the genus (see *G. nigricans*).

Key to the Species

1. Branches terete throughout . . . . *G. intermedius*
1. Branches flattened, at least in the upper parts 2

2. Branches usually from 1-2 mm in width; plants on drying showing little change in colour . . . . *G. tenuis*
2. Branches less than 1mm in width; plants becoming black on drying . *G. nigricans*

Gymnogongrus intermedius Kylin

Kylin, 1938.

Plants somewhat bushy, from 2-3 cm in height, cartilaginous in texture; branches repeatedly dichotomously divided and occasionally polychotomous in the apical region, fastigiate distally, terete and to about 0.5 mm or more in diameter; cystocarps to about 400 μm in diameter, projecting on both sides of a branch.

For figures see Kylin, 1938, pl. 4, fig. 10.

Occurs as "tufts on rocks in the surf zone" according to Fox (1957).

Distribution. World: in warm temperate and tropical parts of the eastern Atlantic Ocean. West Africa: Côte d'Ivoire: ?John 1977a. Nigeria: Fox 1957, John 1977a.

Gymnogongrus nigricans P. Dangeard Pl. 32, fig. 5.

Dangeard, 1952.

Plants bushy, to about 5 cm in height, cartilaginous in texture and black on drying; branches usually regularly dichotomously divided throughout, terete below and flattened in the upper parts, from 0.4-0.8 mm in width; cystocarps to about 0.5 mm in diameter, projecting more on one side of a branch than the other.

Plate 32: Fig. 1. *Gracilaria verrucosa:* Portion of the sparingly divided and filiform plant bearing prominent cystocarps. Figs 2, 3. *Gracilaria ferox:* 2. Plant showing the close branching giving the upper portions a fan-like outline; 3. Terminal portion of a branch with the apices obtuse to acute. Fig. 4. *Gracilaria foliifèra:* Plant growing on a *Turritella* shell and bearing prominent cystocarps on the flattened branches. Fig. 5. *Gymnogongrus nigricans:* Plant showing the regularly dichotomously divided and somewhat flattened branches. Fig. 6. *Gymnogongrus tenuis:* Plant with the flattened branches gradually tapering towards their tips. Fig. 7. *Gigartina acicularis:* Plant with irregularly distichous or pinnate branching, becoming secund and patent towards tips.

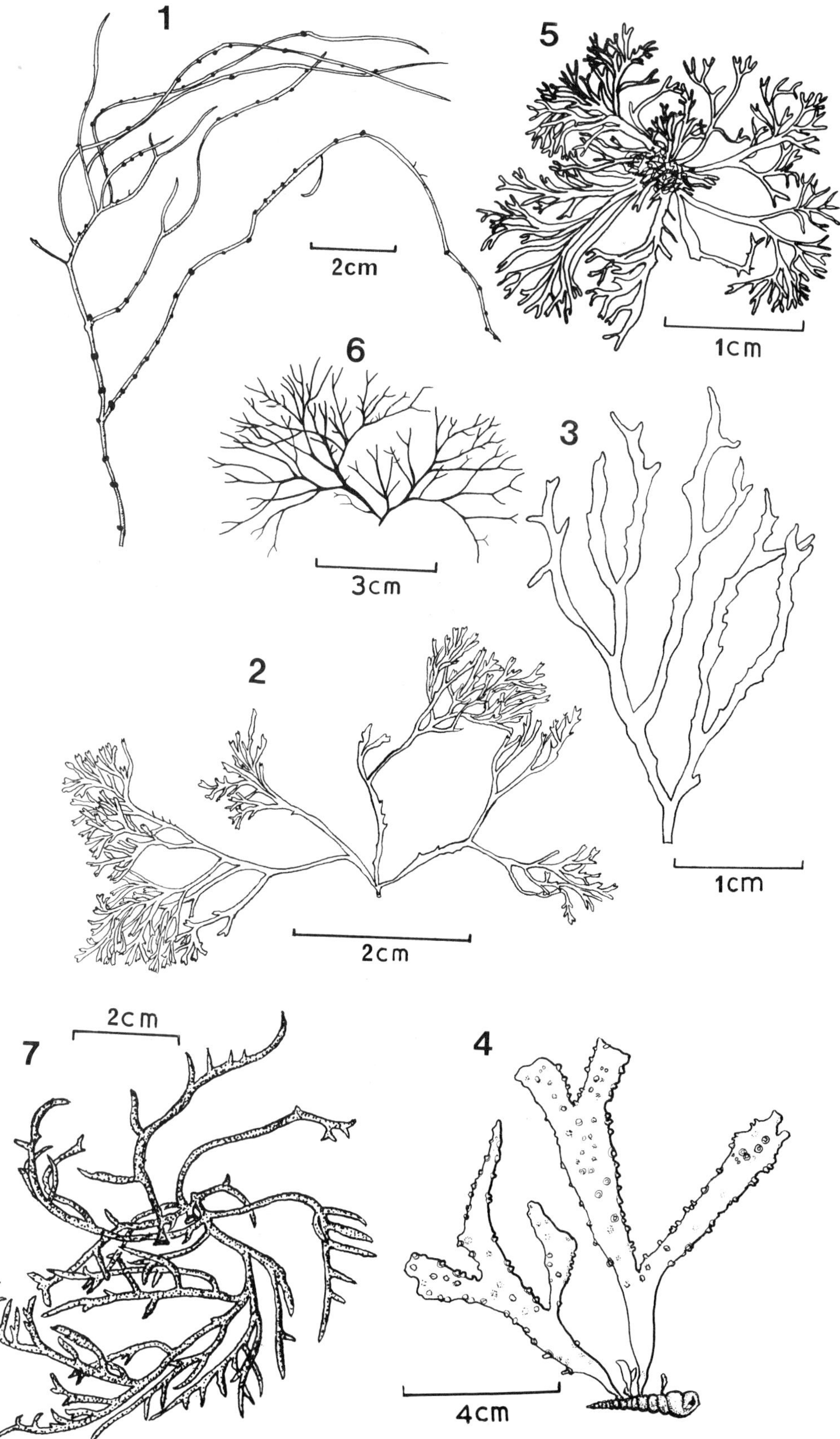
1
2cm
5
1cm
6
3cm
3
1cm
2
2cm
2cm
7
4
4cm

For additional figures see Dangeard, 1952, p. 319, pl. 18, figs a-c.

Occurs as solitary plants or with other small algae in the turf growing in the lower eulittoral subzone over a wide range of exposure to wave action.

Distribution. World: in warm temperate and tropical parts of the eastern Atlantic Ocean. West Africa: Benin: John 1977a, John & Lawson 1972b. Bioko: unpublished. Cameroun: Hoppe 1969, John 1977a, John & Lawson 1972a, Lawson 1955, Stephenson & Stephenson 1972. Côte d'Ivoire: John 1977a. Ghana: John 1977a, John & Lawson 1972a. Liberia: De May *et al.* 1977, John 1977a.

Remarks. We have found growing along a breakwater in Benin plants most probably referable to this species even though they appear to be rather different in gross habit. The plants on the wave-exposed side were compact and closely branched whilst those on the sheltered side were larger and looser.

Gymnogongrus tenuis (J. Agardh) J. Agardh Pl. 32, fig. 6.

J. Agardh, 1850; *Chondrus tenuis* J. Agardh, in Kützing, 1849.

Plants erect and bushy, from 2-4 cm in height, cartilaginous to nearly membranaceous; branches dichotomously divided and occasionally polychotomously in the apical region, usually flattened, to about 2 mm in width, linear in shape, and narrower towards the tips; cystocarps spherical to oval, from 0.5-0.8 mm in diameter.

For additional figures see Børgesen, 1919, p. 357, fig. 352.

Occurs on rocks in wave-exposed situations in the eulittoral zone.

Distribution. World: widespread in warm temperate and tropical seas. West Africa: Côte d'Ivoire: John 1977a. Gambia: John 1977a, John & Lawson 1977b. Liberia: De May *et al.* 1977, John 1977a. Nigeria: Fox 1957, John 1977a.

Doubtful Record

Gymnogongrus griffithsiae (Turner) Martius

The only reports of this species from the tropical West African coast is the very early one as "*F.* [*Fucus*] *Griffithsiae*" by Hornemann (1819) from "Danish Guinea"(present-day Ghana) and recently by Aleem (1978) from the Sierra Leone peninsula. Though known as far south as Mauritanie (see Lawson and John, 1977) these two isolated reports are regarded with doubt until the material on which they are based can be examined. It might prove impossible to verify the Hornemann record since part of the Isert collection was destroyed by fire in 1807.

Family Gigartinaceae

Chondrus Stackhouse 1797

Doubtful Record

Chondrus crispus Stackhouse

We regard with doubt the very early record of this species (as *Fucus crispus* Linnaeus) from tropical West Africa by Hornemann (1819) based on material collected by Isert from "Danish Guinea" (now Ghana). It is widely distributed in colder waters occurring on the eastern side of the Atlantic as far south as southern Spain and possibly Morocco and the Cape Verde Islands (Dixon and Irvine, 1977a).

Gigartina Stackhouse 1809

Plants consisting of terete to flattened branches, fleshy to cartilaginous, subdichotomously, irregularly or pinnately divided; branches composed of a medulla of anastomosing filaments or of cells of various shapes and a cortex of dichotomously branched anticlinal rows of cells becoming progressively smaller towards the surface; tetrasporangia in flattened to globose sori on papillae, developing from the innermost cortical cells, tetrapartite; spermatangia arising in irregular sori on the surface of papillae; cystocarps very prominent, bulging above the surface or on short papillae.

World Distribution: widespread from boreal-antiboreal to tropical seas.

Gigartina acicularis (Roth) Lamouroux — Pl. 32, fig. 7

Lamouroux, 1813; *Ceramium acicularis* Roth, 1806.

Plants forming clumps or mats, from 3-5 cm in height, dark reddish-purple to brownish, with the branches often bleached to yellowish-green and sometimes having a banded appearance; branching sparingly irregularly dichotomous or pinnate, secund and patent near the branch tips; branches terete or somewhat subterete towards the apices, to about 2 mm in diameter, partly procumbent with the ultimate branches often recurved and forming secondary small attachment discs upon contact with a substratum.

For additional figures see Dixon and Irvine, 1977a, p. 238, fig. 87a, b; Taylor, 1960, p. 783, pl. 60, fig. 6.

Occurs as small cushion-like clumps on very wave-exposed rocks and as extensive, loose mats where exposure is less. Found in the lower part of the upper eulittoral subzone and as often pure stands on steeply sloping rock surfaces in the lithothamnia subzone.

Distribution. World: widespread in warm temperate and tropical seas. West Africa: Bioko: unpublished. Cameroun: Dixon & Irvine 1977a, Dizerbo 1974, Hoppe 1969, Lawson 1955, 1966, Stephenson & Stephenson 1972. Gambia: John & Lawson 1977b. Ghana: Dickinson & Foote 1950, Dizerbo 1974, Hoppe 1969, John & Asare 1975, Lawson 1956, 1957b, 1966, Stephenson & Stephenson 1972; *Fucus acicularis*, Hornemann 1819. Guinée: ?Lawson 1966, ?Sourie 1954, Stephenson & Stephenson 1972. Liberia: De May *et al.* 1977. São Tomé: Carpine 1959. Sierra Leone: Aleem 1978, John & Lawson 1977a.

Order Cryptonemiales* ** · Family Corallinaceae***

Articulated Forms

Amphiroa Lamouroux 1812

Plants forming erect to prostrate clumps, with the erect branches irregularly or di- or trichotomously divided, consisting of heavily calcified, terete or flattened segments jointed by flexible articulations; segments in median section composed of a medulla of one to several transverse rows of long cells alternating with rows of small ovoid cells and corticated by a number of layers of small cells; articulations in section of several, more rarely one, transverse rows of cells, corticated; conceptacles often somewhat projecting and scattered over the surface of the segments, with a terminal pore.

World Distribution: widespread in warm temperate and tropical seas.

Key to the Species

1. Branches terete throughout .... *A. rigida*
1. Branches flattened, at least in the upper parts 2

2. Branches irregularly dichotomously divided, with one of the forks reduced to 2 or 3 segments and the other fully developed .... *A. linearis*
2. Branches regularly dichotomously divided throughout 3

Callophyllis lecomtei Hariot (Kallymeniaceae) is so far only known from the Congo Republic (De Toni, 1897; Hariot, 1895, 1896). To the genus *Callophyllis* has been referred a sterile membranaceous red alga found at about 10 m below low water off Ghana (John *et al.*, 1977, 1980; Lieberman *et al.*, 1979).

**There has been collected off Ghana a lubricous alga having zonate tetrasporangia but no other reproductive organs have been found. It is probable that this is a member of the family Dumontiaceae but without sexual material it is not possible to assign it to a genus. *Dudresnaya* is a member of this family and Bodard (1971) reports a "*Dudresnaya* aff. *crassa*" growing off the coast of Sénégal and Lieberman *et al.*, (1979) a "*Dudresnaya* sp." off Ghana.

****Dermatolithon papillosum* (Zanardini). Foslie var. *cystoseirae* (Hauck) Lemoine is known from the island of Pagalu (Lemoine, *pers. comm.*), and is probably the same plant she published with an unspecified location from the Gulf of Guinea region of tropical West Africa (Lemoine, 1965).

3. Plants low and semi-prostrate; segments from 1.5-2 mm in width in the upper parts and the articulations black and conspicuous . . . . *A. peruana*
3. Plants always erect; segments less than 0.6 mm in width and articulations not particularly conspicuous . . . . *A. beauvoisii*

Amphiroa beauvoisii Lamouroux Pl. 33, fig. 1.

Lamouroux, 1816.

Plants forming erect clumps, up to 5 cm in height; branching regularly dichotomous; segments terete towards the base of the branches, about 400 μm in diameter, becoming flattened above and with surface striations, from 600-650 μm in width; segments composed of a medulla of 2 to 3(-4) rows of long cells, from 50-110 μm long, alternating with a row of short cells to about 35 μm in length; articulations of 2 rows of long cells.

For additional figures see Hamel and Lemoine, 1953, p. 42, fig. 7, pl. 5, figs 3-6.

Occurs on moderately wave-exposed rocks in the lower eulittoral subzone and to a depth of 10 m growing on rocky platforms.

Distribution. World: in warm temperate and tropical parts of the Atlantic Ocean as well as in the Mediterranean. West Africa: Côte d'Ivoire: John 1972c, 1977a. Gambia: John & Lawson 1977b. Ghana: John 1977a, John *et al.* 1980, Lieberman *et al.* 1979. Liberia: De May *et al.* 1977.

Remarks. According to Dr. A.W. Johansen (*pers. comm.*) this plant is probably identical to the species described and illustrated by Pilger (1919) as *Amphiroa annobonensis* from the offshore island of Pagalu.

Amphiroa linearis Kützing Pl. 33, fig. 6.

Kützing, 1858.

Plants usually greater than 5 cm in height; branching irregularly dichotomous, with one of the forks reduced and composed of only 2 to 3 segments while the other fork develops normally and is further subdivided; segments usually somewhat flattened, from 0.5-1.5 mm in width and 2 to several times as long as broad; segments having at least 3 rows of long cells alternating with 1 row of short cells.

For additional figures see Kützing, 1858, pl. 46, fig. 2a-c.

There is no information on the ecology of this plant collected from Gabon.

Distribution. World: so far known only from tropical West Africa. West Africa: Gabon: De Toni 1905, Kützing 1858.

Amphiroa peruana Areschoug Pl. 33, fig. 2.

Areschoug, 1854.

Plants low and semi-prostrate, forming rounded clumps or more extensive patches, to about 3 cm in height; branching irregularly dichotomous and distichous; segments subterete in the lower parts of the branches and becoming above compressed, angular or subcuneate, from 1.5-2 mm in width and 3(-4) mm in length, with the articulations black and conspicuous; segments consisting of 2 rows of long cells from 45-100 μm in length with 1 row somewhat shorter than the other and alternating with a short row usually less than 45 μm in length; articulations consisting of a number of transverse rows of cells.

For additional figures see Taylor, 1945, p. 417, pl. 50, figs 1, 2.

Occurs on moderately sheltered to moderately wave-exposed rocks in the lower eulittoral subzone; occasionally in shallow tidepools.

Distribution. World: in tropical parts of the Atlantic and Pacific Oceans. West Africa: Côte d'Ivoire: John 1972c, 1977a. Ghana: John 1977a; ***A. brevianceps,*** John & Lawson 1972a. Liberia: John 1972c, 1977a.

Amphiroa rigida Lamouroux Pl. 33, fig. 5.

Lamouroux, 1816.

Plants forming a loosely matted turf, from 1-2 cm in height; branching irregular and somewhat lateral; segments terete throughout, from (150-)250-450 μm in diameter and sometimes to about 4 mm in length; segments consisting of 2 (rarely 1) rows of long cells, from 50-100 μm in length and alternating with a row of shorter cells to 40 μm long; articulations made up of 2 transverse rows of cells.

For additional figures see Suneson, 1937, p. 48, 50 to 52, figs 28-32, pl. 4, fig. 13; Hamel and Lemoine, 1953, p. 41, fig. 6, pl. 5, figs 3-6.

Occurs on moderately wave-exposed rocks in lower eulittoral tidepools, often entangled with other small algae.

Distribution. World: in warm temperate and tropical parts of the Atlantic Ocean as well as the Mediterranean. West Africa: Cameroun: unpublished. Ghana: John *et al.* 1977, Lieberman *et al.* 1979; *A. cryptarthrodia,* John & Lawson 1972a.

Plate 33: Fig. 1. *Amphiroa beauvoisii:* Portion of a plant with the conceptacles borne laterally on the subterete segments. Fig. 2. *Amphiroa peruana:* Plant showing conceptacles on the flattened segments and with conspicuous articulations between consecutive segments. Figs 3, 4. *Jania verrucosa:* 3. Habit of plant; 4. Portion of a plant showing the inflated terminal segments containing the conceptacles. Fig. 5. *Amphiroa rigida:* Portion of a plant with the branching somewhat irregular and lateral. Fig. 6. *Amphiroa linearis* (after Kützing, 1858): Portion of a plant showing the irregularly dichotomously divided branches having one fork normally developed and the other reduced to 2 or 3 segments. Fig. 7. *Corallina pilulifera:* Portion of a plant showing the pinnate branching and the terminal conceptacles.

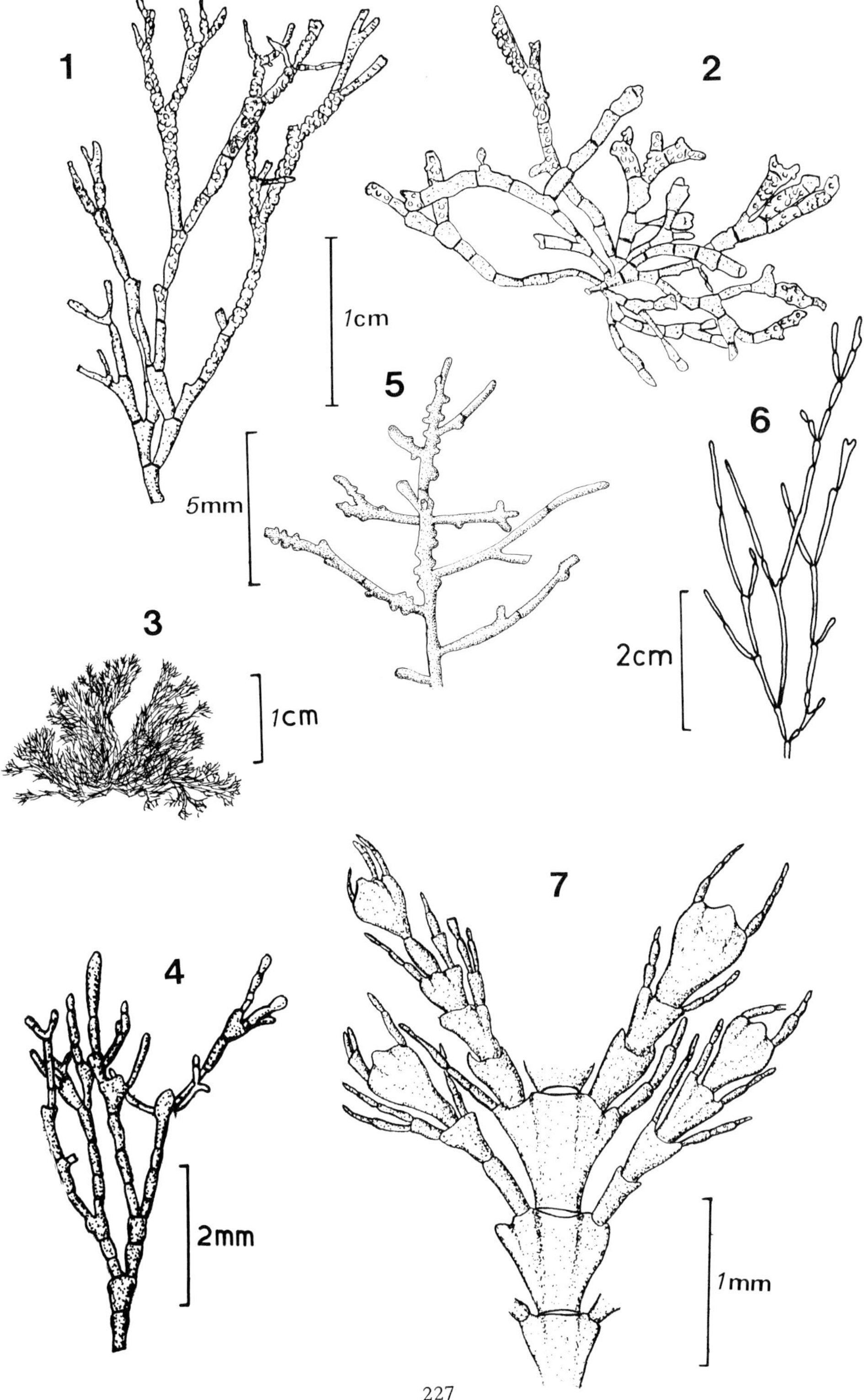
1
2
1cm
5
6
5mm
3
2cm
1cm
7
4
2mm
1mm

Doubtful Record

Amphiroa capensis Areschoug ex J. Agardh

This plant has been reported from the island of São Tomé by Henriques (1886b, 1887, 1917) but Steentoft (1967), who has examined this material, considers it best to leave it without specific determination as it consists of just a few fragments.

Corallina Linnaeus 1759

Plants usually erect, with the branches generally distichous and oppositely pinnate, consisting of terete or flattened and heavily calcified segments, articulated, arising from a crustose basal layer; segments in longitudinal section composed of a medulla of transverse rows of cells and an outer cortication; articulations in longitudinal section composed of a single row of long and thick-walled cells; tetrasporic and carposporic conceptacles having a broad fertile area on the floor and the former containing about thirty to forty mature sporangia per conceptacle; spermatangial conceptacles with the fertile area confined to the lower portion of a bulbose chamber having a long exit canal.

World Distribution: cosmopolitan.

Key to the Species

1. Plants often having the lower branches somewhat prostrate and irregularly divided; conceptacles usually having forked branchlets arising from the adjacent segment. . . . . *C. pilulifera*
1. Plants with the erect branches arising directly from a crustose base; conceptacles without flanking branchlets. . . . . *C. mediterranea*

Corallina mediterranea Areschoug

Areschoug, in J. Agardh, 1852.

Plants in bushy clumps, generally from 3-4 cm in height; branches often having a somewhat pyramidal outline, distichous and pinnate, with the segments subterete to compressed, ovate or subtriangular, about 0.5-1.5 mm broad and to 1 mm or more in length; branchlets arising at an angle from the upper corners of the flattened segments, simple and usually 1 to 3 segments long; conceptacles terminal on the branchlets, pyriform or ovoid or elongated at the apex, bearing 1 or more horn-like projections about 2 segments long.

For figures see Hamel and Lemoine, 1953, pl. 1, fig. 3, pl. 2, figs 1, 2, 6.

There is no information on the ecology of this species in the region.

Distribution. World: from warm temperate to tropical parts of the eastern Atlantic Ocean as well as the Mediterranean. West Africa: Sénégal (Casamance): Chevalier 1920.

Corallina pilulifera Postels & Ruprecht Pl. 33, fig. 7.

Postels and Ruprecht, 1840.

Plants forming erect clumps or occasionally forming extensive carpets, from 2-4(-10) cm in height; lower branches somewhat prostrate, entangled and irregularly divided, with the segments terete to subterete, about 0.5 mm in width and to about 1 mm in length; erect branches distichous and pinnate, with the compressed segments cuneate and to about 1.5(-3) mm in width; branchlets arising at acute angles from the upper corners of the flattened segments, simple or furcate and 2 to 3 segments long; conceptacles terminal on the branchlets, ovoid and having horn-like projections of from 2 to 3 segments, with forked branchlets sometimes arising on each side of a conceptacle.

For additional figures see Yendo, 1902, pl. 3, figs 14-16, pl. 7, figs 14-16.

Occurs in the lower eulittoral subzone and particularly abundant on gently sloping rock surfaces subject to wave surge or in the outflow channels of tidepools where there is current surge.

Distribution. World: probably pantropical. West Africa: Côte d'Ivoire: John 1972c, 1977a. Gambia: John & Lawson 1977b. Ghana: John 1977a. Liberia: John 1972c, 1977a.

Haliptilon (Decaisne) Lindley 1846

Plants erect, distichous and pinnately branched, with the lateral branches much divided, arising from a crustose or rhizomatous base; branches consisting of terete or somewhat flattened segments, heavily calcified and with flexible articulations at the nodes; segments in longitudinal section composed of a medulla of irregular cells and the articulation in section having a single transverse row of cells; conceptacles immersed and arising terminally on inflated segments, with a single apical pore and sometimes having horned projections; tetrasporic conceptacles having a restricted fertile area on the floor and containing at maturity about four to eight sporangia; spermatangial conceptacles having a fertile area lining most of the elongate chamber and with a short exit canal; cystocarpic conceptacles having a restricted fertile area on the floor.

World Distribution: in warm temperate and tropical seas.

Haliptilon subulata (Ellis & Solander) Johansen Pl. 34, figs 1, 2.

Johansen, 1970; *Corallina subulata* Ellis & Solander, 1786.

Plants bushy, usually to about 4 cm in height; branching distichous and pinnate, with the lateral branches often many times divided and sometimes dichotomous towards the tips; segments of the main branches flattened and broadly deltoid, rarely terete and cylindrical, costate, with the upper ends usually truncate, to about 500 μm in width; branchlets arising at acute angles from the upper corners of the upper flattened segments, up to several segments long, simple or forked, terete and attenuate or somewhat spathulate; conceptacles ovate to elliptical, with or without horn-like projections that are often longer than the conceptacles.

For additional figures see Taylor, 1960, p. 763, pl. 50, figs 1, 2 (as *Corallina subulata).*

Occurs on larger algae or more rarely on beach rocks exposed to moderate wave action in the upper part of the lower eulittoral subzone.

Distribution. World: as for genus. West Africa: Ghana: unpublished. Liberia: De May *et al.* 1977.

Jania* Lamouroux 1812

Plants usually erect, with slender and terete branches regularly dichotomously divided, consisting of calcified segments and flexible articulations, with crustose or rhizomatous bases; segments in longitudinal section composed of irregularly arranged cells and the articulations in section of a single row of large cells; conceptacles immersed and arising terminally on inflated segments, with a single apical pore, sometimes with horned projections present; tetrasporic conceptacles having a restricted fertile area on the floor and containing about four to eight sporangia at maturity; spermatangial conceptacles having a fertile area lining most of the elongate chamber, with a short exit canal; cystocarpic conceptacles having a restricted area on the floor.

World Distribution: widespread in warm temperate and tropical seas.

Key to the Species

1. Segments from 200-400(-450) μm in diameter and 1-3 mm in length .... *J. crassa*
1. Segments from 30-210 μm in diameter and usually less than 1 mm in length ... 2

2. Segments never greater than 85 μm in diameter .... *J. capillacea*
2. Segments usually greater than 100 μm in diameter and if less then only in the upper branches.. 3

3. Some of the lower branches arcuate and attached terminally or laterally to the substratum by discoid holdfasts; branching usually at wide angles (>45°)........ .... *J. adhaerens*
3. All the branches normally erect and basally attached; branching at narrower angles ((45°0.. 4

4. Segments about 200 μm in diameter .... *J. verrucosa*
4. Segments from 80-150 μm in diameter............................ *J. rubens*

**Jania longifurca* Zanardini has been mentioned from the Congo in two papers by Hariot (1895, 1896) but under the name of *Corallina longifurca* in the earlier.

Jania adhaerens Lamouroux Pl. 34, fig. 3.

Lamouroux, 1816.

Plants forming semi-prostrate and spreading clumps, to about 1 cm in height, with some of the lower branches arcuate and attached to the substratum by a flat disc; branches repeatedly dichotomously divided at angles of from 45-75°, with conical apices; segments near the base of the plant from 100-210 μm in diameter and decreasing to 30-60 μm in the upper branches, from 2 to 8 times as long as broad.

For additional figures see Børgesen, 1917, p. 195 to 197, figs 184-187.

Occurs as variously sized clumps growing on rocks in the lower eulittoral subzone and found to a depth of 10 m on rocky platforms.

Distribution. World: as for genus. West Africa: Cameroun: Dangeard 1952, Schmidt & Gerloff 1957; *Corallina adhaerens,* Pilger 1911. Gambia: John & Lawson 1977b. Ghana: Lieberman *et al.* 1979. São Tomé: Carpine 1959, Steentoft 1967. Sierra Leone: Aleem 1978, John & Lawson 1977a.

Jania capillacea Harvey

Harvey, 1853.

Plants more or less erect, less than 1 cm in height, usually basally attached but occasionally discoid holdfasts developing laterally on some of the lower branches; branches repeatedly dichotomously divided at an angle of about 45°, with the tips somewhat conical; segments to 70(-85) μm in diameter throughout, from 4-10(-12) times as long as broad.

For figures see Børgesen, 1917, p. 199, fig. 188.

Occurs growing in clumps or dense mats of algae in the lower eulittoral subzone in moderately wave-sheltered situations.

Distribution. World: in tropical parts of the Atlantic and Pacific Oceans. West Africa: Gabon: John & Lawson 1974a.

Jania crassa Lamouroux Pl. 35, fig. 1.

Lamouroux, 1821.

Plants forming small clumps or occasionally extensive mats, to about 5 cm or more in height; branches regularly dichotomously divided at a narrow angle (<45°), with the upper branches occasionally recurved and the apices acute to conical; segments from 200-400(-450) μm in diameter and 1-3 mm in length.

For additional figures see Dawson, 1953, p. 227, pl. 27, figs 1, 2 (as *J. natalensis*).

Occurs over the lower eulittoral subzone as small clumps on wave-exposed rocks and forming large, extensive patches on moderately sheltered to moderately wave-exposed rocks, particularly where there is wave surge.

Distribution. World: as for genus. West Africa: Côte d'Ivoire: John 1977a. Gabon: John 1977a; *J. natalensis*, Schmidt 1929, Steentoft 1967. Ghana: John 1977a; *J. natalensis*, Lawson 1956, Stephenson & Stephenson 1972.

Jania rubens (Linnaeus) Lamouroux Pl. 34, figs 4, 5.

Lamouroux, 1816; *Corallina rubens* Linnaeus, 1767.

Plants forming entangled cushions or subcorymbose clumps, from 0.5-5 cm in height; branches regularly dichotomously divided at an acute angle; segments from (60-)80-150(-200) μm in diameter, short and barrel-shaped in the lower parts of the branches and somewhat cylindrical above, to 1-1.5 mm in length towards the apices.

For additional figures see Hamel and Lemoine, 1953, p. 38, fig. 4, pl. 3, figs 2, 2.

Occurs from the upper part of the lower eulittoral subzone down to a depth of about 10 m mixed together with other small algae or occasionally on larger algae, and growing in greatest abundance in moderately wave-sheltered situations.

Distribution. World: as for genus. West Africa: Cameroun: Fox 1957, Pilger 1911, Schmidt & Gerloff 1957; *Corallina rubens*, Goor 1923. Gambia: John & Lawson 1977b. Ghana: John *et al.* 1977, 1980, Lawson 1956, 1957b, 1966, Lieberman *et al.* 1979. Nigeria: Fox 1957. São Tomé: Steentoft 1967.

Jania verrucosa Lamouroux Pl. 33, figs 3, 4.

Lamouroux, 1816.

Plants forming dense clumps or mats, usually less than 2 cm in height; branches regularly dichotomously divided at an acute angle, fastigiate, with the apices rounded or somewhat club-shaped; segments about 200 μm in diameter throughout, less than 1 mm in length and usually 2 to 3 times as long as broad.

For additional figures see Harvey, 1863, pl. 251, figs 1-4 (as *J. fastigiata*).

Occurs in the algal turf or on larger algae growing in moderately sheltered to moderately wave-exposed situations in the lower eulittoral subzone.

Distribution. World: as for genus. West Africa: Côte d'Ivoire: *J. fastigiata*, John 1972c, 1977a. Ghana: Lieberman *et al.* 1979; *J. fastigiata*, John 1977a. Liberia: *J. fastigiata*, De May *et al.* 1977, John 1977a. Sierra Leone: John & Lawson 1977a.

Plate 34: Figs 1, 2. *Haliptilon subulata:* 1. Habit of plant; 2. Portion of plant showing the pinnate branching and terminal conceptacles sometimes with horn-like projections. Fig. 3. *Jania adhaerens:* Portion of a plant showing the arcuate lower branches bearing at intervals flat attachment discs. Figs 4, 5. *Jania rubens:* 4. Plant showing the dichotomous pattern of branching; 5. Portion of a branch showing the terete segments and the uncalcified articulations.

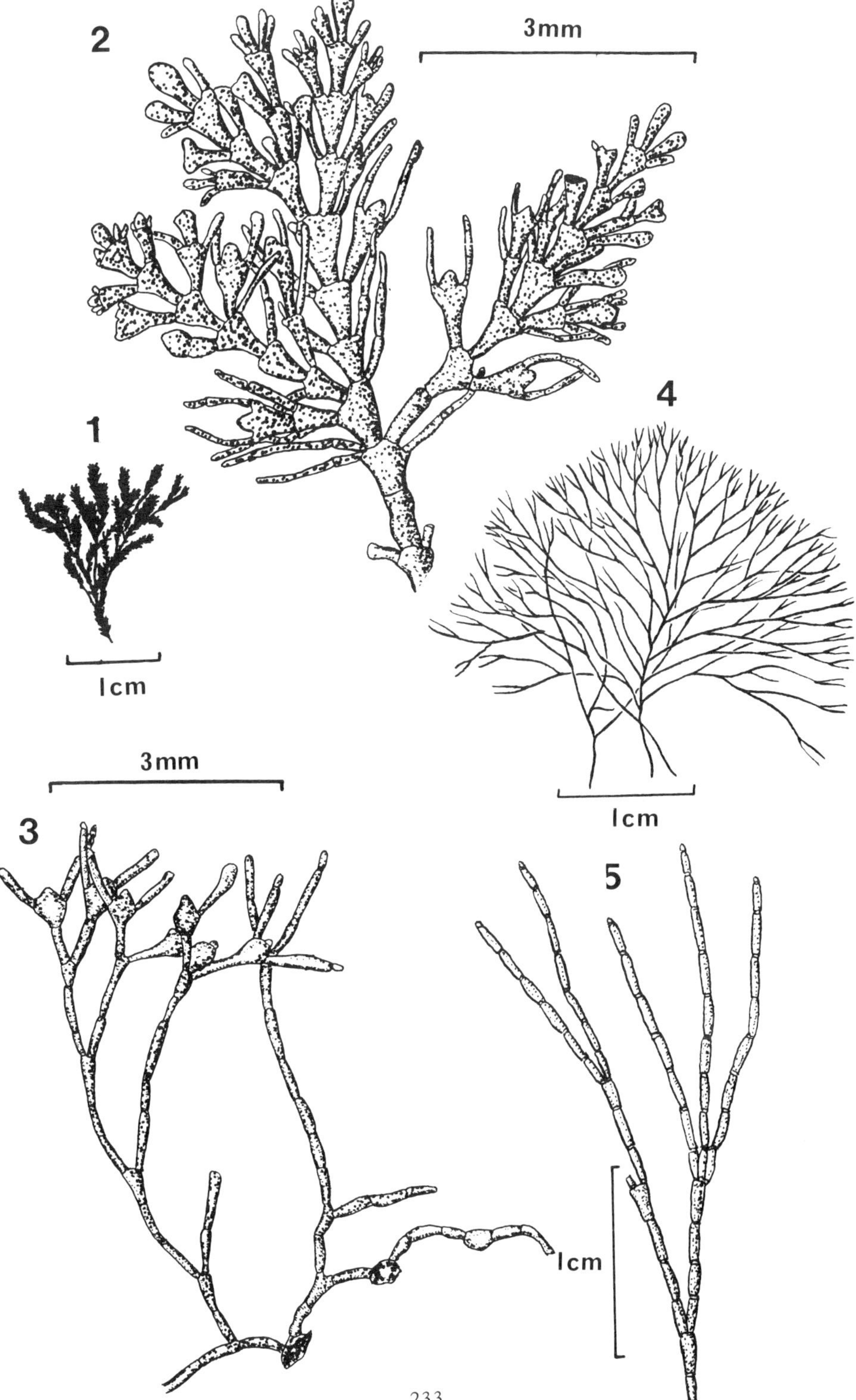
2
3mm
1
1cm
4
3mm
3
1cm
5
1cm

Crustose Forms*

Fosliella Howe 1920

Plants crustose, forming lightly calcified expansions; thallus consisting of a single-layered hypothallus and epithallus, always polystromatic in the vicinity of the conceptacles, with the filaments of the hypothallus radially arranged and usually closely united, often having hair-bearing swollen cells (trichocytes); intercell fusions present; asexual conceptacles conical or oblong-conical, with a single pore; cystocarpic conceptacles similar to the sporangial but slightly smaller.

World Distribution: cosmopolitan.

Key to the Species

1. Cells in surface view about 8-27 μm in length; trichocytes arising terminally on the filaments. . . . . *F. farinosa*
1. Cells in surface view from 6-10 μm in length; trichocytes usually absent . . . . *F. lejolisii*

Fosliella farinosa (Lamouroux) Howe — Pl. 35, figs 2, 3.

Howe, 1920; *Melobesia farinosa* Lamouroux, 1816.

Plants forming disc-like to irregular crusts, often somewhat crowded; thallus in surface view composed of irregularly disposed filaments of cells elongated radially; cells from (4-)6-10(-12) μm in diameter and (8-)13-17(-27) μm in length, usually bearing a smaller cell distally on their upper side; trichocytes arising terminally on the filaments, from 15-20 μm in diameter and 22-33 μm long; asexual conceptacles from (66-)100-110(-150) μm in diameter.

For additional figures see Lemoine, in Børgesen, 1917, p. 172, fig. 165 (as *Melobesia farinosa*).

var. **solmsiana** (Falkenberg) W.R. Taylor

Taylor, 1939; *Melobesia solmsiana* (non Crouan frat.) Falkenberg, 1879.

Plants forming a net-work of meandering filaments, sometimes growing together but not forming a regular disc, only attached occasionally at points of contact.

For figures see Solms-Laubach, 1881, pl. 1, figs 9, 12, 13 (as *Melobesia solmsiana*).

Occurs as pink spots on the surface of larger algae such as *Sargassum* growing in the eulittoral zone and down to a depth of 15 m. The species and variety developed

*Several species are reported by Lemoine (1964) simply as from the "Golfe de Guinée". See Price *et al.* (1983) for details concerning these records.

within a few days on plexiglass surfaces attached to the sea bed at a depth of about 15 m below the low water.

Distribution. World: as for genus. West Africa: Gabon: John & Lawson 1974a. Gambia: John & Lawson 1977b. Ghana: John & Lawson 1972a, Lieberman *et al.* 1979; *F. farinosa* var. *solmsiana,* unpublished.

Remarks. The confused nomenclatural status of the variety is discussed in some detail by Taylor (1939).

Fosliella lejolisii (Rosanoff) Howe

Howe, 1920; *Melobesia lejolisii* Rosanoff, 1866.

Plants usually forming disc-like crusts, variously sized and occasionally crowded together; thallus in surface view composed of radially arranged filaments, with the cells about 7 μm in diameter and from 6-10 μm in length; thallus near the margin 1 cell thick or with a smaller cell on the upper side of each basal cell; thallus towards the centre of the disc with the original basal cell layer subdivided; asexual conceptacles from 150-300 μm in diameter.

For figures see Hamel and Lemoine, 1953, p. 104, 105, figs 65-67 (as *Melobesia lejolisii*).

So far only found at a depth of about 15 m growing on roughened plexiglass plates attached by a framework to the sea bed.

Distribution. World: widespread in boreal-antiboreal to tropical seas. West Africa: Ghana: unpublished.

Remarks. We have provisionally assigned to this species Ghanaian plants found growing only on an artificial substratum though outside the Gulf of Guinea it is usually reported as confined to the leaves of sea grasses.

Hydrolithon (Foslie) Foslie 1909

Plants forming crusts and with or without erect branches; hypothallus monostromatic and non-coaxial, usually of quadrate cells; perithallus of irregularly sized and arranged cells, only occasionally layered; epithallus monostromatic; intercell fusions common and heterocysts present; conceptacles all having a single pore.

World Distribution: cosmopolitan.

Hydrolithon boergesenii (Foslie) Foslie

Foslie, 1909; *Goniolithon boergesenii* Foslie, 1901.

Plants forming a crust having wart-like prominences on the surface; hypothallus of nearly quadrate cells, somewhat elongated at right angles to the surface, from 5-13 μm in diameter and 12-22(-33) μm in height; perithallial cells irregularly arranged, ovoid to elongate and from 5-20(-30) μm in size, with heterocysts rectangular and consisting of 4 to 8 cells, to about 15 μm in diameter and 13-33 μm high; epithallus of subrectangular cells; asexual conceptacles from 300-400 μm in diameter and 100-125(-140) μm in height; spermatangial conceptacles from 50-75 μm in diameter and 40-65 μm high; cystocarpic conceptacles larger, from 125-225 μm in diameter and 90-125 μm high.

For figures see Masaki, 1968, pl. 27, figs 1-3, pl. 28, figs 1-8, pl. 67, figs 1-7, pl. 68, figs 1-4 (as *Porolithon boergesenii*).

There is no information on the ecology of this plant in the region other than it was found growing together with *Lithophyllum retusum*. In other parts of its range it is commonly reported from wave-exposed situations in the eulittoral zone.

Distribution. World: in warm temperate and tropical parts of the Atlantic Ocean. West Africa: São Tomé: *Goniolithon boergesenii* var. *africana*, Foslie 1907, Hariot 1908, Henriques 1917; *G. boergesenii* f. *africana* De Toni 1889; *Porolithon boergesenii*, De Toni 1924, Lemoine 1917, Steentoft 1967.

Lithophyllum* Philippi 1837

Plants calcareous, consisting of a crust, with or without erect coralloid branches, sometimes growing completely unattached; hypothallus monostromatic or coaxial and polystromatic; perithallus weakly to strongly layered and with secondary pit connections; epithallus mono- to polystromatic; asexual conceptacles single-pored, sporangia confined to the periphery of the floor and bi- or tetrasporic.

World Distribution: cosmopolitan.

There is no information on the ecology of any of the species belonging to this genus which have so far been reported from the region.

Key to the Species

1. Plants wholly encrusting, thin and with age developing wart-like excrescences and wavy ridges on the surface .... *L. subtenellum*
1. Plants consisting of an encrusting layer bearing erect and subdichotomously divided branches 2

**Lithophyllum kaiseri* (Heydrich) Heydrich is reported by Pilger (1919) from the island of Pagalu as *L. kotschyanum*; see discussion in Price *et. al.* (1983).

2. Branches somewhat flattened and markedly inflated towards the tips, with the apices usually truncate and depressed in the centre; hypothallial cells from (9-) 11-18(-22) μm in height and the extended perithallial cells from 9-11(-14) μm in height . . . . *L. retusum*
2. Branches terete or slightly flattened, nodulose, with the apices obtuse and little inflated; hypothallial cells from 18-32 μm in height and the extended perithallial cells from 9-20 μm in height. . . . . *L. simile*

Lithophyllum retusum Foslie — Pl. VII, figs C-E*

Foslie, 1897.

Plants consisting of branches free or arising from a thin crust, more or less erect and from 2.5-3 cm in height; branches anastomosing and more or less fastigiate, up to 3 or more times subdichotomously divided, flattened, about 2 mm thick below and increasing towards the often inflated tips which are usually truncate and somewhat depressed in the centre; hypothallial cells from 7-11 μm in diameter and (9-) 11-18(-22) μm in height; perithallus with some cells subquadrate and from (6-)7-11 μm in diameter, with other cells elongated at right angles to the surface, from 7-9 (-11) μm in diameter and from 9-11(-14) μm in height; epithallial cells elongated at right angles to the surface, from 9-11 μm in diameter; asexual conceptacles cone-shaped but not particularly prominent, from 150-200 μm in diameter in surface view and about 95 μm in height.

Distribution. World: in tropical parts of the eastern Atlantic Ocean. West Africa: Ghana: Adey & Lebednik 1967, Foslie 1909, Steentoft 1967. São Tomé: Adey & Lebednik 1967, De Toni 1905, Foslie 1897, 1909, Foslie & Printz 1929, Steentoft 1967.

Remarks. The reports of *Lithophyllum retusum* from São Tomé by Hariot (1908) and Henriques (1917) are believed to be misidentifications of a plant which might be an as yet undescribed species (see Steentoft, 1967).

Lithophyllum simile Foslie — Pl. 35, fig. 8.

Foslie, 1909.

Plants consisting of erect and subglobose clumps, to about 5 cm in height; branches congested and subdichotomously divided, usually terete and rarely slightly flattened, from 2-4 mm in width and often nodulose, with the apices obtuse and occasionally a little inflated; hypothallial cells from 9-11 μm in diameter and 18-32 μm in height; perithallial cells subquadrate and from (6-)7-9 μm diameter, occasionally extended at right angles and more rarely parallel to the surface, from (7-)9-12 μm in diameter and 9-20 μm in height; ?conceptacles in section from 170-230 μm in diameter and 75-95 μm in height.

*after Foslie and Printz, 1929, pl. 64, figs 12-14.

For additional figures see Foslie, 1929, pl. 63, fig. 22.

Distribution. World: in the tropical parts of the eastern Atlantic Ocean. West Africa: São Tomé: Adey & Lebednik 1967, De Toni 1924, Foslie 1909, Foslie & Printz 1929, Steentoft 1967.

Lithophyllum subtenellum (Foslie) Foslie.

Foslie, 1900c; *Goniolithon subtenellum* Foslie, 1898.

Plants forming thin crusts, usually to about 0.5 mm in thickness, initially orbicular with a lobed or dentate margin and becoming with age confluent, thrown up into wavy ridges or develops wart-like excrescences to about 2 mm in height; hypothallial cells elongated parallel to the surface, from 9-21 μm in diameter and from 5-13 μm in height; perithallus weakly developed, with the cells in older plants quadrate to somewhat rectangular, from 5-10 μm in diameter and 6-14 μm in height, somewhat shorter in deep water plants; asexual conceptacles subhemispherical to conical, often numerous and crowded together, from 200-300 μm in diameter in surface view, with the sporangia about 40 μm in diameter and 75 μm in height.

For figures see Foslie and Printz, 1929, pl. 53, figs 1-3; Hamel and Lemoine, 1953, p. 98, fig. 63, pl. 20, fig. 1, 2 (as *Lithothamnium subtenellum*).

Distribution. World: widespread in warm temperate and tropical parts of the eastern Atlantic Ocean as well as in the Mediterranean. West Africa: São Tomé: Adey & Lebednik 1967, J. Feldmann 1942, Hariot 1908, Henriques 1917; *Lithothamnium subtenellum*, Steentoft 1967.

Doubtful Record

Lithophyllum tortuosum (Esper) Huvé

See the remarks under *Bifurcaria bifurcata* regarding the element of doubt that attaches to Sourie's (Sourie, 1954) report of this plant as *Tenarea tortuosa* from Guinée.

Plate 35: Fig. 1. *Jania crassa:* Portion of a plant showing the terete segments and uncalcified articulations. Figs 2, 3. *Fosliella farinosa:* 2. Crustose colonies growing as discs on the surface of a foliar appendage of *Sargassum;* 3. Surface view of a colony showing the branched cell rows. Figs 4-7. *Porolithon africanum* (after Pilger, 1919): 4. Section of a branch showing conceptacles containing zonately divided tetrasporangia; 5. Outermost layers of cells; 6. Middle layers of often anastomosing cells; 7. Innermost layers of cells elongated at right angles to the surface. Fig. 8. *Lithophyllum simile:* Portion of a plant showing the congested and nodulose branches. Fig. 9. *Porolithon aequinoctiale:* Plant showing the erect branches to be congested, anastomosing and often tapering or narrowing abruptly towards the obtuse apices.

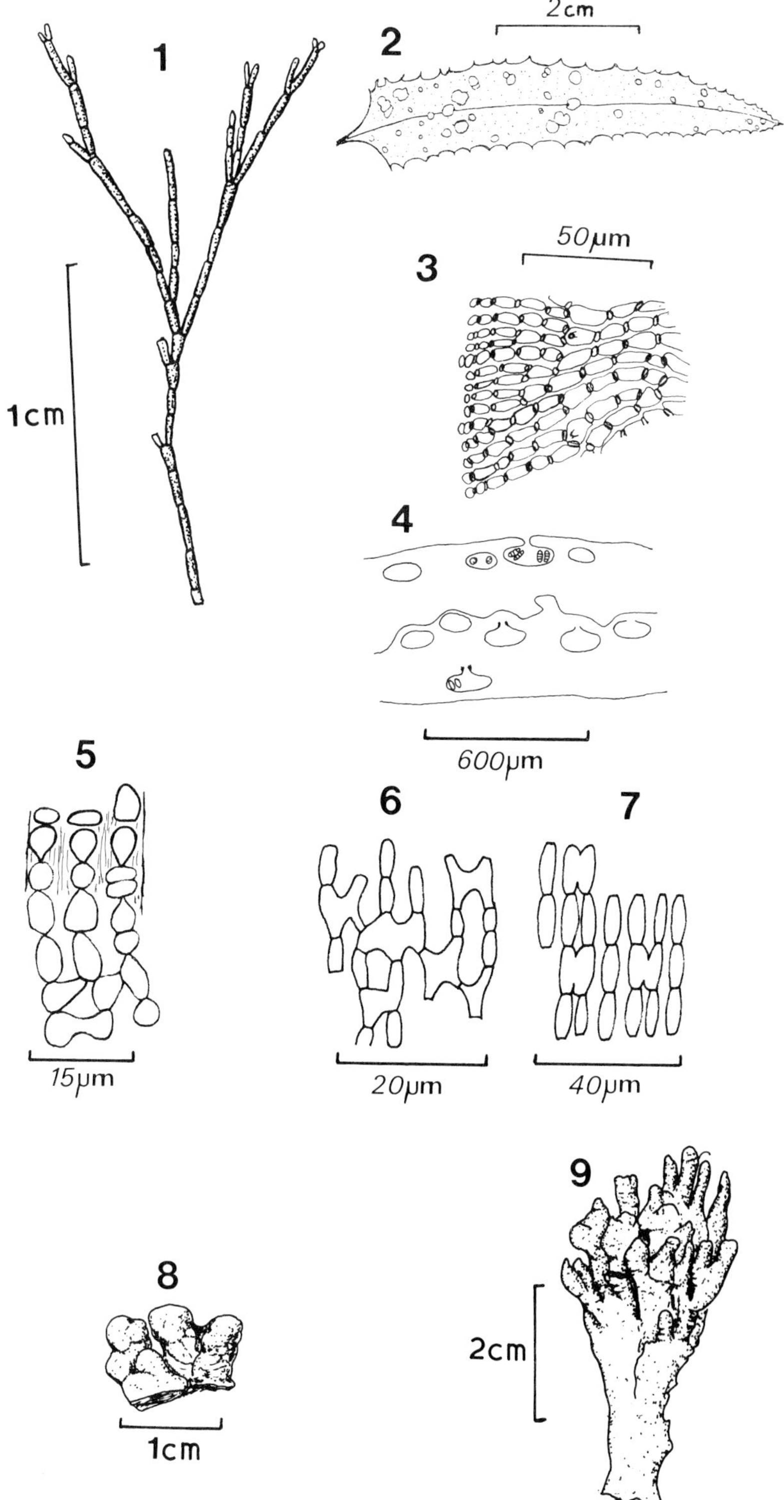
1
2
2cm
1cm
3
50μm
4
600μm
5
15μm
6
20μm
7
40μm
8
1cm
9
2cm

Mesophyllum Lemoine 1928

Plants calcareous, forming thin expansions or occasionally as thick crusts, with or without coralloid branches; hypothallus parallel to the substratum and with cells in concentric rows especially in the lower parts, coaxial; perithallus weakly layered and intercell fusions common; epithallus of a layer of rectangular cells; asexual conceptacles having a number of pores in the roof and sexual conceptacles unipored.

World Distribution: widespread in warm temperate and tropical seas.

Key to the Species

1. Branches from 3-6 mm in diameter and markedly nodulose, with the apices thickened and somewhat inflated; asexual conceptacles from 500-700 μm in diameter in surface view. . . . . *M. brachycladum*
1. Branches from 1.5-2.5 mm in diameter and more or less smooth, usually inflated below and tapering above; asexual conceptacles from 200-400 μm in diameter in surface view. . . . . *M. canariense*

?**Mesophyllum brachycladum** (Foslie) Adey

Adey, 1970; *Lithothamnion brachycladum* Foslie, 1900a.

Plants consisting of rounded clumps of radiating branches; branches irregularly dichotomous, fastigiate and usually anastomosing, terete, from 3-6 mm in diameter, markedly nodulose, with the apices thickened and somewhat inflated; branches showing little differentiation into distinct tissues but having cup-shaped layers of cells present; cells of branches frequently rectangular, from 7-8 μm in diameter and 9-14 μm in height, sometimes smaller and rounded at the corners; asexual conceptacles scattered or crowded in the upper parts of the branches, slightly prominent and a little flattened in the centre, in surface view from 500-700 μm in diameter, with the roof perforated by about 40 pores.

For figures see Foslie and Printz, 1929, pl. 12, fig. 19 (as *Lithothamnion brachycladum*).

Distribution. World: in warm temperate and tropical parts of the eastern Atlantic Ocean. West Africa: Príncipe: *Lithothamnion brachycladum*, Børgesen 1929, De Toni 1905, Foslie 1900a, Steentoft 1967.

Remarks. There is some uncertainty regarding this record since Foslie (1900a) did not verify his determination by examining its structure. An element of doubt exists regarding the taxonomic position of this species since the holotype material does not possess a basal hypothallus (Adey, 1970). It is on the nature of the secondary hypothallus that this genus is separated from other closely related genera.

Mesophyllum canariense (Foslie) Lemoine

Lemoine, in Børgesen, 1929; *Lithothamnion canariense* Foslie, 1906b.

Plants forming cushion-like clumps, to about 10 cm in diameter and from 1.5-2.5 (-3.5) cm in height; branches erect and anastomosing, irregularly dichotomously divided and fastigiate, terete and from 1.5-2.5 mm in diameter, inflated below and tapering above; hypthallial cells of the crust from 15-25 μm in diameter and to about 30 μm in height; perithallial cells of the crust from 7-8 μm in diameter and 4-6 μm in height; inner cells of the branches from 10-15 μm in diameter and 4-12 μm in height, with the outer cells from 7-10 μm in diameter and 3-7 μm in height; asexual conceptacles hemispherical, from 200-400 μm in diameter in surface view; cystocarpic conceptacles conical and in surface view from 250-500 μm in diameter.

For figures see Børgesen, 1929, p. 33, 34, figs 8, 9, pl. 1, figs 1, 2.

Distribution. World: as for genus. West Africa: São Tomé: Steentoft 1967; *Lithophyllum marlothii,* Hariot 1908, Henriques 1917.

Doubtful Record

Mesophyllum floridanum (Foslie) Adey.

This plant has been reported as *Lithothamnion floridanum* from São Tomé by Adey and Lebednik (1967), Foslie (1906a) and Steentoft (1967). It has been decided to regard these São Tomé records as being uncertain since according to Foslie (1906a) the specimens on which they are based "stand very near to *L. floridanum,* but in that they are sterile, they cannot be definitely determined.".

Neogoniolithon Setchell & Mason 1943

Plants calcareous, crustose or branching, occasionally completely free; hypothallus parallel to the substratum, coaxial or non-coaxial; perithallial cells generally not layered, internal fusions present, with heterocysts single or grouped in short vertical rows; epithallus usually single-layered; asexual sporangia single-pored.

World Distribution: widespread in warm temperate and tropical seas.

Neogoniolithon mamillare (Harvey) Setchell & Mason

Setchell and Mason, 1943; *Melobesia mamillaris* Harvey, 1847.

Plants crustose, thin, surface extensively covered by mammilliform tubercles in young crusts, gradually lengthening into erect terete branches which become divided and interlaced; hypothallus several cells thick, composed of rectangular cells from 18-36 μm in length, with long axis of cells at right angles to surface; perithallus of regular rows of quadrate to subquadrate cells from 7-15 μm in size; epithallus usually single-layered, with cells rectangular and from 11-15 μm in breadth; asexual and female conceptacles similar, conical and gradually tapering to summit, with cavity ranging from 400-700 μm in breadth; male conceptacles less prominent and smaller, with cavity from 300-400 μm in breadth.

For figures see Pilger, 1919, p. 406, 407, 409, 410, 412, figs 1-23 (as *Goniolithon mamillare*).

There is no information on the ecology of this species in the region.

Distribution. World: pantropical. West Africa: Bioko: Schmidt & Gerloff 1957. Cameroun: Schmidt & Gerloff 1957.

Undetermined Species

Neogoniolithon sp.

An unidentified species has been reported from São Tomé by Carpine (1959) and gets a secondary mention by Steentoft (1967) in her revision of the marine algae of some of the islands in the Gulf of Guinea. The apparent loss (Huvé, *pers. comm.*) of the Carpine collections makes it impossible to examine this plant.

Phymatolithon Foslie 1898

Plants calcareous, completely crustose or also bearing erect branches; hypothallus parallel to the substratum and non-coaxial; perithallial cells not layered and intercell fusions common; epithallus monostromatic or absent; conceptacles developing deep in the perithallus, with the asexual conceptacles having many pores in the roof.

World Distribution: probably widespread from boreal-antiboreal to tropical seas.

Phymatolithon tenuissimum (Foslie) Adey.

Adey, 1970; *Lithothamnion tenuissimum* Foslie, 1900a.

Plants initially forming small delicate crusts and often having a crenulate margin, with age increasing to 100-250(-400) μm in thickness and the originally smooth surface developing scaly thickenings and sometimes indistinct concentric striations towards the margin; hypothallus of 2 to 3 layers of cells elongated parallel to the substratum and to about 10 μm in diameter; perithallus weakly developed, having the cells rounded or quadrate and thick-walled, from 3-5 μm in diameter; asexual conceptacles scattered or somewhat grouped together and not very prominent, from 180-200 μm in diameter in surface view, with about 10 to 15 pores in the roof.

For figures see Foslie and Printz, 1929, pl. 1, figs 1-3; Lemoine, 1924, p. 116, fig. 1, pl. 4, fig. 5 (as *Lithothamnion tenuissimum*).

Occurs according to Foslie (1900a) "at St. Thomé, where it seems to be of pretty common occurrence.".

Distribution. World: in warm temperate and tropical parts of the eastern Atlantic Ocean. West Africa: São Tomé: *Lithothamnion tenuissimum,* Adey & Lebednik 1967, De Toni 1905, J. Feldmann 1939, 1942, Foslie 1900a, Foslie & Printz 1929, Steentoft 1967.

Porolithon (Foslie) Foslie 1909

Plants calcareous, forming often massive horizontal crusts, with or without erect branches; hypothallus polystromatic with the cells elongated parallel to the substratum, non-coaxial; perithallial cells polystromatic and often variable in form and dimensions, with intercell fusions common, without secondary pit connections, having heterocysts single or grouped in transverse discs; epithallus mono- or polystromatic; asexual conceptacles having a single pore in the roof.

World Distribution: cosmopolitan.

Key to the Species

1. Plants wholly encrusting, with the surface uneven and often light in colour...... .... *P. onkodes*
1. Plants consisting in part of erect and anastomosing branches 2

2. Branches terete to subcompressed, about 2 mm in width just below the apex and often increasing somewhat abruptly lower down............... *P. aequinoctiale*
2. Branches compressed, either short and wart-like or longer, anastomosing and divided, often giving rise to folds or winding ridges .... *P. africanum*

Porolithon aequinoctiale Foslie Pl. 35, fig. 9; Pl. VII, fig. F.*

Foslie, 1909.

Plants forming erect clumps, about 4 cm in height and 2.5 cm in diameter; branches congested, anastomosing and subdichotomously divided, terete to subcompressed and to about 2 mm in width just below the obtuse apex, increasing somewhat abruptly lower down; branches having rows of short irregular cells from 5-7 μm in diameter and 7-10 μm in height, alternating with rows of longer cells from 14-20 μm in height; cells towards the surface subquadrate, from 5-7 μm in diameter and 6-7(-9) μm in height, with the outermost layers of cells extended parallel to the surface, from 4 to 6 times broader than high and 5-7 μm in diameter; spermatangial conceptacles from 115-250 μm in diameter and 20-30 μm in height in section.

There is no information on the ecology of this plant in the region.

*after Foslie and Printz, 1929, pl. 70, figs 4, 5, as *Lithophyllum aequinoctiale.*

Plate VII (after Foslie and Printz, 1929): A. ***Porolithon africanum:*** Part of an old specimen collected from São Tomé (from pl. LXVIII, fig. 2, as ***Lithophyllum africanum*** f. ***intermedia***). B. ***Pseudolithophyllum irregulare:*** Specimen loosened from some hard object and collected from São Tomé (from pl. XII, fig. 22, as ***Lithothamnion irregulare***). C-E. ***Lithophyllum retusum:*** Young and old specimens collected from São Tomé (from pl. LXIV, figs 12-14). F. ***Porolithon aequinoctiale:*** Part of a specimen seen from the side and collected from São Tomé (from pl. LXX, fig. 5 as ***Lithophyllum aequinoctiale***).

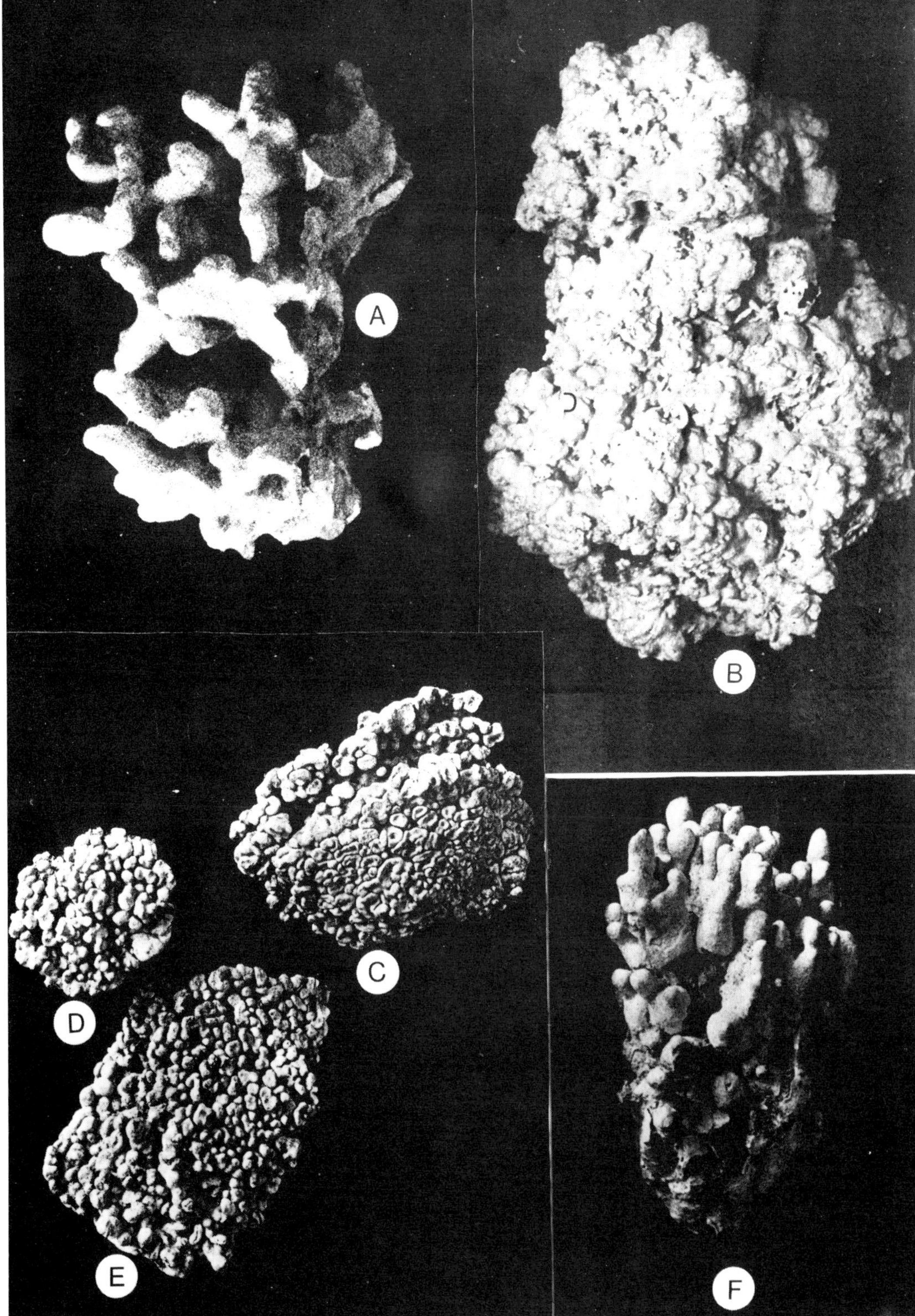
A
B
C
D
E
F

Distribution. World: in tropical parts of the eastern Atlantic Ocean. West Africa: São Tomé: De Toni 1924, Foslie 1909; *Lithophyllum aequinoctiale*, Adey & Lebednik 1967, Foslie & Printz 1929, Steentoft 1967.

Porolithon africanum (Foslie) Foslie Pl. 35, figs 4-7, Pl. VII, fig. A*

Foslie, 1909; *Lithophyllum africanum Foslie*, 19006.

Plants forming subhemispherical masses, about 5 cm in diameter, consisting of branches borne on obpyramidal or somewhat compressed and confluent lobes; branches angular or compressed and rarely subterete, from 3-7 mm in width and to about 10 mm in height, either simple, short and wart-like or else longer, anastomosing, sometimes palmate, occasionally folded or forming winding ridges, with the apices obtuse; inner cells of the branches elongated at right angles to the surface and from 7-11(-14) μm in height, with occasionally groups of cells from 9-12 μm in diameter and 14-25 μm high also present; outermost cells of branches flattened, from 4-9 μm in diameter and 4 to 9 times broader than high; asexual conceptacles immersed and only slightly convex, in section from 100-200 μm in diameter and 80-90 μm in height.

For additional figures see Pilger, 1919, p. 417, 418, figs 24-28 (as *Lithophyllum africanum*).

There is no significant information on the ecology of this plant in the region.

Distribution. World: in warm temperate and tropical parts of the eastern Atlantic Ocean. West Africa: Bioko: *Lithophyllum africanum*, Schmidt & Gerloff 1957. Cameroun: *L. africanum*, Schmidt & Gerloff 1957. São Tomé: Foslie 1909; *L. africanum*, Adey & Lebednik 1967; *L. africanum* f. *intermedia*, Foslie 1909, Foslie & Printz 1929, Steentoft 1967; *Lithophyllum ponderosum*, Foslie 1897; *Lithophyllum incrustans*, De Toni 1905; *Lithothamnion ponderosum* Henriques 1917.

Porolithon onkodes (Heydrich) Foslie Pl. 36, figs 1, 2.

Foslie, 1909; *Lithothamnion onkodes* Heydrich, 1897.

Plants usually forming overlapping crusts having an uneven surface, from 2-4 mm in thickness; cells of the hypothallus several layers thick, subquadrate to elongated at right angles to the surface, from 2-5 μm in diameter and from 4-12 μm in height, with groups of 4 to 5(-10) larger cells usually 6-12 μm in diameter and from 16-25 μm in height; cells of epithallus in one or occasionally 2 layers, irregularly spherical to somewhat flattened, up to 6 μm in diameter and from 2-3 μm in height; asexual conceptacles in section about 165 μm in diameter and occasionally to 210 μm, from 68-85 μm in height.

*after Foslie and Printz, 1929, pl. 68, fig. 2, as *Lithophyllum africanum* f. *intermedia*.

For additional figures see Heydrich, 1897, pl. 1, figs 11a, b (as *Lithothamnion onkodes*).

Occurs as light pink crusts growing over steeply sloping rocks in the lower eulittoral subzone where there is moderate to severe wave action.

Distribution. World: probably pantropical. West Africa: Ghana: unpublished. São Tomé: unpublished (Lemoine, *pers. comm.*).

Pseudolithophyllum* Lemoine 1913

Plants calcareous, consisting of a well-developed crust, with or without erect branches; hypothallus mono- to polystromatic, non-coaxial; perithallus usually well-developed and having intercell fusions common, with large heterocysts absent; epithallus one to five layers thick; asexual conceptacles having a single pore in the roof.

World Distribution: cosmopolitan.

Key to the Species

1. Surface of crust covered by wart-like excrescences from 1-3 mm in diameter .... *P. irregulare*
1. Surface of individual crusts smooth........................... *P. mildbraedii*

Pseudolithophyllum irregulare (Foslie) Adey Pl. VII, fig. B**.

Adey, 1970; *?Lithothamnion irregulare* Foslie, 1907.

Plants forming irregular crusts from 4-8 cm in diameter, with the surface covered by wart-like excrescences which are from 1-3 mm in diameter and especially abundant in the centre where they often coalesce; cells of hypothallus ovoid to spherical, from 6-14 μm in diameter and 11-29 μm in height; cells of perithallus indistinctly layered and subquadrate, from 6-9 μm in diameter and to 11(-14) μm in height; ?asexual conceptacles arising on the excrescences, not prominent, from 300-375 μm in diameter in surface view; cystocarpic conceptacles conical, to about 750 μm in diameter in surface view and to 600 μm in height.

There is no information on the ecology of this plant in the region.

Distribution. World: in warm temperate and tropical parts of the eastern Atlantic Ocean. West Africa: São Tomé: *Lithothamnion irregulare*, Adey & Lebednik 1967, De Toni 1924, Foslie 1907, Foslie & Printz 1929; *Lithophyllum irregulare*, Steentoft 1967; ***Tenarea irregularis,*** Børgesen 1929.

**Pseudolithophyllum leptothalloideum* (Pilger) De Toni was one of two species of *Lithophyllum* described by Pilger (1919) from Pagalu.

**after Foslie and Printz, 1929, pl. 12, fig. 22, as *Lithothamnion irregulare.*

Remarks. Adey (1970) has some reservations in placing this plant in *Pseudolithophyllum* since sporangial conceptacles have not been positively identified.

Pseudolithophyllum mildbraedii (Pilger) De Toni

De Toni, 1924; *Lithophyllum mildbraedii* Pilger, 1919.

Plants forming thin, flat, almost circular crusts to about 3.5 cm in diameter, with the margin undulate and crenate, often a ridge or branch-like excrescence formed at the edge where two crusts come into contact; hypothallus indistinct, consisting of a layer of cells giving rise to curved or straight rows of cells; perithallus of laterally united rows of cells, often cells more regularly arranged above than below, with cells rounded to somewhat rectangular having their long axes at right angles to the surface, from 8-18 μm in length; epithallus of 1 to several layers of flat, closely arranged cells, often this covering layer(s) of cells irregular or partly removed in older parts of the crust; asexual conceptacles hemispherical, only slightly projecting above the surface, from 150-160 μm in diameter (internal cavity), with roof 1 to 3 cell-layers thick.

For figures see Pilger, 1919, p. 425, 426, figs 40-47 (as *Lithophyllum mildbraedii*).

There is no significant information on the ecology of this species in the region.

Distribution. World: only known from the eastern parts of the tropical Atlantic. West Africa: Bioko: Schmidt & Gerloff 1957. Cameroun: Schmidt & Gerloff 1957.

Remarks. Originally described from material collected from Pagalu (=Annobon) by J. Mildbraed in October 1911.

Family Corynomorphaceae

Corynomorpha J. Agardh 1872a

Plants erect, terete, clavate, simple or forked, arising from a short stipe and attached by a discoid holdfast; branches composed of a medulla of anastomosing filaments and stellate cells, with a cortex of anticlinal rows of dichotomously divided filaments; tetrasporangia in nemathecia at the tips of the branches; sporangia tetrapartitely, irregularly or almost zonately divided; spermatangia arising in chains; cystocarps immersed in terminal nemathecia.

World Distribution: in tropical parts of the Atlantic and Indian Oceans.

Corynomorpha prismatica (J. Agardh) J. Agardh Pl. 36, figs 3-6.

J. Agardh, 1872a; *Dumontia prismatica* J. Agardh, 1841.

Plants usually gregarious and forming extensive patches, from 5-15 cm in height, purplish to reddish-brown and pale at the apex, firm and cartilaginous; branches terete and clavate, to about 5 mm in diameter, simple or bifurcate, occasionally arising in a whorl from a collar-like swelling at the top of a short stipe; tetrasporangial nemathecia at the tips of the branches and forming a cap about 2 cm in length; cystocarpic nemathecia somewhat ovoid, to about 1 cm in length.

For additional figures see Balakrishnan, 1962, p. 79, 80, 83, 84, figs 1-2.

Occurs most commonly growing on sand-covered rocks at a depth of about 10 m and occasionally epiphytic on *Thamnoclonium*.

Distribution. World: as for genus. West Africa: Ghana: John & Lawson 1972a, John *et al.* 1977, Lieberman *et al.* 1979.

Family Cryptonemiaceae*

Cryptonemia** J. Agardh 1842

Plants consisting of flattened, lobed or palmately divided fronds, with or without a partial midrib, occasionally having proliferations from the margin or on the midrib when present, arising from a short stipitate base; thallus having a medulla of periclinally directed filaments and stellate cells present, with an inner cortex of large colourless cells and an outer cortex of short anticlinal rows of cells; tetrasporangia scattered in the cortex or in sori, sporangia tetrapartite; cystocarps immersed, scattered or in special segments.

World Distribution: widespread in warm temperate and tropical seas.

Key to the Species

1. Plants having a short midrib and an entire margin. . . . . *C. luxurians*
1. Plants without a midrib and the margin bearing forked crenulations. . . . . *C. crenulata*

Cryptonemia crenulata (J. Agardh) J. Agardh — Pl. 37, figs 1-3.

J. Agardh, 1851; *Phyllophora crenulata* J. Agardh, 1841.

Plants often solitary or as small patches, about 10 cm in height, reddish-brown; fronds sparingly irregularly dichotomously divided especially towards the base, leathery and curled, to about 2 cm in width, with forked crenulations along the margin which are particularly well-developed around the broadly rounded and sometimes terminally indented apex; tetrasporangia in irregularly rounded sori just below the apex of the frond.

*For discussion concerning the correct family name see Silva (1980, p. 82).

**A sterile *Cryptonemia* having a conspicuous midrib has been reported from Gambia (John and Lawson, 1977b) and this might either be *C. luxurians* or *C. seminervis* (see remarks under *C. luxurians*).

For additional figures see Taylor, 1960, p. 779, pl. 58, fig. 4.

Occurs in the sublittoral fringe and shallow sublittoral on steeply sloping rock gullies and ledges in moderately wave-exposed situations; occasionally to a depth as great as 12 m.

Distribution. World: in warm temperate and tropical parts of the Atlantic Ocean. West Africa: Ghana: Guiry & Irvine 1974, John *et al.* 1977, 1980, Lieberman *et al.* 1979.

Remarks. The margin of the frond is inflated mainly by the expansion of the medulla and this thickening is apparent to the naked eye by its deeper colour especially in young plants. This character is also developed to a lesser extent in *C. luxurians* and is a phenomenon found in a large number of species in this group (M. Guiry, *pers. comm.*). According to De Toni (1905) this species belongs to the genus *Acrodiscus* because the tetrasporangia are grouped together in sori.

Cryptonemia luxurians (C. Agardh) J. Agardh Pl. 36, figs 7-9.

J. Agardh, 1851; *Sphaerococcus lactuca* var. *luxurians* C. Agardh, 1823.

Plants usually solitary, about 10 cm or more in height, bright reddish-purple or pinkish-red (deep water plants); fronds irregularly dichotomously divided, often with numerous marginal proliferations, membranaceous to leathery, to about 3(-5) cm wide, with the margin entire and a conspicuous midrib extending to about half way to the broadly rounded apex; tetrasporangia confined to small marginal proliferations.

For additional figures see Taylor, 1960, p. 779, pl. 58, fig. 3.

Occurs as small plants on the underside of rock overhangs and steeply sloping surfaces in moderately wave-exposed situations in the sublittoral fringe, often much of its surfaced covered by epizoans. The largest and most delicate plants have been found growing on calcareous cobbles down to a depth of 28 m.

Distribution. World: widespread in warm temperate and tropical parts of the Atlantic Ocean. West Africa: Côte d'Ivoire: ?John 1977a; *C. seminervis*, John 1972c. Ghana: Guiry & Irvine 1974, John 1977a, John *et al.* 1977, Lieberman *et al.* 1979; *C. seminervis*, Dickinson & Foote 1950, Lawson 1956, Stephenson & Stephenson 1972. Liberia: De May *et al.* 1977.

Plate 36: Figs 1, 2. *Porolithon onkodes:* 1. Section through the upper part of the crust showing the single-layered epithallus and intercell fusions common in the perithallus; 2. Section of the crust showing a row of large columnar heterocyst cells in the perithallus. Figs 3-6. *Corynomorpha prismatica:* 3. Plants showing the club-shaped branches occasionally arising in whorls; 4. Cystocarpic plant with the ovoid nemathecium at the tip of the branch; 5. Tetrasporic plant with the elongate apical nemathecium; 6. Section showing the dichotomously branched rows of cortical cells arising from the loosely filamentous medulla. Figs 7-9. *Cryptonemia luxurians:* 7. Plant with the flattened fronds having a midrib in the lower third; 8. Section of the frond showing the wide filamentous medulla, the inner cortex and the outermost layer of subrectangular cell; 9. A lateral proliferation containing tetrasporangia.

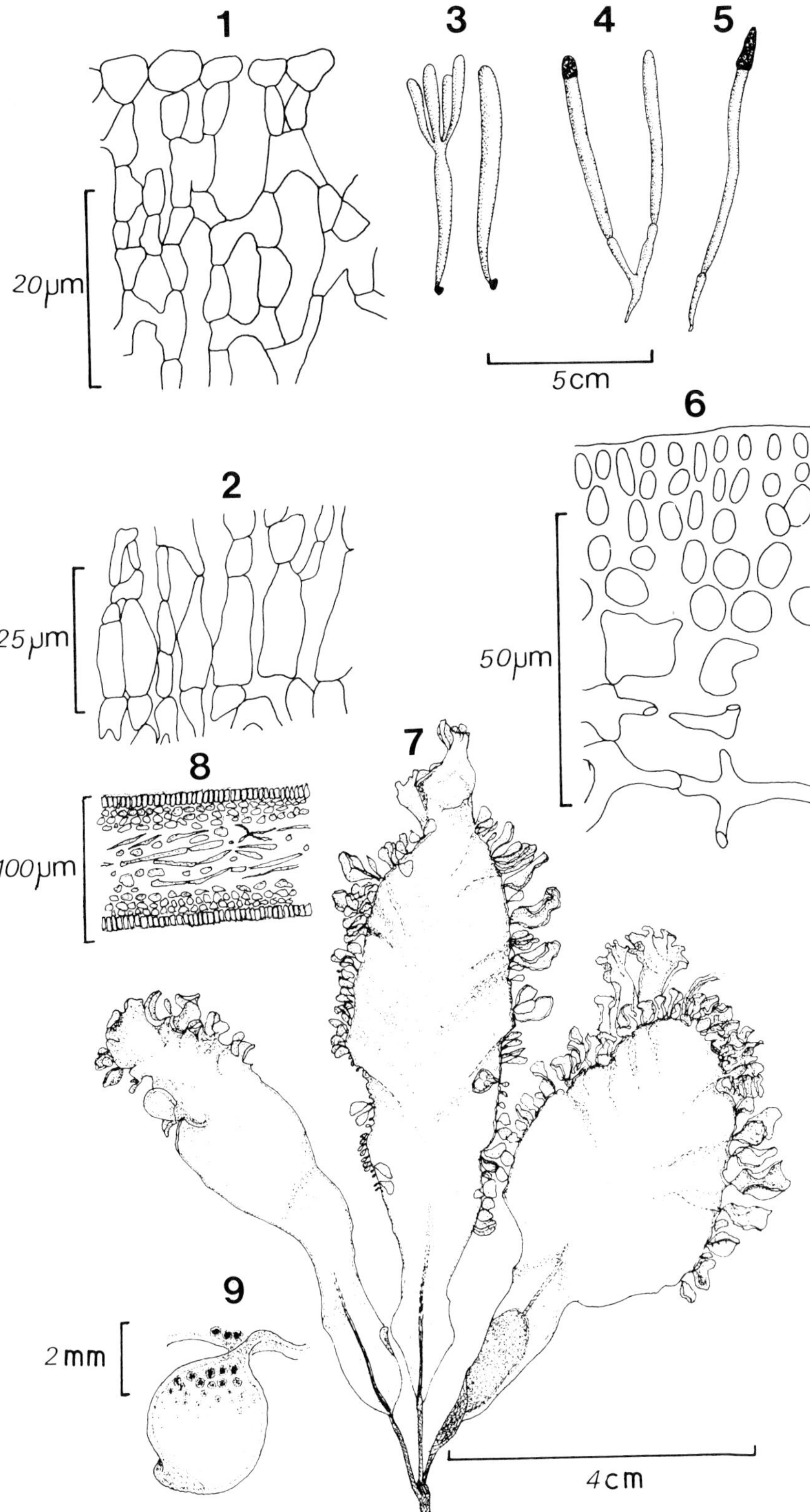
1
3
4
5
20µm
5cm
6
2
25µm
50µm
8
7
100µm
9
2mm
4cm

Remarks. There is a distinct possibility that *Cryptonemia luxurians* and *C. seminervis* are conspecific as the only doubtful distinguishing feature seems to be the confinement of the tetrasporangia in the former to marginal proliferations whilst they are scattered throughout the frond in the latter (M. Guiry, *pers. comm.*).

Grateloupia C. Agardh 1823

Plants fleshy to membranaceous, often gelatinous, consisting of terete to compressed branches, dichotomously to pinnately divided and occasionally proliferous; thallus made up of a medulla and cortex; medulla composed of a net-work of anastomosing filaments periclinally directed and one or two layers of stellate cells; cortex consisting of a layer of stellate cells and dichotomously divided rows of cells becoming smaller towards the surface; tetrasporangia scattered in the cortex, sporangia tetrapartite; spermatangia in sori on the surface; cystocarps small and inconspicuous, scattered and embedded in the cortex, with a pore.

World Distribution: widespread in warm temperate and tropical seas.

Key to the Species

1. Plants many times pinnately, bipinnately or irregularly divided, usually less than 3mm in width .... *G. filicina*
1. Plants simple or often with a few irregular divisions, sometimes proliferous, usually more than 5 mm wide .... *G. doryphora*

Grateloupia doryphora (Montagne) Howe — **Pl. 37, figs 6, 7.**

Howe, 1914; *?Halymenia doryphora* Montagne, 1839.

Plants solitary or forming large mats, to about 40 cm in length or on rare occasions reaching 80 cm, violet to reddish-purple and often bleached to green in the upper parts; branching simple or with a few divisions towards the base, occasionally proliferous from the margin and surface, sometimes finely pinnate; branches from 5-20 mm or more in width, attenuate at the base and the apex.

For additional figures see Abbott and Hollenberg, 1976, p. 433, fig. 383; Ardré and Gayral, 1961, p. 43, figs a-g, p. 44, figs h-l, pl. 1-3 (as *G. lanceola*).

Plate 37: Figs 1-3. *Cryptonemia crenulata:* 1. Plant showing the frond with an undulate margin bearing forked crenulations; 2. Section of the frond showing the filamentous medulla, the large inner cortical cells, and the smaller outer cortical cells forming a very compact outermost layer; 3. Section towards the margin of the frond showing the distinct marginal thickening due to the increased development of the medullary tissues. Figs 4, 5. *Grateloupia filicina:* 4. Plant many times divided with the branches often attenuate at the base and apex; 5. Structure of the cortex showing the short dichotomously divided rows of cells arising from the filamentous medulla. Figs 6, 7. *Grateloupia doryphora:* 6. Plant showing the flattened branches to be simple or a few times irregularly divided; 7. Tetrapartitely divided tetrasporangia immersed between the short rows of cells of the outer cortex.

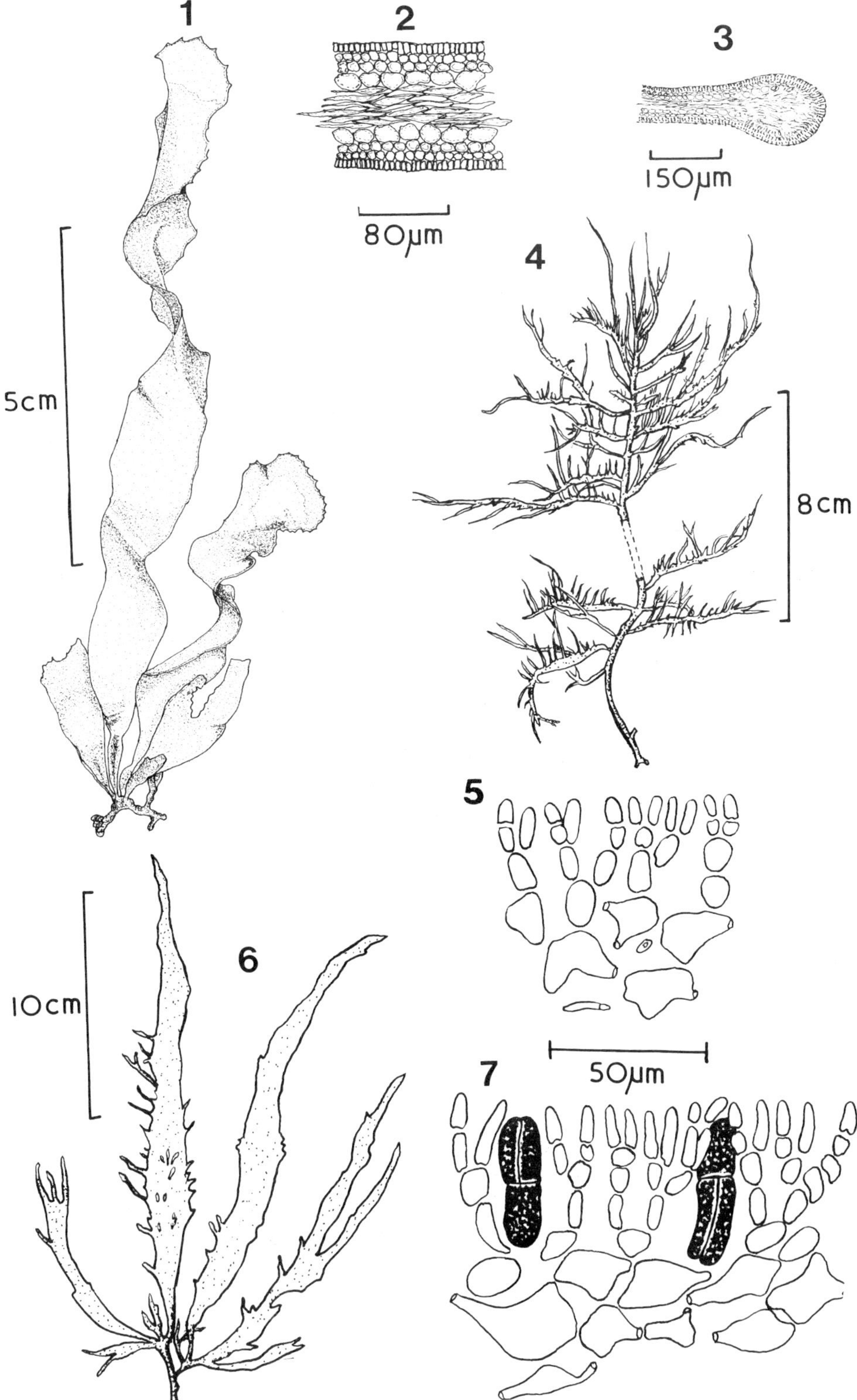
1
2
3
5cm
80µm
150µm
4
8cm
5
6
10cm
50µm
7

Occurs often on sand-covered beach rocks in the lower eulittoral subzone and very occasionally forming a band many metres in extent on breakwaters; found commonly in shallow tidepools in the eulittoral zone.

Distribution. World: as for genus. West Africa: Côte d'Ivoire: John 1977a. Gambia: John & Lawson 1977b. Ghana: John 1977a, John *et al.* 1977, Lieberman *et al.* 1979; *G. gibbesi*, Dickinson & Foote 1950. Liberia: De May *et al.* 1977, John 1977a.

Grateloupia filicina (Lamouroux) C. Agardh Pl. 37, figs 4, 5.

Agardh, 1823; *Delesseria filicina* Lamouroux, 1813.

Plants solitary or forming loose mats, to about 25 cm in height, usually reddish-purple; branching pinnate, with the lower branches often bipinnate or irregularly divided, rarely with proliferations developing from the surface; branches attenuate at the base and the apex, with the lower ones usually longer than the upper branches, to about 3 mm in breadth.

For additional figures see Dixon, 1966, p. 52, figs 3, 2a-k.

Occurs most commonly on beach rocks and in tidepools in moderately sheltered to moderately wave-exposed situations in the eulittoral zone. The most extensive patches of this plant are to be found in brackish-water habitats such as river estuaries, whilst solitary and little branched plants grow in very wave-exposed places.

Distribution. World: as for genus. West Africa: Benin: John 1977a, John & Lawson 1972b. Cameroun: Fox 1957, John 1977a, Pilger 1911, Schmidt & Gerloff 1957; *G. filicina* f. *filiformis*, De Toni 1924, Pilger 1911. Côte d'Ivoire: Bodard 1966c, John 1972c, 1977a. Gambia: John & Lawson 1977b. Ghana: Dickinson & Foote 1950, Fox 1957, John 1977a, John & Asare 1975, Lawson 1966. Liberia: De May *et al.* 1977. Nigeria: Fox 1957, John 1977a. Togo: John 1977a, John & Lawson 1972b, Pilger 1911.

Remarks. Some of the large and well-developed plants found in the Gulf of Guinea are close to variety *luxurians* Gepp & Gepp which recently has been reported from the south coast of the British Isles (Farnham, 1980).

Doubtful Record

Grateloupia dichotoma J. Agardh

Schmidt (1929) records the distribution of this species in the Atlantic as "...von der englischen und normannischen Küste südwärts bis nach Kamerun.". This plant is known from the Canary Islands but this is the only report of it from tropical West Africa and so we regard its mention from the region as being most doubtful.

Halymenia C. Agardh 1817

Plants very variable in form, fleshy or gelatinous, consisting of foliaceous or terete or angular branches, entire or variously lobed and divided; medulla of periclinally directed filaments and some crossing from cortex to cortex, with stellate cells present; cortex of cells becoming smaller towards the surface and not obviously filamentous; tetrasporangia scattered and embedded in the cortex, tetrapartite; cystocarps having a pericarp formed of a few medullary filaments, with a distinct pore.

World Distribution: widespread in warm temperate and tropical seas.

Key to the Species

1. Plants terete and slightly flattened just below the axils of the dichotomously divided branches. . . . . *H. agardhii*
1. Plants of another form. 2

2. Plants having an erose-dentate margin and projections on the surface. . . . . *H. duchassaingii*
2. Plants having an entire and sinuate or slightly dentate margin and a smooth surface. 3

3. Plants very delicate and gelatinous; fronds irregularly orbicular to elliptical-obovate, usually more than 10 cm in width and length . . . . *H. actinophysa*
3. Plants firm and only slightly gelatinous; fronds ovate and lobed, usually less than 10 cm in width and length. . . . . *H. vinacea*

Halymenia actinophysa Howe — Pl. 38, fig. 1.

Howe, 1911.

Plants delicate, membranaceous and very gelatinous, dull pink to rose red and occasionally purplish-red, attached by a small stipe less than 1 cm long; fronds irregularly orbicular to elliptical-obovate, rarely lobed, to about 15 cm or more in width and to 30 cm in length, with the margin usually sinuate or even slightly dentate; cystocarps immersed and from 160-230 μm in diameter.

For additional figures see Dawson, 1954b, p. 317, pl. 4, figs 29-34, p. 339, pl. 15, p. 341, pl. 16, fig. 61.

Occurs on rocky platforms and calcareous cobble at depths from 10 to 28 m and most abundant at the lower end of this range; occasionally found in the drift.

Distribution. World: in tropical parts of the Pacific and Atlantic Oceans. West Africa: Ghana: John & Lawson 1972a, John *et al.* 1977, 1980, Lieberman *et al.* 1979.

Halymenia agardhii De Toni — Pl. 40, fig. 8.

De Toni, 1905.

Plants solitary and bushy, from 4-12 cm or more in height, rose red, somewhat fleshy; branches consisting of 4 to 8(-12) dichotomies in one plane, sometimes having a few short dichotomous proliferations, terete and slightly flattened just below the branch axils, from 2-6 mm in width.

For additional figures see Taylor, 1960, p. 765, pl. 51, figs 1, 2.

Occurs on moderately sheltered to moderately wave-exposed rocks in the immediate sublittoral zone and occasionally in tidepools in the lower eulittoral subzone; found also to depths as great as 25 m.

Distribution. World: as for genus. West Africa: Ghana: John & Lawson 1972a, John *et al.* 1977, Lieberman *et al.* 1979.

Halymenia duchassaingii (J. Agardh) Kylin Pl. 38, fig. 2.

Kylin, 1932; *Meristotheca duchassaingii* J. Agardh, 1872.

Plants solitary, to about 18 cm in length, pink to purplish-red, with the foliaceous fronds arising from a stipitate base; fronds irregularly divided from near the base into a few lanceolate segments, from 1-5 cm in width, with the margin irregularly lobed and erose-dentate and having few to many spine-like projections on the surface.

For additional figures see Taylor, 1960, p. 767, pl. 52, fig. 2.

Occurs on moderately sheltered to moderately wave-exposed rocks in the immediate sublittoral and down to a depth of about 6 m.

Distribution. World: in tropical parts of the Atlantic Ocean. West Africa: Cameroun: unpublished. Gabon: John & Lawson 1974a.

Halymenia vinacea Howe & W.R. Taylor Pl. 38, fig. 3.

Howe and Taylor, 1931.

Plants solitary, to about 10 cm in height, rose red, membranaceous and slightly gelatinous, attached by a small stipe; fronds simple or more usually divided into ovate lobes, less than 5 cm in width and from 50-80 μm in thickness, with the margin always entire; inner cortical cells elongate, from 7-14 μm in diameter and about 7 μm in height; outer cortical cells quadrate or polygonal, from 6-10 μm in diameter.

For additional figures see Howe and Taylor, 1931, p. 31, fig. 16, pl. 1, figs 1, 2, pl. 2, fig. 6.

Plate 38: Fig. 1. *Halymenia actinophysa:* Plant shortly stipitate with the membranaceous frond having a sinuate margin. Fig. 2. *Halymenia duchassaingii:* Plant membranaceous and irregularly divided, with the margin erose-dentate and the surface bearing a few spine-like projections. Fig. 3. *Halymenia vinacea:* Plant membranaceous and divided into ovate lobes.

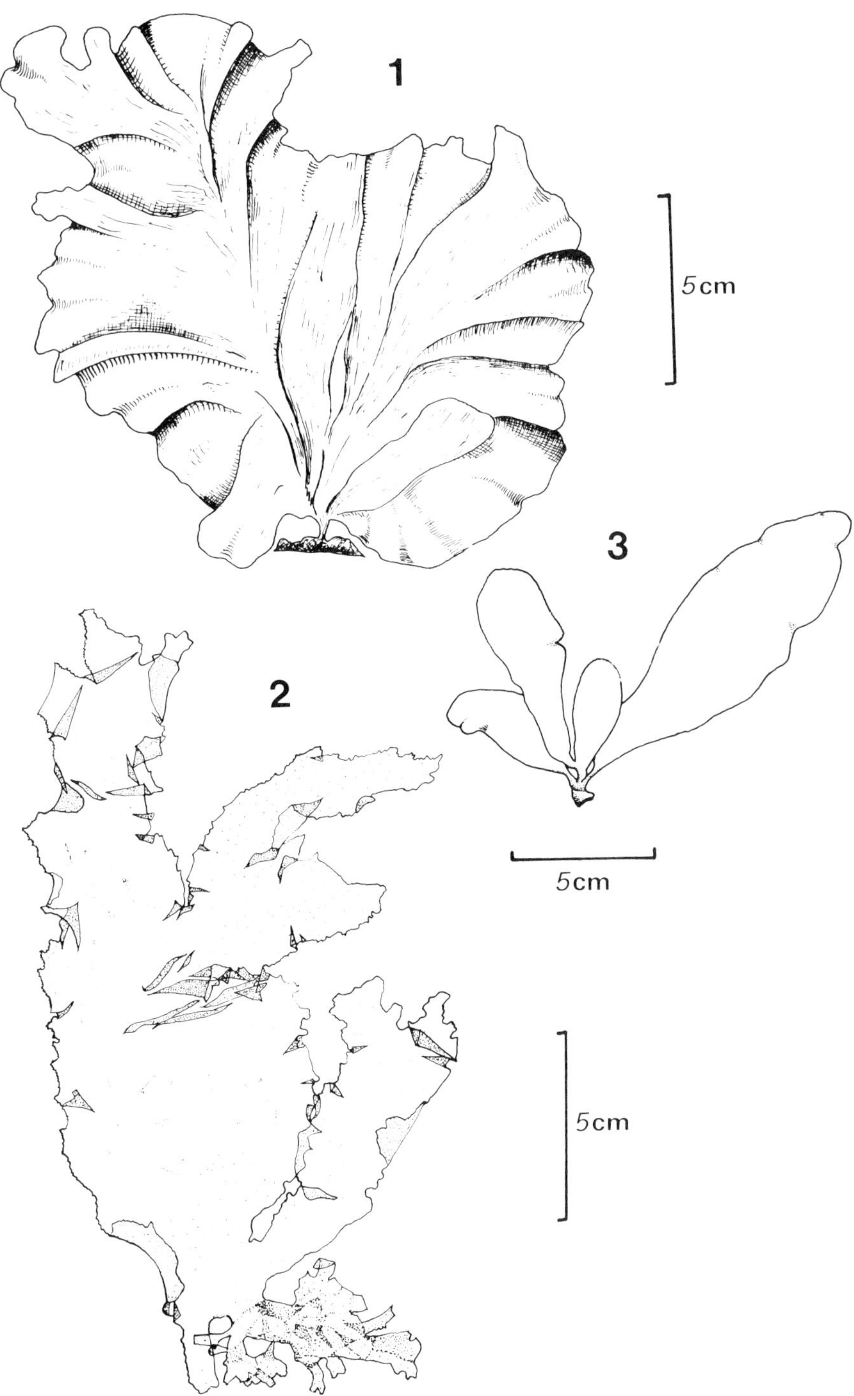
1
5cm
3
2
5cm
5cm

Found only on rare occasions in the region growing on rocky platforms or calcareous cobbles at depths between 20 and 25 m.

Distribution. World: in tropical parts of the Atlantic Ocean. West Africa: Ghana: John *et al.* 1977, 1980.

Remarks. Our plants closely correspond to the description of Howe and Taylor (1931) of the type collection from Brazil though even at the present time little is known regarding its variability in vegetative morphology.

Doubtful Record

Halymenia floresia (Clemente) C. Agardh

Aleem (1978) reports this plant dredged from a depth of 20 m over the continental shelf of the Sierra Leone peninsula. We regard this to be a doubtful record of a plant not previously reported from the mainland of West Africa though known from the Canary Islands.

Thamnoclonium Kützing 1843b.

Plants erect, consisting of terete to flattened branches, dichotomously to laterally divided, with the surface rough and irregular being covered by numerous short spinuous tubercules, sometimes intimately associated with a sponge; medulla of dense interwoven filaments; cortex pseudoparenchymatous, composed of rounded to angular assimilatory cells becoming smaller towards the surface; reproductive organs borne in lateral appendages scattered or in clusters on the branches; tetrasporangia scattered in the thickened outer cortex of the lateral appendages, sporangia tetrapartite.

World Distribution: in warm temperate and tropical seas.

Thamnoclonium claviferum J. Agardh Pl. 40, figs 1, 2.

J. Agardh, 1876.

Plants usually solitary, from 15-25 cm in height, rigid and horny, dark reddish-brown, with one to several erect branches arising from a broad discoid holdfast; branches several times alternately divided and arising at acute angles, often simple for a number of centimetres before the first division, terete or slightly compressed above; peltate warts in whorls of 4, oval to oblong multifid-foliated or echinate, attached to the main axis by a slender neck, covering plant except for the lower parts; reproductive organs borne in small roundish or slightly flattened lateral appendages.

For additional figures see Harvey, 1863, pl. 293.

Occurs at depths ranging from 10 to 20 m growing on rocky platforms or more usually found on sand-covered calcareous cobbles. It affords a suitable substratum for a number of other organisms and on occasion several plants of *Corynomorpha* have been found growing on a single individual.

Distribution. World: as for genus. West Africa: Ghana: John *et al.* 1977, 1980; *Thamnoclonium* sp., John & Lawson 1972a.

Family Peyssonneliaceae

Peyssonnelia Decaisne 1841

Plants prostrate, more or less calcified, attached directly to the substratum or by unicellular rhizoids, occasionally free towards the margin; thallus consisting of a layer of relatively uniform hypothallial cells, rarely with a subhypothallus, and a perithallus of straight or somewhat curved rows of cells which are simple or a few times dichotomously divided; tetrasporangia borne in superficial nemathecia, tetrapartite; spermatangia grouped in superficial sori; carpogonia developing in nemathecia.

World Distribution: cosmopolitan.

Key to the Species

1. Plants pink or light red, membranaceous and often free towards the margin; calcification confined to the hypobasal region. *P. inamoena*
1. Plants usually dark reddish-purple, hard and attached throughout; calcification well-developed except for the last few layers of cells towards the surface . . . . *P. polymorpha*

Peyssonnelia inamoena Pilger — Pl. 39, figs 3-5.

Pilger, 1911.

Plants often light red in the centre and pinkish towards the margin, membranaceous and smooth surfaced, becoming wrinkled on drying, attached to the substratum loosely and occasionally completely free towards the margin; thallus composed of hypothallial cells about 33 μm in diameter from which arise straight or very occasionally curved rows of perithallial cells from 8-10 μm in diameter; perithallial cells elongated at right angles to the surface below and becoming quadrate to somewhat flattened near the surface; calcification confined to the hypobasal region.

For additional figures see Denizot, 1968, p. 98 to 101, figs 78-84, P. 74, fig. 69.

Occurs in the lower eulittoral zone as prostrate and largely attached plants whilst in more sheltered situations such as in deeper water (10-28 m) it grows with the margins completely free.

Distribution. World: in warm temperate and tropical seas. West Africa: Cameroun: De Toni 1924, Pilger 1911, Schmidt & Gerloff 1957, Womersley & Bailey 1970. Ghana: John *et al.* 1977, 1980, Lieberman *et al.* 1979. Príncipe: *P. rubra*, Barton 1887, 1901, Steentoft 1967.

Remarks. Denizot (1968), who has revised the non-coralline encrusting red algae, has pointed out the very close resemblance between this taxon and *P. rubra*. The latter species is differentiated only by the presence of large calcified cells (cystoliths). Many of the earlier reports of this species from West Africa have been found on re-examination of the material on which they were based to be *P. inamoena*.

Peyssonnelia polymorpha (Zanardini) Schmitz Pl. 39, fig. 6.

Schmitz, in Falkenberg, 1879; *Lithymenia polymorpha* Zanardini, 1860.

Plants dark reddish-purple and often with swollen excrescences on the surface, completely crustose; thallus composed of a hypothallus of cells from 16-34 μm in diameter and 28-32 μm in height, from which arise straight and erect rows of perithallial cells about 10-20 μm in diameter; perithallial cells elongated at right angles to the surface below and becoming quadrate to flattened above; calcification well-developed and to within a few cells of the surface.

For additional figures see Denizot, 1968, p. 130, fig. 109.

Occurs in the lower part of the eulittoral zone on moderately wave-exposed rocks.

Distribution. World: in warm temperate and tropical seas, probably widespread. West Africa: Côte d'Ivoire: John 1977a. ?Sierra Leone: John & Lawson 1977a.

Doubtful Record

Peyssonnelia rosenvingii Schmitz

The only report of this species from West Africa is the one by Aleem (1978) who mentions it growing on stones and rocks at various localities around the Sierra Leone peninsula. We regard this West African record as doubtful until the material on which it is based is examined and the identification verified.

Plate 39: Figs 1, 2. *Halichrysis depressa:* 1. Cystocarpic plant showing the frond irregularly lobed or lacerated along the margin and with the surface papillose; 2. Section showing the large-celled medulla interspersed with smaller cells and the imprecisely delimited cortex. Figs 3-5. *Peyssonnelia inamoena:* 3. Undersurface of plant showing the single point of attachment with the substratum; 4. Upper surface bearing hemispherical nemathecia; 5. Section through a nemathecium showing developing carposporangia and the paraphyses arising superficially on the upper surface of the thallus. Fig. 6. *Peyssonnelia polymorpha:* Section through the upper part of the crust showing the rows of perithallial cells becoming somewhat flattened towards the surface. Figs 7, 8. *Hildenbrandia rubra:* 7. Crust in section showing the rows of nearly quadrate cells; 8. Tetrasporangia obliquely and irregularly divided. Fig. 9. *Rhodophysema africana:* Crust in section showing the tetrapartitely divided sporangia borne on a stalk and surrounded by paraphyses.

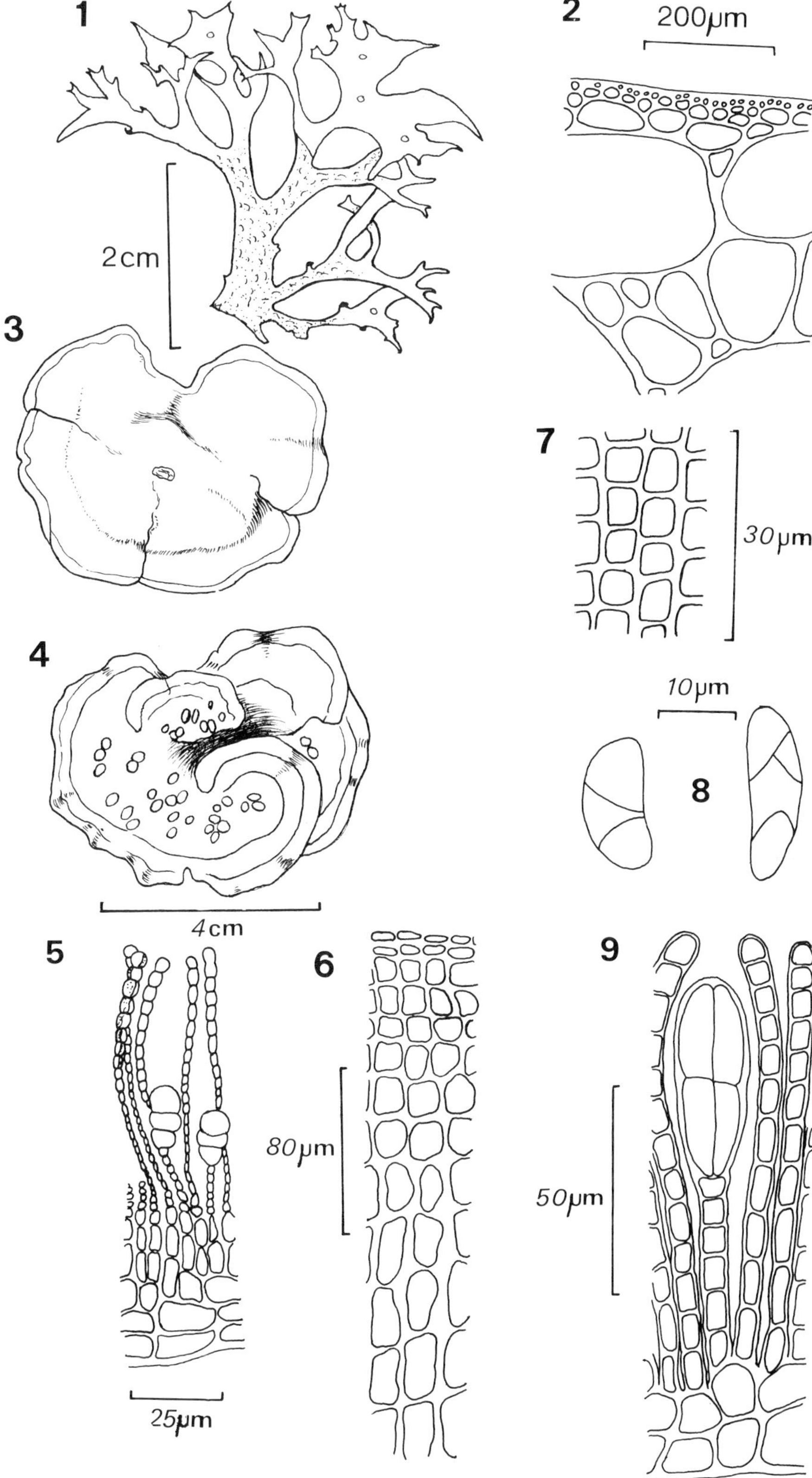
1
2
200µm
2cm
3
7
30µm
4
10µm
8
4cm
5
6
9
80µm
50µm
25µm

Family Hildenbrandiaceae

Hildenbrandia Nardo 1834

Plants forming closely adherent crusts of indefinite extent, without rhizoids or calcification; thallus homogeneous, composed of erect and closely united rows of quadrate cells or more rarely somewhat flattened cells; tetrasporangia arising in scattered subcylindrical to flask-shaped conceptacles which are occasionally elongated at right angles to the surface, sporangia tetrapartite or irregularly zonately divided; sexual reproduction unknown.

World Distribution: cosmopolitan.

Hildenbrandia rubra (Sommerfelt) Meneghini Pl. 39, figs 7, 8.

Meneghini, 1841; *Verrucaria rubra* Sommerfelt, 1826.

Plants forming dark red or nearly violet-red crusts, to about 500 μm in thickness; thallus consisting of little-branched rows of compact and nearly quadrate cells, from 4-7 μm in diameter; tetrasporangial conceptacles to about 120 μm in breadth and having a broad pore; tetrasporangia obliquely tetrapartite to somewhat irregularly divided, from 8-10 μm in diameter and 10-24 μm in length.

For additional figures see Newton, 1931, p. 297, fig. 184 (as *H. prototypus*).

Occurs as small encrusting discs to large expansions many metres in extent and is most common on moderately wave-exposed rocks subject to much surge of water; found in the upper eulittoral subzone and extends down into the sublittoral fringe.

Distribution. World: as for genus. West Africa: Côte d'Ivoire: *H. prototypus*, John 1977a. Gambia: *H. prototypus*, John 1977a, John & Lawson 1977b. Ghana: *H. prototypus*, John 1977a, Lawson 1957b, 1966. Liberia: *H. prototypus*, De May *et al.* 1977, John 1977a. São Tomé: *H. prototypus*, John 1977a; *H. rosea*, Henriques 1886b, 1887, 1917; *Hildenbrandia* sp., Steentoft 1967. Sierra Leone: *H. prototypus*, Aleem 1978, John 1977a, John & Lawson 1977a.

Remarks. Steentoft (1967) has examined the São Tomé collection but was unable to identify it to species as it was sterile. Henriques (1886b, 1887, 1917) reports this material as *H. rosea*, which is now accepted as a synonym of *H. rubra*, and is included above for the sake of completeness.

Order Palmariales · Family Palmariaceae

Palmaria Stackhouse 1801

Doubtful Record

Palmaria palmata (Linnaeus) O. Kuntze

This is the accepted nomenclatural equivalent of the plant cited as "*F.* [*Fucus*] *sobiliferus fl. dan.* cum varietatibus..." by Hornemann (1819) from "Danish Guinea" (present-day Ghana). We regard this record from the tropical coast of West Africa of a cold water species as most unlikely. It is possible that the original collection of Isert was destroyed by fire in 1807. For further comments on the Hornemann record see Price *et al.* (1983).

Rhodophysema Batters 1900

Plants encrusting and uncalcified, attached to the substratum directly, rhizoids absent; thallus with a monostromatic basal layer, cell fusions common, erect filaments little branched and laterally united; tetrasporangia in sori on the surface, associated with paraphyses, sessile or stalked, tetrapartitely divided; spermatangia develop superficially; carpogonia single-celled, terminal on erect filaments, dividing to form a tetrasporangium and a generative stalk cell on fertilization; carposporophyte absent.

World Distribution: from boreal-antiboreal to tropical parts of the Atlantic Ocean.

Rhodophysema africana John & Lawson — Pl. 39, fig. 9.

John and Lawson, 1974a.

Plants forming dark reddish-purple crusts; thallus composed of 2 layers of almost quadrate cells, rarely more, with cells from 7-14(-17) μm in diameter; tetrasporangia from 16-21μm in diameter and 35-42(-59) μm in length, borne on a stalk of 5 to 10(-14) cells; paraphyses somewhat club-shaped, from 110-145(-180) μm in length and consisting of from 12 to 16(-20) cells, from 3.5-7(-10) μm in diameter and cells 2 to 3 times longer than broad below and becoming shorter than broad above.

Occurs in the lower part of the barnacle subzone on rocks and also on the shells of gastropods in the immediate sublittoral.

Distribution. World: in tropical parts of the eastern Atlantic Ocean. West Africa: Gabon: Ganesan & West 1975, John & Lawson 1974a, South & Whittick 1976.

Champia Desvaux 1809

Plants erect or semi-prostrate, consisting of terete to flattened branches, alternately or oppositely divided, hollow, regularly constricted and having transverse septa at the constrictions; branches having an interrupted outer layer of cortical cells, an inner continuous layer of large thin-walled cells, and a few longitudinal gland-bearing filaments running along the outer part of the central cavity; tetrasporangia scattered amongst the cortical cells, tetrahedrally divided; cystocarps sessile and ovate, surrounded by small filaments, having an apical pore.

World Distribution: widespread from boreal-antiboreal to tropical seas.

Key to the Species

1. Plants semi-prostrate, with the branches often attached to one another and to the substratum by discoid holdfasts; branches markedly compressed and the segments less than half as long as broad .... *C. vieillardii*
1. Plants erect and basally attached; branches terete or subterete and segments usually greater than half as long as broad.................................. 2
2. Plants obviously tripinnately divided; branches having very indistinct septa .... *C. tripinnata*
2. Plants usually bipinnately divided; branches having distinct septa 3
3. Plants usually less than 5 cm in height; branches terete to subterete and from 0.5-1.5 mm in width.. *C. parvula*
3. Plants usually from 5-12 cm in height; branches terete and from 1.5-2.5 mm in diameter... *C. salicornoides*

Champia parvula (C. Agardh) Harvey — Pl. 40, fig. 6.

Harvey, 1853; *Chondria parvula* C. Agardh, 1824.

Plants forming small clumps, to about 5 cm in height; branches distichous and alternate or occasionally opposite, sometimes intricate and coalescing, terete or subterete, having obvious septa at the constrictions; segments from 0.5-1.5 mm in width and 0.5 to 2 times as long as broad.

For additional figures see Newton, 1931, p. 440, fig. 263.

Occurs together with other small algae growing on moderately wave-exposed rocks in the lower eulittoral subzone and on cobbles and rocky banks at depths down to 12 m.

Distribution. World: as for genus. West Africa: Cameroun: Lawson 1955, John 1977a, Stephenson & Stephenson 1972. Côte d'Ivoire: John 1977a. Ghana: John *et al.* 1977, Lieberman *et al.* 1979. Príncipe: Carpine 1959, John 1977a, Steentoft 1967. São Tomé: Carpine 1959, John 1977a, Steentoft 1967.

Champia salicornoides Harvey Pl. 40, fig. 5.

Harvey, 1853.

Plants solitary and bushy, to about 10 cm in height, pale rose red, somewhat gelatinous; branches irregularly opposite, alternate or occasionally verticillate, terete, with obvious constrictions at the septa; segments from 2-3 mm in diameter and 1.5 to 2.5 times as long as broad.

For additional figures see Børgesen, 1920, p. 410, fig. 394.

Occurs on rocky platforms and calcareous cobbles at depths of from 8 to 15 m.

Distribution. World: widespread in many warm temperate and tropical seas. West Africa: Ghana: John & Lawson 1972a, John *et al.* 1977, 1980, Lieberman *et al.* 1979.

Champia tripinnata Zanardini

Zanardini, 1851.

Plants loose and bushy, from 4-5 cm in height, light red and almost translucent; branches distichous and conspicuously tripinnate, alternate, subopposite or occasionally verticillate, terete to subterete, with inconspicuous septa; segments from 1-2.5 mm in width and 1 to 2 times as long as broad.

For figures see Zanardini, 1858, pl. 11, figs 2, 2a.

There is no information on the ecology of this species in the region.

Distribution. World: in the eastern tropical Atlantic Ocean and the Red Sea. West Africa: Príncipe: Steentoft 1967.

Champia vieillardii Kützing Pl. 40, fig. 4.

Kützing, 1886.

Plants semi-prostrate, to about 3 cm in height, forming patches as much as 30 cm in diameter, reddish-purple and having a bluish iridescence when fresh, with the ultimate branches often arcuate and attached both to the lower branches and the substratum by discoid holdfasts; branches distichous and usually alternate, rarely opposite and tripinnate, markedly compressed; segments from 2-4 mm in width and less than half as long as broad.

For additional figures see Kützing, 1866, pl. 37, figs e, f.

Occurs on rocky platforms and over large polyzoans at depths ranging from 10 to 28 m though in other parts of its range it is known from the eulittoral zone.

Distribution. World: probably widespread in tropical seas. West Africa: Ghana: John 1972a, John & Lawson 1972a, John *et al.* 1977, Lieberman *et al.* 1979.

Remarks. The confusion that has existed between this species and *Champia compressa* Harvey has been discussed by Dawson (1954a) and the present situation is also referred to by John & Lawson (1972a).

Chylocladia Greville ex Hooker 1833

Doubtful Record

Chylocladia verticillata (Lightfoot) Bliding

This is usually considered to be a cold temperate species though Aleem (1978) reports collecting a single specimen (as *Chylocladia kaliformis* (Goodenough & Woodward) Hooker) from the lower littoral at Lakka on the Sierra Leone peninsula. Until the material on which this West African record is based can be examined and the determination verified it seems best for now to regard it as a doubtful.

Lomentaria Lyngbye 1819

Plants consisting of terete or compressed branches, hollow and tubular, usually constricted at intervals into slender segments tapering towards both ends, with branching irregular or somewhat lateral; branches having a medulla of a few gland-bearing longitudinal filaments in the central cavity, with the cortex composed of one or more layers of small roundish or angular cells; tetrasporangia scattered or grouped together in cavities on the surface of the branches; spermatangia in superficial sori; cystocarps scattered and prominent, globose or subconical, with a single pore.

World Distribution: widespread from boreal-antiboreal to tropical seas.

Lomentaria articulata (Hudson) Lyngbye Pl. 40, fig. 7.

Lyngbye, 1819; *Ulva articulata* Hudson, 1762.

Plants soft and bushy, to about 5 cm in height, reddish-brown and sometimes iridescent when fresh; branches dichotomous below and oppositely or pinnately divided above, lateral branches sometimes in whorls at the nodes, terete to subterete, with the apices often attenuate; segments often cylindrical below and becoming elliptical to moniliform above, from 0.5-2 mm in diameter and up to 6 mm in length.

For additional figures see Lyngbye, 1819, pl. 30, figs 1-4.

Pilger (1911) reports this plant as growing in luxuriance over rocks protruding through sand.

Distribution. World: as for genus. West Africa: Cameroun: Goor 1923, Pilger 1911, Schmidt & Gerloff 1957.

Plate 40: Figs 1, 2. *Thamnoclonium claviferum:* 1. Plant several times alternately divided and towards the base devoid of spinuous tubercles; 2. Portion of a branch showing the tubercles. Fig. 3. *Chrysymenia enteromorpha:* Plant irregularly divided into saccate branches tapering towards base and apex. Fig. 4. *Champia vieillardii* (after John and Lawson, 1972a): Plant with segmented and flattened branches and attached by a basal holdfast and marginal haptera. Fig. 5. *Champia salicornoides:* Portion of the seg-

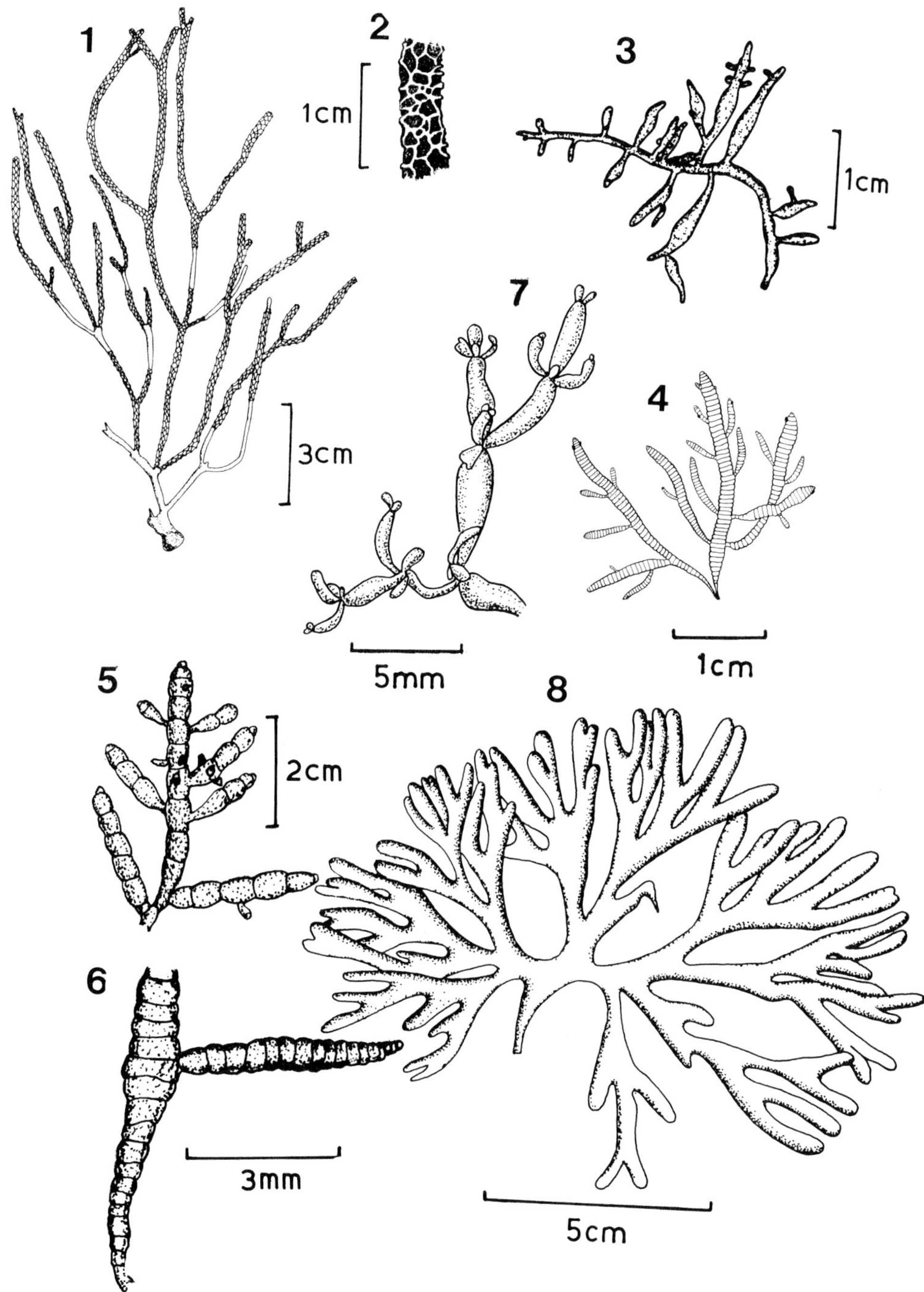

mented plant bearing cystocarps. Fig. 6. *Champia parvula:* Portion of plant showing the branches regularly constricted into short segments. Fig. 7. *Lomentaria articulata:* Portion of a plant showing the branches constricted into cylindrical to somewhat elliptical or fusiform segments. Fig. 8. *Halymenia agardhii:* Plant showing the terete branches slightly flattened at the dichotomies.

Family Rhodymeniaceae

Botryocladia (J. Agardh) Kylin 1931

Plants consisting of terete and dichotomously or alternately divided branches, bearing branchlets in the form of variously sized and shaped vesicles, attached to the substratum by discoid holdfasts; branchlets having gland cells projecting into a mucilage-filled central cavity, with the cells of the cortex becoming smaller towards the outside and the outermost layer not always continuous; tetrasporangia scattered and embedded in the cortex, tetrapartitely divided; spermatangia in small sori on the surface; cystocarps scattered, prominent, with a single pore.

World Distribution: widespread from warm temperate to tropical seas, less common in cold temperate and boreal-antiboreal seas.

Key to the Species

1. Plants semi-prostrate and spreading, with the robust branches from 10-25 cm in length .... *B. botryoides*
1. Plants erect, with the slender branches less than 3 cm in length 2

2. Vesicles 2 to 3 layers thick, with the outermost layer or layers forming a discontinuous cortex arranged in a net-like manner .... *B. lawsonii*
2. Vesicles 4 layers thick, with a continuous outer cortical layer ...... *B. guineensis*

Botryocladia botryoides (Wulfen) J. Feldmann — Pl. 41, fig. 1.

J. Feldmann, 1942; *Fucus botryoides* Wulfen, 1789.

Plants forming semi-prostrate and spreading clumps, to about 3 cm in height, dark reddish-brown; branches terete, alternately to irregularly divided, from 10-25 cm in length, bearing vesicles at intervals on the upper side; vesicles ovoid to pyriform, from 3-8 mm in length, with a continuous outer layer of assimilatory cells.

For additional figures see Børgesen, 1920, p. 403, fig. 388 (as *Chrysymenia uvaria*).

Occurs in amongst other algae in the turf or on sand-covered rocks in moderately sheltered to moderately wave-exposed situations and extends from the lower eulittoral subzone to a depth of about 10 m.

Distribution. World: in warm temperate and tropical parts of the eastern Atlantic Ocean as well as in the Mediterranean. West Africa: Cameroun: John & Lawson 1972a; *Chrysymenia uvaria*, Pilger 1911. Gabon: John & Lawson 1974a. Ghana: John & Lawson 1972a, John *et al.* 1977, Lieberman *et al.* 1979.

Botryocladia guineensis John — Pl. 41, figs 2-4.

John, 1972a.

Plants erect, from 2-5 cm in height, rose red to reddish-brown, with a single branch arising from a small discoid holdfast; branches terete, from 1-1.5 mm in diameter, somewhat irregularly divided, bearing from 2 to 6(-10) vesicles; vesicles spherical to elongate-pyriform, to about 16(-20) mm in greatest diameter and to 25(-30) mm in length, composed of 4 layers of cells with the outermost cell layer continuous; gland cells spherical to pyriform, from 4 to 8(-12) on inner face of cells lining cavity.

For additional figures see John, 1972a, p. 33 to 35, figs 1-7.

Occurs on calcareous cobbles lying on sand at depths ranging from 12 to 25 m though in greatest abundance at the lower end of this depth range.

Distribution. World: in the tropical eastern Atlantic Ocean. West Africa: Ghana: John 1972a, John *et al.* 1977, 1980, Lieberman *et al.* 1979.

Botryocladia lawsonii John Pl. 41, figs 5-7.

John, 1980.

Plants erect, from 1-4 cm in height, reddish-brown, often several branches arising from a discoid holdfast; branches terete, from 300-700 μm in diameter, irregularly divided, bearing from 5 to 20(-24) shortly stipitate vesicles; vesicles initially sub-spherical becoming elongate and cylindrical, from 1-8(-10) mm in greatest diameter and 4-20(-25) mm in length, from 2 to 3 cells thick, with the outermost layers arranged in a net-like manner over the large innermost cells; gland cells spherical to ovoid, from 2 to 8(-10) occasionally present on inner face of cells lining cavity.

For additional figures see John, 1980, p. 91, 92, figs 1-7.

Occurs on calcarous cobbles at a depth of 13 to 15 m.

Distribution. World: tropical West Africa. West Africa: Ghana: John 1980; *Botryocladia* sp., John *et al.* 1977, Lieberman *et al.* 1979.

Chrysymenia J. Agardh 1842

Plants consisting of soft, terete or somewhat flattened branches, hollow and sac-like or tubular, usually alternately or irregularly divided; medulla absent or composed of just a few filaments; inner cortex of large and closely packed cells, sometimes bearing one to several gland cells; outer cortex of one to three layers of small cells; tetrasporangia scattered between the cortical cells, tetrapartite; cystocarps scattered and slightly projecting, hemispherical, opening by a single pore.

World Distribution: widespread in warm temperate and tropical seas.

Chrysymenia enteromorpha Harvey Pl. 40, fig. 3.

Harvey, 1853.

Plants solitary, to about 3 cm in length, light reddish-purple and usually iridescent; branches somewhat irregularly divided, terete to slightly compressed, to 4 mm in width, saccate and oblong, tapering towards the base, rounded or slightly tapering towards the apex; septa at the constriction of the branches and consisting of large polygonal cells; cystocarps bluntly conical, about 180 μm in diameter.

For additional figures see Børgesen, 1920, p. 397 to 399, figs 381-383.

Found on a few occasions in the region growing on rocky platforms at a depth of about 10 m.

Distribution. World: as for genus. West Africa: Ghana: John & Lawson 1972a, John *et al.* 1977.

Halichrysis* (J. Agardh) Schmitz 1889

Plants prostrate or semi-prostrate and dorsiventral; medulla consisting of large cells interspersed with small cells or filaments, delimited imprecisely from the cortex of a few layers of cells with the outermost cells small and assimilatory; tetrasporangia produced on a one-celled stalk laterally on the cortical cell rows, tetrapartite; cystocarps forming wart-like projections on the surface and opening by a single pore, with the gonimolobes developed on the top of a prominent gonimoblast stalk arising from a large fusion cells.

World Distribution: in the Mediterranean and warm temperate and tropical parts of the Atlantic and Pacific Oceans.

Halichrysis depressa (J. Agardh) Schmitz — Pl. 39, figs 1, 2.

Schmitz, 1889; *Chrysymenia depressa* J. Agardh, 1851.

Plants consisting of a number of foliaceous fronds, irregularly lobed or lacerated along the margins, overlapping and sometimes united by localised fusions, attached to the substratum by shortly stalked haptera often arising marginally as finger-like projections; fronds pink to dark reddish-purple, sometimes mottled and iridescent with a metallic hue, fleshy, from 600-800 μm thick; tetrasporic plants irregularly or regularly forked to form linear or cuneate segments, from 1-10 mm in width; cystocarpic plants simple, rarely forked, to about 20 mm in width, with the surface papillose.

*See discussion in Price *et al.* (1983).

Plate 41: Fig. 1. *Botryocladia botryoides:* Plant consisting of several small vesicles borne on a well-developed system of semi-prostrate branches. Figs 2-4. *Botryocladia guineensis:* 2. Plant having a little developed system of erect branches bearing large vesicles; 3. Section through a mature cystocarp (after John, 1972a); 4. Section of a vesicle showing the inner layers of large colourless cells and the smaller outer layers of assimilatory cells (after John, 1972a). Figs 5-7. *Botryocladia lawsonii:* (after John, 1980): 5. Plant showing several elongated vesicles borne on an irregularly divided erect axis; 6. Surface view of a vesicle with tetrapartitely divided tetrasporangia arising amongst the net-like arrangement of cortical cells; 7. Section of a vesicle showing the layer of large inner colourless cells and the discontinuous outer layer or layers of assimilatory cells.

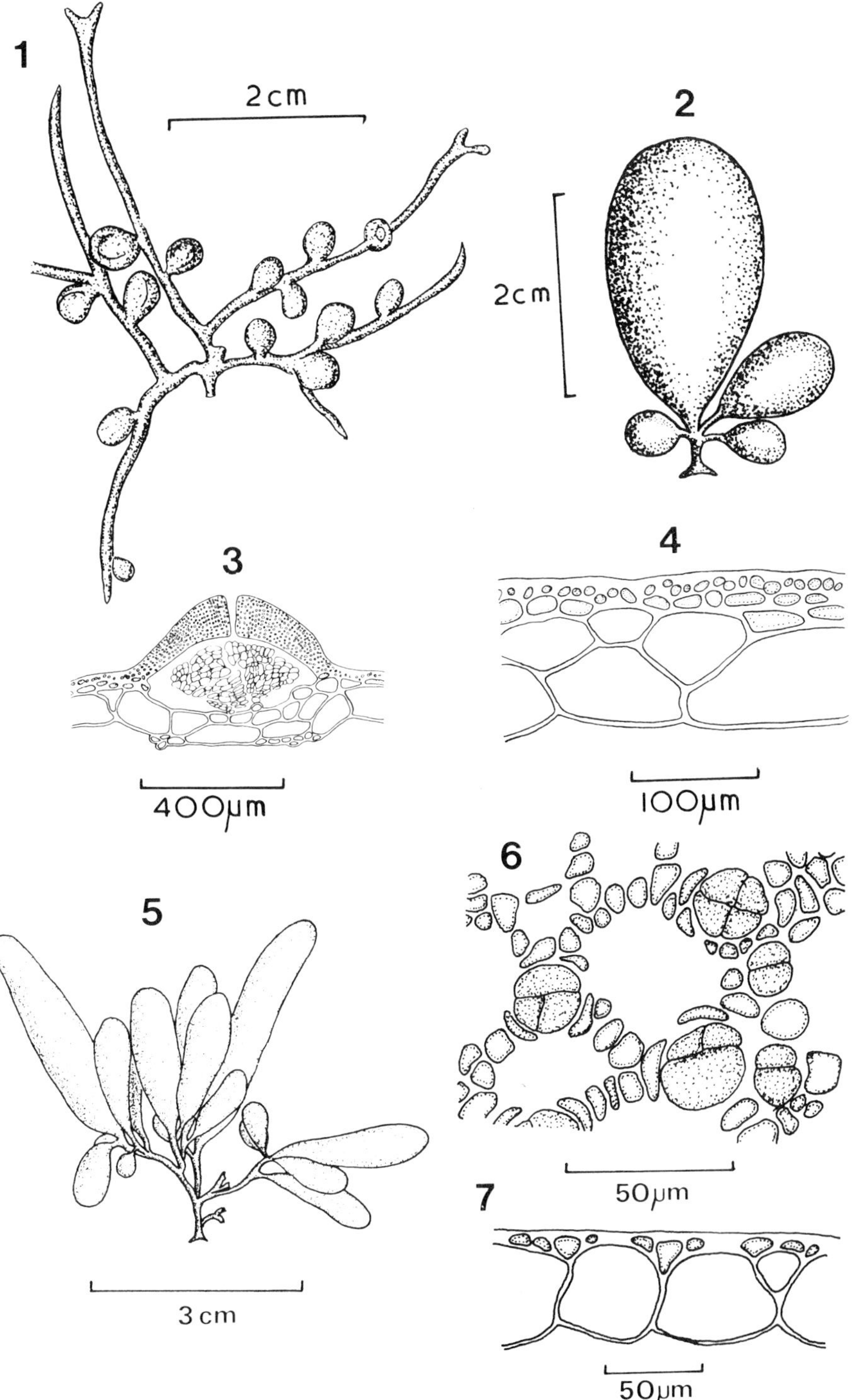
1
2 cm
2
2 cm
3
400 µm
4
100 µm
5
3 cm
6
50 µm
7
50 µm

For additional figures see Huvé & Huvé, 1976, p. 378 to 380, 382, 384, 386, 388, 390, figs 1-27.

Occurs on rocky platforms and on rare occasions growing on calcareous cobbles at depths ranging from about 8 to 16 m.

Distribution. World: in the Mediterranean and tropical parts of the eastern Atlantic. West Africa: ?Bioko: unpublished. Ghana: unpublished.

Remarks. Plants having a similar colouration, habit and vegetative structure have been found in shallow subtidal habitats in Cameroun (unpublished) and Gabon. The Gabon material has zonate tetraspores and has tentatively been referred to the genus *Meristotheca* (John & Lawson, 1974a), whilst sterile Ghanaian plants closely resemble *Meristotheca coacta* Okamura (John *et al.*, 1977). Recent collections of cystocarpic plants have all been found on detailed examination to be *Halichrysis depressa*.

Rhodymenia Greville 1830

Plants erect and flattened, firm and fleshy, consisting of dichotomously, irregularly or palmately divided fronds, sometimes proliferous from the margin, arising from a stipitate or stoloniferous base and having discoid holdfasts; medulla composed of large isodiametric cells; cortex of a few layers of small assimilative cells, particularly thick in fertile parts due to the development of anticlinal rows of cells; tetrasporangia scattered over the entire frond or in sori, embedded in cortex or specialised nemathecia, sporangia tetrapartite; cystocarps scattered or confined to branch tips, very prominent and hemispherical, with a distinct pore.

World Distribution: cosmopolitan.

The members of this genus exhibit a great deal of morphological plasticity and yet in the past species separation has often been largely based on vegetative features. This genus is in need of revision on a global scale.

Rhodymenia pseudopalmata (Lamouroux) Silva Pl. 42, figs 1-3.

Silva, 1952; *Fucus pseudopalmatus* Lamouroux, 1805.

Plants erect, up to 6 cm in height, cartilaginous, reddish-purple to purplish-brown, often fan-shaped in appearance, expanding from a stipitate base with a small discoid holdfast or occasionally from a stoloniferous base; branches flattened, cuneate or rounded towards base, usually dichotomously divided, from 2-6(-8) mm in breadth, occasionally tapering towards the rounded or spathulate apices; tetrasporangia in small rounded sori near the tips of the branches.

For additional figures see Guiry, 1977, p. 387, fig. 1, p. 388, figs 2, 3, p. 391, figs 14a-h, p. 393, figs 15-22, p. 395, figs 23-28, p. 402, figs 31-34, p. 404, figs 35a-f, p. 407, figs 36-40.

Occurs most commonly on the underside of overhanging rock ledges in the sublittoral fringe usually growing together with *Cryptonemia luxurians;* occasionally to a depth of 25 m on rocky platforms and calcareous cobbles.

Distribution. World: in warm temperate and tropical seas. West Africa: Gabon: ?John & Lawson 1974a. Ghana: John *et al.* 1977, Lieberman *et al.* 1979; *R. palmetta,* Dickinson & Foote 1950, Lawson 1956, Stephenson & Stephenson 1972. Sierra Leone: John & Lawson 1977a.

Remarks. Many plants from the region appear to correspond closely with variety *pseudopalmata,* but a few of those with narrow fronds often tapering towards the rounded apices show some resemblance to *R. holmesii* Ardissone (= *R. pseudopalmata* variety *elisiae* (Duby) Guiry in Guiry & Hollenberg).

Order Ceramiales · Family Ceramiaceae*

Antithamnion Nägeli 1847

Plants filamentous, consisting of an erect and prostrate system of branches, with the erect branches alternately divided below and often opposite or whorled above or occasionally secund, bearing two or four branches of determinate growth which may be alternately, subdichotomously or unilaterally divided; gland cells often present, sessile on the determinate branches or branchlets; chloroplasts numerous, elliptical to rounded, banded or sometimes elongate; tetrasporangia commonly borne on the adaxial side of the determinate branches, often replacing branchlets, sporangia tetrapartite; gonimblast formed towards the base of the determinate branches.

World Distribution: cosmopolitan.

Key to the Species

1. Determinate branches bearing opposite or secund branchlets and often terminating in a prominent gland cell .... *A. butleriae*
1. Determinate branches of another form 2

2. Determinate branches alternately or subdichotomously divided....... *A. elegans*
2. Determinate branches bearing adaxially from each cell short branchlets........ .... *Antithamnion* sp.

Antithamnion butleriae Collins — Pl. 42, fig. 5.

Collins, 1901.

*Further plants believed to be referable to this family have been found in Gabon, Bioko and Ghana but the lack of reproductive structures has prevented definitive determinations. The genus *Lasiothalia* is reported with reservations from Liberia (see De May *et al.*, 1977).

Plants found growing entangled with other small red algae, less than 1 mm in height; main branches dichotomous, alternate or sometimes opposite; determinate branches absent near the base or sometimes unilateral below, opposite above, with cells up to 30 μm in diameter and 2 to 6 times longer than broad, bearing a series of simple branchlets on the adaxial surface, from 2 to 3 cells in length; gland cell formed terminally on the determinate branches, touching the basal cell of the branchlet on each side.

For additional figures see Børgesen, 1920, p. 466, fig. 425a-c.

Found in the lower eulittoral subzone in amongst an uniseriate sandbinding red alga which is possibly a species of *Spermothamnion.*

Distribution. World: tropical parts of the Atlantic Ocean. West Africa: Sierra Leone: John and Lawson 1977a.

Antithamnion elegans Berthold Pl. 43, fig. 1.

Berthold, 1882.

Plants growing together with other small algae, or as felty patches on the surface of larger algae and polyzoans, to about 3 mm in height; branches of the erect filaments alternately divided and bearing oppositely a pair of determinate branches arising from the upper part of a cell, with cells 16-25 μm in diameter and 1 to 2.5 times longer than broad; determinate branches alternately or subdichotomously divided, from 5-12 μm in diameter at the base and cells from 1 to 4 times longer than broad, with apical cells tapering and blunt; gland cells oblong, from 7-10 μm in diameter and to about 16 μm in length, borne on the first or second cell above the division of a determinate branch; tetrasporangia sessile and often on the first or second cell of a determinate branch, about 25 μm in diameter and to 35 μm in length.

For additional figures see Børgesen, 1930a, p. 57 to 59, figs 21-23.

Occurs in the algal turf in the lower eulittoral subzone where wave-action is moderate and found growing on larger algae and polyzoans in the shallow sublittoral down to a depth of about 15 m.

Distribution. World: in warm temperate and tropical parts of the Atlantic Ocean. West Africa: Ghana: John *et al.* 1977, 1980, Lieberman *et al.* 1979. Togo: John & Lawson 1972b.

Plate 42: Figs 1-3. *Rhodymenia pseudopalmata:* 1, 2. Variation in form of the flattened and dichotomously divided plant; 3. Section showing the large medullary cells and the smaller outer cortical cells. Fig. 4. *Antithamnion* sp. (after John and Lawson, 1977a): Terminal portion of a branch showing the determinate branches decreasing in length towards the tip of the main axis and bearing short adaxial branchlets. Fig. 5. *Antithamnion butleriae* (after John and Lawson, 1977a): Terminal portion of a branch with many of the determinate branches terminating in a large gland cell. Figs 6, 7. *Centroceras bellum* (after John and Lawson, 1977a): 6. Portion of a plant showing a prostrate axis bearing erect branches and tetrasporangia immersed in the swollen tip of one of these branches; 7. Portion of a branch showing the vertical rows of cortical cells.

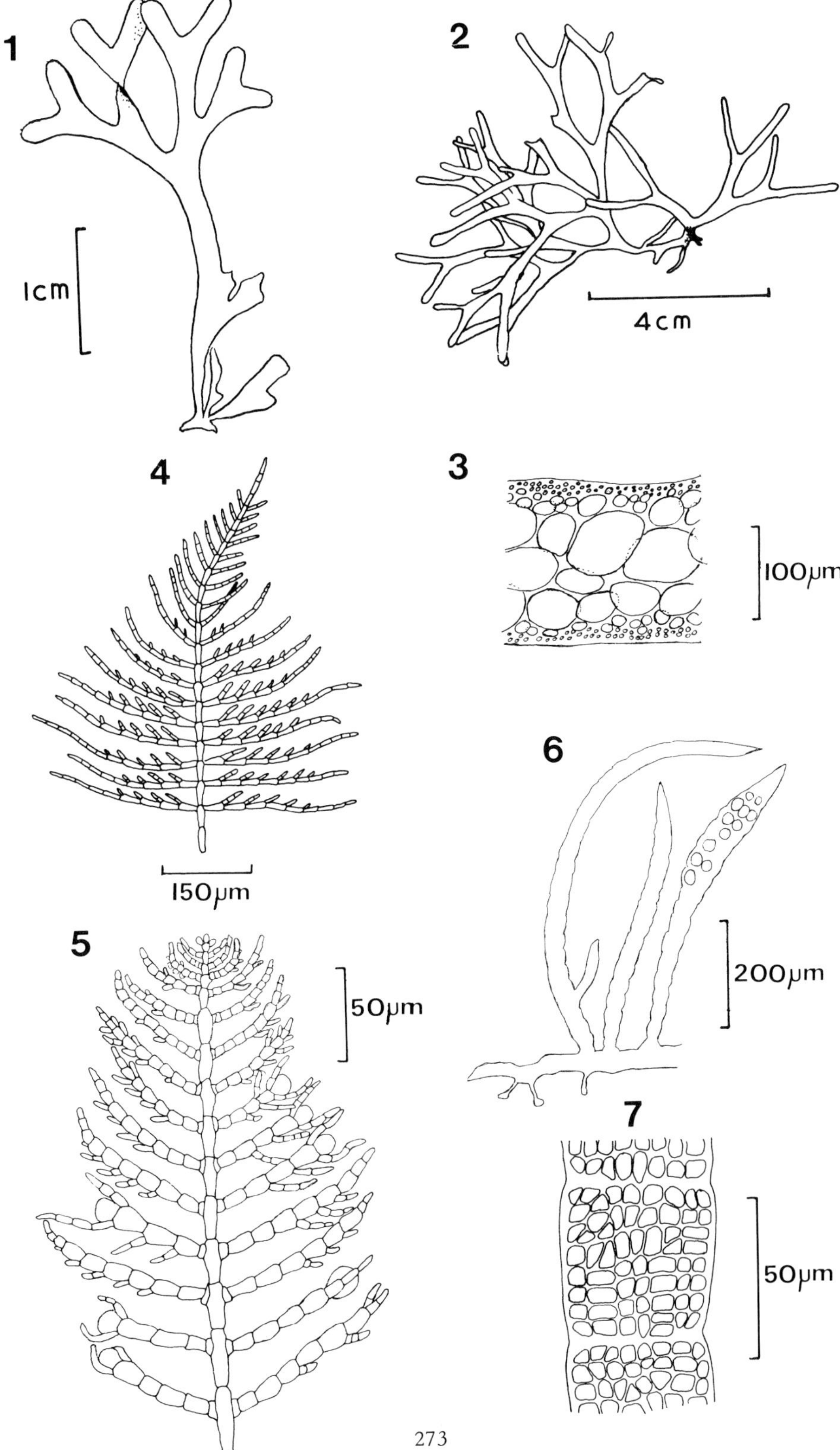
1
2
1cm
4cm
4
3
100µm
6
150µm
5
50µm
200µm
7
50µm

Remarks. The plants from the region closely agree with the detailed description and figures provided by Børgesen (1930a) of Canary Island specimens although our plants have somewhat smaller dimensions.

Undetermined Species

Antithamnion sp. Pl. 42, fig. 4.

Plants about 500 μm in height, found as a few branches entangled with other small algae; branches of the erect system borne on a short cell of a prostrate branch opposite to a multicellular rhizoid-like branch, pinnately divided; cells of main branches about 20 μm in diameter and up 4 times longer than broad; determinate branches decreasing in length upwards and giving the plant a pyramidal outline, bearing adaxially at an acute angle one to several-celled branchlets; secondary indeterminate branches arising opposite to a determinate branch; gland cells rare, arising on the branchlets.

Occurs together with *Antithamnion butleriae* in the lower eulittoral subzone.

Distribution. West Africa: Sierra Leone: John & Lawson 1977a.

Remarks. This plant bears a close resemblance to *Antithamnion defectum* Kylin but differs in having a determinate branch arising opposite a pinnate secondary indeterminate branch (see Hardy-Halos, 1968, p. 176, fig. 6, pl. 1.)

Bornetia Thuret 1855

Plants filamentous, subdichotomously to unilaterally divided and usually consisting of large cylindrical to globose cells; tetrasporangia and spermatangia arising in dense clusters, borne on short, incurved, lateral branches terminating in forcipate involucral cells; tetrasporangia tetrahedrally divided; carposporangia attached to whorled involucral cells, without a gelatinous covering.

World Distribution: widespread from boreal-antiboreal to tropical seas.

Bornetia secundiflora (J. Agardh) Thuret Pl. 43, fig. 2.

Thuret, 1855; *Griffithsia secundiflora* J. Agardh, 1841.

Plate 43: Fig. 1. *Antithamnion elegans:* Terminal portion of a plant with the opposite or occasionally whorled branchlets of determinate growth sometimes bearing clusters of spermatangia. Fig. 2. *Bornetia secundiflora:* Portion of a plant showing the regularly dichotomously divided branches of large cells. Figs 3, 4. *Centroceras clavulatum:* 3. Terminal portion of a branch showing the whorls of spines at the nodes; 4. Portion of a branch with the cortical cells in longitudinal rows and spines at the nodes. Figs 5, 6. *Grallatoria tingitana:* 5. Portion of a branch showing the spirally arranged branchlets of determinate growth; 6. Portion of a creeping branch with a rhizoid growing from the basal cell of a branchlet.

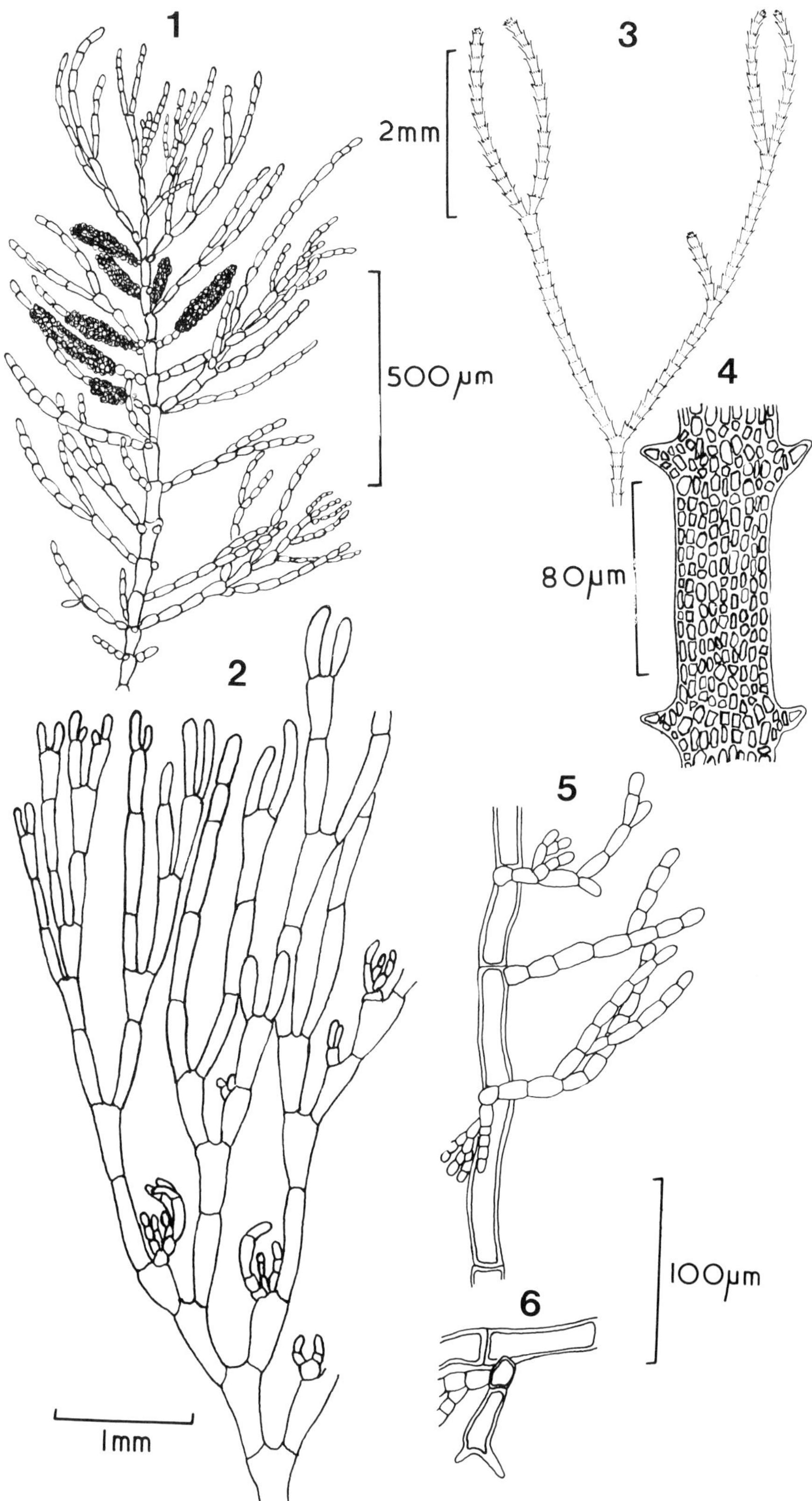
1
2mm
3
500 µm
4
80 µm
2
5
100 µm
6
1mm

Plants forming coarse, erect and bushy tufts, from 6-12 cm in height, bright pinkish-red; branches regularly dichotomously divided at intervals of about 2 to 5 cells; cells of the branches large and cylindrical to pyriform, from 280-420(-700) μm in diameter and 2 to 6 times longer than broad, with the apical cell blunt; tetrasporangia clustered on short stalks, borne on the inner side of curved involucral branchlets; spermatangia sessile, borne on recurved and bifurcate involucral branchlets.

For additional figures see Baldock and Womersley, 1968, p. 198, figs 1-4, pl. 1, fig. 1.

Occurs on rocky platforms and calcareous cobbles at depths of about 8 to 10 m.

Distribution. World: as for genus. West Africa: Ghana: John & Lawson 1972a, John *et al.* 1977, Lieberman *et al.* 1979.

Callithamnion* ** Lyngbye 1819

Plants filamentous, monosiphonous, corticate by downgrowing rhizoidal filaments or ecorticate, arising from a rhizoidal holdfast, variously but usually copiously branched; branching monopodial, but sometimes of spreading appearance from pseudodichotomous development; adventitious branching irregular (from the cortical matrix in some species) giving laterals of form similar to the main system branches; branching pattern alternate, spiral or distichous, sometimes regularly or irregularly pseudodichotomous; active main apices or major lateral apices surpassed by lateral development, or clearly exceeding it according to species; determinate apical cells gradually to abruptly domed, of variable length, or conical; apical or subapical unicellular hyaline hairs occasional in younger growth regions of some species, commonly and suddenly deciduous; vegetative cell length-to-diameter ratio 1:1 to 16:1; mature cells multinucleate, with uninucleate apices, or uninucleate throughout; chloroplasts irregular and discoid (young), developing to irregular masses or ribbons from division and coalescence with age; chromosome content usually (but not always) n=28-33; gametangial phase commonly dioecious, less often monoecious; spermatangia on outer laterals in (coalescent) cushions, or in adaxial secund rows, sometimes luxuriantly spreading to other surfaces and to higher order laterals; with paired (occasionally only unilateral) support cells; carpogonial branch four-celled (including carpogonium), straight, 'U'-shaped, or zig-zag in divergent planes; auxiliary cell generating after fertilisation two, less often three, gonimoblast filaments; gonimoblasts producing rounded or conical gonimolobes, subsidiarily lobed or not; carpogonia commonly 25-40 μm × 15-30 μm; carposporophyte normally diploid; sporophytes with tetrasporangia, less often (in some species) with bisporangia or polysporangia; tetrasporangia tetrahedral, rounded, ellipsoid or

*We are indebted to Mr. J. H. Price for kindly providing the account of this genus.

**Pilger (1920) has described *Callithamnion mildbraedii* from the island of Pagalu.

ovoid, less often nearly spherical, variable in size and shape, commonly one per cell, adaxially sessile on peripheral laterals, sometimes spreading to other surfaces; parasporangia in some species.

World Distribution. probably pan-Atlantic (except in coldest areas of Arctic; Antarctic) and Mediterranean; greatest abundance in North Eastern Atlantic and Western Mediterranean; sparsely present in Pacific and Indian Oceans.

Material reported as *Callithamnion tenuissimum*, as *C. roseum sensu* Harvey and (by Pilger) as *C. mildbraedii* may overlap in morphology. Only small amounts of *C. tenuissimum* and *C. roseum* are so far known for West Africa; this material is lacking one or other reproductive aspect and is largely in poor condition. Similar problems arise with the reported *C. hookeri*, although the few tetrasporangia present are identical with the usual form and position in *C. hookeri* of the Northern Hemisphere. The most fully recorded *Callithamnion* is *C. granulatum*, detected only in very restricted locations in western Ghana, Liberia, and Côte d'Ivoire. Sandy-bottomed pools bear fine filamentous forms such as *C. tenuissimum*, *C. roseum* (and probably *C. mildbraedii*); these are not prolific and more comparative material is required for study of the taxa.

Certain other taxa in the genus (e.g. *C. corymbosum; C. decompositum*) are firmly recorded from areas north and south of the Gulf of Guinea, although they are not yet detected within the region covered here. Contrasting cool and warmer water temperature regimes are certainly part of the explanation, but it would not be unreasonable to expect at least the predominantly subtidal *C. decompositum* (known in Mauritanie; Sénégal; Angola) eventually to be found in cool upwelling areas of western Ghana or the deeper subtidal rock reefs off the Ghanaian coast.

Key to the Species

1. Robust plants of at least semi-exposed wave conditions; some adventitious weft developed from thick cortication; primary branching pseudodichotomous in aspect .... *C. granulatum*
1. More delicate plants of conditions protected from wave-action (pools; lower eulittoral in shelter); some cortication on axial and major lateral bases, but no adventitious weft; primary branching aspect monopodial, alternate 2

2. Indeterminate apices not overtopped by adjacent young subordinate laterals from the same axis [the rare occurrence of overtopping in *C. hookeri* is easily dismissed by scanning rest of plant concerned] .... *C. hookeri*
2. Indeterminate apices almost always overtopped and somewhat ensheathed by rapid growth of immediate subordinate laterals from the same axis [rare absence of overtopping is easily dismissed by scanning rest of plant concerned] 3

3. Most tetrasporangia isolated, oblong-ellipsoid, 75-85 $\times$ 35-45 μm; plants isolated, epiphytic .... *C. tenuissimum*
3. Most tetrasporangia in series, globose-rounded, 45-85 $\times$ 40-50 μm; plants tufted, chiefly epilithic in detrital conditions .... *C. roseum sensu* Harvey

Callithamnion granulatum (Ducluzeau) C. Agardh Pl. 45, figs 7-9.

C. Agardh, 1828; *Ceramium granulatum* Ducluzeau, 1805.

Dense spongy tufts, dull purplish- or brownish-red, with roughly pyramidal plant outline, becoming straggly; branching monopodial-alternate, with peripheral tufts of pseudodichotomous aspect; peripheral laterals abruptly to gradually tapered towards their apices; axes and major laterals densely corticate by downgrowing rhizoids, the latter producing recognisable adventitious laterals; basal axis gross diameter 300-500(-700) μm; ecorticate upper branch diameters 60-70(-100) μm; cell length-diameter ratios 1 to 3(-4); mature cells multinucleate; plants commonly dioecious (female, male, cystocarpic and tetrasporangial combinations known); spermatangia in hemispherical tufts distal on smaller laterals; cystocarps paired, near branch apices, usually of a single rounded gonimolobe; carposporangia 25-35 μm; tetrasporangia tetrahedral, rounded to ovoid, in adaxial peripheral series or in peripheral axils, 45-60 × 50-60 μm.

For additional figures see Feldmann-Mazoyer, 1941, p. 480, fig. 191, p. 144, fig. 49; Price, 1978, p. 265, fig. 2, p. 280, fig. 11.

Growing on other algae, or less often on rock, in lower eulittoral subzone and sublittoral fringe, in moderately to considerably wave-beaten or current-washed situations. Not common.

Distribution. World: North Eastern Atlantic, Shetland to Ghana; Western Mediterranean; Adriatic. West Africa: Côte d'Ivoire: John 1977a. Ghana: John 1977a. Liberia De May *et al.* 1977, John 1977a.

Remarks. Material from Ghana, Côte d'Ivoire and Liberia is close in form to North-Eastern Atlantic and Mediterranean *Callithamnion granulatum.* This is reflected in habitats exploited (unusual for *Callithamnion* in the Gulf of Guinea area), i.e. granite boulders and retained pools in areas of moderate to considerable exposure to wave-action. Plants are habitually epiphytic, spreading to become epilithic. Differences from European material include (i) less robust form; (ii) dense lower axial cortication with less frequent and less dense adventitious weft; (iii) minor variations in the early gametangial stages. The few available plants or the less extreme exposure of habitats in West Africa may have caused the less bushy appearance and spongy texture of the sample as compared with mature European plants. Peripheral lateral tufts are less robust in form and cause more straggly growth. Occasional larger individuals (e.g. No. 6790, Bushua Bay, Ghana) are more similar to smaller British specimens; size reduction may possibly be due to presence at range periphery or to clinal morphological variation between northern Atlantic forms and those from further south. Work on this proceeds, the present application of *granulatum* being provisional.

Callithamnion hookeri (Dillwyn) S.F. Gray

Gray, 1821; *Conferva hookeri* Dillwyn, 1809.

Bushy, straggly, or sparsely branched, sometimes even reduced to single small axes with few laterals; major axes 300-500 μm broad, commonly somewhat corticate but sometimes sparsely so, cortication lacking in juvenile forms; adventitious laterals usual from cortication further north but not noted in Ghanaian fragments; branching pattern variable, alternate-spiral to highly irregular in axes/major laterals, distichous-alternate or spiral in peripheral systems, the latter not normally overtopping indeterminate apices on most of plant; bearing laterals usually with strong axial zigzag; apical cells rounded, with basal breadth 5-10 μm; cells (very variable in number and size) in peripheral laterals cylindrical to slightly tapered; mature cells varying between 100 μm maximum diameter (near base of thallus) to ca 40 μm diameter (near apices); cell length/breadth ratios between 0.5:1 and 5.0:1. West African (Ghanaian) material lacks reproductive stages other than tetrasporangia. Tetrasporangia near globose, 45-55 × 45-50 μm, adaxial on peripheral laterals, one per cell in all known cases (sometimes two, one adaxial, one abaxial, known for more northerly areas).

For figures see Fletcher, 1980, pl. 19, figs 4, 5.

In rock pools and consistently damp places, lower littoral of wave-exposed shores; commonly epiphytic, but also epizoic and/or epilithic with less frequency. Uncommon and sparse; enmeshed with other lower layer "turf" forms and difficult to detect.

Distribution (see Dixon and Price, 1981, for a fuller overall statement). World: Atlantic, Iceland and northern Norway to Canaries and Morocco; Baltic Sea coast of Germany; Newfoundland southwards (Hoek, 1982; South, 1976; South and Hooper, 1980); in areas of cold upwelling in Gulf of Guinea (Price in Edwards, 1979; Price in Hoek, 1982); perhaps Angola (Hoek, 1982; Lawson *et al.*, 1975). Taxonomic clarification required for records from the Mediterranean and Pacific Ocean. West Africa: Ghana (details previously unpublished).

Remarks. For analysis of morphological variation in relation to differences in environmental conditions of habitats tolerated, see Price (1978).

Callithamnion roseum *sensu* Harvey Pl. 44, figs 1-3.

Harvey, 1849, pl. 230; determinations are usually based on this plate.

Plants forming soft and slightly iridescent tufts, 1-2 cm high; branching monopodial, alternate (spiralled), sometimes peripherally approaching pseudodichotomous in aspect; peripheral laterals gradually tapering, basal diameter 15-50 μm, apical diameter 5-15 μm; peripheral lateral apices blunt or pointed, larger laterals normally rounded to blunt; main axis and major laterals basally corticate, with diameters 140-170(-180) μm; cell length-to-diameter ratios (1.5:1 to)5:1; tetrasporangia in adaxial distal series on peripheral lateral cells, globose to rounded, sessile, 40-50 × 45-85 μm; other reproductive stages not known in West Africa.

For additional figures see Harvey, 1849, pl. 230; Price, 1978, p. 282, fig. 14B, C, p. 295, fig. 21; Fletcher, 1980, pl. 19, fig. 6.

Mostly in sandy or detrital-bottomed sublittoral fringe and lower eulittoral tide pools, sheltered from wave-action.

Distribution. World: typically North East Atlantic, possibly also eastern coast of North America. Reports from elsewhere need clarification. West Africa: Cameroun: unpublished. Gabon: *C. gallicum*, John & Lawson 1974a.

Callithamnion tenuissimum (Bonnemaison) Kützing

Kützing, 1849; *Ceramium tenuissimum*, Bonnemaison, 1828.

Plants isolated, rose-red, small, 1-2 cm high, epiphytic on various algae; branching monopodial, alternate, pseudodichotomous in aspect towards peripheral regions; tetrasporangia tetrahedral, usually isolated, occasionally in sessile, distal, adaxial series on peripheral laterals, oblong-ellipsoid, generally 75-85 × 35-45 μm, some smaller and ovoid in shape; other reproductive organs not known for West Africa.

For figures see Feldmann-Mazoyer, 1941, p. 448, fig. 186 (as *Aglaothamnion tenuissimum*).

Epiphytic on various algae; lower eulittoral subzone, often in red algal turf.

Distribution. World: North Eastern Atlantic and Mediterranean. West Africa: Ghana: unpublished. Sierra Leone: ?John & Lawson 1977a.

Remarks. The determination is tentative only, based mainly on the shape and position of tetrasporangia. Only small amounts of material were available, all dried on paper and in poor condition. Resuscitation was poorly possible and few dependable data were therefore obtained. The whole poorly-known taxon may simply represent a spasmodically appearing growth form of *C. byssoides* Arn. ex Harv. in Hook. or of *C. roseum sensu* Harvey. The only characteristic really likely to separate the former two on common morphological grounds is tetrasporangial shape, a rather variable basis for specific distinction. Dimensions of the material accord with those given by Feldmann-Mazoyer, 1941: 469. The report by Aleem (1978) of *C. byssoides* growing epiphytically at one locality on the Sierra Leone peninsula, and previous reports from the Canary Islands, are all likely to be based on forms of this species.

Doubtful Record

Callithamnion byssoides Arnott ex Harvey in Hooker

See remarks under *C. tenuissimum*.

Plate 44: Figs 1-3. *Callithamnion roseum:* 1. Portion of a branch showing the tetrasporangia arising in a secund series on the adaxial side of the ultimate divisions; 2. Apical portion of a branch; 3. Terminal portion of a branch showing the alternate to pseudodichotomous branching. Figs 4, 5. *Ceramium codii:* 4. Terminal portion of a branch; 5. Portion of a branch showing the very simple nodal structure. Figs 6, 7. *Ceramium strobiliforme:* 6. Terminal portion of a plant showing the tetrasporangia borne in the swollen and corticated tip of a branch; 7. Structure of the upper nodes of a branch.

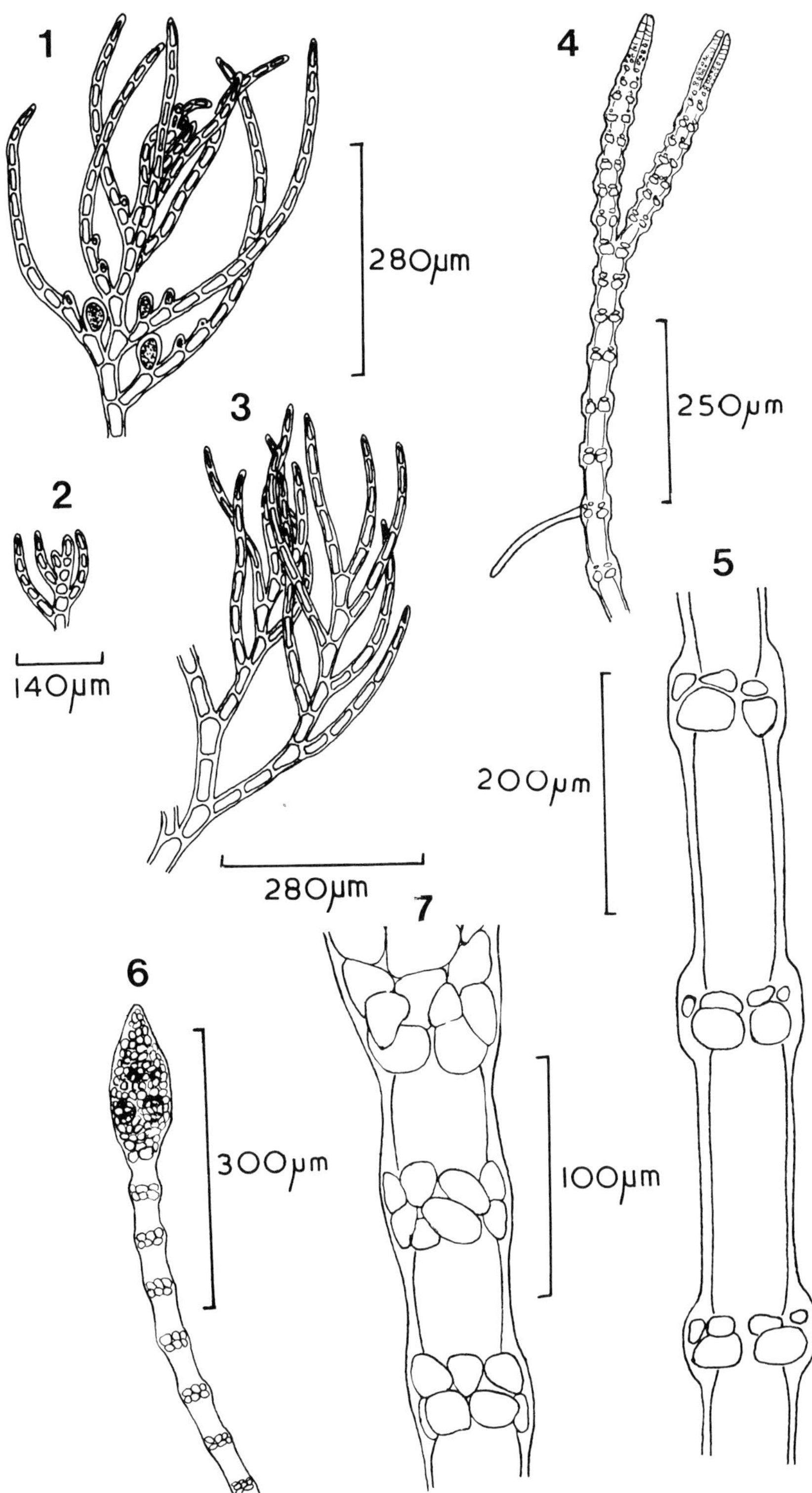
1
280μm
2
140μm
3
280μm
4
250μm
5
200μm
6
300μm
7
100μm

Centroceras Kützing 1841

Plants filamentous, erect or prostrate, dichotomously, subdichotomously or occasionally irregularly divided, with the apices simple or forcipate; branches consisting of an axial row of cells completely corticated by quadrate to rectangular cells arranged in vertical rows, cortical cells cut-off from upper end of axial cell, with a ring of spines at the nodes in some species; tetrasporangia arising as a transverse ring around the nodes, sometimes on specialized branches, sporangia tetrahedrally divided; spermatangia clustered, terminal on branchlets; gonimoblast bilobed, surrounded by a few incurved branchlets.

World Distribution: widespread in warm temperate and tropical seas.

Key to the Species

1. Plants from 2-6 cm in height; branches bearing a whorl of 1- or 2-celled spines at many of the nodes .... *C. clavulatum*
1. Plants minute, usually less than 1 mm in height; branches devoid of spines .... *C. bellum*

Centroceras bellum Setchell & Gardner Pl. 42, figs 6, 7.

Setchell and Gardner, 1924.

Plants minute, to about 1 mm high, with the erect branches arising from prostrate axes; branches of the erect system pseudodichotomously divided, from 70-120 μm in diameter, with the internodes up to twice as long as broad; cortication consisting of rows of cells, with 2 to 3 rows of somewhat disorganized and isodiametric or quadrate cells at the nodes; cells of the cortex from 3-4 μm wide and 12-16 μm in length at the larger internodes; tetrasporangia embedded in swollen ultimate divisions of the branches, with 3 to 4 sporangia per node.

For additional figures see Setchell and Gardner, 1924, p. 825, pl. 26, fig. 48, p. 853, pl. 40c, p. 929, pl. 782.

Occurs in the low algal turf covering moderately wave-exposed rocks in the lower eulittoral subzone.

Distribution. World: in tropical parts of the Pacific and Atlantic Oceans. West Africa: Sierra Leone: *Centroceras* sp., John & Lawson 1977a.

Remarks. Børgesen (1953) suggested that *Centroceras bellum* ought to be placed in his newly created genus *Ceramiella* but this tranfer was not formally made until 16 years later (Diaz-Piferrer, 1969). Nevertheless, we have decided to follow Hommersand (1963) who after a thorough investigation of this species decided to retain it in the genus *Centroceras* until further information was forthcoming on its sexual reproduction.

Centroceras clavulatum (C. Agardh) Montagne Pl. 43, figs 3, 4.

Montagne, 1846-1849; *Ceramium clavulatum* C. Agardh, in Kunth, 1822.

Plants forming dense tufts or mats of entangled filament, from 2-6 cm in height, dark purplish-red or sometimes bleached to pink or greenish-yellow; branching dichotomous, each dichotomy equally developed, at narrow angles; branches from 70-200 μm in diameter, with the internodes 350-500 μm in length and decreasing near the tips of the branches; spines arising at the nodes, 1- or 2-celled, particularly abundant towards the branch apices.

For additional figures see Abbott and Hollenberg, 1976, p. 605, fig. 547; Feldmann-Mazoyer, 1941, p. 338, 340, figs 128, 129.

Occurs as dense cushion-like tufts on very wave-exposed rocks and as loose mats or mixed in the algal turf in more wave-sheltered situations. This is one of the most common algae along the West African coast, usually found in greatest abundance over the lower part of the upper eulittoral subzone and the lithothamnia subzone below.

Distribution. World: as for genus. West Africa: Benin: John 1977a, John & Lawson 1972b. Cameroun: Fox 1957, John 1977a, Lawson 1955, 1966, Richardson 1969, Schmidt & Gerloff 1957, Stephenson & Stephenson 1972; *Ceramium clavulatum*, Pilger 1911. Bioko: unpublished. Côte d'Ivoire: John 1972c, 1977a. Gabon: John & Lawson 1974a; *Centroceras inerme*, De Toni 1903, Kützing 1849; *Ceramium clavulatum*, Hariot 1895. Gambia: John & Lawson 1977b. Ghana: Dickinson & Foote 1950, Fox 1957, John 1977a, John & Pople 1973, Lawson 1953, 1956, 1957b, 1966, Stephenson & Stephenson 1972. Guinée: Lawson 1966, Marchal 1960, Sourie 1954, Stephenson & Stephenson 1972. Liberia: Askenasy 1888, De May *et al.* 1977, De Toni 1903, Fox 1957, John 1972c, 1977a. Nigeria: Fox 1957, John 1977a, Lawson 1966, Stephenson & Stephenson 1972. São Tomé: Carpine 1959, Fox 1957, John 1977a, Lawson 1966, Steentoft 1967; *Centroceras hyalacanthum*, Henriques 1886b, 1887; *Ceramium clavulosum*, Hariot 1908, Henriques 1917. Sierra Leone: Aleem 1978, Fox 1957, John & Lawson 1977a, Lawson 1954b, 1957a, 1966, Stephenson & Stephenson 1972. Togo: John 1977a, John & Lawson 1972b.

Ceramium* Roth 1797

Plants filiform, often profusely branched, regularly dichotomously divided and sometimes alternate or subpinnate, with the apices frequently forcipate; branches consisting of an axial row of large colourless cells, with bands of assimilatory cells restricted to the nodes or occasionally extending over the internodal region; gland cells present or absent on the nodes; tetrasporangia sessile, naked or partly or completely covered by an involucre of cortical cells, tetrahedrally divided; spermatangia sessile, usually on the nodes near the tips of the branches, surrounded by short and

**Ceramium gracillimum* (Kützing) Griffiths & Harvey is reported from the Congo (Hariot, 1895, 1896) and *C. leptosiphon* Pilgèr from Pagalu (De Toni, 1924; Pilger, 1920).

incurved branchlets; gonimoblast borne at the upper nodes, often surrounded by incurving branchlets.

World Distribution: cosmopolitan.

Key to the Species

1. Plants beset with many adventitious branchlets and the apices always divaricate; nodal bands usually greater than 200 μm in diameter . . . . *C. cornutum*
1. Plants rarely or never bearing adventitious branchlets, with the apices either straight or forcipate; nodal bands nearly always less than 200 μm in diameter . . 2

2. Branches simple or having a single dichotomy and terminating in a corticated cone-shaped tip . . . . *C. strobiliforme*
2. Branches usually repeatedly dichotomously divided and the apices of another . . . form 3

3. Nodal bands having a clear space separating the lowermost cell row from those rows above. 4
3. Nodal bands with no clear space existing between one cell row and another 5

4. Nodal bands usually greater than 60 μm in diameter and the lowermost row of cells quadrate to angular . . . . *C. taylorii*
4. Nodal bands less than 60 μm in diameter and the lowermost row of cells transversely elongated. . . . . *C. gracillimum* var. *byssoideum*, in part

5. Nodal bands having a row of large and longitudinally elongate cells. 6
5. Nodal bands of variously shaped cells but never longitudinally elongated 7

6. Branches in upper parts sometimes conspicuously incurved and chain-like; axial cells from 1 to 3 times longer than broad . . . . *C. mazatlanense*
6. Branches straight or only slightly curved; axial cells from 1 to 12 times longer than broad . . . . *C. tenerrimum*

7. Nodal bands less than 45 μm in diameter and consisting of 1 or 2 rows of cells. . . . . *C. codii*
7. Nodal bands usually greater than 45 μm in diameter and consisting of more than 2 rows of cells, at least in the older branches 8

8. Plants normally greater than 5 cm in height; nodal bands composed of several irregular rows of small cells all of similar dimensions. . . . . *C. tenuissimum*
8. Plants less than 5 cm in height; nodal bands of up to 5 rows of cells, with at least 1 row consisting of cells larger than the rest 9

9. Nodal bands usually less than 60 μm in diameter and having the lowermost row of cells transversely elongated . . . . *C.gracillimum* var. *byssoideum*, in part
9. Nodal bands always greater than 60 μm in diameter and never with the cells of the lowermost row transversely elongated . . . . *C. ledermannii*

Ceramium codii (Richards) Mazoyer Pl. 44, figs 4, 5.

Mazoyer, 1938; *Ceramothamnion codii* Richards, 1901.

Plants to about 1 mm in height, usually epiphytic, consisting of erect branches arising from a creeping system; branches subdichotomously divided, with a straight or slightly incurved and often simple apex; nodal bands prominent, from 25-33(-43) μm in diameter and from 16-23 μm in length, consisting of a row of large and slightly transversely elongate cells about 16 μm in diameter, sometimes with an upper row of smaller cells present; axial cells about 2 to 3 times as long as broad, with striations normally present in the wall; tetrasporangia solitary and partly covered by an involucre of cells.

For additional figures see Richards, 1901, pl. 21, 22 (as *Ceramothamnion codii*).

Occurs epiphytically on larger algae growing in moderately wave-sheltered situations in the lower eulittoral subzone and to a depth of about 10 m.

Distribution. World: probably widespread in warm temperate and tropical seas. West Africa: Ghana: Lieberman *et al.* 1979. Togo: ?John & Lawson 1972b.

Ceramium cornutum P. Dangeard Pl. 45, fig. 4.

Dangeard, 1952.

Plants forming epiphytic tufts, from 2-5 cm in height; branches regularly dichotomously divided, beset with many forked adventitious branchlets, with the tips of the branches and branchlets divaricate; nodal bands to about 300 μm in diameter and 140 μm in length, consisting of a row of large and nearly quadrate cells about 70 μm in diameter, with a single row of small irregularly shaped cells below and 1 to several rows above; axial cells about 1.5 times as long as broad in the older parts and shorter in younger branches, with striations in wall; tetrasporangia semi-prominent, usually somewhat secund on the abaxial side of the branches and branchlets, 1 to 3 sporangia per node, covered by an involucre of cells.

For additional figures see Dangeard, 1952, p. 317, pl. 17, figs c-i.

Occurs on larger algae growing on cobbles and rocky platforms at depths ranging from 10 to 25 m, also known from the lower eulittoral subzone where it is found in the algal turf.

Distribution. World: in warm temperate and tropical parts of the eastern Atlantic Ocean. West Africa: Ghana: unpublished.

Ceramium gracillimum (Kützing) Griffiths & Harvey Pl. 45, figs 1, 2.

Griffiths and Harvey, in Harvey, 1849; *Hormoceras gracillimum* Kützing, 1841.
var. **byssoideum** (Harvey) Mazoyer
Mazoyer, 1938; *Ceramium byssoideum* Harvey, 1853.

Plants forming epiphytic tufts or felty patches, from 2-5(-30) mm in height; branches dichotomously or occasionally pinnately divided, with the apices forcipate; nodal bands prominent, to about 60 μm in diameter and 40 μm in length in older branches, consisting below of 1 or 2 rows of transversely elongate or occasionally quadrate cells and above of up to several rows of smaller irregularly arranged angular cells; gland cells present; axial cells to about 4 to 5 times longer than broad, without striations in wall; tetrasporangia in the ultimate branches, naked or having a few-celled involucre.

For additional figures see Feldmann-Mazoyer, 1941, p. 294, fig. 109.

Occurs on various larger algae growing in moderately wave-exposed situations in the lower eulittoral subzone and has been dredged from a depth of 11 m.

Distribution. World: probably widespread in warm temperate and tropical seas. West Africa: Cameroun: unpublished. Gabon: John & Lawson 1974a. Ghana: unpublished. Liberia: De May *et al.* 1977. São Tomé: Carpine 1959, Steentoft 1967. Sierra Leone: John & Lawson 1977a; *C. byssoideum,* Aleem 1978.

Remarks. We have found specimens in Cameroun and Ghana in which the nodal structure is similar to *Ceramium masonii* Dawson in having a clear space between the lowermost row of cells and those above. Nevertheless, in all other respects these plants resemble *C. gracillimum* variety *byssoideum* particularly in having naked or almost naked sporangia. Mazoyer (1938) has reduced *Ceramium byssoideum* to the status of a variety of *C. gracillimum* whilst Taylor (1960) has retained the former species for the plant on the western side of the Atlantic.

Ceramium ledermannii Pilger Pl. 45, fig. 6.

Pilger, 1911.

Plants forming small tufts or felty patches on the surface of larger algae, from 1-10 mm in height; branches dichotomously divided, with the apices forcipate; nodal bands from 90-120 μm in diameter and to about 30 μm in length, consisting in young branches of a row of large cells below and a row of smaller cells above, in older parts often having a few rows of smaller irregular or angular cells above the row of larger ones and occasionally also below; axial cells often somewhat barrel-shaped, from 1 to 2 or rarely more times longer than broad, with striations in wall; tetrasporangia in the upper branches, with a few-celled involucre.

For additional figures see Pilger, 1911, p. 309, 310, figs 18-23.

Occurs as an epiphyte on larger algae such as species of *Laurencia* and *Gracilaria* and sometimes mixed in with the algal turf; found over a wide range of wave-exposure and extends from the upper eulittoral subzone down into the immediate sublittoral.

Distribution. World: in the tropical parts of the eastern Atlantic Ocean. West Africa: Cameroun: De Toni 1924, John 1977a, Pilger 1911. Côte d'Ivoire: John 1977a. Gabon: John & Lawson 1974a. Liberia: De May *et al.* 1977, John 1977a.

Ceramium mazatlanense Dawson Pl. 45, fig. 3.

Dawson, 1950a.

Plants forming tufts, from 1-3 cm in height, often growing entangled with other small algae; branches dichotomously divided, sometimes conspicuously incurved and chain-like in appearance, beset occasionally with a few short spine-like branchlets, with forcipate apices; nodal bands from 100-210 μm in diameter and to about 70 μm in length, consisting of a row of large longitudinally elongated cells from 12-20 μm in diameter and 25-40 μm in length, with from 1 to 3 rows of small irregular cells above the larger cell row and sometimes in older branches a layer of small cells developing below the large cells; axial cells from 1 to 3 times longer than broad, with striations in the wall; tetrasporangia prominent and covered with an involucre of cells, with 1 or more rarely 2 sporangia per node, borne in regular series on the abaxial surface of strongly curved branches.

For additional figures see Jaasund, 1970a, p. 66, fig. 1a; Dawson 1950a, p. 125, pl. 2, figs 14, 15.

Occurs epiphytically or mixed in the algal turf growing in the lower eulittoral subzone and the sublittoral fringe.

Distribution. World: probably pantropical. West Africa: Gabon: John & Lawson 1974a. Ghana: unpublished.

Ceramium strobiliforme nov. sp. Pl. 44, figs 6, 7.

Plantae epiphyticae solitariae, usque ad 1 mm altae; rami simplices vel aliquando singulatim dichotomi; apex rami strobiliformis, omnino corticatus, 100-250 μm longitudine, 70-100 μm diametro; nodis basi ca. 30 μm diametro, sub apicem 70 μm diametro, 20-35 μm longitudine, supra ex cellulis irregularibus 1-3 seriatis et infra cellulis transverse elongatis uniseriatis constans; cellulae axiales 1.5-3-plo longiores quam latae, parietibus striatis; tetrasporangia ca 40 μm diametro, in apicem immersa, itaque ad apicem *Equiseti* strobilum simulans.

Plants solitary and epiphytic, up to 1 mm in height; branches simple or occasionally with a single dichotomy; apex of branches cone-shaped, completely corticate, from 100-250 μm in length and 70-100 μm in diameter; nodal bands about 30 μm in diameter towards the base of a branch and increasing to about 70 μm just below the corticated apex, from 20-35 μm in length, consisting of 1 to 3 rows of irregular cells above and a lower row of transversely elongated cells; axial cells 1.5 to 3 times longer than broad, with striations in the wall; tetrasporangia about 40 μm in diameter, embedded in the swollen branch tip which resembles the apical strobilus of *Equisetum*.

Holotype: 8791, Vernon Bank, Ghana, 11 April 1978. Deposited in the British Museum (Natural History), London (B M).

Occurs as an epiphyte on species of *Chondria* and *Gelidium* growing on rocky platforms at a depth of about 10 m.

Distribution. World: in tropical West Africa. West Africa: Ghana: unpublished.

Remarks. This plant bears a superficial resemblance to *Ceramium cingulatum* Weber-van Bosse and *C. equisetoides* Dawson, both of which have a continuous cortical layer over the often swollen branch tip. The Ghanaian plant differs from these two species in its overall dimensions, in the actual shape of the tip of a branch, and in the lesser degree of cortication.

Ceramium taylorii Dawson Pl. 45, fig. 5.

Dawson, 1950a.

Plants forming small tufts, to about 10 cm in height; branches alternately or dichotomously divided, with the apices forcipate; nodal bands up to 130 μm in diameter and to about 70 μm in length, consisting of a lowermost row of quadrate or more rarely angular cells, about 14 μm in diameter, separated by a clear space from a row of large and somewhat transversely elongated cells that have a few irregular rows of smaller cells above; gland cells often present; axial cells from 3 to 4 times longer than broad, with striations in the wall; tetrasporangia about 40 μm in diameter, often prominent and whorled, up to 4 sporangia per node, with an involucre of cells.

For additional figurers see Dawson, 1950a, p. 125, pl. 2, fig. 13, p. 135, pl. 4, figs 31-33.

Occurs with other species of *Ceramium* growing on lithothamnia in the lower eulittoral subzone.

Distribution. World: probably cosmopolitan. West Africa: Cameroun: unpublished.

Remarks. Dawson (1950a) has drawn attention to the close resemblance between this species and *Ceramium fastigiatum* Harvey, the main feature separating the two species being the presence of a clear space between the lowermost row of cortical cells and those above in *C. taylorii.* It is possible that some of the earlier records of *Ceramium fastigiatum* from warm temperate parts of the eastern Atlantic Ocean may prove on re-examination of the material on which they are based to be *C. taylorii.*

Plate 45: Figs 1, 2. *Ceramium gracillimum* variety *byssoideum:* 1. Terminal portion of a branch showing the prominent nodes; 2. Structure of two of the upper nodes. Fig. 3. *Ceramium mazatlanense:* Portion of a plant showing the strongly incurved branches and the distinctive arrangement of the tetrasporangia. Fig. 4. *Ceramium cornutum:* Portion of a plant showing the common adventitious branchlets and the tetrasporangia arising in series on the branches and branchlets. Fig. 5. *Ceramium taylorii:* Structure of the node with the lower row of quadrate cells separated from those above by a clear space. Fig. 6. *Ceramium ledermannii:* Structure of a node in the lower part of a branch. Figs 7-9. *Callithamnion granulatum:* 7. Bushy habit of plant; 8. Terminal portion of a loosely branched peripheral lateral; 9. Terminal portion of a denser peripheral lateral showing the alternate branching pattern.

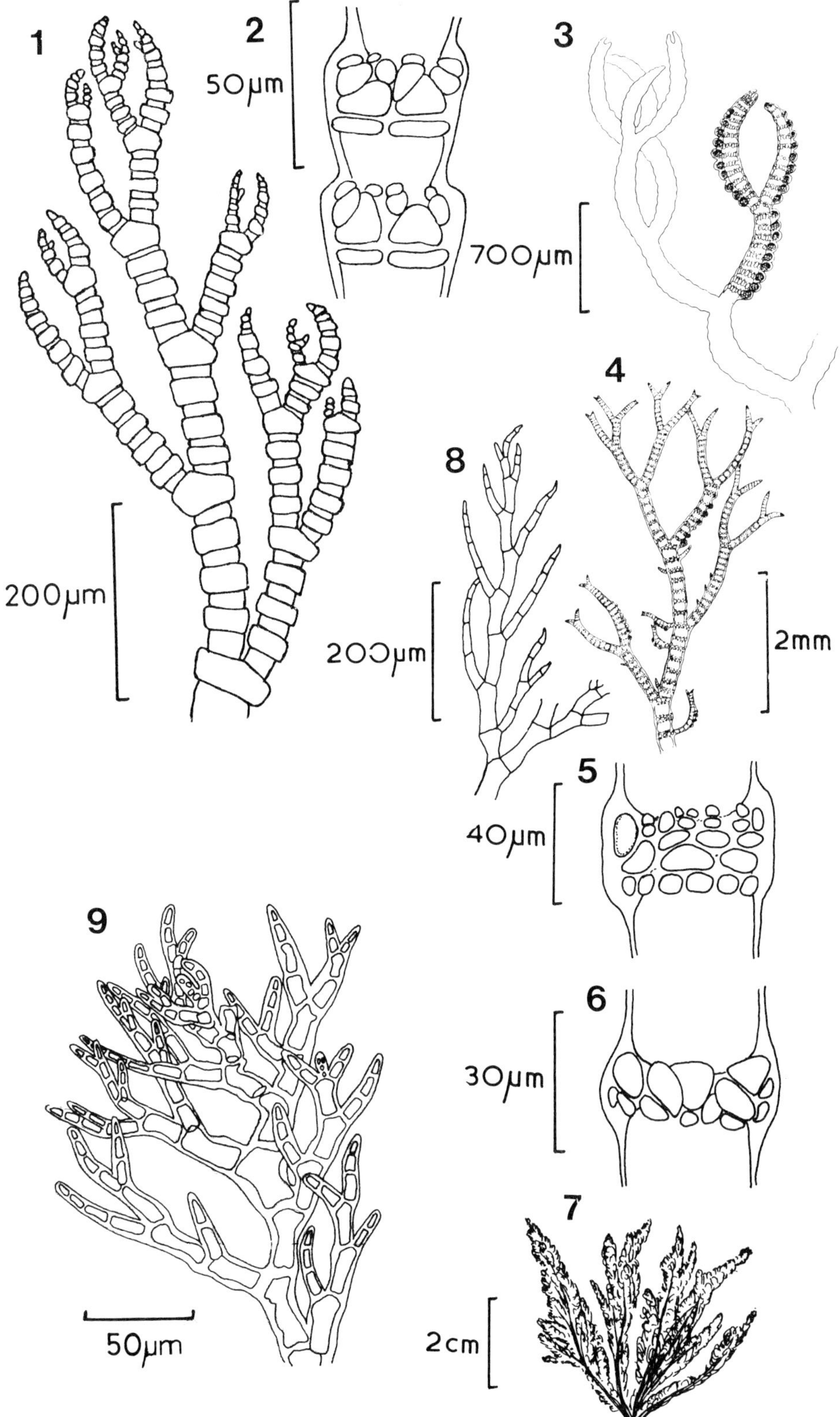
1
2
50µm
3
700µm
200µm
4
8
200µm
2mm
5
40µm
9
6
30µm
7
50µm
2cm

Ceramium tenerrimum (Martens) Okamura

Okamura, 1921; *Hormoceras tenerrimum* Martens, 1866.

Plants forming tufts, about 3-4 cm in height, reddish-purple; branches dichotomously divided and fastigiate, with the apices forcipate and sometimes strongly incurved; nodal bands up to 150 μm in diameter and from 20-50 μm in length, consisting of a lowermost row of longitudinally elongate cell, about 20-25 μm in diameter and 35-40 μm in length, with 1 to 3 rows of small, irregularly arranged and shaped cells, from 6-10 μm in diameter above; axial cells up to 12 times longer than broad in older parts of the branches and decreasing towards the branch tips; tetrasporangia developed on the upper branches, little emmersed and naked.

For figures see Feldmann-Mazoyer, 1941, p. 291, 292, figs 107, 108.

There is no information on this species in the region although in other parts of its range it is commonly found as an epiphyte.

Distribution. World: in warm temperate and tropical parts of the Atlantic. West Africa: Guinée: Piccone 1900, 1901, Sourie 1954.

Ceramium tenuissimum (Roth) J. Agardh

J. Agardh, 1851; *Ceramium diaphanum* var. *tenuissimum* Roth, 1806.

Plants forming tufts from 10-12 cm in height or a low mat when mixed with other algae; branches dichotomously divided, sometimes bearing a few short divaricate lateral branchlets; nodal bands from 70-175 μm in diameter and 50-120 μm in length, strongly projecting in the upper branches, consisting of several rows of small and somewhat irregularly arranged cells, from 6-12 μm in diameter; gland cells sometimes present; axial cells up to 8 times longer than broad, with striated walls; tetrasporangia secundly arranged on the branches or occasionally whorled, with an involucre of a few cells.

For figures see Feldmann-Mazoyer, 1941, p. 299, 301, figs 113, 114.

Occurs on sheltered to moderately wave-exposed rocks in the lower eulittoral subzone and in the sublittoral fringe.

Distribution. World: widespread in warm temperate and tropical seas. West Africa: Cameroun: John 1977a, Lawson 1955, Stephenson & Stephenson 1972. Côte d'Ivoire: John 1977a. Sierra Leone: John & Lawson 1977a.

Doubtful Record

Ceramium fastigiatum (Roth) Harvey form **flaccida** H.E. Petersen

The report of this plant from the "lower littoral" on the Sierra Leone peninsula (Aleem, 1978, 1980b) is the first from the eastern side of the Atlantic Ocean. This is considered as a doubtful record until the material on which it is based is examined and the identification verified. See remarks under *Ceramium taylorii.*

Dohrniella Funk 1922

Plants filamentous, consisting of a prostrate system of branches from which arise sparingly alternately divided erect branches bearing branchlets; cells at the base of the branchlets small, larger above and each with one to three papilliform cells; chloroplasts numerous and spherical to elongate; tetrasporangia stalked and arising on the upper cells of the branchlets, tetrahedrally divided; spermatangia stalked and clustered towards the tips of the branches; gonimoblasts formed on the erect branches and four-lobed, with the two upper lobes larger than the two lower ones.

World Distribution: in warm temperate and tropical parts of the Atlantic Ocean and the Mediterranean.

Dohrniella antillarum (W.R. Taylor) Feldmann-Mazoyer — Pl. 46, figs 1, 2.

Feldmann-Mazoyer, 1941; *Actinothamnion antillarum* Taylor, in Taylor and Arndt, 1929.

Plants forming soft and delicate tufts, up to 10 mm in height, attached by multicellular rhizoids developed from the basal cells of the prostrate branches; branching sparingly alternate and branchlets spirally arranged; branches to about 50 μm in diameter, with the lower cells to 8 times as long as broad and decreasing above to 1.5 times as long as broad; branchlets having 1 to 2 small spherical cells at the base, becoming cylindrical above and eventually the cells inflated and almost in moniliform rows towards the apex, to about 22 μm in diameter, and bearing 1 or occasionally 2 papilliform cells.

For additional figures see Taylor and Arndt, 1929, p. 659, 661, figs 1-10 (as *Actinothamnion antillarum*).

Occurs on calcareous cobbles, larger algae as well as on polyzoans growing at depths from 10 to 15 m.

Distribution. World: in warm temperate and tropical parts of the Atlantic Ocean. West Africa: Ghana: John & Lawson 1972a, John *et al.* 1977, Lieberman *et al.* 1979.

Grallatoria Howe 1920

Plants filamentous, dorsiventral, consisting of a prostrate system of branches usually becoming secondarily erect, bearing alternately and in a double spiral the branches of determinate growth, attached by uni- or multicellular holdfasts arising from nearly every cell of the creeping axes; chloroplasts rounded or elongate to fusiform; tetrasporangia sessile or stalked and arising on the inner side of determinate branches, tetrapartitely or tetrahedrally divided; gonimoblasts terminal on branches of indeterminate growth, with slender involucral branchlets present.

World Distribution: in warm temperate and tropical seas.

Grallatoria tingitana (Schousboe ex Bornet) Abbott Pl. 43, figs 5, 6.

Abbott, 1976; *Callithamnion tingitanum* Schousboe, in Bornet, 1892.

Plants usually forming small felty patches, rarely greater than 1 mm in height, with the prostrate branches attached by holdfasts growing out from the basal cells of the determinate branches; branching sparingly alternate and the determinate branches spirally arranged at a divergence of about one-quarter to one-half; cells of the indeterminate branches cylindrical to barrel-shaped, to about 30 μm in diameter and at least 1.5 times longer than broad; cells of the determinate branches narrower and often barrel-shaped particularly in the upper parts, with the apical cells blunt or occasionally slightly tapering.

For additional figures see Feldmann-Mazoyer, 1941, p. 125, fig. 38, p. 430, fig. 169, pl. 4 (as *Callithamniella tingitana*).

Occurs mixed with other small algae growing on rocks on the floor of moderately wave-sheltered tidepools in the eulittoral zones and also found with other algae on rocky platforms at a depth of about 10 m.

Distribution. World: as for genus. West Africa: Ghana: Lieberman *et al.* 1979; *Callithamniella tingitana*, John 1977a. Liberia: *Callithamniella tingitana*, De May *et al.* 1977, John 1977a.

Griffithsia* C. Agardh 1817

Plants filamentous, consisting of laterally or subdichotomously divided branches of noticeably large cells which often bear distally di- to polydichotomous hairs; tetrasporangia in whorls at the nodes, with or without large involucral cells, sporangia tetrahedrally divided; spermatangia in sori at the nodes of the upper cells of the branches or at the distal end of the terminal cell; carposporangia often having a gelatinous covering, arising laterally on a branch, partly surrounded by involucral cells.

World Distribution: widespread from boreal-antiboreal to tropical seas.

*This genus is possibly represented in Ghana (John *et al.*, 1980; Lieberman *et al.*, 1979) by a second species which, unlike *Griffithsia schousboei*, has filaments made up of narrow, elongate cells. Unfortunately the absence of reproductive structures has prevented positive identification even to genus.

Plate 46: Figs 1, 2. *Dohrniella antillarum:* 1. Terminal portion of a branch showing the presence of papilliform cells on some of the upper inflated cells of the spirally arranged branchlets; 2. Portion of a prostrate axis with an erect branch and rhizoids arising from the same cell. Fig. 3. *Griffithsia schousboei:* Portion of a branch showing the lateral branching and the large spherical to pyriform cells. Figs 4, 5. *Gulsonia ecorticata:* 4. Portion of a plant with the whorled branchlets arising along the axial filament; 5. Portion of a branch with the whorl of branchlets arising from the upper part on an axial cell and a tetrasporangium borne on the second division of the branchlet. Figs 6, 7. *Spermothamnion investiens:* 6. Portion of a plant showing a prostrate axis attached by elongate holdfasts and bearing erect axes; 7. Portion of a branch with the tetrasporangia borne on 1-celled stalks.

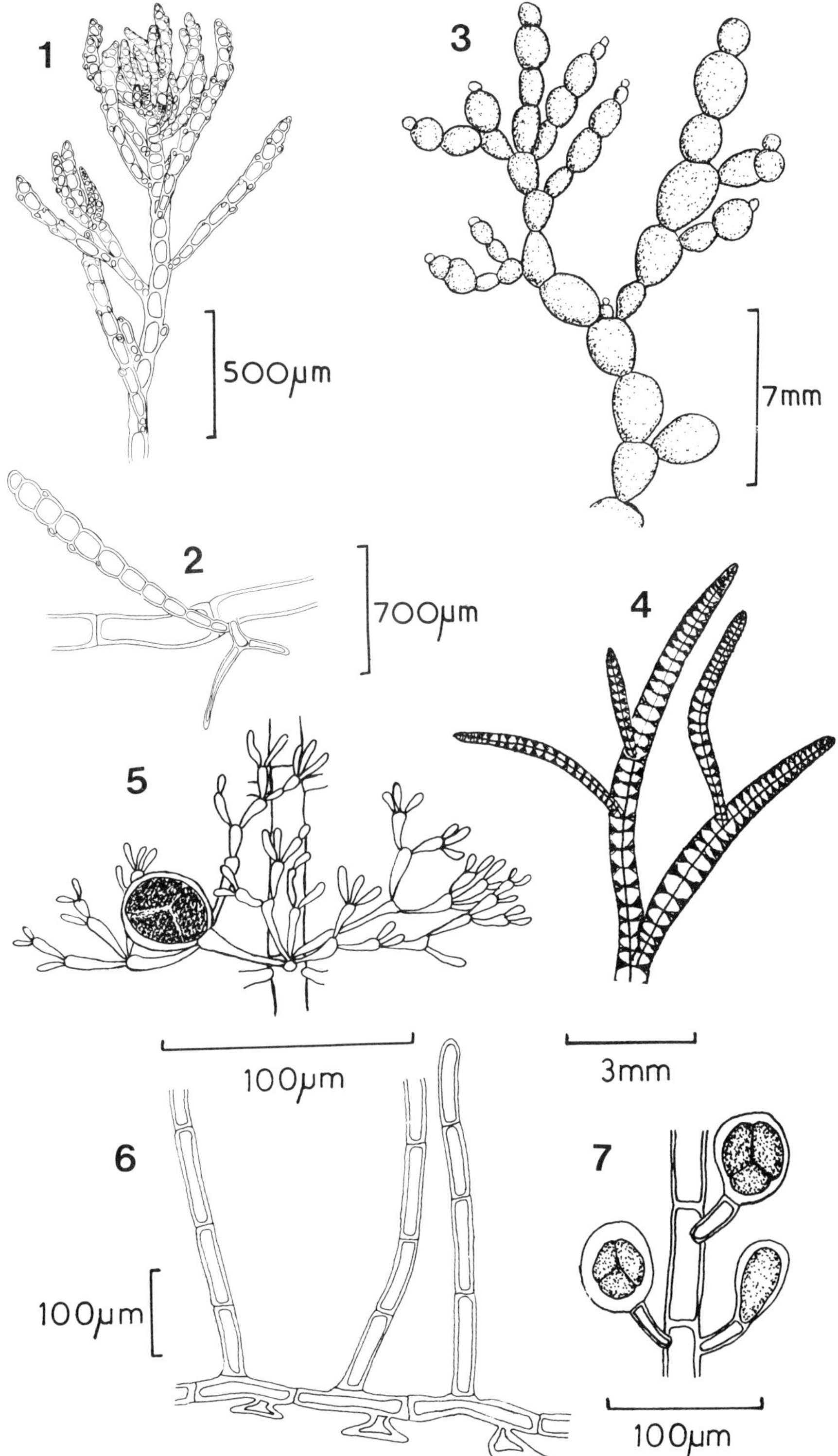
1
500µm
3
7mm
2
700µm
4
5
100µm
3mm
6
100µm
7
100µm

Griffithsia schousboei Montagne Pl. 46, fig. 3.

Montagne, 1839-1841.

Plants forming globose tufts on larger algae, from 1-4 cm in height, bright red; branches borne laterally and with 2, 3 or occasionally more arising from a single cell, moniliform; cells of the branches spherical, ovoid or pyriform, to about 1.5 mm in diameter and to 3 mm in length, corticated by a few filaments towards the base; tetrasporangia at the nodes and associated with involucral cells, sporangia to about 50 μm in diameter; spermatangia in clusters at the nodes of the upper cells of a branch.

For additional figures see Feldmann-Mazoyer, 1941, p. 416, 417, figs 164, 165.

Found on larger algae such as species of *Laurencia* and *Gelidiopsis* as well as on articulated corallines growing in moderately sheltered to moderately wave-exposed situations in the lower eulittoral subzone.

Distribution. World: in warm temperate and tropical parts of the Atlantic Ocean as well as in the Mediterranean. West Africa: Côte d'Ivoire: John 1977a. Ghana: John 1977a.

Doubtful Record

Griffithsia barbata (J.E. Smith) C. Agardh

Drift plants are recorded by Aleem (1978) from Sierra Leone during the dry season and represent the first mention of this species from the mainland of West Africa. The source of the Sierra Leone material remains unknown and specific attribution is only possible if the plants are fertile.

Gulsonia Harvey 1855

Plants consisting of an axial row of cells, bearing from the upper part of each cell a whorl of four short branchlets, with the lateral branches of indeterminate growth arising from basal cells of branchlets or very rarely from axial cells, ecorticate or sometimes having rhizoidal cortication; tetrasporangia sessile and borne on second or third divisions of the branchlets, sporangia tetrahedrally divided; carpogonial branches four-celled and arising on branches of indeterminate growth, with an involucre of filaments at the base of the carposporophyte having developed from the axial cell.

World Distribution: in warm temperate and tropical seas.

Gulsonia ecorticata nov. sp. Pl. 46, figs 4, 5.

Plantae molles gelatinosae caespitosae, epiphyticae, roseae, 1-3 cm altae; axis principalis basi 100 μm diametro, apicem versus gradatim attenuatus, cellulis 3-5-plo longioribus quam latis; rami principalis sparsi aut e axe principali aut e cellulis basalibus ramulorum verticillatorum emergentes; ramuli verticillati di- trichotomi, 4-6-plo divisi, 1-3 cellulis terminalibus, ca. 7-16 μm longitudine, ad apicem angustatis, haud transverse divisis, etiam in cellulis glandularibus muniti; tetrasporangia in ramulis ordinis secundi vel tertii, ex apicibus cellulorum exorta, 33-77 μm diametro; carposporangia in ramis lateralibus singulatim vel binatim exorta, maturitate ca. 230 μm, carposporis ca. 24 μm longis, 16-18 μm diametro.

Plants forming soft gelatinous tufts, epiphytic, light rose red, from 1-3 cm in height; main axis irregularly divided, to about 100 μm in diameter and decreasing gradually towards the apex, formed from cells 3 to 5 times as long as broad; branches arising laterally from the basal cell of the whorled branchlets or from the main axis; branchlets 4 to 6 times di- to trichotomously divided, with 1 to 3 tapering terminal cells, from 7-16 μm in length, having gland cells present; tetrasporangia arising on the distal parts of the cells of the second or third divisions of the branchlets, sporangia from 33-77 μm in diameter; carposporangia arising on the lateral branches and usually single or in pairs, about 230 μm in diameter when mature, with the carpospores to about 24 μm in length and from 16-18 μm in diameter.

Holotype: 7230, Vernon Bank, Ghana, 28 January 1975. Deposited in the British Museum (Natural History), London (BM).

Occurs as an epiphyte on larger algae such as species of *Gracilaria* and *Dictyurus fenestratus* growing on rocky platforms at a depth of about 10 m.

Distribution. World: in the tropical eastern Atlantic Ocean. West Africa: Ghana: *Gulsonia* sp., John *et al.* 1977, Lieberman *et al.* 1979.

Remarks. The Ghanaian plants fall well within the accepted limits of this genus especially in having the carposporangia arising on the lateral branches of indeterminate growth (Wollaston, 1968). This new species differs from the other two in this genus, *Gulsonia annulata* Harvey and *G. mediterranea* Kylin, in the absence of cortication, the tapering and short length of the apical cells of the branchlets, and the delicate nature of the plant due to the smaller dimensions of many of its parts. Other features in which it differs relate to the absence of hairs, presence of gland cells, and the origin of the indeterminate branches. So far very few fertile individuals have been found and so it has not been possible to follow the development of the carposporophyte.

Microcladia (Turner) Greville 1830

Doubtful Record

Microcladia glandulosa (Solander ex Turner) Greville

This plants has been reported by Kützing (1863) as having been collected from the "...embouchure de la riviere de Gabon, Guinée". It is not clear to which West African country is being referred to in this statement which Kützing presumably copied from the Lenormand plant (No. 289) which he cites. This plant cannot be located in the Caen Herbarium (C), where much of the Lenormand collection is deposited, and it is possible that it might have been destroyed in the last war (Gayral, *pers. comm.*). There is no other mention of this species from tropical West Africa and so at present we prefer to regard it as a doubtful record.

Spermothamnion* Areschoug 1847

Plants consisting of an erect system of monosiphonous filaments laterally or occasionally oppositely branched, arising from a creeping system attached by unicellular and long holdfasts; cells having a single nucleus and many small chloroplasts; sporangia in clusters on the side of the lateral branchlets or terminal on them, polysporic or tetrahedrally divided; carposporangia arising towards the ends of the lateral branchlets.

World Distribution: widespread from boreal-antiboreal to tropical seas.

Key to the Species

1. Cells from 14-18(-20) μm in diameter and (3-)4 to 6 times longer than broad. . . . . *S. investiens*
1. Cells from 28-35 μm in diameter and (2-)3 to 4 times longer than broad . . . . *S. speluncarum*

Spermothamnion investiens (Crouan frat.) Vickers — Pl. 46, figs 6, 7.

Vickers, 1905; *Callithamnion investiens* Crouan frat., in Mazé and Schramm, 1870-1877.

Plants forming a felty covering over larger algae and sedentary animals, rarely epilithic, rose red; erect filaments simple or occasionally alternately branched, with

*There may be at least another representative of this genus in Sierra Leone (John and Lawson, 1977a), but since all the specimens found have been sterile positive identification even to genus has not been possible.

the apical cell rounded; cells from 14-18(-20) μm in diameter and from (3-)4 to 6 times longer than broad, with the wall up to 5 μm thick especially in the cells of the prostrate filaments; tetrasporangia secund or on occasion opposite, borne terminally on a 1-celled lateral branchlets, sporangia ovate or round.

For additional figures see Børgesen, 1917, p. 200, 201, figs 189, 190a-c (as *S. investiens* var. *cidericola*).

Occurs as a red felty covering to larger algae such as old plants of *Galaxaura* as well as on sponges and barnacles growing in the sublittoral fringe and to a depth of about 10 m; rarely mixed with other algae when found growing on rocks in the eulittoral zone.

Distribution. World: in warm temperate and tropical parts of the Atlantic Ocean. West Africa: ?Cameroun: unpublished. Gabon: John & Lawson 1974a. Ghana: *Spermothamnion* sp., Lieberman *et al.* 1979.

Remarks. We have tentatively referred to this species sterile plants found in Cameroun and Ghana similar in size and vegetative structure to the material from Gabon.

Spermothamnion speluncarum (Collins & Hervey) Howe

Howe, 1920; *Rhodochorton speluncarum*, Collins and Hervey, 1917.

Plants forming a dense mat, epilithic, with the attaching haptera and erect filaments emerging from the middle of the cells of the prostrate filaments; erect filaments simple, occasionally alternately or secundly branched, usually slightly tapering above, with the apical cell terminally blunt; cells from 28-35 μm in diameter and from (2-)3 to 4 times longer than broad; tetrasporangia solitary, opposite or occasionally in secund series, sessile or on stalks of 1 or 2 cells, from 50-60 μm in diameter.

For figures see Børgesen, 1930a, p. 17, 18, figs 5, 6.

Occurs on sand-covered rocks in the lower eulittoral subzone and occasionally in tidepools.

Distribution. World: in warm temperate and tropical parts of the Atlantic Ocean. West Africa: Ghana: unpublished.

Spyridia Harvey 1833

Plants erect, consisting of irregularly divided branches, often bearing branchlets of determinate growth; branches corticated by alternating series of transverse rows of small cells at the nodes and large longitudinally elongate cells at the internodes, with rhizoidal filaments in older parts; branchlets of an axial row of cells and cortication restricted to the nodes, deciduous, terminating in a short few-celled spine; tetrasporangia single or in groups at the nodes of the branchlets, tetrahedrally divided; spermatangia forming cylindrical patches on branchlets, covering nodes and internodes; gonimoblasts having two or more irregular lobes and surrounded by an involucre of branchlets.

World Distribution: widespread in warm temperate and tropical seas.

Key to the Species

1. Ultimate branches usually distinctly swollen towards the tips and slightly compressed . . . . *S. clavata*
1. Ultimate branches cylindrical and always terete 2

2. Branchlets with only a single terminal spine. . . . . *S. filamentosa*
2. Branchlets initially with recurved spines on the terminal or 2 distal nodes. . . . . *S. hypnoides*

Spyridia clavata Kützing Pl. 47, fig. 4.

Kützing, 1841.

Plants usually bushy and to about 16 cm in height; branches alternately or occasionally oppositely divided and distichous, terete below and often somewhat compressed above, from 1-2 mm in diameter towards base, with the ultimate division often swollen and clavate, having obtuse or acuminate apices; branchlets usually absent from the swollen ultimate branches, to about 60 μm in diameter and with the segments from 1 to 3 times as long as broad, curved upwards and terminating in a hooked spine.

For additional figures see Børgesen, 1917, p. 236, fig. 227.

Occurs on rocky platforms and calcareous cobbles at depths ranging from 10 to 12 m and occasionally cast-up in the drift.

Distribution. World: as for genus. West Africa: Ghana: John & Lawson 1972a. Gambia: John & Lawson 1977b, Steentoft 1967. São Tomé: Hariot 1908, Henriques 1917, John & Lawson 1972a, Steentoft 1967. Sénégal (Casamance): Chevalier 1920.

Spyridia filamentosa (Wulfen) Harvey

Harvey, in Hooker, 1833; *Fucus filamentosus* Wulfen, 1803.

Plants bushy, to about 20 cm in height; branches alternately divided, radial, terete throughout, from 1-2 mm in diameter, often devoid of branchlets below; branchlets radially arranged, from 20-45 μm in diameter and segments about 2 to 4 times longer than broad, terminating in a straight spine.

Plate 47: Figs 1-3. *Spyridia hypnoides* variety *disticha* form *inermis:* 1. Plant showing the distichous arrangement of the branches and the branchlets; 2. Portion of a plant showing the corticated branches bearing uniseriate branchlets; 3. Terminal portion of a branchlet with the spine-like tip and the nodal cortication. Fig. 4. *Spyridia clavata:* Portion of a plant showing the ultimate branches usually distinctly swollen.

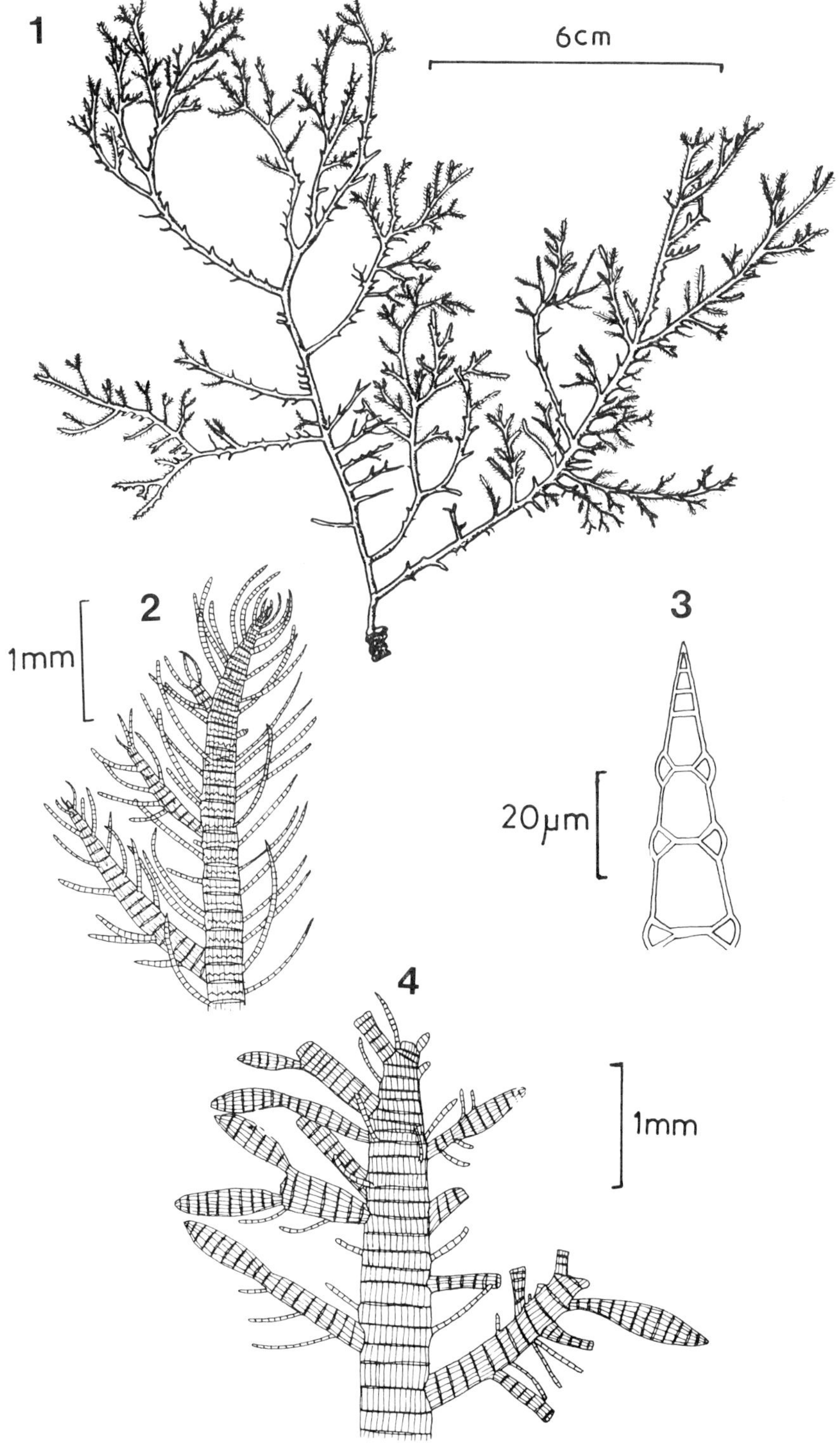
1
6cm
2
1mm
3
20µm
4
1mm

For figures see Børgesen, 1917, p. 233 to 235, figs 222-225.

Frequently reported in shallow water though there is no information on its ecology in the Gulf of Guinea.

Distribution. World: in warm temperate and tropical parts of the Atlantic Ocean. West Africa: Sierra Leone: Aleem 1978.

Remarks. Many of the early reports of this species from the offshore islands in the Gulf of Guinea have been discounted by Steentoft (1967) who believes them to be misdeterminations of *Spyridia aculeata* (now *S. hypnoides*).

Spyridia hypnoides (Bory) Papenfuss

Papenfuss, 1968a; *Thamnophora hypnoides* Bory, in Belanger, 1834.

Plants solitary and to about 25 cm in height; branches alternately divided, radial or tending to be distichous, terete throughout, often tapering towards the base and apex, with the tips sometimes revolute; branchlets up to 60 μm in diameter and segments to about 1.5 times as long as broad, terminating in a spine and having 1 to 3 hooked spines on the first and second nodes.

For figures see Børgesen, 1917, p. 237, fig. 228 (as *S. aculeata* var. *typica*).

var. **disticha** (Børgesen) nov. comb. Pl. 47, figs 1-3.
Spyridia aculeata var. *disticha* Børgesen, 1917, p. 238.
f. **inermis** (Børgesen) nov. comb.
S. aculeata var. *disticha* f. *inermis* Børgesen, 1917, p. 239.

Branches and branchlets distichously arranged and the latter having hooked spines near the tips rare or absent.

For additional figures see Børgesen, 1917, p. 237, 239, 240, figs 228-230 (as *S. aculeata* var. *disticha* f. *inermis*).

Occurs as the variety in some abundance on calcareous cobbles at depths from 10 to 25 m and has been dredged to over 30 m. The typical form has been reported dredged at 11 m by Carpine (1959) from around one of the offshore islands.

Distribution. World: as for genus. West Africa: Gabon: *S. hypnoides* var. *disticha* f. *inermis*, John & Lawson 1974a. Ghana: *S. hypnoides*, John *et al.* 1977, 1980, Lieberman *et al.* 1979; *S. hypnoides* var. *disticha* f. *inermis*, John & Lawson 1972a. Príncipe: *S. hypnoides* var. *disticha* f. *inermis;* as *S. aculeata* f. *inermis*, Carpine 1959, Steentoft 1967. São Tomé: *S. hypnoides;* as *S. aculeata*, Steentoft 1967; as *S. aculeata* f. *typica* (= *f. aculeata*), Carpine 1959, Steentoft 1967; as *S. filamentosa,* Henriques 1886b, 1887, 1917.

Wrangelia C. Agardh 1828

Plants consisting of variously divided branches, often distichously arranged; branches ecorticated or corticated by filaments arising from the nodes and from the base of the branchlets, sometimes well-developed and pseudoparenchymatous; reproductive organs borne on the branchlets or on the ultimate branches, often surrounded by an involucre of filaments; tetrasporangia shortly stalked, tetrahedrally divided.

World Distribution: widespread in warm temperate and tropical seas.

Key to the Species

1. Plants usually solitary, stiff and erect, to about 10 cm in height; branches to about 650 μm in diameter and with pseudoparenchymatous cortication .... *W. penicillata*
1. Plants forming soft and delicate tufts or mats, from 1-3 cm high; branches less than 260 μm in diameter and cortication absent or of a few irregular filaments .. .. *W. argus*

Wrangelia argus (Montagne) Montagne — Pl. 48, fig. 6.

Montagne, 1856; *Griffithsia argus* Montagne, 1839-1841.

Plants forming soft tufts and occasionally mats, to about 3 cm in height, dark purplish-red or sometimes greenish-red and gelatinous when producing tetrasporangia, occasionally iridescent; branching of the main axes dichotomous and of the branchlets subdichotomous; branches to about 260 μm in diameter, having the cells from 2 to 4 times as long as broad, ecorticate or with a few corticating filaments arising from the lower cells of the branchlets or from the nodes; branchlets having the basal cells from 70-100 μm in diameter and 2 to 10 times as long as broad, often terminating in 1 or 2 short acute spines.

For additional figures see Børgesen, 1916, p. 116, figs 125, 126.

Occurs in the lower eulittoral subzone and down to a depth of about 2 m, growing as a component of the algal turf in sheltered situations and in more wave-exposed places as plumose tufts, mats or felty ring-like patches on rocks and lithothamnia.

Distribution. World: as for genus. West Africa: Cameroun: John 1977a. Gambia: John & Lawson 1977b. Ghana: Dickinson & Foote 1950, Gordon 1972, John 1977a, John & Pople 1973, Lawson 1956, 1957b, 1966, Stephenson & Stephenson 1972. Liberia: De May *et al.* 1977, John 1977a. Togo: John 1977a, John & Lawson 1972b.

Wrangella penicillata (C. Agardh) C. Agardh — Pl. 48, figs 4, 5.

C. Agardh, 1828; *Griffithsia penicillata* C. Agardh, 1824.

Plants erect and solitary, to about 10 cm in height, red to reddish-orange; branching alternate and distichous; branches to about 650 μm in diameter and the internodes half to twice as long as broad, with well-developed cortication often pseudoparenchymatous; branchlets with basal cells from 60-70 μm in diameter and 0.5 to 2 times as long as broad, terminating in a blunt cell or an acute spine.

For additional figures see Børgesen, 1916, p. 120, figs 131, 132.

Occurs on rocky platforms or calcareous cobbles at a depth of about 10 m and sometimes cast-up in the drift.

Distribution. World: as for genus. West Africa: Ghana: John & Lawson 1972a, John *et al.* 1977, Lieberman *et al.* 1979.

Family Delesseriaceae*

Caloglossa J. Agardh 1876

Plants consisting of leaf-like branches, constricted at intervals and dichotomously divided, with the secondary branches usually arising from the midrib; branches membranaceous, with a monostromatic lamina having oblique series of subhexagonal cells running to the margin, and a prominent midrib having the axial rows of cells obscured by elongate superficial cells; tetrasporangia in sori on each side of the midrib and near the tips of the branches; cystocarps prominent, sessile on the midrib.

World Distribution: widespread from boreal-antiboreal to tropical seas.

Key to the Species

1. Branches usually from 1-2 mm in width .... *C. leprieurii*
2. Branches never greater than 0.5 mm wide .... *C. ogasawaraensis*

*On a number of occasions a flattened, foliaceous plant having a network of macroscopic veins and a crenulate margin, has been collected at depths in excess of 10 m off Ghana where it is sometimes one of the most common algae particularly on *Turritella* shells strewn over sand or mud bottoms (>30 m depth). This plant may be a species of *Hymenena* (John *et al.*, 1977, 1980; Lieberman *et al.*, 1979) or *Cryptopleura*, genera whose limits are not clearly defined and which are in need of critical revision (Abbott and Hollenberg, 1976).

Plate 48: Figs 1, 2. *Caloglossa leprieurii:* 1. Terminal portion of a plant showing the flattened branches with a midrib and constricted into segments; 2. Portion of a branch tip with tetrasporangia grouped together in sori. Fig. 3. *Caloglossa ogasawaraensis:* Portion of a plant with the secondary branches arising from the constrictions and not from the midrib. Figs 4, 5. *Wrangelia penicillata:* 4. Plant erect and the branching distichous and alternate; 5. Portion of a plant showing the monosiphonous branchlets borne on the corticated main axis. Fig. 6. *Wrangelia argus:* Portion of a branch bearing tetrasporangia on involucrate filaments and with corticating filaments growing down over the monosiphonous axis.

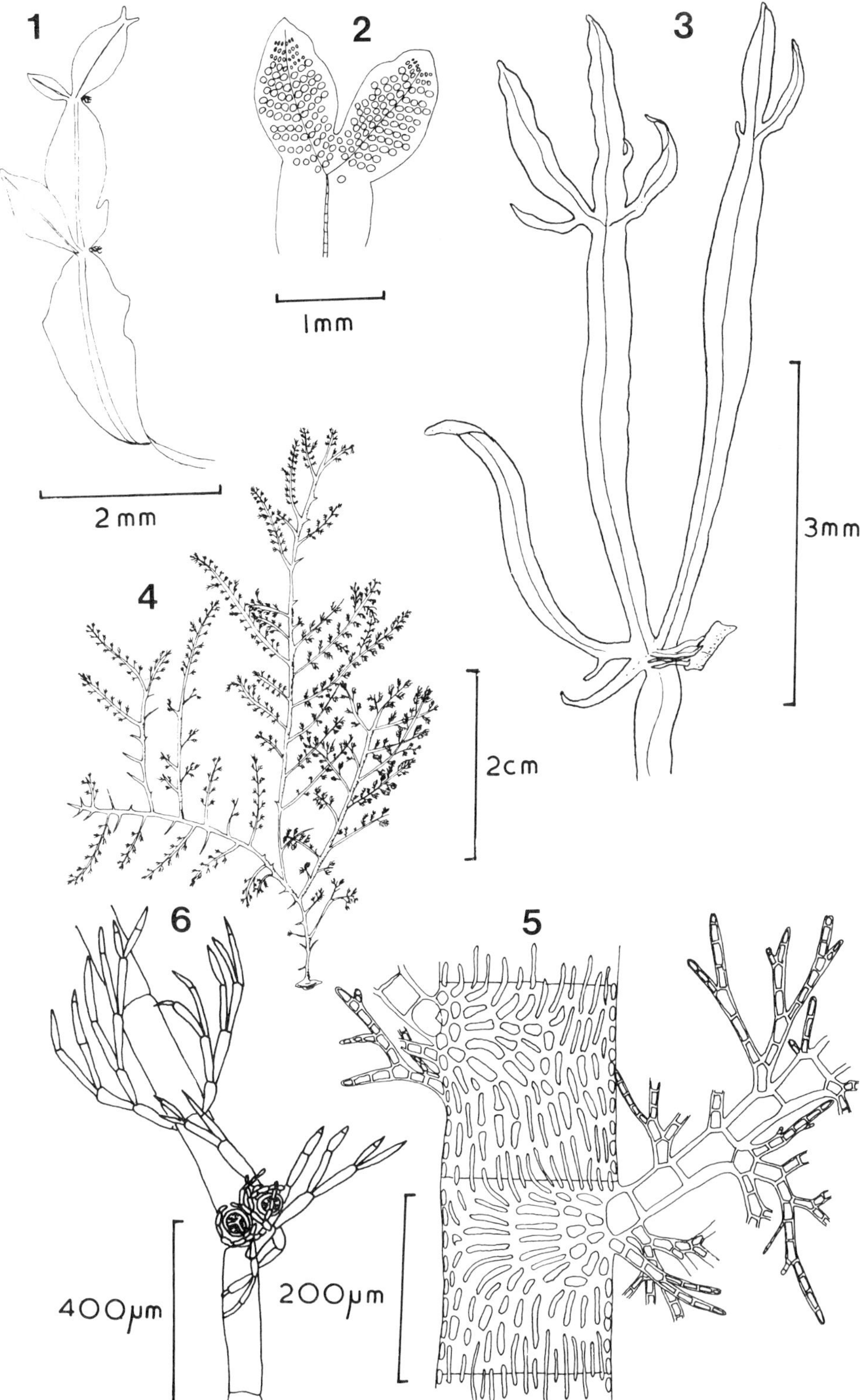
1
2
3
2 mm
1mm
3mm
4
2cm
6
5
400µm
200µm

Caloglossa leprieurii (Montagne) J. Agardh Pl. 48, figs 1, 2.

J. Agardh, 1876; *Delesseria leprieurii* Montagne, 1840.

Plants usually forming semi-prostrate mats, to about 1 cm in height, brownish-black, attached by rhizoids arising from the underside of the branches at the forkings; branches constricted into somewhat lanceolate segments, from 1-2 mm in width, with the secondary branches arising at the narrow constrictions or directly from the midrib.

For additional figures see Papenfuss, 1961, p. 10, 11, 13, 14, 16, 17, figs 1-30.

Occurs on rocks, logs or the roots of mangroves and is particularly abundant in cracks, crevices or in shaded situations in brackish-water habitats, growing in the eulittoral zone and occasionally extending into the littoral fringe.

Distribution. World: as for genus. West Africa: Bioko: unpublished. Cameroun: John 1977a, Lawson 1955, Pilger 1911, Post 1957a, 1957b, 1963b; *C. leprieuri* var. *hookeri,* Post 1936, 1955a, 1963b, Stephenson & Stephenson 1972. Côte d'Ivoire: John 1972c, 1977a. Gabon: John & Lawson 1974a. Ghana: Dickinson & Foote 1950, Hariot 1896, John 1977a, Lawson 1960c, Post 1957a, 1957b, 1963b, Stephenson & Stephenson 1972; *C. leprieuri* var. *hookeri,* Post 1955a, 1963b, 1965. Guinée: Post 1955a, 1955b, 1957a, 1957b; *C. leprieuri* f. *continua,* Post 1936, 1955b; *C. leprieuri* var. *hookeri,* Marchal 1960, Post 1936, 1955a, 1959. Liberia: De May *et al.* 1977. Nigeria: John 1977a, Post 1936, 1957a; *C. leprieuri* var. *hookeri,* Post 1963b. São Tomé: Post 1966a. Sierra Leone: Aleem 1978, John & Lawson 1977a, Post 1955a, 1955b, 1965, 1968.

Remarks. There is some confusion regarding the typification of the variety *hookeri* since Post's (Post, 1936) description and illustrations (p. 54, 55, figs 2, 3) of this plant are based on material which had previously been made the type of *C. mnoides* by J. Agardh (1876). It seems best at the present to include the West African plants which conform to *C. mnoides* (=*C. leprieurii* var. *hookeri*) in *C. leprieurii* until more is known of this genus regarding its developmental morphology and the range of phenotypic variation shown by plants growing in different environments. For similar reasons we have chosen to omit f. *continua* (Okamura) Post reported by Post (1936, 1955b) from "Dakar-Guinea".

Caloglossa ogasawaraensis Okamura Pl. 48, fig. 3.

Okamura, 1908.

Plants usually prostrate and growing entangled with other small algae, less than 1 cm in height, attached by rhizoids arising from the undersurface; branches constricted at intervals into linear to lanceolate segments, about 0.5 mm in width, with the secondary branches arising from the constrictions and never from the midrib.

For additional figures see Okamura, 1908, pl. 37, figs 1-11.

Occurs often together with *Caloglossa leprieurii* on rocks in river estuaries and on the roots of *Rhizophora* in mangrove swamps; found also along open parts of the coast growing with other small red algae in the littoral fringe.

Distribution. World: probably pantropical. West Africa: Côte d'Ivoire: John 1972c, 1977a, Post 1963b, 1966a, 1966b. Nigeria: unpublished. São Tomé: John 1977a, Post 1966a, 1968.

Hypoglossum Kützing 1843b

Plants consisting of terete branches bearing foliar appendages, some orders of branching arising from the midrib of the appendages, occasionally proliferous; foliar appendages having a monostromatic lamina and a thick and corticated midrib; tetrasporangia and spermatangia in discontinuous sori on or along the midrib, sporangia tetrahedrally divided; cystocarps sessile, arising on the midrib, with a produced lateral pore.

World Distribution: widespread from boreal-antiboreal to tropical seas.

Hypoglossum guineense nov. sp. Pl. 49, figs 1-3.

Plantae molles subtiles, plerumque singulatim crescentes, 4-10 cm altae, pallidissime roseae; pars basalis teres, irregulariter divisa, frondes emittens; frons linearis vel lanceolata, minute serrata, basi obtusa, apice subacuta, 3-10 mm lata ad 6 cm longa, costa prominente; pericarpium hemisphaericum, poro conspicuo buccinato munitum, 420-770 μm diametro, sessile, costa frondis diminutae fere delaminatae insidente.

Plants soft and delicate, usually growing singly, from 4-10 cm in height, very pale pink; branches terete and irregularly divided, giving rise to foliar appendages; foliar appendages linear to lanceolate and having minute marginal serrations, with the base obtuse and the apex subacute, from 3-10 mm in width and to about 6 cm in length, having a prominent midrib; cystocarps sessile and with a conspicuous trumpet-shaped margin to pore, from 420-770 μm in diameter, arising on the midrib of a reduced foliar appendage having the lamina little developed.

Holotype: 6036, Vernon Bank, Ghana, 8 November 1968. Deposited in the British Museum (Natural History), London (B M).

Occurs on rocky platforms and on calcareous cobbles at depths between 10 to 28 m.

Distribution. World: in the tropical eastern Atlantic Ocean. West Africa: Ghana: *Hypoglossum* sp., John *et al.* 1977, 1980, Lieberman *et al.* 1979.

Remarks. There are but few species in this genus in which the margins of the foliar appendages are serrated. The Ghanaian plants show a certain resemblance to *Hypoglossum serrulatum* (Harvey) J. Agardh known from Australia. Nevertheless, this new species differs from the Australian plant in having more regular and considerably smaller marginal serrations, subacute rather than blunt apices, and is more delicate in form due to its generally smaller dimensions.

Platysiphonia Børgesen 1931

Plants consisting of an erect system of flattened, alternately or unilaterally divided branches arising from a prostrate system of terete branches, ecorticate or corticated only towards the base, attached by rhizoids or branched haptera; flattened branches usually six cells broad, with the flanking cells half the length of the lateral pericentral cells; tetrasporangia arising in two longitudinal rows on either side of the axial filament, tetrahedrally or tetrapartitely divided; spermatangia borne in groups on flattened branches; cystocarps adaxial on ultimate branches, sessile, usually urn-shaped, with a terminal pore.

World Distribution: widespread in warm temperate and tropical seas.

Platysiphonia miniata (C. Agardh) Børgesen — Pl. 49, figs 4, 5.

Børgesen, 1931; *Hutchinsia miniata* C. Agardh, 1828.

Plants forming delicate tufts or small patches, to about 2 cm in height, light reddish-purple, attached by an intricate system of rhizoids; branching irregular and lateral branches diverging obliquely from the adaxial surface of the main axis, with the central basal cell of the lateral branch embedded at the point of attachment; branches usually flattened and each flanking cell half the length of the central cells; tetrasporangia usually confined to the ultimate lateral branches.

For additional figures see Womersley and Shepley, 1959, p. 196, 198, figs 83-99.

Occurs as patches on the sandy floor of rocky tidepools and in the algal turf on moderately wave-sheltered rocks in the eulittoral zone. Found also at a depth of about 10 m growing on other algae and over large polyzoans.

Distribution. World: as for genus. West Africa: Côte d'Ivoire: John 1972c, 1977a. Ghana: John 1977a, John & Lawson 1972a, John *et al.* 1977, Lieberman *et al.* 1979.

Plate 49: Figs 1-3. *Hypoglossum guineense:* 1. Portion of a plant showing the distinct midrib and regularly serrated margin to the membranaceous frond; 2. Portion of a frond with one of the marginal teeth; 3. Cystocarp borne on the midrib of a reduced branch and having an extended pore. Figs 4, 5. *Platysiphonia miniata:* 4. Portion of a plant having the tetrasporangia arising in two rows on either side of the axial filament in a flattened lateral branch; 5. Portion of a branch showing each flanking cell to be half the length of the central cells.

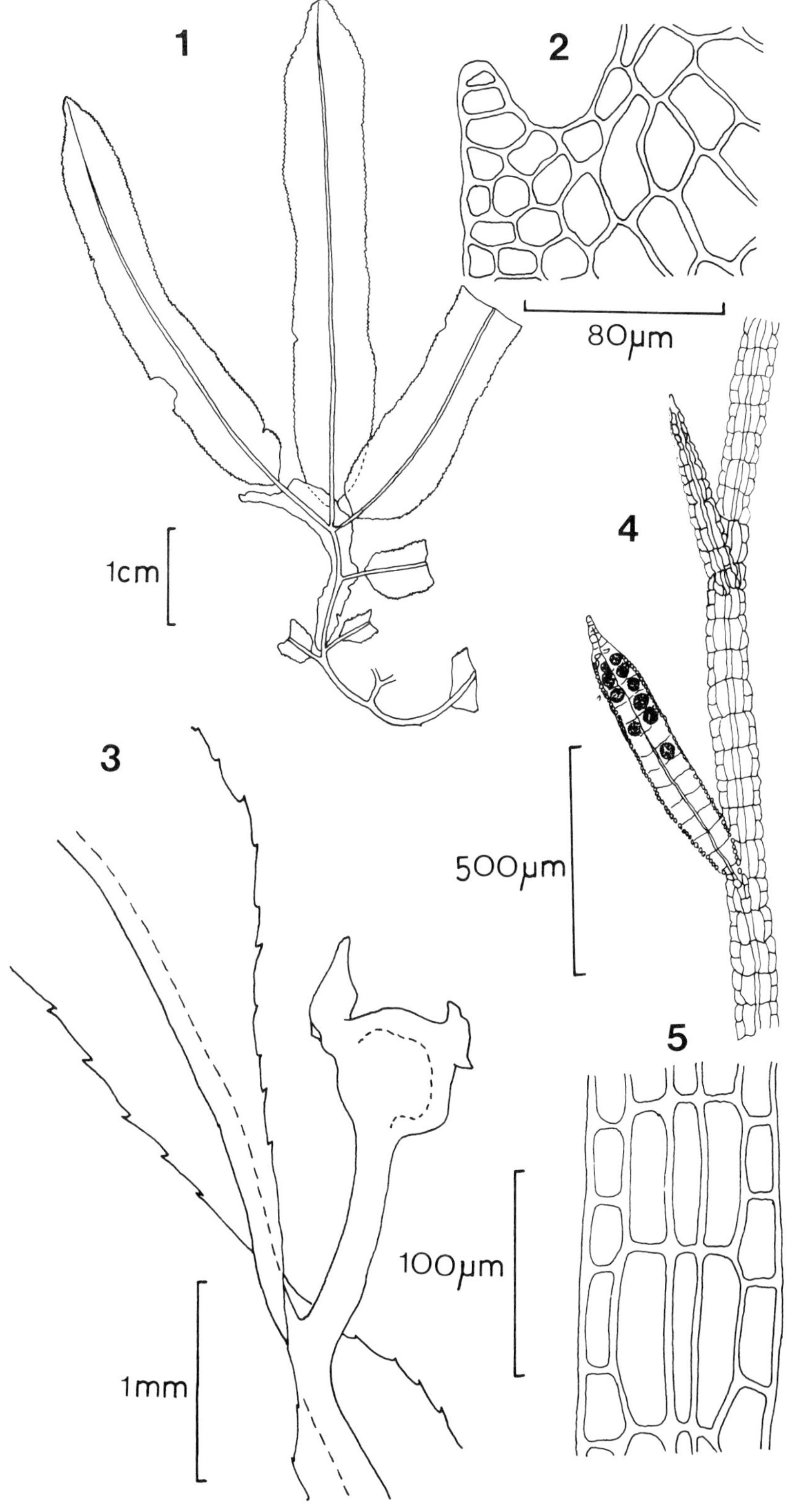
1
2
80µm
1cm
4
3
500µm
5
100µm
1mm

Taenioma J. Agardh 1863

Plants consisting of erect terete branches bearing flattened branchlets and arising from a prostrate system of alternately divided branches, attached by unicellular rhizoids; terete branches having four pericentral cells; branchlets of determinate growth flattened and five cells in width, having the marginal cells divided and half the length of the inner cells of each segment, terminating in two or more long hairs; tetrasporangia borne in the flattened branchlets, sporangia tetrapartite; spermatangia on specialised branchlets; cystocarps urn-shaped and replacing branchlets.

World Distribution: probably widespread in warm temperate and tropical seas.

Key to the Species

1. Branches from 46-80 μm in diameter and branchlets invariably terminating in 2 long hairs.. *T. nana*
1. Branches from 70-150 μm in diameter and branchlets terminating in 3, more rarelv 2, hairs .... *T. perpusillum*

Taenioma nana (Kützing) Papenfuss

Papenfuss, 1952; *Polysiphonia nana* Kützing, 1863.

Plants growing together with other small algae, to about 1 mm in height, greenish-red, with an erect system of branches arising at intervals of from 2 to 9 segments from a prostrate system; branching alternate, with a branch and a branchlet usually borne on every third segment of the erect system; terete branches from 46-80 μm in diameter; branchlets from 46-71 μm in width, from 18 to 23 segments long, terminating invariably in 2 long hairs.

For figures see Lawson, 1960b, p. 364, fig. 1, 1 m, pl. 2i, j; Børgesen, 1919, p. 339, fig. 337 (as *T. perpusillum*).

Occurs on moderately sheltered to moderately wave-exposed rocks in the lower eulittoral subzone.

Distribution. World: as for genus. West Africa: Bioko: unpublished. Sierra Leone: John & Lawson 1977a, Lawson 1960b; *T. perpusillum*, Lawson 1954b, 1957a.

Plate 50: Figs 1, 2. *Dictyurus fenestratus:* 1. Plant showing the alternately divided erect branches arising from a creeping stoloniferous base; 2. Portion of a branch showing the elaborate sac-like network of anastomosing branchlets. Figs 3, 4. *Taenioma perpusillum:* 3. Portion of a branch showing the three terminal monosiphonous hairs; 4. Cystocarp urn-shaped and replacing a branchlet. Fig. 5. *Acanthophora ramulosa* (after Steentoft, 1967): Range of form of the cystocarp, usually subtending a spine.

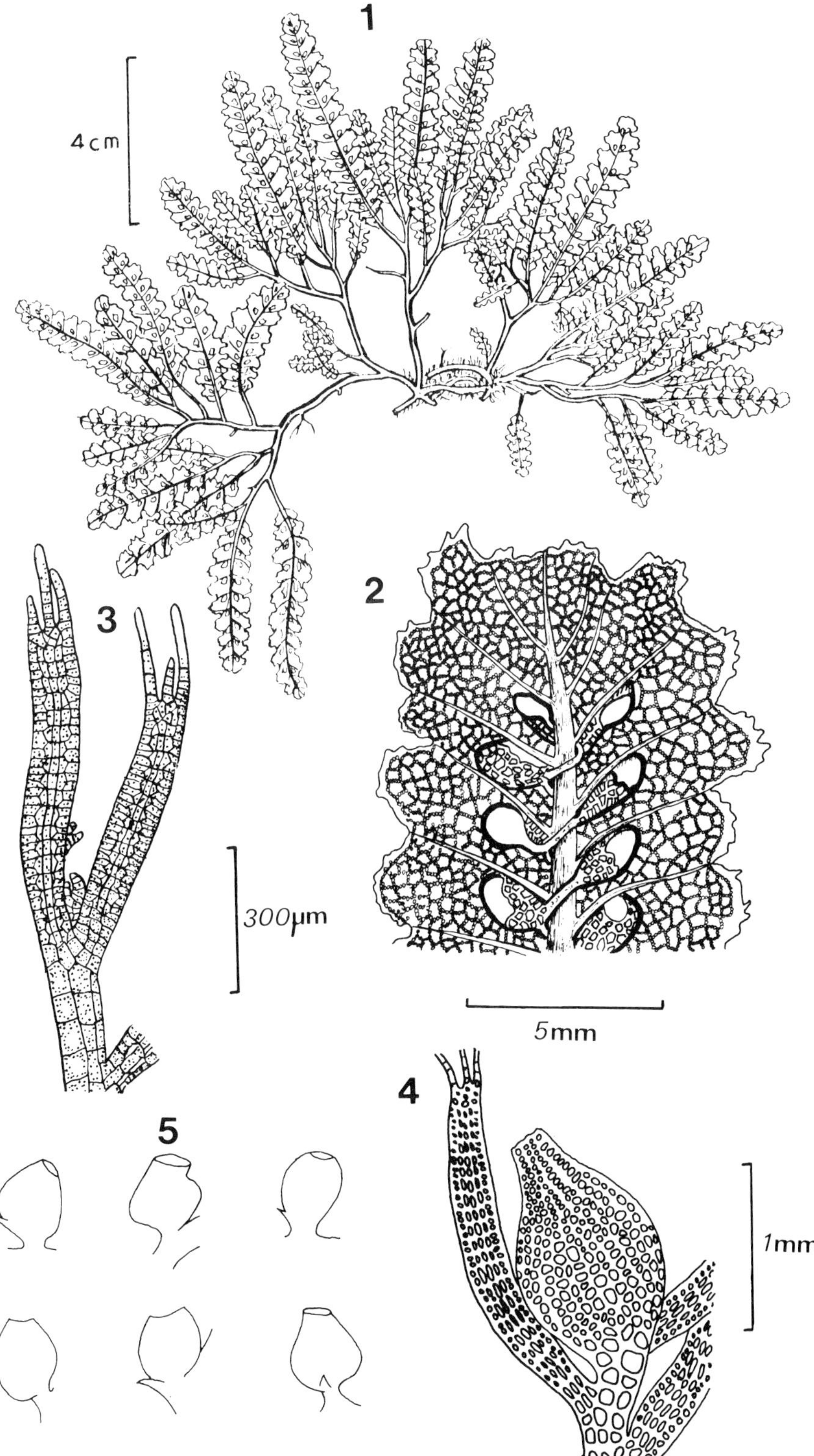
1
4cm
2
3
300µm
5mm
4
5
1mm

Taenioma perpusillum (J. Agardh) J. Agardh Pl. 50, figs 3, 4.

J. Agardh, 1863; *Polysiphonia perpusilla* J. Agardh, 1848a.

Plants forming felty patches or mixed with other small algae, from 1-3 mm in height, reddish-purple, with the erect branches arising at intervals of 6 to 7 segments from the prostrate system of branches; branching alternate and somewhat fastigiate, with the branches and branchlets given off from every third or fourth segment; branches from 70-150 μm in diameter; branchlets from 50-98 μm in width, from 20 to 29 segments long, terminating in 2 or 3 long hairs.

For additional figures see Lawson, 1960b, p. 364, fig. 1a-k, pl. 2a-h.

Occurs as felty ring-like patches or tufts on rocks or lithothamnia in moderately wave-exposed situations and associated with other algae where wave-exposure is less.

Distribution. World: as for genus. West Africa: Côte d'Ivoire: John 1972c, 1977a, John & Lawson 1977b. Ghana: Dickinson & Foote 1950, John 1977a, John & Lawson 1977a, Lawson 1954b, 1956, 1960b, Stephenson & Stephenson 1972. Liberia: De May *et al.* 1977, John 1972c, 1977a, John & Lawson 1977a.

Family Dasyaceae

Dasya C. Agardh 1824

Plants erect, consisting of irregularly or alternately divided branches bearing dense whorls of pseudodichotomously divided branchlets; branches having four or five pericentral cells and rhizoidally corticated in older parts; branchlets ramellate in having a polysiphonous base and monosiphonous above, occasionally monosiphonous throughout; tetrasporangia in stalked and lanceolate to ovate-lanceolate stichidia, four to five per segment, tetrahedrally divided; spermatangia in lanceolate-conical clusters on the ramelli; cystocarps sessile or stalked on smaller branches.

World Distribution: widespread from boreal-antiboreal to tropical seas.

Key to the Species

1. Plants forming bushy epiphytic tufts, from 1-1.5 cm in height ...... *Dasya* sp. A
1. Plants solitary, epilithic, normally greater than 2 cm in height 2

2. Branchlets from 15-40 μm in diameter at base and gradually narrowing towards apices... *D. baillouviana*
2. Branchlets from 5-12 μm in diameter throughout .... *Dasya* sp. B

Dasya baillouviana (Gmelin) Montagne

Montagne, 1839-1841; *Fucus baillouviana* Gmelin, 1768

Plants solitary, to about 5 cm in height, bright red; branching sparingly to profusely alternate; branches corticated, up to 1 mm in diameter, often naked below and bearing many ramellate branchlets in upper parts giving them a plumose appearance; branchlets pseudodichotomously divided, from 15-40 μm in diameter at base, narrowing towards apex; tetrasporangial stichidia borne near the base of the branchlets, to about 120 μm in diameter and to 400 μm or more in length.

For figures see Falkenberg, 1901, pl. 18, figs 5-17 (as *D. elegans*); Jaasund, 1976, p. 120, fig. 243 (as *D. pedicellata*).

This plant is of rare occurrence in the region having been found on only few occasions growing on cobbles and rocky banks at 10 to 15 m, dredged also from a depth of about 10 fathoms (18 m).

Distribution. World: probably widespread in warm temperate and tropical seas. West Africa: Ghana: John *et al.* 1977, Lieberman *et al.* 1979; *D. pedicellata*, Dickinson 1952, den Hartog 1959.

Undetermined Species

Dasya sp. A. Pl. 51, fig. 4.

Plants forming bushy tufts on larger algae, from 1-1.5 cm in height; branches alternately divided, from 100-140(-200) μm in diameter in lower parts, ecorticate throughout; branchlets arising alternately on branches, irregularly dichotomously divided, from 25-50 μm in diameter, with the cells to about 3(-5) times as long as broad and decreasing towards the tips to about equal in length and breadth, tapering above and terminating in an obtuse apex; tetrasporangial stichidia ovate to nearly conical, from 70-100 μm in diameter and about 200 μm in length.

Occurs on larger algae in the lower eulittoral subzone and to a depth of about 10 m.

Distribution. West Africa: Ghana: unpublished.

Dasya sp. B. Pl. 51, figs 5, 6.

Plant 15 cm in height; branches irregularly alternately and occasionally oppositely divided, spreading, to about 2 mm in diameter, with rhizoidal cortication well-developed; branchlets borne only on the upper branches and arising irregularly, irregularly dichotomously divided, monosiphonous throughout, to about 1 mm long, with the cells from 5-12 μm in diameter and 3 to 10 times as long as broad, having an obtuse apex; tetrasporangial stichidia arising on distinct ramelli, lanceolate, from 60-80 μm in diameter and up to 1.5 mm in length; cystocarps shortly stalked, from 400-500 μm in diameter, urceolate, with a beaked ostiole.

Collected on just one occasion at a depth of about 10 m growing on a rocky platform.

Distribution. West Africa: Ghana: unpublished.

Dictyurus Bory 1834

Plants erect, consisting of terete and alternately divided branches, giving rise to a regular network or veil formed of anastomosing branchlets; branches having four pericentral cells surrounded by a thick pseudoparenchymatous cortex; branchlets ecorticate and ramellate, polysiphonous below and monosiphonous above; tetrasporangia in stichidia arising from the polysiphonous bases of the branchlets.

World Distribution: probably widespread in many warm temperate and tropical seas.

Dictyurus fenestratus Dickinson Pl. 50, figs 1, 2.

Dickinson, 1951.

Plants bushy, solitary or gregarious, from 6-14 cm in height, reddish-purple; branches arising from a creeping basal system, from 1-2 mm in diameter, naked below and bearing above an elaborate sac-like network of anastomosing branchlets forming a partially open spiral veil, with the free edges of the net forming fan-shaped structures; reproductive organs unknown.

For additional figures see Dickinson, 1951, p. 294, pl. 3, p. 296.

Occurs often as small single plants on the underside of rock ledges in moderately wave-exposed situations in the sublittoral fringe or covering extensive areas at about 10 m depth where it is often growing on sand-covered rocks. This plant is found occasionally as solitary individuals down to about 22 m.

Distribution. World: in warm temperate and tropical parts of the eastern Atlantic Ocean. West Africa: Ghana: Aregood & Hackett 1972, Dickinson 1951, John *et al.* 1977, 1980, Lawson 1956, Lieberman *et al.* 1979.

Remarks. Reproductive structures have yet to be found in this species.

Plate 51: Figs 1, 2. *Acanthophora spicifera:* 1. Plant sparingly irregularly branched and beset with lateral branches of determinate growth; 2. Terminal portion of a plant showing the spines confined to the lateral branches. Fig. 3. *Heterosiphonia wurdemanni* (after John and Lawson, 1972a): Portion of a branch showing the dichotomously divided branchlets arising at two segment intervals, and a tetrasporangial stichidium terminating an ultimate division of a branchlet. Fig. 4. *Dasya* sp. A: Portion of a branch with the tetrasporangial stichidia arising on the dichotomously divided branchlets. Figs 5, 6. *Dasya* sp. B: 5. Upper portion of a plant showing the irregularly divided branches covered with branchlets; 6. Portion of a branch showing the alternately divided branchlets and stalked tetrasporangial stichidia borne on the corticated main axis. Figs 7, 8. *Acanthophora muscoides:* 7. Portion of a plant with branches beset with lateral branches of determinate growth; 8. Portion of a branch with the spines present on both the main branches and the lateral branches.

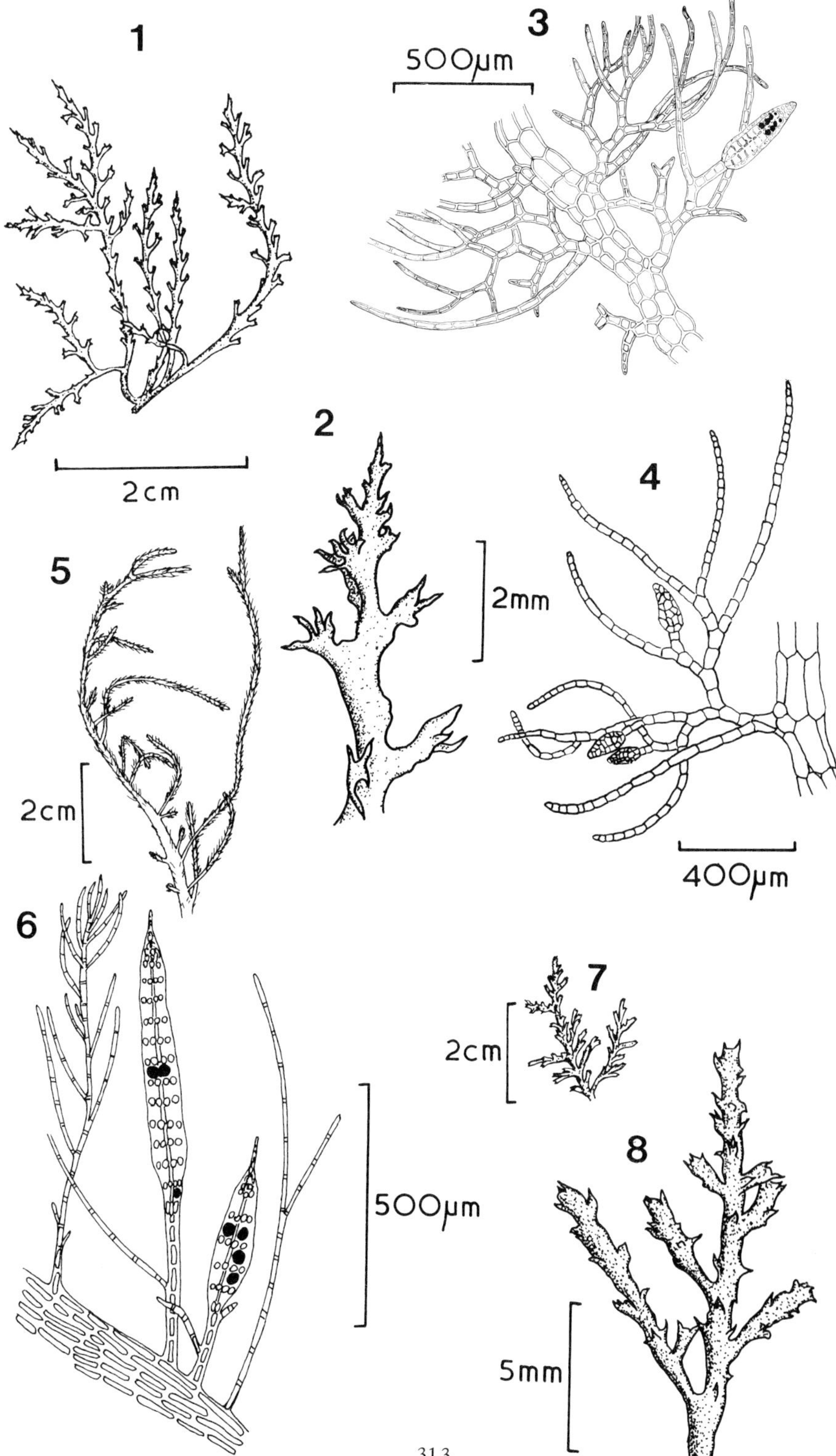
1
2cm
3
500µm
2
2mm
4
400µm
5
2cm
6
500µm
7
2cm
8
5mm

Heterosiphonia Montagne 1842

Plants erect or sometimes dorsiventral consisting of terete or flattened branches, more or less dichotomously divided or distichous and pinnate, bearing alternately divided and often ramellate branchlets; branches having four or more pericentral cells and rhizoidal cortication; branchlets monosiphonous throughout or more often polysiphonous towards the base, separated by two to several internodes; tetrasporangial stichidia cylindrical and terminal on a ramellus, with from four to six tetrahedrally divided tetrasporangia per segment; cystocarps arising near the base of the short branchlets, often ovoid to urn-shaped, with a terminal pore.

World Distribution: widespread from boreal-antiboreal to tropical seas.

Heterosiphonia wurdemanni (Bailey ex Harvey) Falkenberg Pl. 51, fig. 3.

Falkenberg, 1901; *Dasya wurdemanni* Bailey, in Harvey, 1853.

Plants solitary or sometimes entangled with other algae, to about 3 cm or more in height; branching dichotomous or somewhat irregular, with the branchlets arising alternately on the branches from every second segment; branches to about 180 μm in diameter and having from 4 to 6 pericentral cells; branchlets dichotomously divided, monosiphonous, to about 56 μm in diameter at the base, with the cells equal in length and breadth in the lower parts, either short and squarrose or long and tapering with the cells increasing to 6 times as long as broad towards the apices.

For additional figures see Børgesen, 1919, p. 325 to 327, figs 326-328.

Occurs on rocky platforms and calcareous cobbles at depth of about 8 to 10 m and also in the sublittoral fringe, occasionally cast-up in the drift.

Distribution. World: as for genus. West Africa: Ghana: John & Lawson 1972a, John *et al.* 1977, Lieberman *et al.* 1979.

Remarks. The Ghanaian plants are somewhat intermediate between the varieties *laxa* and *squarrosa* as recognised by Børgesen (1919) and orginally given form rank by Falkenberg (1901).

Micropeuce J. Agardh 1899

Plants usually erect and bushy, with the alternately divided branches bearing lateral branchlets arising from the basal cells of alternately or subdichotomously divided monosiphonous ramelli; branches havings five pericentral cells and rhizoidal cortication; tetrasporangia borne in little-modified ultimate branchlets, spirally arranged and one per segment.

World Distribution: widespread from boreal-antiboreal to tropical seas.

Micropeuce mucronata (Harvey) Kylin Pl. 57, figs 1, 2.

Kylin, 1956; *Dasya mucronata* Harvey, 1853.

Plants bushy, to about 10 cm or more in height, light red; branches alternately or somewhat irregularly divided, to about 1 mm in diameter, often naked below but bearing spirally arranged branchlets in upper parts; ramelli from 1 to 3 times subdichotomously divided, from 40-60 μm in diameter at base and cells about 3 times as long as broad below and increasing towards the apex to 8 times as long as broad, with the terminal cell subulate; tetrasporangia developing in upper parts of little modified branchlets, up to 15 in series.

For additional figures see Taylor, 1928, pl. 27, figs 5, 6 (as *Brongniartella mucronata*).

Found on but few occasions growing on rocky platforms and calcareous cobbles at depths ranging from 10 to 27 m.

Distribution. World: probably widespread in many warm temperate and tropical seas. West Africa: Ghana: John *et al.* 1977, Lieberman *et al.* 1979.

Family Rhodomelaceae

Acanthophora Lamouroux 1813

Plants usually coarse and erect, consisting of alternately or irregularly divided terete branches of indeterminate growth and similar but shorter branches of determinate growth, partly or completely beset with spirally disposed spine-like branchlets; branches in section showing five obvious pericentral cells only towards the apices, with parenchymatous cortication; tetrasporangia tetrahedral, arising in the short branches of determinate growth; spermatangia forming flattened and disc-like clusters; cystocarps developing in the axils or near the basal parts of the spine-like branchlets.

World Distribution: in warm temperate and tropical seas.

Key to the Species

1. Plants sparingly branched; spine-like branchlets if present confined to the short branches of determinate growth .... *A. spicifera*
1. Plants densely branched and bushy; spine-like branchlets present on all the branches 2
2. Branches of determinate growth either covered by spine-like branchlets or naked and having a characteristic ovoid to cylindrical shape, without a subtending spine .... *A. ramulosa*
2. Branches of determinate growth always covered by spine-like branchlets and usually subtended by a spine .... *A. muscoides*

Acanthophora muscoides (Linnaeus) Bory Pl. 51, figs 7, 8.

Bory, 1827-29; *Fucus muscoides* Linnaeus, 1753.

Plants erect and bushy, to about 16 cm in height; branches to about 2.5 mm in diameter, irregularly divided with the divisions often arising close together in the upper parts, bearing spirally disposed spine-like branchlets; determinate branches often subtended by a spine; tetrasporangia borne on short and very spiny determinate branches; cystocarps urn-shaped and on short and somewhat swollen branchlets.

For additional figures see Kützing, 1865, pl. 77, figs a-c; Taylor, 1960, p. 807, fig. 3.

Occurs in the lower eulittoral subzone on gently sloping and moderately wave-sheltered rocks, often associated with a good deal of silt; found only on one occasion at a depth of 11 m.

Distribution. World: probably pantropical. West Africa: Bioko: unpublished. Cameroun: unpublished. Gabon: John & Lawson 1974a. Gambia: John & Lawson 1977b. São Tomé: Carpine 1959, Steentoft 1967.

Acanthophora ramulosa Kützing Pl. 50, fig. 5.

Kützing, 1843b.

Plants erect and very bushy, to about 20 cm in height; branches many times irregularly divided and covered by spine-like branchlets, with some of the short determinate branches in older parts naked and ovoid or cylindrical in shape; tetrasporangia in the terminal portions of short, wrinkled and spineless, determinate branches; cystocarps on short stalks and urn-shaped to ovoid.

For additional figures see Kützing, 1865, pl. 76, figs a, b.

There is no information available on the ecology of this plant in the region.

Distribution. World: in tropical parts of the eastern Atlantic Ocean. West Africa: São Tomé: Steentoft 1967; *A. muscoides* (pro parte), Hariot 1908, Henriques 1917.

Acanthophora spicifera (Vahl) Børgesen Pl. 51, figs 1, 2.

Børgesen, 1910; *Fucus spiciferus* Vahl, 1802.

Plants erect and lax, to about 20 cm in height; branches from 2-3 mm in diameter, sparingly irregularly divided, with the ultimate divisions often long and arcuate; branches of determinate growth short and dense in upper parts, beset with spine-like branchlets especially towards the tips; cystocarps arising in axils of the spine-like branchlets, shortly stalked and urn-shaped.

For additional figures see Børgesen, 1918, p. 260 to 263, figs 253-258; Taylor, 1960, p. 805, fig. 3, p. 807, figs 1, 2.

Occurs in the lower eulittoral subzone and sublittoral fringe on wave-sheltered rocks, often associated with beds of *Sargassum*.

Distribution. World: probably pantropical. West Africa: Cameroun: Lawson 1954b, 1955, Schmidt & Gerloff 1957, Steentoft 1967; *A. thierii*, Pilger 1911. Gabon: *A. thierii*, Hariot 1896. Príncipe: Steentoft 1967. São Tomé: Steentoft 1967; *A. muscoides*, Hariot 1908 (pro parte), Henriques 1886b, 1887, 1917 (p. p.). Sierra Leone: Aleem 1978, John & Lawson 1977a, Lawson 1954b, Steentoft 1967.

Bostrychia Montagne 1842

Plants dorsiventral, consisting of a creeping and erect system of branches attached either by rhizoidal branchlets or rhizoids growing out from the pericentral cells; branches terete and filiform, sometimes more or less flattened, usually alternately divided and distichous, mostly polysiphonous with pericentral cells equal in length to the axial cells or transversely divided, corticated or ecorticated; branchlets often ramellate being partly monosiphonous, simple or sometimes divided; tetrasporangia borne in stichidia and several per segment, sporangia tetrahedrally divided; cystocarps arising on the tips of the branchlets, with a conspicuous pore.

World Distribution: widespread in warm temperate and tropical seas.

Key to the Species

1. Branches ecorticate throughout; branchlets sometimes transformed into primary holdfasts 2
1. Branches corticated, at least in part; primary holdfasts formed from out growths from the pericentral cells. 3

2. Branchlets polysiphonous at base and largely monosiphonous above. . . . . *B. moritziana*
2. Branchlets largely polysiphonous and with up to 3 terminal segments monosiphonous (non f. *moniliforme*) . . . . *B. radicans*

3. Branches rhizoidally corticated . . . . *B. calliptera*
3. Branches parenchymatously corticated 4

4. Branches showing little difference in size and degree of branching throughout plant; branchlets having less than 5 monosiphonous segments at the tip. . . . . *B. montagnei*
4. Branches long and bearing distinctly shorter lateral ones; branchlets monosiphonous throughout or with 5 or more monosiphonous terminal segments 5

5. Branchlets of the last order monosiphonous throughout. . . . . *B. tenella*
5. Branchlets polysiphonous towards base and monosiphonous for the last 5 to 15 segments. . . . . *B. binderi*

Bostrychia binderi Harvey — Pl. 52, fig. 7.

Harvey, 1847.

Plants forming tufts or mats, to about 4 cm in height, dark reddish-brown or often bleached to pale brown or greenish-brown; branching alternate and pinnate, with the lateral branches crowded, relatively short and often tripinnate below but somewhat longer and bipinnate in upper parts; branches with 6 to 7 pericentral cells and pseudoparenchymatous cortication; branchlets polysiphonous towards base and monosiphonous for the last 5 to 15 segments; tetrasporangial stichidia linear to lanceolate, with 4 to 5 sporangia per segment.

For additional figures see Richardson, 1975, pl. 23, fig. 2, pl. 26, fig. 5.

Occurs in the littoral fringe on the undersurface of stranded logs, in cracks and crevices in rocks, and sometimes on open rocks provided they are shaded by overhangs and trees: often associated with *B. tenella.*

Distribution. World: as for genus. West Africa: Bioko: unpublished. Cameroun: Cordeiro-Marina 1978, Lawson 1955, Post 1936, 1963b, Steentoft 1967. Gabon: John & Lawson 1974a. Ghana: Lawson 1960c, Post 1955a, 1955b, 1959, 1963b, Steentoft 1967, Stephenson & Stephenson 1972; *B. tenella* (pro parte), Dickinson & Foote 1950. Guinée: Post 1955a, Steentoft 1967. Liberia: De May *et al.* 1977. Nigeria: unpublished. São Tomé: Carpine 1959, Post 1966a, 1968, Steentoft 1967. Sierra Leone: John & Lawson 1977a, ?Lawson 1957a, Levring 1969, Post 1955a, 1959, 1965, 1968, Steentoft 1967; *B. tenella,* Lawson 1954b.

Bostrychia calliptera Montagne Pl. 52, figs 1-4.

Montagne, 1840.

Plants more or less prostrate, usually as a low mat to about 2 cm in height, attached by rhizoids arising at or near the origins of the branches; branching di- or polydichotomous and the smaller divisions pinnate; branches having 6 pericentral cells and rhizoidally corticated, with the branchlets having 4 pericentral cells throughout; tetrasporangial stichidia stalked and lanceolate, with the sporangia in 2 distinct rows.

For additional figures see Kützing, 1865, pl. 19, figs d-g.

Occurs in mangrove swamps together with other small algae growing on the breathing roots of *Avicennia;* also on shaded rocks in the littoral fringe in river estuaries.

Distribution. World: probably widespread in tropical seas. West Africa: Côte d' Ivoire: John 1972c, 1977a. Ghana: John 1977a, Lawson 1960c, Post 1963b, 1965. Guinée: Post 1936, 1955a, 1955b, 1959, 1963b, Schnell 1950. Sierra Leone: John & Lawson 1977a, Post 1965, 1968.

Remarks. Some plants found in the Niger Delta have the morphology of this species but the cortication does not appear to be clearly rhizoidal.

Bostrychia montagnei Harvey

Harvey, 1853.

Plants forming tufts, to about 5 cm in height; branching alternate and distichous, repeatedly pinnate in the upper parts; branches having 6 to 7 pericentral cells and a pseudoparenchymatous cortex of two to several layers of cells with the outer ones small and irregularly arranged; branchlets often absent below, incurved and polysiphonous except for short monosiphonous tips.

For figures see Harvey, 1853, p. 55, pl. 14, fig. B, 1-4; Taylor, 1960, p. 811, fig. 1.

Occurs together with other small red algae growing on the stilt roots of the mangrove *Rhizophora*.

Distribution. World: probably widespread in warm temperate and tropical parts of the Atlantic Ocean. West Africa: Sierra Leone: John & Lawson 1977a; *B. scorpioides* var. *montagnei,* Post 1965, 1968.

Remarks. It is not clear from Post's earlier publication (Post, 1965) whether the typical form of *Bostrychia scorpioides* was present along with the variety *montagnei* Harvey, this latter is now considered to be a synonym of *B. montagnei* Harvey (see Sluiman, 1979). Nevertheless, in a later work (Post, 1968) she repeats the collection information given in the earlier one and there is no ambiguity as only the variety is referred to.

Bostrychia moritziana (Sonder) J. Agardh Pl. 52, fig. 5.

J. Agardh, 1863; *?Polysiphonia moritziana* Sonder, in Kützing 1849.

Plants forming small tufts, to about 2 cm in height; branching dichotomous and the smaller divisions repeatedly pinnate especially in the upper parts; branches having 7 to 8 pericentral cells, ecorticate; branchlets polysiphonous towards the base, incurved and monosiphonous above; tetrasporangial stichidia elongate to lanceolate, with 3 to 4 sporangia per segment.

For additional figures see Kützing, 1865, pl. 24, figs a-c (as *B. monosiphonia*).

Occurs on the roots of mangroves and also known from rocks in freshwater streams, rare in shaded situations in the littoral fringe on open parts of the coast.

Distribution. World: as for genus. West Africa: Cameroun: Post 1936. Gambia: John & Lawson 1977b. Liberia: De May *et al.* 1977. São Tomé: Post 1963b. Sierra Leone: unpublished.

Remarks. This species has also been found in Ghana and Nigeria but so far only from freshwater habitats.

Bostrychia radicans (Montagne) Montagne

Montagne, 1850b; *Rhodomela radicans* Montagne, 1840.

Plants forming a loose mat, to about 1.5 cm in height; branching alternate and distichous; branches having 6 to 8 pericentral cells, often curved, ecorticate; branchlets polysiphonous below and monosiphonous for up to 3 segments at their very tips;

tetrasporangial stichidia borne on long stalks, ovate and oblong, with 3 to 4 series of tetrasporangia, 2 often very prominent.

For figures see Kützing, 1865, pl. 20, figs a-c; Pilger, 1911, p. 305, fig. 13.

f. **depauperata** Montagne

Montagne, 1850b.

Main branches sparingly and irregularly divided.

f. **hapteromanica** Post

Post, 1936.

Branchlets usually all transformed into haptera.

f. **moniliforme** Post Pl. 52, fig. 6.

Post, 1936.

Branchlets ramellate, occasionally having up to 18 terminal monosiphonous segments.

Occurs on the open coast in the upper eulittoral subzone and the littoral fringe, usually confined to shaded places such as cracks and crevices in rocks and beneath overhanging trees. This plant is also found in brackish-water habitats being at similar levels on estuarine rocks and on the roots of mangroves.

Distribution. World: probably widespread in warm temperate seas and pantropical. West Africa: Bioko: unpublished. Cameroun: Cordeiro-Marino 1978, John 1977a, Lawson 1955, Pilger 1911, Post 1936, 1955a, 1957a, 1963a, 1963b, Schmidt & Gerloff 1957, Steentoft 1967, Stephenson & Stephenson 1972. *B. radicans* f. *depauperata,* Post 1936. *B. radicans* f. *hapteromanica,* John 1977a, Post 1936. *B. radicans* f. *moniliforme,* Codeiro-Marina 1978, Post 1936. Côte d'Ivoire: *B. radicans* f. *hapteromanica,* John 1972c, 1977a, Post 1963a. Gabon: John & Lawson 1974a, Post 1963a, Steentoft 1967. Ghana: John 1977a, Lawson 1960c, 1966, Post 1963a, 1963b, Steentoft 1967. *B. radicans* f. *moniliforme,* Lawson 1960c, Post 1963a,

Plate 52: Figs 1-4. *Bostrychia calliptera:* 1. Portion of a plant showing the pinnate branching; 2. Tip of a branchlet monosiphonous; 3. Lower portion of a branchlet polysiphonous; 4. Branch with complete cortication. Fig. 5. *Bostrychia moritziana:* Portion of a plant showing the monosiphonous tips to the branchlets and polysiphonous below. Fig. 6. *Bostrychia radicans* form *moniliforme:* Portion of a plant with only the apical segments of a branchlet monosiphonous. Fig. 7. *Bostrychia binderi:* Portion of a plant showing the branchlets to be polysiphonous only towards the base and with the tetrasporangial stichidia borne near the branch tip.

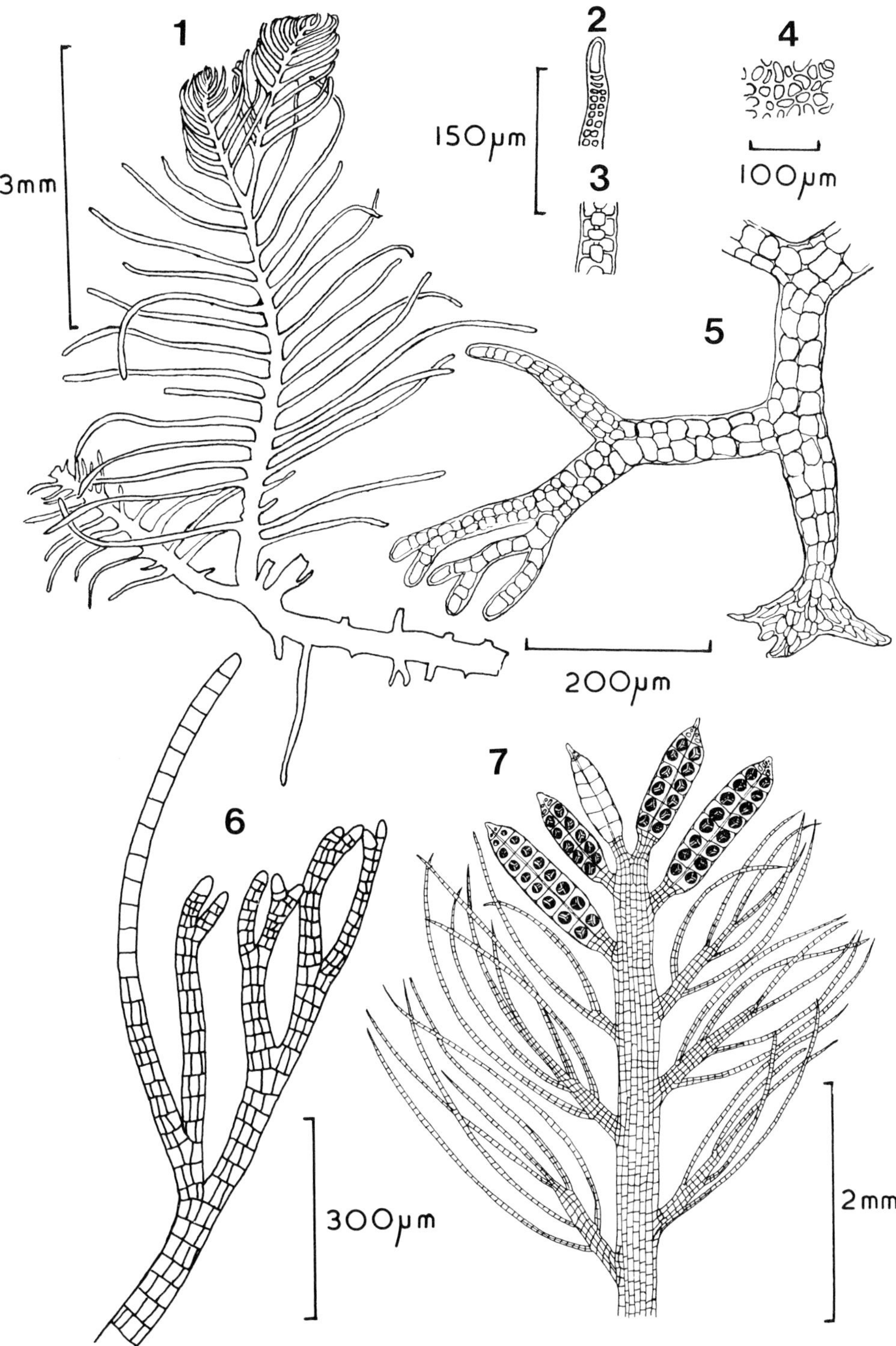
1
3mm
2
150µm
3
4
100µm
5
200µm
6
300µm
7
2mm

1963b. Guinée: Marchal 1960, Post 1955a, 1959, 1963a, Steentoft 1967. *B. radicans* f. *moniliforme,* Post 1955a, Schnell 1950. Liberia: Post 1965. Nigeria: Cordeiro-Marino 1978, John 1977a, Post 1936, 1963a, 1963b, Steentoft 1967; *B. simpliciuscula,* Fox 1957. São Tomé: John 1977a, Post 1963a, 1963b, Steentoft 1967. *B. radicans* f. *depauperata,* Post 1963b, 1966a. Sierra Leone: Aleem 1978, Cordeiro-Marino 1978, John & Lawson 1977a, Lawson 1954b, 1957a, 1966, Levring 1969, Longhurst 1958, Post 1955a, 1963a, 1965, 1968, Steentoft 1967. *B. radicans* f. *depauperata,* Post 1965. *B. radicans* f. *moniliforme,* Post 1965.

Remarks. This species has been reported elsewhere in tropical West Africa but only from freshwater habitats (see Post, 1968). The report of *Bostrychia simpliciuscula* from Nigeria is based on a misidentification.

Bostrychia tenella (Vahl) J. Agardh.

J. Agardh, 1863; *Fucus tenellus* Vahl, 1802

Plants forming tufts or mats, from 3-5 cm in height, yellowish-green or often dark red; branching alternate and distichous, repeatedly pinnate; branches having 6 to 8 pericentral cells and fewer in the lateral branches, with 3 to 4 layers of cortical cells in the older parts; branchlets monosiphonous throughout; tetrasporangial stichidia stalked and linear to lanceolate, with usually 4 sporangia in each segment.

For figures see Børgesen, 1918, p. 300 to 302, figs 299-303; Pilger, 1911, p. 306, figs 14, 15.

Occurs in the littoral fringe growing in cracks and crevices in rocks and sometimes over tree-shaded boulders, often together with other *Bostrychia* species.

Distribution. World: as for genus. West Africa: Bioko: unpublished. Cameroun: Lawson 1955, Pilger 1911, Post 1936, 1955a, 1963b, Stephenson & Stephenson 1972. Ghana: Dickinson & Foote 1950 (pro parte), Lawson 1956, 1960c, Post 1955a, 1957, 1963b, 1968, Stephenson & Stephenson 1972; *B. binderi* f. *terrestre,* Post 1963b. Guinée: Post 1955a. Sierra Leone: Aleem 1978, Lawson 1954b, Post 1959.

Remarks. *Bostrychia binderi* f. *terrestre* (Harvey) Post may be considered to be probably no more than a depauperate specimen. Post (1936) based this form on a Harvey collection (no. 22) from the Friendly Islands which was called *B. terrestre* in his exsiccata. This Harvey collection was earlier described by J. Agardh (1863) as *B. tenella* var. *terrestris* and later raised to specific rank by J. Agardh (1897), only to be reduced by De Toni (1924) to synonymy under *B. tenella.* Harvey never provided a diagnosis for his plant even though Post (1963b) acknowledges him as the authority for the form. According to the rules of botanical nomenclature this form has never been validly described.

Bryocladia Schmitz 1897

Plants consisting of alternately divided branches bearing branchlets, arising from a creeping and matted base; branchlets stiff and often recurved, spirally disposed on short branches of determinate growth, sometimes bearing hairs; branches consisting of from six to sixteen pericentral cells, ecorticate; tetrasporangia borne along the outermost side of a branchlet, with one sporangium per segment; cystocarps stalked and urn-shaped, arising on the lateral branches.

World Distribution: probably widespread in warm temperate and tropical seas.

Key to the Species

1. Branches usually having 8 pericentral cells, with the branchlets simple or subsimple . . . . *B. cuspidata*
1. Branches usually having 10 pericentral cells, with the branchlets alternately divided and pinnate . . . . *B. thyrsigera*

Bryocladia cuspidata (J. Agardh) De Toni — Pl. 53, fig. 9.

De Toni, 1903; *Polysiphonia cuspidata* J. Agardh, 1848a.

Plants forming erect and loose tufts or mats, to about 8 cm in height, dark purplish-black; branches having 8 pericentral cells, simple or very sparingly divided towards the base; branchlets simple or occasionally subdivided near the base, tapering markedly towards the apex.

For figures see Taylor, 1960, p. 805, fig. 2.

Occurs on beach rocks in the lower eulittoral subzone in moderately wave-exposed situations.

Distribution. World: in tropical parts of the Atlantic Ocean. West Africa: Cameroun: Pilger 1911. Gambia: John & Lawson 1977b. Liberia: De May *et al.* 1977.

Bryocladia thyrsigera (J. Agardh) Schmitz — Pl. 53, figs 1, 2.

Schmitz, in Falkenberg, 1901; *Polysiphonia thyrsigera* J. Agardh, 1848a.

Plants forming dense, coarse tufts or mats, to about 10 cm in height, brownish-green or purplish-black; branches having 10 pericentral cells, sparingly divided, beset with lateral branches bearing dense, stiff and pinnately divided branchlets; tetrasporangia in series in densely clustered recurved branchlets.

For additional figures see Kützing, 1864, pl. 33, figs d-g (as *Polysiphonia thyrsigera*).

Occurs in the algal turf or as tufts or large extensive mats on moderately sheltered to moderately wave-exposed rocks in the lower eulittoral subzone; often common on sand-scoured beach rocks where few other algae can exist.

Distribution. World: in warm temperate and tropical parts of the Atlantic Ocean. West Africa: Côte d'Ivoire: John 1972c, 1977a. Gambia: John & Lawson 1977b. Ghana: Dickinson & Foote 1950, Fox 1957, John 1977a, Lawson 1956, 1966, Stephenson & Stephenson 1972. Liberia: De May *et al.* 1977, John 1972c, 1977a. Nigeria: Fox 1957, John 1977a, Lawson 1966, Saenger 1974, Stephenson & Stephenson 1972. Sierra Leone: John & Lawson 1977a. Togo: John 1977a, John & Lawson 1972b.

Bryothamnion Kützing 1843b

Plants usually erect and bushy, consisting of firm, terete to angled or compressed branches, alternately divided and bearing simple or forked branchlets; branchlets having six to eight pericentral cells and a thick cortex; tetrasporangia in stichidia arising at the axils of the branches, spirally disposed and one per segment; cystocarps stalked and subglobose, borne on short modified branchlets.

World Distribution: probably widespread in warm temperate and tropical seas. tropical seas.

Key to the Species

1. Branches triangular in section; branchlets spirally disposed....... *B. triquetrum*
1. Branches terete or compressed; branchlets arising marginally or in 3 or 4 rows .. .. *B. seaforthii*

Bryothamnion seaforthii (Turner) Kützing Pl. 53, fig. 4.

Kützing, 1843b; *Fucus seaforthii* Turner, 1809.

Plants forming bushy clumps, from 8-20 cm in height, purplish-red, membranaceous to subcartilaginous; branching sparingly pinnate below, dense and fastigiate in upper parts; branchlets arising bilaterally or sometimes in 3 or 4 rows; branches terete to compressed, with the branchlets bearing spine-like ramelli and to about 4(-6) times forked.

Plate 53: Figs 1, 2. *Bryocladia thyrsigera:* 1. Plants arising from a creeping base; 2. Terminal portion of a plant with the branches beset with pinnately divided branchlets. Fig. 3. *Bryothamnion triquetrum:* Upper portion of a branch showing the spirally disposed branchlets. Fig. 4. *Bryothamnion seaforthii:* Portion of a branch with the branchlets arising bilaterally. Fig. 5. *Chondria tenuissima* (after John and Lawson, 1972a): Portion of a plant with the branches having a somewhat pyramidal outline due to the branchlets becoming shorter towards the branch tips. Figs 6, 7. *Chondria bernardii:* 6. Plant with the branches alternately divided; 7. Terminal portion of a branch showing a cystocarp subtended by a spine. Fig. 8. *Digenia simplex:* Portion of a plant showing its bushy appearance due to the main axes being beset with simple branchlets. Fig. 9. *Bryocladia cuspidata:* Portion of a plant with the erect branch beset with simple or subsimple branchlets.

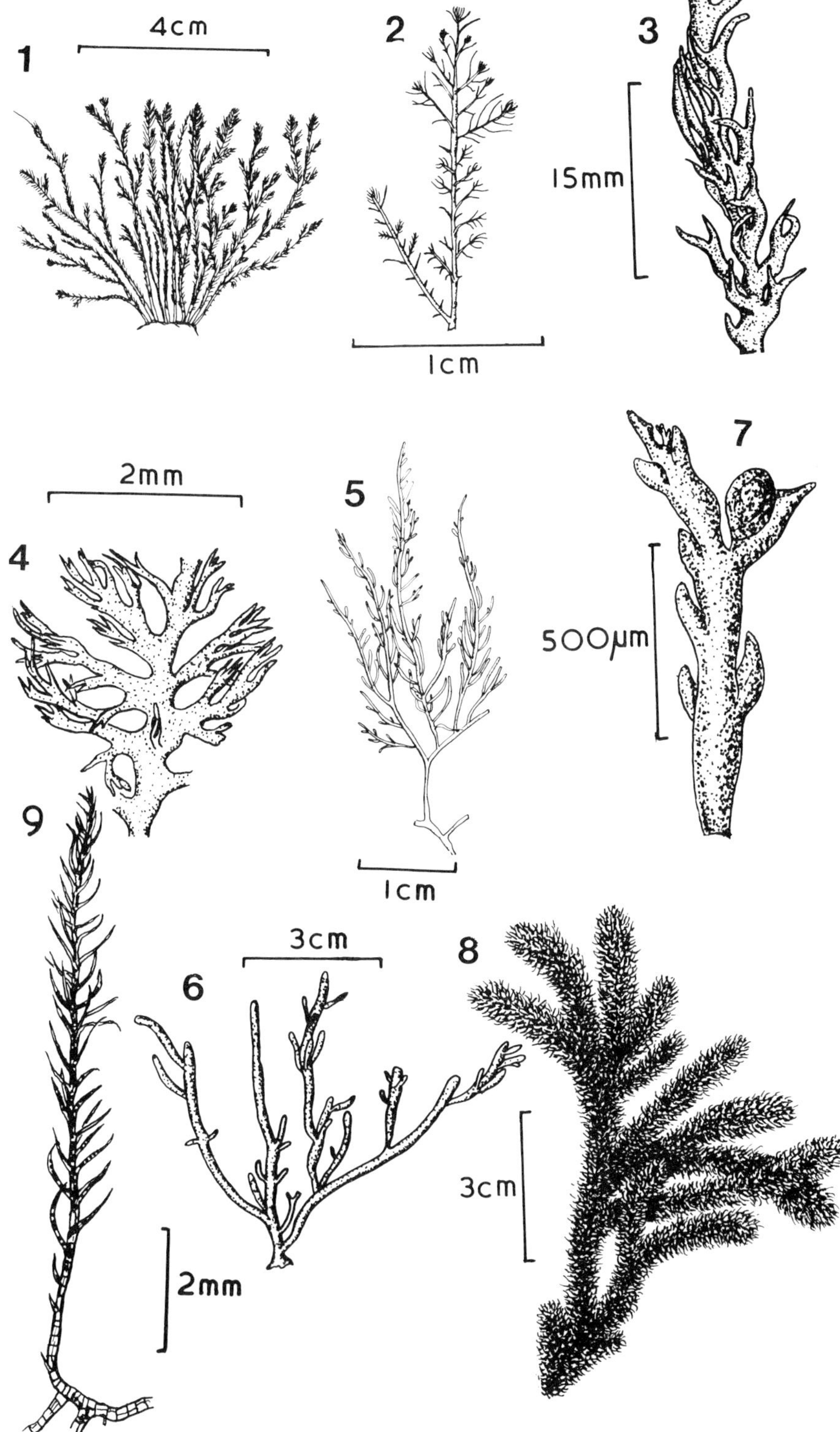

1
4cm
2
1cm
3
15mm
4
2mm
5
1cm
7
500μm
9
6
3cm
8
3cm
2mm

For additional figures see Børgesen, 1918, p. 284, 285, figs 284-286.

There is no information on the ecology of this species in the region but in other parts of its range it is found as a deep water plant.

Distribution. World: probably widespread in tropical seas. West Africa: Guinée: Børgesen 1918. São Tomé: De Toni 1903, Falkenberg 1901, Hariot 1908, Henriques 1886b, 1887, 1917, Steentoft 1967.

Bryothamnion triquetrum (Gmelin) Howe Pl. 53, fig. 3.

Howe, 1915; *Fucus triqueter* Gmelin, 1768.

Plants erect and to about 25 cm in height, purplish-brown, fleshy to subcartilaginous; branching irregularly alternate; branchlets spirally disposed along angles of branches, widely spaced below and crowded in upper parts; branches terete towards base and triangular above; branchlets very short and spine-like, simple below and 1 or 2 times divided in upper parts, often trifid and bearing subulate ramelli.

For additional figures see Børgesen, 1918, p. 283, 284, figs 282, 283.

Occurs as an apparently rare plant in the region having been dredged on only one occasion from a depth of 11 m whilst in other parts of its range it is known from still greater depths.

Distribution. World: probably widespread in tropical seas. West Africa: São Tomé: Carpine 1959, Steentoft 1967; *B. triangulare,* De Toni 1903, Falkenberg 1901, Hariot 1908, Henriques 1886b, 1887, 1917.

Chondria C. Agardh 1817

Plants consisting of terete or somewhat flattened branches, fleshy to cartilaginous, irregularly to alternately divided, with the branchlets often club-shaped or spindle-shaped, somewhat constricted at the base and terminating in a cluster of hairs; apical cell prominent or at the base of a depression; branches in section having five pericentral cells and a several-layered cortex; tetrasporangia developed from the pericentral cells and restricted to the ultimate branchlets, tetrahedrally divided; spermatangia arising towards the tips of the branchlets as dense clusters, usually forming flat or curved plates; cystocarps sessile and usually prominent, with the pericarp more than one cell thick, having a single pore.

World Distribution: widespread from boreal-antiboreal to tropical seas.

Key to the Species

1. Branches subterete below and obviously flattened above .... *C. confusa*
1. Branches terete throughout... 2

2. Branches from 1-2.5 mm in diameter and branchlets soft and spindle-shaped ... . *C. tenuissima*
2. Branches usually less than 1 mm in diameter and branchlets usually club-shaped 3

3. Branches normally greater than 500 μm in diameter........................ 4
3. Branches less than 500 μm in diameter.................................... 5

4. Branchlets spine-like towards the tips of the branches; cystocarps characteristically subtended by a spine .... *C. bernardii*
4. Branchlets never appearing spine-like; cystocarps unknown ..... *Chondria* sp. B

5. Branches often arcuate and usually bearing a secund series of obtuse or truncate branchlets .... *Chondria* sp. A
5. Branches straight, with the cylindrical to truncate branchlets opposite or verticillate .... *Chondria densa*

Chondria bernardii P. Dangeard — Pl. 53, figs 5, 6.

Dangeard, 1951a.

Plants usually solitary, to about 4 cm in height, purplish-red, somewhat cartilaginous; branching alternate and occasionally proliferous towards the apices; branches terete throughout, to about 1 mm in diameter and the upper branches curving upwards, somewhat constricted at the base; branchlets small and club-shaped, appearing spine-like towards the branch tips; cystocarps to about 600 μm in diameter, subtended by a spine.

For additional figures see Dangeard, 1951a, p. 19, pl. 2, figs d-g.

Occurs on sand-covered rocks in tidepools in the lower eulittoral subzone subject to moderate wave action, occasionally found at a depth of about 10 m growing on calcareous cobbles.

Distribution. World: in warm temperate and tropical parts of the eastern Atlantic Ocean. West Africa: Ghana: John *et al.* 1977, Lieberman *et al.* 1979.

Chondria confusa nom. nov. — Pl. 56, figs 8, 9.

C. platyclada P. Dangeard (1952, p. 306, 307) non W.R. Taylor (1945, p. 295).

Plants forming intricately woven cushion-like clumps or looser clumps in deep water, from 2-3 cm in height, reddish-purple and sometimes iridescent, with the branches attached to one another and to the substratum by rhizoids; branching distichous or subdistichous, with the branchlets arising alternately or rarely subopposite; branches and branchlets subterete below and becoming flattened above, usually about 1 mm in breadth, attenuate towards the base and apex, usually having acute tips; cells in surface view generally longer than broad, from 70-80 μm in length and from 10-12 μm in breadth.

For figures see Dangeard, 1952, p. 325, pl. 21, figs A-I (as *C. platyclada*).

Occurs on calcareous cobbles at depths between 10 and 15 m although in Sénégal it is reported growing in the eulittoral zone.

Distribution. World: in warm temperate and tropical parts of the Atlantic Ocean. West Africa: Ghana: *C. platyclada*, John *et al.* 1977, 1980, Lieberman *et al.* 1979.

Remarks. The name *Chondria platyclada* given by Dangeard (1952) to material from Sénégal is illegitimate and must be rejected as it is a later homonym for a plant described by Taylor (1945) from the Pacific Ocean. We have been unable to obtain Dangeard's type specimen collected from Sénégal (no number given) but have renamed his plant *Chondria confusa* because of the confusion caused by his unfortunate choice of specific epithet.

Chondria densa P. Dangeard Pl. 56, fig. 5.

Dangeard, 1951a.

Plants forming small compact cushion-like clumps, from 2-3 cm in height, purplish-red, with the erect branches arising from a creeping base; branching of the main axes irregular, with the branchlets usually borne opposite one another or verticillate and particularly well-developed towards the branch tips; branches terete, to about 500 μm in diameter; branchlets short, cylindrical and truncate at the apex; cells in surface view longer than broad, from 40-70 μm in length and 30-45 μm in breadth, becoming smaller and almost quadrate towards the tips of the branchlets.

For additional figures see Dangeard, 1951a, p. 19, pl. 2a-c.

There is no information on the ecology of this plant in the region other than that it was collected in the eulittoral zone; it has been found on only one occasion.

Distribution. World: in warm temperate and tropical parts of the eastern Atlantic Ocean. West Africa: Ghana: unpublished.

Remarks. Bodard (1968), when referring to *Laurencia microcladia* in Sénégal, places the following species in quotation marks as follows: "= *C. densa* J. Feldm., = *Chondria densa* Dangeard" which seems to imply that he believes *Chondria densa* to be conspecific with it. No further information is given to justify his reason for believing this to be the case.

Chondria tenuissima (Goodenough & Woodward) C. Agardh Pl. 53, fig. 5.

C. Agardh, 1821a; *Fucus tenuissimus* Goodenough and Woodward, 1797.

Plants solitary, to about 12 cm or more in height, reddish-brown to straw coloured, with the main branches somewhat cartilaginous and the lower order branches soft and flaccid; branching irregular below and alternate and widely spaced above, with the secondary branches usually decreasing gradually in length in upper parts and so giving plant a broad pyramidal outline; branches terete, from 1-2.5 mm in diameter; branchlets to about 0.5 mm in diameter, spindle-shaped.

For additional figures see Thuret and Bornet, 1878, pls 43-48.

Occurs on rocky platforms and small calcareous cobbles at depths ranging from 22 to 25 m.

Distribution. World: probably widespread in warm temperate and tropical seas. West Africa: Gambia: John & Lawson 1977b. Ghana: John & Lawson 1972a, John *et al.* 1980, Lieberman *et al.* 1979.

Undetermined Species

Chondria sp. A. Pl. 56, figs 6, 7.

Plants forming loose tufts, from 3-4 cm in height, purplish-red, with erect branches arising from an irregularly divided system of creeping branches; erect branches terete, to about 450 μm in diameter, simple or rarely divided, often arcuate; branchlets cylindrical and having obtuse or truncate apices, borne in series on the arcuate branches and sometimes a secund series developing on the branchlets; surface cells very slightly elongate, from 35-45 μm in length and 30-40 μm in width, nearly quadrate towards tips of the branchlets; tetrasporangia from 80-90 μm in diameter, embedded towards the tips of the branchlets.

Occurs on fairly sheltered rocks in the lower eulittoral subzone.

Distribution: West Africa: Ghana: unpublished.

Chondria sp. B.

Plants forming loose tufts, from 2-4 cm in height, purplish-red, fleshy to cartilaginous, with the erect branches arising from a poorly developed creeping base; branches sparingly and irregularly divided, to about 1 mm in diameter; branchlets borne irregularly or alternately on the branches, abruptly tapering towards the apices; surface cells elongate throughout branch system, up to 60-70 μm in length and 10-20 μm in width.

There is no information on the ecology of this plant in the region and only one collection of it has been made.

Distribution: West Africa: Ghana: unpublished.

Digenea C. Agardh 1823

Plants erect, coarse and bushy, consisting of terete and dichotomously or irregularly divided branches, beset with simple and short branchlets; branches having six to eight pericentral cells surrounded by a thin cortex; branchlets lacking cortication in the upper parts; tetrasporangia arising towards the tips of the branchlets, spirally disposed and one per segment; spermatangia forming small, flat and ovoid discs at the apices of the branchlets; cystocarps ovoid, borne on short stalks on the branchlets.

World Distribution: apparently widespread in warm temperate and tropical seas.

Digenea simplex (Wulfen) C. Agardh Pl. 53, fig. 8.

C. Agardh, 1823; *Conferva simplex* Wulfen, 1803.

Plants usually bushy, brownish-red, somewhat cartilaginous especially in upper parts; branches irregularly or dichotomously divided, beset with radially arranged branchlets; branchlets simple and stiff, from 3-5(-15) mm in length.

For additional figures see Falkenberg, 1901, pl. 9, figs 25-29.

There is no information on the ecology of this plant in the region but it is likely that it was collected from the eulittoral zone, though in other parts of the Atlantic Ocean it has been dredged to a depth of 20 m (Taylor, 1960).

Distribution. World: as for genus. West Africa: Ghana: Dickinson & Foote 1950, Steentoft 1967. São Tomé: Hariot 1908, Henriques 1886b, 1887, 1917, Steentoft 1967.

Halodictyon Zanardini 1843

Doubtful Record

Halodictyon mirabile Zanardini

Aleem (1978) reports this net-like plant growing on hydroids in the lower littoral at one locality on the Sierra Leone peninsula. This is the first mention of this species from the mainland of West Africa though it is known from the Canary Islands. We regard this record as doubtful until it can be verified by examination of the material on which it is based.

Herposiphonia* Nägeli 1846

Plants dorsiventral, consisting of a main axis attached by unicellular rhizoids and bearing simple branches of determinate growth and compound branches of indeterminate growth arising adaxially at regular intervals from the nodes; branches having from eight to twenty pericentral cells, ecorticate throughout; branches often terminating in a cluster of hairs; apices of the main axes and compound branches incurved; tetrasporangia in a single series, with one sporangium per segment, tetrahedrally or obliquely tetrapartitely divided; spermatangia in dense lanceolate or somewhat elliptical clusters, arising in series on simple branches and developed from rudimentary hairs; cystocarps as in *Polysiphonia* and arising in a similar position to the spermatangia.

**Herposiphonia brachyclados* Pilger has been reported from the island of Pagalu (De Toni, 1924; Pilger, 1920).

World Distribution: probably widespread in warm temperate and tropical seas.

Key to the Species

1. Plants from 1-4 cm in height, with the branches of indeterminate growth secondarily erect. . . . . *H. guineensis*
1. Plants to about 1 cm in height, with the indeterminate branches always prostrate . . . . *H. tenella*

Herposiphonia guineensis nov. sp. Pl. 54, figs 1, 2.

Plantae atroferrugineae, caespitosae aut tegetes formantes, 1-4 cm altae; rami principales indeterminati initio prostrati cum rhizoideis, demum erecti sine rhizoideis, tres ramos simplices determinatos inter ramos indeterminatos successivos ferentes; rami omnes 9-10 cellulas pericentrales continentes; rami principales usque ad 200 μm diametro, e segmentis aeque longis ac latis vel brevioribus compositi; rami determinati clavati obtusi, tum valde curvati et dense fasciculati cum ramis erectis, plerumque 70-140 μm diametro et segmentis brevioribus quam latis apicem versus, 56-70 μm diametro et segmentis 1-1.5-plo longioribus quam latis basem versus; trichoblastae pseudodichotome 3-4 ramosae; cellulae ca. 14 μm diametro vel 2-10-plo longioribus quam latis basem versus; chromatophorae vulgo zonatae; tetrasporangia in ramis determinatis curvatis seriatim formata.

Plants forming dark reddish-brown tufts or mats, from 1-4 cm in height; main indeterminate branches initially prostrate and attached by rhizoids, later secondarily erect and without rhizoids, bearing 3 simple determinate branches between successive branches of indeterminate growth; branches having 9 to 10 pericentral cells; main compound branches about 200 μm in diameter, with the segments equal in length and breadth or shorter than broad; branches of determinate growth club-shaped, obtuse, when arising on the erect branches strongly curved and densely clustered, usually from 70-140 μm in diameter and segments from 1 to 1.5 times as long as broad towards the base; hairs 3 to 4 times pseudodichotomously divided, with the cells about 14 μm in diameter and from 2 to 10 times as long as broad near the base; chloroplasts frequently zonate; tetrasporangia formed in series in curved determinate branches.

Holotype: A1478, Christiansborg, Ghana, 29 December 1959. Deposited in the British Museum (Natural History), London (BM).

Occurs in moderately wave-exposed situations as ring-like patches on lithothamnia, whilst where wave action is less it grows either in the algal turf or as an epiphyte on larger algae. It is found in the lower eulittoral subzone and extends into the sublittoral fringe.

Distribution. World: in tropical parts of the eastern Atlantic Ocean. West Africa: Côte d'Ivoire: *H. densa*, John 1972c, 1977a. Ghana: *H. densa*, John 1977a, John &

Pople 1973, Lawson 1956, Stephenson & Stephenson 1972. Liberia: *H. densa*, John 1972c, 1977a. Togo: *H. densa*, John 1977a, John & Lawson 1972b.

Remarks. This new species resembles *Herposiphonia hollenbergii* Dawson (1963) in having the main axes semi-erect and the determinate branches somewhat club-shaped but is generally a much more robust plant (see Abbott and Hollenberg, 1976, p. 16, fig. 665). Formerly we have mistakenly attributed collections of this plant to *Herposiphonia densa* Pilger described from Cameroun. See remarks under *H. tenella*.

Herposiphonia tenella (C. Agardh) Ambronn Pl. 54, figs 3-5.

Ambronn, 1880; *Hutchinsia tenella* C. Agardh, 1828.

Plants forming greenish to light reddish-purple patches, occasionally with other small algae, less than 1 cm in height; main axes prostrate, with 3 simple branches lying between each compound branch or its rudiment; branches having 8 to 14 pericentral cells; prostrate axes from (80-)100-140(-200) μm in diameter, with the segments from 0.6 to 1.5(-2) times as long as broad; branches of determinate growth from 50-84(90) μm in diameter at the base and tapering above, with segments 0.8 to 2 times as long as broad.

For additional figures see Børgesen, 1918, p. 287 to 289, figs 287-289; Børgesen, 1920, p. 473, fig. 430.

var. **densa** (Pilger) nov. stat.

Herposiphonia densa Pilger, 1911, p. 307.

Plants forming a low very thick and dense mat, about 7-8 mm in height; branch arrangement similar to the typical form; main prostrate axes 90-100 μm in diameter and segments about equal in length and breadth; simple branches of determinate growth to about 70 μm in diameter and the segments often up to 3 times as long as broad, tapering to an acute apex; cystocarps urn-shaped, from 320-490 μm in diameter and 450-520 μm in length, borne near the apex of a branch.

For figures see Pilger, 1911, p. 307, 308, figs 16, 17 (as *H. densa*).

var. **secunda** Hollenberg Pl. 54, fig. 6.

Hollenberg, 1968b.

Plants usually growing entangled with other small algae; branches of indeterminate growth or its rudiment arising from every fifth or sixth node and preceded by a simple determinate branch.

For additional figures see Børgesen, 1920, p. 470, fig. 428 (as *H. secunda*).

Occurs often as circular or ring-like felty patches on rocks or lithothamnia in moderately wave-exposed situations and with other algae in the algal turf in less exposed places. The variety *densa* usually grows on low beach rocks and is associated with a good deal of sand. Found in the lower eulittoral subzone and occasionally to a depth of about 15 m on rocky banks and calcareous cobbles.

Distribution. World: as for genus. West Africa: Cameroun: *H. tenella* var. *densa;* as *H. densa,* Dangeard 1952, De Toni 1924, John 1977a, ?Lawson 1955, Pilger 1911, Schmidt & Gerloff 1957, Stephenson & Stephenson 1972. *H. tenella* var. *secunda;* as *H. secunda,* Lawson 1955. Gambia: John & Lawson 1977b. Ghana: John 1977a, Lawson 1956. *H. tenella* var. *secunda;* as *H. secunda,* John & Lawson 1972a. Liberia: De May *et al.* 1977, John 1977a. São Tomé: Carpine 1959. Sierra Leone: Aleem 1978, John & Lawson 1977a. *H. tenella* var. *secunda,* John & Lawson 1977a, as *H. secunda,* Aleem 1978.

Remarks. An examination of the type material from Cameroun (Kribi, no. 526) of *Herposiphonia densa* Pilger (1911) has shown it to closely resemble *H. tenella.* Nevertheless, it does differ from this latter species in a number of minor respects such as its dense growth habit, and having slightly narrower and often longer segments in the branches of determinate growth. We have therefore proposed to retain a varietal status for it as *H. tenella* var. *densa.*

Laurencia* Lamouroux 1813

Plants consisting of terete or occasionally compressed branches, with branching radial or more rarely distichous; apical cells in pit-like depressions on the blunt branch tips, often accompanied by rudimentary hairs; branches in section composed of large colourless medullary cells, with or without lenticular thickenings in the cell walls, and an outermost layer of quadrate to palisade-like cells sometimes projecting near the tips of the ultimate branchlets; tetrasporangia embedded in the ultimate branchlets, with no relationship between sporangia and pericentral cells, sporangia tetrahedrally divided; spermatangia oval to oblong and borne on hairs near the apical cell; cystocarps prominent, ovate to spherical or urn-shaped, with a thick pericarp and a single pore.

World Distribution: cosmopolitan.

Key to the Species

1. Branches markedly compressed, with the branchlets distichous and pinnate. . . . 2
1. Branches terete or subterete, with the branchlets radially arranged 3

2. Plants often blackish, to olive-green when bleached; outermost layer of cells palisade-like . . . . *L. pinnatifida*
2. Plants usually yellowish-red; outermost cells quadrate. . . . . *L. brongniartii*

*_Laurencia brachyclados_ has been described by Pilger (1920) from the island of Pagalu and secondary citations are made by De Toni (1924) and Yamada (1931). See remarks under *L. perforata.*

3. Plants usually bushy, forming erect and lax tufts, rarely below 4 cm in height . . 4
3. Plants forming compact cushions, dense mats or prostrate clumps, usually less than 4 cm in height 10

4. Branches usually becoming gradually shorter above and giving plant a pyramidal outline 5
4. Branches showing no such regular decrease in branch length towards the apex . 6

5. Branches all erect and free, with the branchlets often swollen . . . . *L. obtusa*
5. Branches of erect system arising from one of entangled and prostrate branches, with the branchlets usually cylindrical . . . . *L. intricata*

6. Branches narrow, from 0.3-0.7 mm in diameter; medullary cells with lenticular thickenings in walls . . . . *L. galostoffii*
6. Branches wider and from 1-2 mm in diameter; medullary cells without lenticular thickenings in walls 7

7. Outermost cells palisade-like. 8
7. Outermost cells quadrate 9

8. Plants not densely beset with swollen branchlets. . . . . *L. intermedia*
8. Plants beset with crowded and swollen branchlets . . . . *L. papillosa*

9. Plants adhering well to herbarium paper; outer cells projecting in the ultimate branchlets . . . . *L. majuscula*
9. Plants not adhering to herbarium paper; outer cells never projecting. . . . *L. poitei*

10. Branches less than 0.6 mm in diameter; medullary cells having lenticular thickenings in the walls . . . . *L. nidifica*
10. Branches usually greater than 0.6 mm in diameter; medullary cells without thickenings in the walls 11

11. Outermost cells palisade-like and projecting near the apices of the ultimate branchlets. . . . . *L. perforata*
11. Outermost cells quadrate and never projecting. . . . . *L. tenera*

Plate 54: Figs 1, 2. ***Herposiphonia guineensis:*** 1. Portion of a plant showing some of the axes becoming secondarily erect; 2. Erect portion of an axis showing the arrangement of the compound and simple club-shaped branches. Figs 3-5 ***Herposiphonia tenella:*** 3. Portion of a plant with the tetrasporangia arising in series in the erect branches; 4. Spermatangia arising in clusters towards the tips of the erect branches; 5. Terminal portion of a creeping axis showing the curved apex and the clusters of hairs at the tips of some of the erect branches. Fig. 6. ***Herposiphonia tenella*** variety ***secunda:*** Portion of a plant showing the characteristic arrangement of the compound and simple branches along the creeping axis. Figs 7, 8. ***Laurencia galostoffii:*** 7. Portion of a plant showing the irregularly to subdichotomously divided branches; 8. Portion near the apex of a branch.

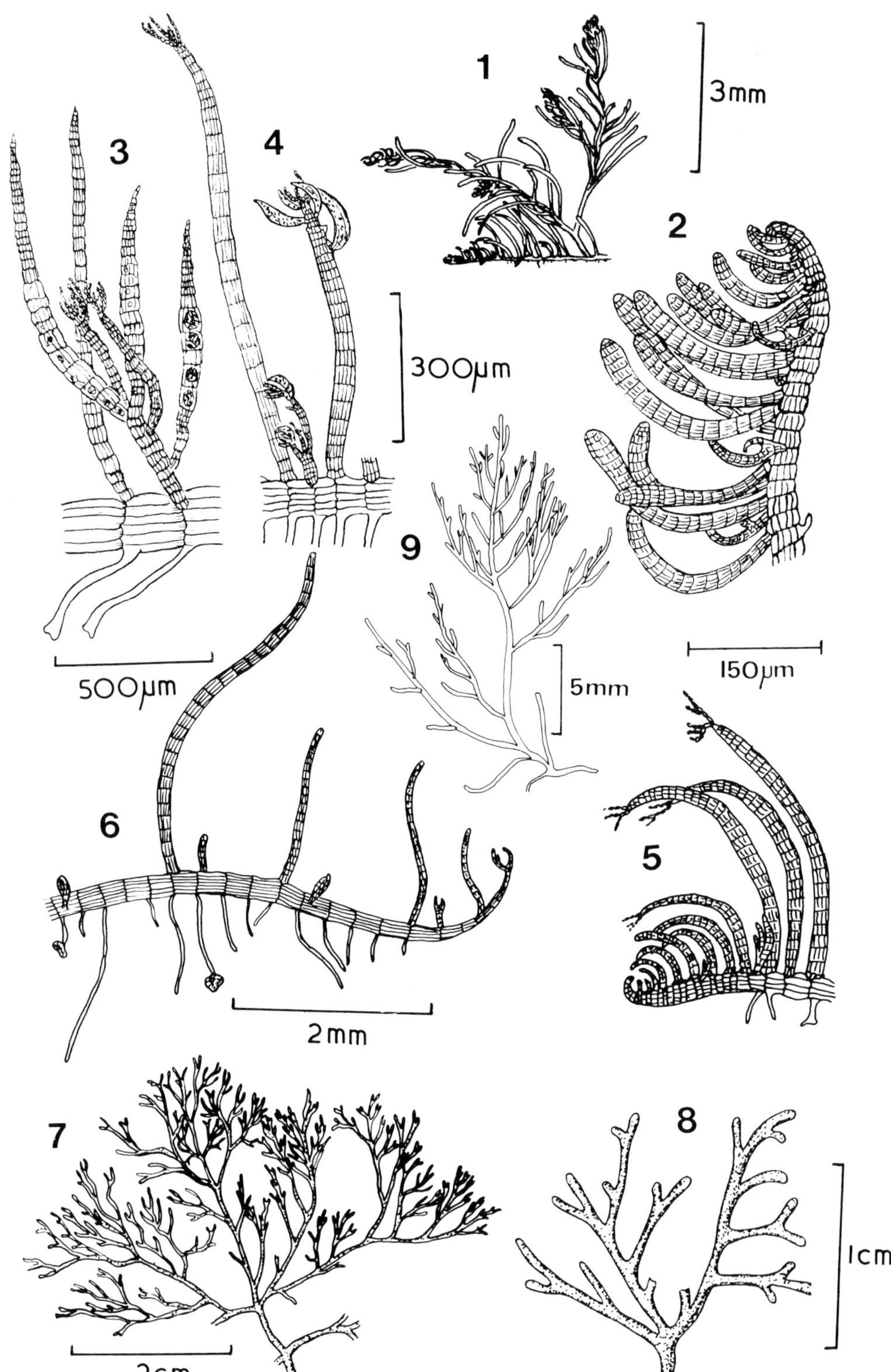
1
3mm
2
3
4
300µm
500µm
9
5mm
150µm
6
5
2mm
7
2cm
8
1cm

Laurencia brongniartii J. Agardh Pl. 55, fig. 1.

J. Agardh, 1841.

Plants forming loose and erect tufts, to about 10 cm in height, usually yellowish-red; branches subdichotomously divided, flattened, from 3-5 mm in width; branchlets distichous and subopposite, terete and cylindrical to clavate or truncate, always simple; medullary cells with lenticular thickenings in the walls uncommon.

For additional figures see Yamada, 1931, p. 301, pl. 25.

Occurs on rocky platforms and calcareous cobbles at depths from 10 to 25 m. The largest and best developed plants are to be found at the lower end of this range.

Distribution. World: probably widespread in tropical seas. West Africa: Ghana: John *et al.* 1977, Liebermann *et al.* 1979.

Laurencia galostoffii Howe Pl. 54, figs 7, 8.

Howe, 1934.

Plants forming compact tufts, from 4-8 cm in height, bright reddish-purple; branches subdichotomously to irregularly divided, dense in upper parts and corymbose, terete or subterete and from 0.3-0.7 mm in diameter; branchlets having truncate or obtuse apices; outermost cells nearly quadrate and walls projecting near the apices of the branchlets; medullary cells with lenticular thickenings in the walls.

For additional figures see Howe, 1934, p. 40, fig. 5.

Occurs on moderately sheltered to moderately wave-exposed rocks in the lower eulittoral subzone and occasionally found in tidepools.

Distribution. World: in tropical parts of the Atlantic and Pacific Oceans. West Africa: Gabon: John & Lawson 1974a. Ghana: John 1977a. Liberia: De May *et al.* 1977, John 1977a.

Remarks. This is often the only soft alga in many tidepools containing otherwise calcareous forms and it is possible that it is resistant to grazing by herbivorous fish.

Laurencia intermedia Yamada Pl. 55, figs 2, 5.

Yamada, 1931.

Plants forming coarse and erect tufts, to about 10 cm or more in height, reddish-purple and occasionally bleached to yellowish-brown; branches oppositely, sub-verticillately or alternately divided, terete, from 1-1.5 mm in diameter, older plants bearing many small and often swollen clavate or truncate branchlets; outermost cells palisade-like and with walls never projecting; medullary cells without lenticular thickenings in the walls.

For additional figures see Saito, 1967, p. 41 to 45, figs 31-35, pls 12, 13.

Occurs as one of the most common littoral plants in some parts of the region growing on moderately sheltered to moderately wave-exposed rocks in the lower eulittoral subzone.

Distribution. World: in tropical parts of the Atlantic and Pacific Oceans. West Africa: Côte d'Ivoire: John 1977a. Ghana: John 1977a; *L. papillosa*, Dickinson & Foote 1950, Lawson 1956, Stephenson & Stephenson 1972. Liberia: De May *et al.* 1977, John 1977a.

Remarks. This plant becomes almost black on drying and will not adhere to herbarium paper. In Ghana it was originally misidentified as *L. papillosa* (Forsskål) Greville from which it is separated by the absence of lenticular thickenings in the walls of the medullary cells.

Laurencia intricata Lamouroux Pl. 55, figs 6, 7.

Lamouroux, 1813.

Plants forming loose, somewhat decumbent and cushion-like clumps, to about 10 cm in height, purplish-pink to olive-green, consisting of an entangled system of prostrate branches giving rise to erect branches; branching irregular and branches coalescing in the lower parts, becoming alternate, opposite or subverticillate above, with the lateral branches in erect portions decreasing in length above and giving the plant a narrowly pyramidal outline; branches terete, to about 1 mm in diameter, with the branchlets clavate and to 0.4 mm in diameter; outermost cells nearly quadrate and slightly projecting in the upper parts of the ultimate branchlets; lenticular thickenings absent from the walls of the medullary cells.

For additional figures see Saito, 1967, p. 13, 14, figs 6, 7, pl. 3, figs 1-3, pl. 4, figs 1-4.

There is no information on the ecology of this plant in the region other than it was found growing entangled with *Bryothamnion triquetrum*. It is known elsewhere in the Atlantic Ocean from shallow water and is believed to be particularly abundant in deep water having been dredged down to a depth of 36 m (Taylor, 1960).

Distribution. World: in warm temperate and tropical parts of the Atlantic and Pacific Oceans. West Africa: São Tomé: Steentoft 1967. ?Sierra Leone: Aleem 1978.

Remarks. Some doubt attaches to the report by Aleem (1978) of this species growing as "tufty cushions" in the littoral on the Sierra Leone peninsula (no locality given). From our knowledge of the marine algae of the peninsula, it seems likely that this may be a misdetermination of *Laurencia tenera* which commonly forms cushion-like clumps and occupies this position on the shore.

Laurencia majuscula (Harvey) Lucas Pl. 55, figs 4, 5.

Lucas, in Lucas and Perrin, 1947; *L. obtusa* (Hudson) Lamouroux var. *majuscula* Harvey, 1863.

Plants erect and bushy, to about 20 cm or more in height, bright reddish-purple and soft in texture; branches alternately divided and sometimes somewhat distichous, terete, to about 1.5 mm in diameter, with the branchlets clavate and from 0.1-0.5 mm in diameter; outermost cells quadrate and clearly projecting in the ultimate branchlets; lenticular thickenings absent from the walls of the medullary cell.

For additional figures see Saito and Womersley, 1974, p. 822, fig. 1a, p. 853, fig. 6.

Occurs as small plants in the immediate sublittoral and as large and well-developed plants in deeper water down to a depth of about 28 m growing on calcareous cobbles.

Distribution. World: probably pantropical. West Africa: Cameroun: unpublished. Gabon: John & Lawson 1974a. Gambia: John & Lawson 1977b. Ghana: Edmunds & Edmunds 1973, John *et al.* 1977, 1980, Lieberman *et al.* 1979.

Remarks. This species was at one time considered to be a variety of *L. obtusa* (see Harvey, 1863) and the two are almost identical anatomically. It appears to be morphologically similar to *L. intricata* but is readily separated from it by having palisade-like cortical cells.

Laurencia nidifica J. Agardh Pl. 56, fig. 1.

J. Agardh, 1863.

Plants forming small cushions or mats, to about 6 cm in height, purplish-red to olive-green, consisting of erect branches arising from an entangled system of prostrate branches; branching opposite or verticillate and only rarely alternate or secund; branches terete and from 0.1-0.6 mm in diameter, with the branchlets to about 0.4 mm in diameter and cylindrical or clavate to truncate; outermost cells quadrate and projecting in the ultimate branchlets; lenticular thickenings in the walls of the medullary cells.

For additional figures see Børgesen, 1945, p. 48, 49, figs 21-24.

Occurs on moderately wave-exposed rocks in the lower part of the eulittoral zone and extending to a depth of about 10 m where it grows on rocky platforms.

Distribution. World: probably widespread in warm temperate and tropical seas. West Africa: Côte d'Ivoire: ?John 1977a. Ghana: John 1977a. Liberia: De May *et al.* 1977, John 1977a.

Plate 55: Fig. 1. *Laurencia brongniartii:* Plant with the compressed branches bearing suboppositely swollen branchlets containing tetrasporangia. Figs 2, 3. *Laurencia intermedia:* 2. Portion of a plant showing swollen fertile branchlets borne on the main axes; 3. Section showing the palisade-like outermost layer of cells. Figs 4, 5. *Laurencia majuscula:* 4. Plant bushy with the branches alternately divided and sometimes distichous; 5. Portion of a branch with the branchlets short and club-shaped. Figs 6, 7. *Laurencia intricata:* 6. Terminal portion of a branch with the branchlets decreasing in length towards the branch tip and so giving it a pyramidal outline; 7. Section showing the quadrate outermost layer of cells.

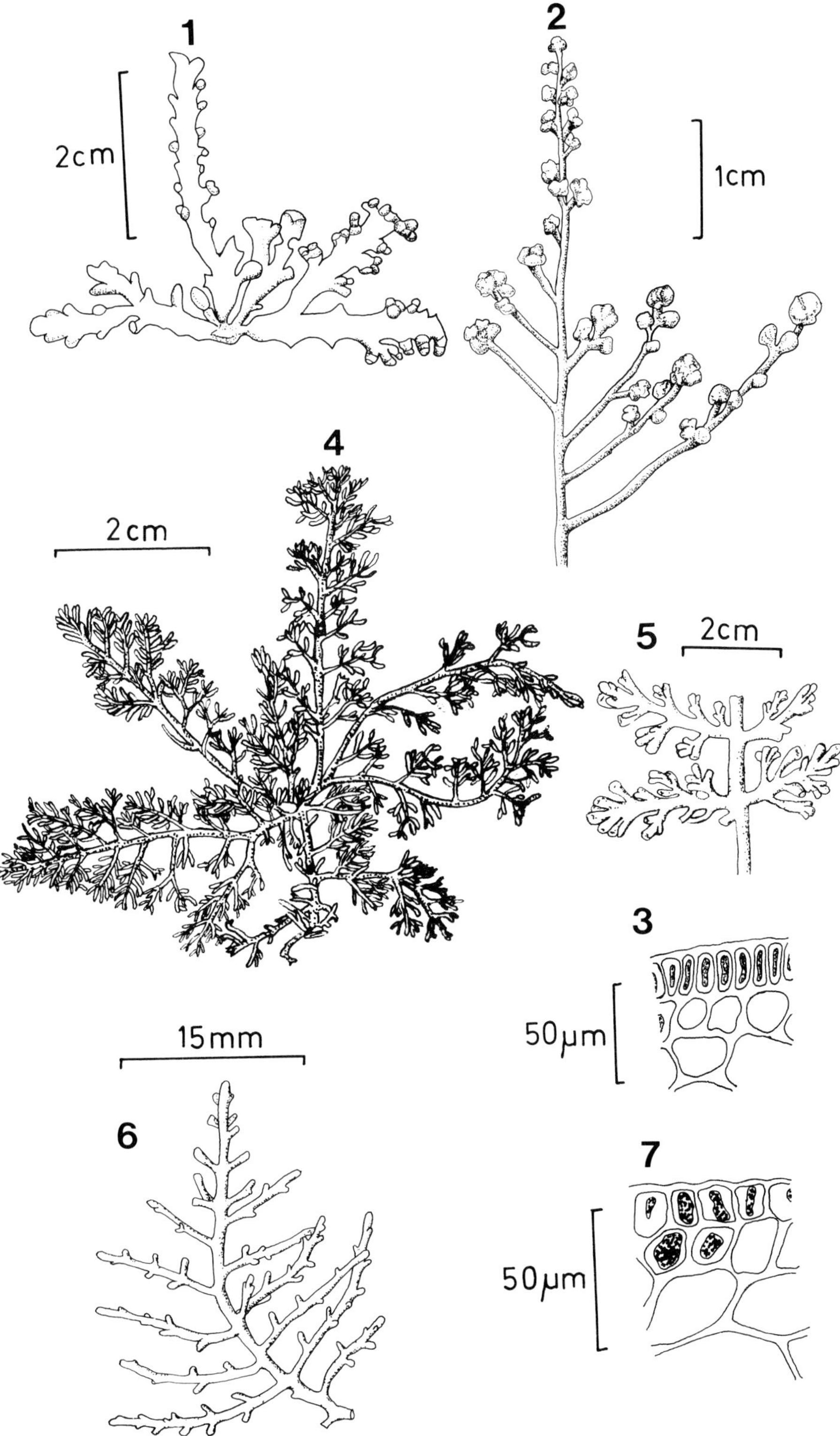
1
2cm
2
1cm
4
2cm
5
2cm
3
50µm
15mm
6
7
50µm

Laurencia obtusa (Hudson) Lamouroux

Lamouroux, 1813; *Fucus obtusus* Hudson, 1778.

Plants forming loose clumps, from 3-8(-17) cm in height, reddish-brown to pale green and firm in texture; branching opposite or sometimes irregularly alternate or subverticillate, arising at wide angles and the lateral branches becoming progressively shorter above so giving the plant a broadly pyramidal outline; branches terete and from 0.8-2.1 mm in diameter, with the branchlets cylindrical to clavate; outermost cells quadrate and with the outer walls only slightly projecting towards the tips of the branchlets; lenticular thickenings absent from the walls of the medullary cells.

For figures see Saito, 1967, p. 6 to 8, 10, 11, figs 1-5, pls 1, 2.

Occurs often associated with beds of *Sargassum* on wave-exposed rocks in the sublittoral fringe and it has also been dredged at a depth of 12 m.

Distribution. World: widespread from boreal-antiboreal to tropical seas. West Africa: Cameroun: Lawson 1955, Steentoft 1967, Stephenson & Stephenson 1972. Gambia: John & Lawson 1977b. Príncipe: Carpine 1959, Steentoft 1967. São Tomé: Carpine 1959, Hariot 1908, Henriques 1886b, 1887, 1917, Steentoft 1967. Sierra Leone: Aleem 1978.

Remarks. Steentoft (1967) considers the São Tomé plants to be close to two varieties, one var. *natalensis* (Kylin) Børgesen is more delicate, smaller and more irregularly branched than the type, whilst the other var. *rigidula* Grunow also is smaller, more rigid and has the erect branches denser than the type. It is doubtful if much useful purpose is served by recognising varieties based on such comparatively minor differences in gross morphology in species which show a very marked degree of morphological plasticity. The main distinguishing feature between *Laurencia obtusa* and *L. intricata* is the presence of a prostrate system of entangled branches in the latter. In fact, Yamada (1931) considers *L. intricata* to be no more than a variety of *L. obtusa*.

Laurencia papillosa (Forsskål) Greville

Greville, 1830; *Fucus papillosus* Forsskål, 1775.

Plants loose and usually to about 10 cm or more in height, purplish to olive-green; branching alternate and sometimes somewhat distichous, with the branchlets crowded together above and appearing to be subcorymbose; branches terete and occasionally slightly subterete, from 1-1.5 mm in diameter, with the branchlets to about 1 mm in diameter, short and truncate, often somewhat swollen and clavate; outermost cells palisade-like and never having projecting walls; lenticular thickenings absent from the walls of the medullary cells.

For figures see Cribb, 1958, p. 185, pl. 7, figs 6-8.

Occurs on wave-exposed rocks or in tidepools in the lower part of the eulittoral zone.

Distribution. World: widespread in warm temperate and tropical seas. West Africa: Cameroun: Pilger 1911, Richardson 1969, Schmidt & Gerloff 1957. Gabon: John & Lawson 1974a.

Remarks. Anatomically this plant and *Laurencia intermedia* are very similar and morphological differences between them are not always well-defined. Yamada (1931), who described the latter species, expressed the opinion that *L. papillosa* and another species (*L. paniculata* J. Agardh) may represent "the extreme forms of one very variable species in which *L. intermedia* may be included.".

Laurencia perforata Montagne Pl. 56, fig. 4.

Montagne, 1839-1841.

Plants forming cushion-like tufts, to about 2 cm in height, olive-green, consisting of an erect system of branches arising from a lower one of intricate and often arcuate branches; branching of the erect system somewhat irregular and the branches bearing a few alternately divided branchlets that are often secundly arranged and curved; branches terete and to about 1 mm in diameter, with the branchlets slightly narrower and cylindrical to clavate; outermost cells palisade-like and never projecting; lenticular thickenings absent from the walls of the medullary cells.

For additional figures see Cribb, 1958, p. 181, pl. 3, figs 1, 2.

Occurs on moderately wave-exposed rocks in the eulittoral zone and commonly found in tidepools.

Distribution. World: widespread in warm temperate and tropical seas. West Africa: Gabon: Hariot 1896, Steentoft 1967. Gambia: John & Lawson 1977b, Steentoft 1967. São Tomé: Carpine 1959, Hariot 1908, Henriques 1917, Steentoft 1967.

Remarks. This plant dries to almost black and does not adhere to herbarium paper. There is the possibility that *Laurencia brachyclados*, described by Pilger (1920) from the island of Pagalu, is no more than a growth form of this species (see De Toni, 1924). There is some confusion regarding Pilger's plant as according to Steentoft (1967) his published description is at variance with isotype material (Mildbraed - 6719) examined by her in the Herbarium of Børgesen in Copenhagen (C).

?Laurencia pinnatifida (Hudson) Lamouroux

Lamouroux, 1813; *Fucus pinnatifidus* Hudson, 1762.

Plants to about 20 cm or more in height, usually blackish and olive-green when bleached; branching alternate and distichous, with the lateral branches pinnatifid and bearing 2 or 3 series of branchlets; branches subterete or compressed and branchlets also somewhat compressed; outermost cells palisade-like and without projecting outer walls; lenticular thickenings in the walls of the medually cells.

For figures see Greville, 1830, pl. 14, figs 1-12.

The only information on this plant in the region is that it was dredged from a depth of 11 m, but elsewhere in the Atlantic Ocean it is commonly found in the eulittoral zone.

Distribution. World: widespread in boreal-antiboreal seas and less common in tropical seas. West Africa: ?Ghana: *Fucus pinnatifidus*, Hornemann 1819. São Tomé: Carpine 1959, Henriques 1885, 1886a, Steentoft 1967.

Remarks. The record from São Tomé cannot be checked as the plant collected by Newton in 1881 has not been located (Steentoft, 1967) and the more recent collection by Carpine (1959) also appears to be lost (Mme H. Huvé, *pers. comm.*). This species is the nomenclatural equivalent of *Fucus pinnatifidus* Linnaeus recorded by Hornemann (1819) from "Danish Guinea" (present-day Ghana). Clarification of the record requires examination of the original collection by Isert though this might have been destroyed by fire during the bombardment of Copenhagen in 1807. This seems an unlikely record for Ghana since it has not been re-discovered though this country has been intensively investigated phycologically over the past twenty years. It bears a superficial resemblance to *Laurencia brongniartii* and it is just possible that these two somewhat compressed species may on occasion be mistaken for one another. For further comments see Price *et al.* (1983).

Laurencia poitei (Lamouroux) Howe Pl. 56, fig. 3.

Howe, 1905; *Fucus poitei* Lamouroux, 1805.

Plants loose and bushy, to about 15 cm in height, pinkish-red; branching irregularly alternate and distichous, with the branches bearing oppositely or alternately branchlets which are numerous and variable in length; branches terete or slightly subterete, from 1-2 mm in width, with the short branchlets usually simple; outermost cells quadrate and never having projecting walls; lenticular thickenings absent from the medullary cell walls.

For additional figures see Børgesen, 1918, p. 246, 247, figs 234, 235.

There is no information on the ecology of this species in the region but elsewhere it is found in shallow water and is reported from depths down to 16 m.

Plate 56: Fig. 1. *Laurencia nidifica:* Plant arising from a prostrate base and the erect branches bearing cylindrical or clavate to truncate branchlets. Fig. 2. *Laurencia tenera:* Portion of a plant showing the alternately or secundly divided upper branches. Fig. 3. *Laurencia poitei:* Terminal portion of a plant showing the oppositely arising branches and branchlets. Fig. 4. *Laurencia perforata:* Portion of a plant with an arcuate branch. Fig. 5. *Chondria densa:* Terminal portion of a plant with the branches bearing short cylindrical branchlets having truncate apices. Figs 6, 7. *Chondria* sp. A: 6. Portion of a plant showing a series of secund branchlets borne on an arcuate branch; 7. Surface of a branch with the cells almost isodiametric. Figs 8, 9. *Chondria confusa:* 8. Portion of a plant showing the spine-like branchlets; 9. Surface of a branch with the cells elongated at right angles to the long axis. Fig. 10. *Lophosiphonia reptabunda:* Terminal portion of a plant with the tips of the erect branches often curved towards the apex of the creeping axis.

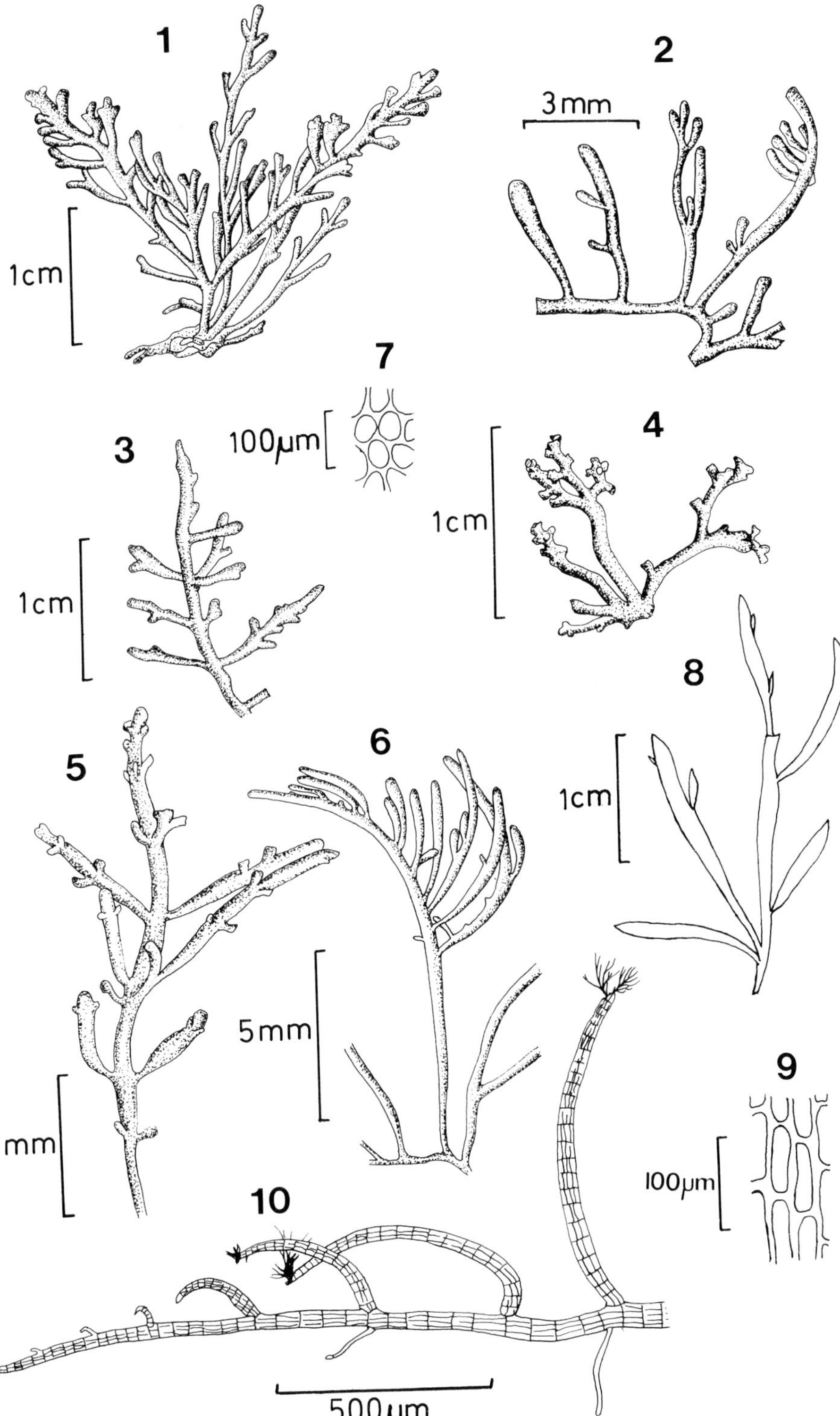
1
2
3mm
1cm
7
100µm
3
4
1cm
1cm
8
5
6
1cm
5mm
9
2mm
100µm
10
500µm

Distribution. World: probably widespread in warm temperate and tropical seas. West Africa: São Tomé: Steentoft 1967; *L. tuberculosa,* Hariot 1908, Henriques 1917; *?Gracilaria poitei,* Henriques 1886b, 1887, 1917.

Laurencia tenera Tseng Pl. 56, fig. 2.

Tseng, 1943.

Plants forming cushion-like tufts or mats, to about 3 cm in height, yellowish- or purplish-brown, with the branches attached to one another and to the substratum by discoid holdfasts; branching pseudodichotomous below at wide angles and the upper branches alternately or secundly divided above at acute angles; branches terete and up to 1 mm in diameter, with the branchlets to about 0.6 mm in diameter and subterete or subturbinate; outermost cells quadrate and only rarely having the walls projecting in the branchlets; lenticular thickenings absent from the walls of the medullary cells.

For additional figures see Cribb, 1958, p. 183, pl. 5, figs 1-10.

Occurs on lithothamnia or rocks in the lower eulittoral subzone and found in tide-pools and extends into the immediate sublittoral, often growing associated with a good deal of sand when in moderately wave-sheltered situations.

Distribution. World: probably pantropical. West Africa: Côte d'Ivoire: John 1972c, 1977a. Gambia: John & Lawson 1977b. Ghana: John 1977a, John & Pople 1973. Liberia: De May *et al.* John 1972c, 1977a. Sierra Leone: John & Lawson 1977a. Togo: John 1977a, John & Lawson 1972b.

Lophocladia Schmitz 1893

Plants usually erect, consisting of repeatedly dichotomously divided branches, bearing spirally arranged and subdichotomously or alternately divided monosiphonous ramelli which are somewhat crowded towards the branch apices; branches having four pericentral cells and a weakly developed rhizoidal cortication; tetrasporangia in stichidia arising from modified ramelli, sporangia spirally disposed and one per segment; spermatangia cylindrical and also borne on the ramelli; cystocarps shortly stalked and usually urn-shaped.

World Distribution: widespread in warm temperate and tropical seas.

Lophocladia trichoclados (C. Agardh) Schmitz Pl. 57, figs 4, 5.

Schmitz, 1893; *?Griffithsia trichoclados* C. Agardh, 1828.

Plants forming small tufts or a creeping mat, from 4-12 cm in height, with an erect system of branches arising from a prostrate system in which occasionally some of the creeping branches become secondarily erect; branching dichotomous at wide angles, with the ramelli similarly divided but at acute angles; branches to about 280 μm in diameter and segments shorter to twice as long as broad, cortication distinctive and developed only in the older parts by rhizoids from the basal ends of the pericentral cells growing down between the pericentral cells; ramelli from 40-60 μm in diameter at the base, with the cells from 80-110 μm in length and up to 8 times as long as broad towards the apices; tetrasporangial stichidia usually arising from the first division of a ramellus.

For additional figures see Børgesen, 1918, p. 302 to 304, figs 304-307; 1919, p. 305 to 307, figs 308-312.

Occurs infrequently on rocky platforms and calcareous cobbles at depths of about 10 to 16 m.

Distribution. World: as for genus. West Africa: Ghana: John & Lawson 1972a, John *et al.* 1977, 1980, Lieberman *et al.* 1979. Príncipe: Carpine 1959. Sierra Leone: Aleem 1978.

Lophosiphonia* Falkenberg 1897

Plants polysiphonous and ecorticate, consisting of a prostrate and erect system of endogenously-produced branches; prostrate branches attached by rhizoidal hold-fasts arising near the distal end of pericentral cells; erect branches sparingly divided, often with recurved apices and frequently bearing a series of deciduous hairs on the convex side; tetrasporangia borne in the ultimate divisions of the erect branches, tetrahedrally divided.

World Distribution: from boreal-antiboreal to tropical seas.

Key to the Species

1. Erect branches usually greater than 500 μm in height, often strongly recurved at the apices and sometimes bearing on the convex side a series of hairs .... *L. reptabunda*
1. Erect branches less than 500 μm in height, with the apices usually straight and the hairs often clustered at the apex....................... *Lophosiphonia* sp.

Lophosiphonia reptabunda (Suhr) Kylin Pl. 56, fig. 9.

Kylin, 1956; *Polysiphonia reptabunda* Suhr, in Kützing, 1843b.

Plants forming loose tufts or extensive low mats, to about 2 cm in height, often dark reddish to purplish-brown; erect branches sparingly divided, with the tips often strongly recurved and sometimes having a series of hairs on the convex side; branch segments composed of 8 to 10 pericentral cells, up to 150 μm in diameter and about equal in length and breadth or slightly longer.

**Lophosiphonia adhaerens* has been described by Pilger (1920) from the island of Pagalu.

For additional figures see Batten, 1923, pl. 23, figs 33-35 (as *Polysiphonia obscura*).

Occurs in the upper eulittoral subzone but most commonly found in the littoral fringe in cracks and crevices in rocks, beneath rock overhangs and occasionally on the underside of large stranded logs.

Distribution. World: in warm temperate and tropical parts of the Atlantic as well as in the Mediterranean. West Africa: Bioko: unpublished. Cameroun: John 1977a; *L. obscura*, Lawson 1955, Stephenson & Stephenson 1972. Côte d'Ivoire: John 1972c, 1977a. Gabon: John & Lawson 1974a. Ghana: John 1977a; *L. obscura*, Dickinson & Foote 1950, Lawson 1956. Sierra Leone: John & Lawson 1977a.

Undetermined Species

Lophosiphonia sp.

Plants reddish-brown in colour, less than 500 μm in height and growing along with other small red algae; prostrate branches from 80-160 μm in diameter and having 8 pericentral cells, attached by thick-walled unicellular rhizoids of about 180 μm in length; erect branches regularly but very sparingly dichotomously divided, to about 120 μm in diameter and gradually tapering towards their tips, with the segments always shorter than broad; hairs long and simple, clustered at the apices of the erect branches.

Occurs in the lower eulittoral subzone growing in an algal turf.

Distribution: West Africa: Sierra Leone: *Lophosiphonia* sp., John & Lawson 1977a.

Remarks. This plant shows a close resemblance to the little-known *Lophosiphonia adhaerens* described by Pilger (1920) from Pagalu.

Murrayella Schmitz 1893

Plants usually erect, with or without a prostrate system of branches, sparingly subdichotomously or alternately divided, bearing ramellate branchlets; branches having four pericentral cells and ecorticate throughout; branchlets polysiphonous below and monosiphonous above the divisions; tetrasporangial stichidia borne on the upper parts of the ramellate branchlets, sporangia usually in whorls of four.

Plate 57: Figs 1, 2. *Micropeuce mucronata:* 1. Plant showing the alternately or somewhat irregularly divided branches bearing spirally disposed branchlets; 2. Terminal portion of a branchlet containing a series of tetrasporangia and beset with subdichotomously divided ramelli. Fig. 3. *Polysiphonia ferulacea:* Portion of a plant showing the sparingly divided and 4 pericentral-celled erect branches arising from prostrate branches bearing rhizoids. Figs 4, 5. *Lophocladia trichoclados:* 4. Portion of a plant with the erect branches covered by monosiphonous ramelli; 5. Portion of a plant with the segmented main axis bearing dichotomously divided monosiphonous ramelli and showing a spermatangial cluster. Figs 6, 7. *Murrayella periclados:* 6. Portion of a plant showing the sparingly divided erect branches arising from a creeping axis; 7. Portion of a branch showing tetrasporangia arising in a ramellate branchlet.

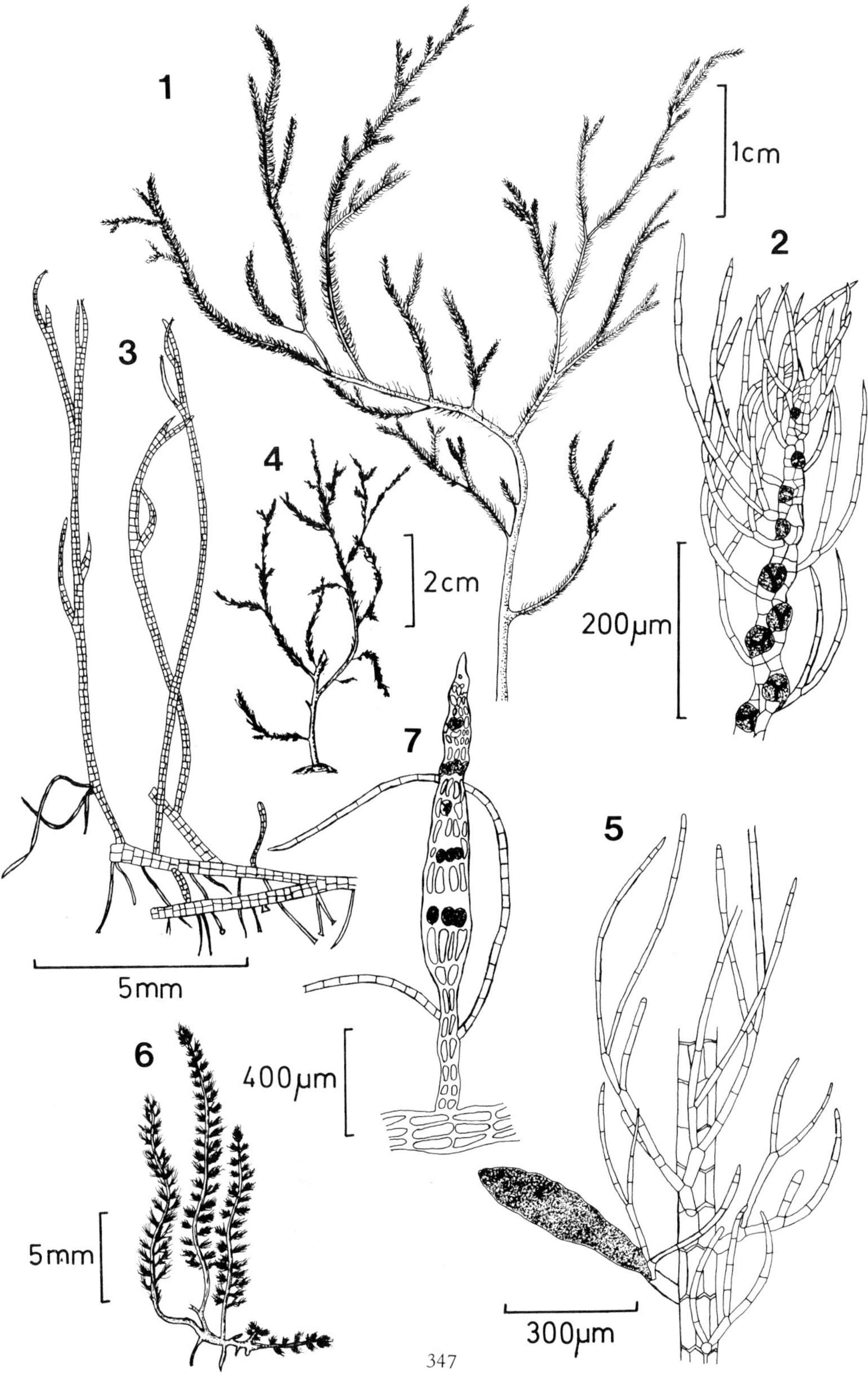
1
1cm
2
200µm
3
4
2cm
5mm
7
5
6
400µm
5mm
300µm

World Distribution: widespread in warm temperate and tropical seas.

Murrayella periclados (C. Agardh) Schmitz Pl. 57, figs 6, 7.

Schmitz, 1893; *Hutchinsia periclados* C. Agardh, 1828.

Plants to about 3 cm in height, dark reddish-brown to purplish; branching of the erect system irregular to alternate and of the creeping system sparingly subdichotomous; branches to about 130 μm in diameter and the segments from 1 to 1.5 times as long as broad; branchlets of two kinds either simple and monosiphonous, with the cells to about 30 μm in diameter and to 60 μm in length, or polysiphonous below and bearing alternately divided monosiphonous ramelli, with intermediate types sometimes present; tetrasporangia in cylindrical stichidia, usually tapering towards the apex, to about 110 μm in diameter and to 560 μm in length.

For additional figures see Børgesen, 1919, p. 414, 315, figs 318-320.

Occurs most commonly entangled with other small algae such as *Bostrychia* spp. growing in cracks and crevices and on rocks shaded by overhangs in wave-exposed situations in the eulittoral zone and occasionally the littoral fringe.

Distribution. World: as for genus. West Africa: Bioko: unpublished. Cameroun: Lawson 1955, Post 1936, 1955a, 1957b, Steentoft 1967, Stephenson & Stephenson 1972. Gambia: John & Lawson 1977b. Ghana: Dickinson & Foote 1950, Lawson 1954b, 1955a, 1966, Post 1936, 1955a, 1957b, 1959, 1963b, 1968, Steentoft 1967. Guineé: Post 1955a, Steentoft 1967. São Tomé: Post 1936, 1966a, 1968, Steentoft 1967. Sierra Leone: Aleem 1978, John & Lawson 1977a, Lawson 1954b, 1957a, Levring 1969, Longhurst 1958, Post 1955a, 1963b, 1965, Steentoft 1967.

Remarks The Ghanaian plants are somewhat unusual in having tufts of monosiphonous ramelli at the apex of the stichidia.

Polysiphonia Greville 1823

Plants consisting of an erect and sometimes also a prostrate system of branches, often dichotomously or laterally divided, attached mostly by unicellular rhizoids; branches polysiphonous, with or without rhizoidal cortication and occasionally pseudoparenchymatous; hairs often present at the apices, mostly spirally arranged, one per segment; tetrasporangia borne singly in successive segments of the ultimate branches, tetrahedrally divided; spermatangia shortly stalked, arising on rudimentary hairs, clustered and lanceolate or long elliptical in shape; cystocarps stalked or sessile, oval or urn-shaped, thin-walled, with a single large pore.

World Distribution: cosmopolitan.

A number of collections of plants believed to be referable to this genus have been made in the region but these have usually been sterile or asexual. Many of these undetermined plants have four pericentral cells whilst fragments of six pericentral-celled *Polysiphonia* spp. have been reported from Benin (John and Lawson, 1972b) and Liberia (De May *et al.*, 1977).

Key to the Species

1. Plants having from 9 to 12 pericentral cells .... *P. camerunensis*
1. Plants having 4 or 6 pericentral cells.. 2

2. Plants having 6 pericentral cells .... *P. denudata*
2. Plants having 4 pericentral cells.. 3

3. Branches to about 350 μm in diameter, with the segments always shorter than broad... *P. ferulacea*
3. Branches to 135 μm in diameter, with the segments often from 2 to 2.5 times longer than broad .... *P. subtilissima*

Polysiphonia camerunensis Pilger Pl. 58, fig. 6.

Pilger, 1911.

Plants to about 10 mm in height, consisting of erect branches arising from a creeping system; branches of the erect system dichotomously divided, having from 9 to 12 pericentral cells, ecorticate; segments of the erect branches to about 110 μm in diameter, nearly equal in length and breadth; segments of the creeping branches to about 130 μm in diameter, shorter than broad; tetrasporangia borne in short lateral branches.

Figures from Pilger, 1911, p. 302, 304, figs 11, 12.

Occurs in the algal turf growing over beach rocks in the lower eulittoral subzone.

Distribution. World: so far only known from the tropical parts of the eastern Atlantic Ocean. West Africa: Cameroun: De Toni 1924, Pilger 1911.

Polysiphonia denudata (Dillwyn) Greville ex Harvey

Greville ex Harvey, in Hooker, 1833; *Conferva denudata* Dillwyn, 1809.

Plants usually forming purplish-brown tufts, up to 25 cm in length; branching pseudodichotomous; branches stiff and widely divergent towards the base of the plant, above becoming flexuous and soft, having usually 6 pericentral cells and corticated near the base; main branches to about 750 μm in diameter, with the segments shorter than broad and increasing to twice as long as broad towards the apex; branchlets absent from the lower branches, from 30-45 μm in diameter and segments up to 2 times longer than broad, formed laterally at the base of hairs when present.

For figures see Børgesen, 1918, p. 269, 270, figs 263-266 (as *P. variegata*).

Found in region growing together with other algae in a sample dredged from a depth of 11 m, although elsewhere in the Atlantic it commonly occurs in sheltered, muddy situations such as are encountered in mangrove swamps and lagoons.

Distribution. World: from boreal-antiboreal to tropical parts of the Atlantic Ocean.

West Africa: São Tomé: Carpine 1959. Sénégal (Casamance): *P. variegata*. Chevalier 1920.

Polysiphonia ferulacea Suhr Pl. 57, fig. 3.

Suhr, in J. Agardh, 1863.

Plants usually dark purplish-brown, forming coarse patches having the appearance of wet fur, from 4-6(-10) cm in height, with an erect and a prostrate system of branches; branching sparingly subdichotomous at an acute angle; branches having 4 pericentral cells, ecorticate; erect branches to about 260 μm in diameter, with the segments shorter than broad; prostrate branches up to 350 μm in diameter, with the segments much shorter than broad.

For additional figures see Børgesen, 1918, p. 278 to 281, figs 277-280.

Occurs most gregariously on moderately wave-exposed rocks or in sheltered situations provided there is considerable wave or current surge. This plant is found in the lower eulittoral subzone, sublittoral fringe and sterile individuals have been found to a depth of about 16 m on calcareous cobbles.

Distribution. World: widespread in warm temperate and tropical seas. West Africa: Côte d'Ivoire: John 1972c, 1977a. Ghana: Dickinson & Foote 1950, Fox 1957, John 1977a, John & Lawson 1972a, John *et al.* 1977, Lawson 1956, Lieberman *et al.* 1979, Stephenson & Stephenson 1972. Liberia: De May *et al.* 1977, John 1977a. Nigeria: Fox 1957, John 1977a. Sierra Leone: John & Lawson 1977a.

Remarks. This plant tends to retain its dark coloration even after prolonged preservation in formalin. Some doubt attaches to the determination of the plants found subtidally off Ghana.

Polysiphonia subtilissima Montagne Pl. 58, fig. 2.

Montagne, 1840.

Plants forming reddish-green, soft and delicate tufts or low mats, from 2-2.5 cm in height, with the erect branches arising at irregular intervals from the prostrate branches; branching often subdichotomous below and alternate above, branches

Plate 58: Fig. 1. *Pterosiphonia pennata:* Portion of an erect branch showing the alternately pinnate branching. Fig. 2. *Polysiphonia subtilissima:* Portion of a 4 pericentral-celled branch containing a mature tetrasporangium. Figs 3-5. *Waldoia antillana:* 3. Prominent apical cell surrounded by rudimentary hairs; 4. Asexual plant with the surface of the flattened branches bearing lanceolate tetrasporangial stichidia and occasionally the tips of the branches flagelliform; 5. Cystocarpic plant with the tips of the flattened branches always obtuse. Fig. 6. *Polysiphonia camerunensis* (after Pilger, 1919): Terminal portion of a branch showing the early stages of branch development. Fig. 7a-k. Terms as used by Drouet (1968) to describe the shape of the end cells and outer wall of the apical cell in his revision of the family Oscillatoriaceae (Cyanophyta): a. rotund, b. hemispherical, c. depressed-hemispherical, d. convex-plate, e. acute-conical, f. obtuse-conical, g. cup-shaped, h. long acute-conical, i. depressed conical, j. truncate-conical, k. attenuate cylindrical.

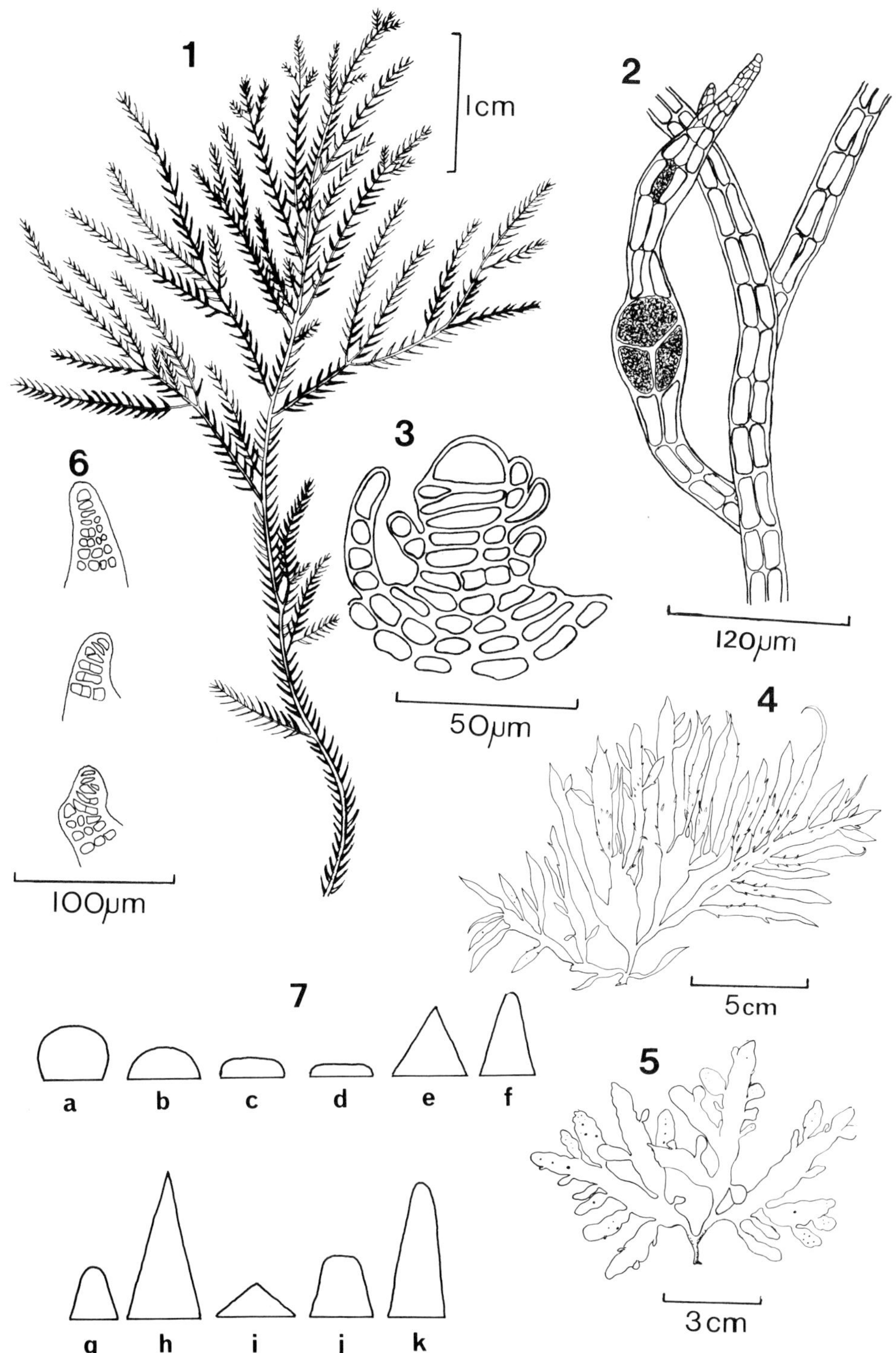
1
1cm
2
120µm
3
50µm
4
5cm
5
3cm
6
100µm
7
a
b
c
d
e
f
g
h
i
j
k

having 4 pericentral cells and ecorticate, to about 70-135 μm in diameter, with the segments shorter than broad below and increasing above to as much as 2 to 2.5 times longer than broad.

For additional figures see Tseng, 1944, pl. 1, figs 1-5.

Occurs most commonly growing through sand lining the floors of somewhat sheltered pools in the lower part of the eulittoral zone, and to a depth of 15 m on rocky banks and calcareous cobbles. This plant is found in other tropical regions of the world often in muddy situations in brackish-water habitats.

Distribution. World: widespread in warm temperate and tropical seas. West Africa: Bioko: unpublished. Cameroun: De Toni 1924, Pilger 1911. Côte d'Ivoire: *Polysiphonia* sp., John 1977a. Gambia: John & Lawson 1977b. Ghana: Lieberman *et al.*, 1979. Liberia: De May *et al.* 1977; *Polysiphonia* sp., John 1977a. Sierra Leone: ?John & Lawson 1977a.

Remarks. Doubt has been expressed as to whether this species is really distinct from *Polysiphonia macrocarpa* Harvey (see Cribb, 1956; Hollenberg, 1968a). The possibility exists that they represent the different ends of a cline of variation with *P. macrocarpa* the colder water form and *P. subtilissima* the warmer water or tropical form.

Pterosiphonia Falkenberg 1897

Plants consisting of erect, alternately pinnately or bipinnately divided branches, arising from a creeping base; branches terete to compressed, having from four to twenty pericentral cells, ecorticate or with parenchymatous cortication in the older branches; hairs not usually terminating branches; tetrasporangia single in successive segments in the upper parts of the ultimate branch divisions, tetrahedrally divided; spermatangia borne in clusters at the tips of the branchlets; cystocarps stalked, ovoid to globular, with a single pore.

World Distribution: widespread in warm temperate and tropical seas.

Pterosiphonia pennata (C. Agardh) Falkenberg — Pl. 58, fig. 1.

Falkenberg, 1901; *Hutchinsia pennatum* C. Agardh, 1824.

Plants forming clumps or loose and extensive mats, to 8 cm or more in height, dark purplish-brown, with the erect branches arising at intervals of about 3 segments from a creeping base; branches slightly compressed and branchlets terete, borne usually at intervals of 2 segments, mostly simple, often incurved towards tip; branches with from 8 to 9 pericentral cells, ecorticate; hairs usually absent; tetrasporangia in series of 8 or more in the branchlets.

For additional figures see Abbott and Hollenberg, 1976, p. 709, fig. 660.

Occurs on moderately sheltered to moderately wave-exposed rocks in the lower eulittoral subzone and the sublittoral fringe, often most gregarious where there is water surge and even sand-scouring such as along the sides of gullies.

Distribution. World: as for genus. West Africa: Cameroun: Lawson 1955. Ghana: Dickinson & Foote 1950, Fox 1957, Lawson 1956, 1966, Stephenson & Stephenson 1972. Nigeria: Fox 1957.

Doubtful Record

Pterosiphonia complanata (Clemente) Falkenberg

Hariot (1896) has with certain reservations reported this plant from Gabon (as *Polysiphonia complanata*) based on just two fragments. We have decided to regard this record as uncertain until more material is discovered.

Ricardia Derbès & Solier 1856

Plants parasitic on the tips of the branches of *Laurencia*, obpyriform to elliptical, with an apical tuft of hairs, penetrating the host by a large, simple or branched, cylindrical rhizoidal cell; thallus polysiphonous at first, becoming parenchymatous with the smallest cells towards the outside, sometimes hollow; tetrasporangia embedded in the outer cortical layers, tetrahedrally divided; spermatangia developing on the hairs; cystocarps immersed and connected with the surface cells.

World Distribution: in warm temperate and tropical parts of the eastern Atlantic Ocean and the Mediterranean.

Ricardia montagnei Derbès & Solier

Derbès and Solier, in Derbès, 1856.

Plants consisting of an ellipsoidal vesicle, to about 1.3 mm in diameter and to 1.4 mm in length, with rounded wart-like elevations on the surface and having simple hairs at the apex, attached to the host by a thick-walled, tapering and rarely branched cell; tetrasporangia developing in sori at the upper end of the vesicle.

For figures see Børgesen, 1930a, p. 76, 78, 79, figs 29-31.

Occurs on plants of *Laurencia obtusa* dredged on one occasion from a depth of 11 m.

Distribution. World: as for genus. West Africa: São Tomé: Carpine 1959, Steentoft 1967.

Waldoia W.R. Taylor 1962b

Plants consisting of flattened and alternately divided branches arising marginally from a central frond; apical cell prominent and with rudimentary hairs; branches having an obscure axial cell surrounded by rhizoidal filaments; tetrasporangia in simple stichidia scattered over the surface of the branches; cystocarps sessile or borne on short stalks, with a slightly produced pore.

World Distribution: in tropical parts of the Atlantic Ocean.

Waldoia antillana W.R. Taylor — Pl. 58, figs 3-5.

Taylor, 1962b.

Plants forming loose and semi-prostrate clumps, to about 10 cm in height, reddish-purple and iridescent when fresh, with the flattened and alternately and often tripinnately divided branches arising on a short stipe; branches irregularly segmented, cuneiform at the base and oblong to linear above, to about 10(-12) mm in width and from 2 to 10 cm in length, with an undulate and crenulate margin; apices obtuse in sexual plants and tapering in asexual plants; tetrasporangia arising towards the acute tips of subconical to subcylindrical stichidia which are to about 6 mm in length; cystocarps spherical to ovate, to about 700 μm in diameter.

For additional figures see Taylor, 1962b, pl. 3, figs 1, 2, pl. 4, fig. 2.

Occurs on rocky platforms and calcareous cobbles usually at depths ranging from 20 to 25 m, rarely found as small plants at a depth as shallow as 10 m.

Distribution. World: as for genus. West Africa: Ghana: John 1972a, John & Lawson 1972a, John *et al.* 1977, 1980, Lieberman *et al.* 1979.

Remarks. For comments regarding this species in the region refer to John & Lawson (1972a).

DIVISION CYANOPHYTA

Order Chroococcales · Family Chroococcaceae

Agmenellum Brébisson 1839

Plants micro- or macroscopic and forming a flat or foliose plate; cells spherical, ovoid or cylindrical, dividing at right angles to the surface and to each other, forming a series of regular rows; sheath gelatinous and hyaline.

Agmenellum thermale (Kützing) Drouet & Daily

Drouet and Daily, 1956; *Merismopedia thermalis* Kützing, 1843b.

Plants initially microscopic and rectangular, becoming eventually elongate and macroscopic, with an irregular margin; cells after division globose, ovoid or cylindrical, from 4-10 μm in diameter and 4-20 μm in length; sheath dense, gelatinous, firm or diffluent.

For figures see Drouet and Daily, 1956, p. 175, 176, figs 134-142.

Occurs together with other blue-green algae floating as a bubbly scum on the surface of brackish-water pools found along the edge of lagoon systems.

Distribution. World: cosmopolitan. West Africa: Ghana: Drouet & Daily 1956, 1957, John & Lawson 1972b.

Anacystis Meneghini 1837

Plants consisting of coccoid cells, irregularly arranged in a gelatinous matrix; cells with the protoplasm homogeneous or a little granular, occasionally containing pseudovacuoles, dividing in three planes perpendicular to each other; sheath hyaline or pigmented.

Key to the Species

1. Cells usually less than 6 μm in diameter . . . . *A. montana*
1. Cells usually 6 μm or more in diameter 2

2. Cells from (8-)12-50 μm in diameter and usually angular after division . . . . *A. dimidiata*
2. Cells usually from 6-12 μm in diameter and becoming spherical after division . . . . *A. aeruginosa*

Anacystis aeruginosa (Zanardini) Drouet & Daily

Drouet and Daily, 1956; *Palmogloea aeruginosa* Zanardini, 1872.

Plants initially microscopic and eventually growing to several centimetres across; cells at division depressed to subspherical, becoming extended and globose, from 6-12 μm in diameter; sheath hyaline or having a gelatinous matrix.

For figures see Drouet and Daily, 1956, p. 174, figs 108-113.

Occurs associated with other blue-green algae as a covering over eulittoral rocks and lithothamnia in usually exposed or very wave-exposed situations.

Distribution. World: probably cosmopolitan. West Africa: Togo: John & Lawson 1972b.

Anacystis dimidiata (Kützing) Drouet & Daily

Drouet and Daily, 1956; *Trochiscia dimidiata* Kützing, 1833.

Plants usually microscopic and consisting of from 2 to 16 cells; cells at division truncate to hemispherical, becoming truncate to globose or very occasionally spherical, from (8-)12-50 μm in diameter; sheath hyaline or having a gelatinous matrix.

For figures see Drouet and Daily, 1956, p. 173, 174, figs 100-107.

Occurs with other blue-green algae as a scum floating on the surface of small brackish-water pools at the edge of lagoons.

Distribution. World: probably cosmopolitan. West Africa: Ghana: Drouet & Daily 1956, 1957, John 1977a. John & Lawson 1972b, Lawson 1960a. Liberia: De May *et al.* 1977, John 1972c, 1977a. São Tomé: John 1977a; *Chroococcus turgidus* var. *submarinus*, Sampaio 1958, 1962a, 1962b. Sierra Leone: *Chroococcus turgidus*, Aleem 1980a.

Anacystis montana (Lightfoot) Drouet & Daily

Drouet and Daily, 1952; *Ulva montana* Lightfoot, 1777.

Plants microscopic or sometimes becoming macroscopic; cells at division each depressed to subspherical or more rarely truncate to hemispherical, becoming globose and to about 6 μm in diameter; gelatinous matrix or a pigmented sheath.

For figures see Drouet and Daily, 1956, p. 169 to 173, figs 16-90A.

Occurs together with other species of blue-green algae as a bluish-grey slime covering barnacles growing in the lower part of the upper eulittoral subzone.

Distribution. World: probably cosmopolitan. West Africa: Ghana: unpublished.

Coccochloris Sprengel 1807

Plants micro- or macroscopic, one or many celled; cells at division subspherical to cylindrical, becoming ovoid to long and cylindrical, dividing in a plane at right angles to the long axis; sheath gelatinous and hyaline or pigmented, homogeneous or lamellated.

Key to the Species

1. Cells to about 3 times as long as broad .... *C. stagnina*
1. Cells usually greater than 3 times as long as broad 2

2. Cells from 1-3 μm in diameter and to about 12 times as long as broad, often curved. .... *C. peniocystis*
2. Cells from 2-6 μm in diameter and to about 8 times as long as broad, always straight. .... *C. elabens*

Coccochloris elabens (Brébisson ex Meneghini) Drouet & Daily

Drouet and Daily 1948; *Micraloa elabens* Brébisson, in Meneghini, 1843.

Cells at division quadrate to spherical to elliptical, becoming cylindrical with the ends truncate to rounded, from 2-6 μm in diameter and to about 8 times as long as broad, always straight; sheath gelatinous and homogeneous.

For figures see Drouet and Daily, 1956, p. 178, figs 164-169.

Occurs with other species of blue-green algae as a lining to the floor of tidepools in the littoral fringe.

Distribution. World: probably cosmopolitan. West Africa: Ghana: unpublished.

Coccochloris peniocystis Drouet & Daily

Drouet and Daily, 1948.

Cells at division cylindrical and becoming longer, with the ends rounded and sometimes tapering to conical, from 1-3 μm in diameter, to about 12 times as long as broad and usually curved; sheath gelatinous and homogeneous or lamellated.

For figures see Drouet and Daily, 1956, p. 178, figs 170-172.

Occurs with a number of other blue-green algae as a spongy mat to about 2 cm thick covering the muddy floor of shallow salt pans as well as on *Sesuvium* plants growing along the edge of such pans.

Distribution. World: probably cosmopolitan. West Africa: Ghana: unpublished.

Coccochloris stagnina Sprengel

Sprengel, 1807.

Cells at division ovoid to elliptical or more rarely elliptical to cylindrical, from (3-)4-8 μm in diameter and to about 3 times as long as broad; sheath gelatinous and homogeneous or conspicuously lamellated.

For figures see Drouet and Daily, 1956, p. 177, 178, figs 145-163.

Occurs in the littoral fringe as a slimy black layer growing on the shaded ends of logs stranded in the upper parts of estuaries.

Distribution. World: cosmopolitan. West Africa: Cameroun: unpublished.

Gomphosphaeria Kützing 1836

Plants usually microscopic and spherical to ovoid, sometimes tuberculate and constricted; cells spherical, ovoid, cylindrical, obovoid or pyriform, radially arranged as a single layer towards the periphery of a gelatinous matrix, dividing in two planes perpendicular to each other; sheath surrounding the cells distinct, sometimes persistent, forming a dichotomously branched structure extending from the centre of the colony.

Gomphosphaeria aponina Kützing

Kützing, 1836.

Plants spherical to ovoid and becoming variously constricted; cells ovoid, cylindrical, obovoid or pyriform and often cordate at division, irregularly arranged or in rows perpendicular to the surface; sheaths firm and sometimes diffluent, forming a branched structure.

For figures see Drouet and Daily, 1956, p. 179, figs 178-180.

Occurs together with other blue-green algae to form a thick spongy covering over the floor of shallow salt pans and over plants of *Sesuvium* growing into these pans.

Distribution. World: cosmopolitan. West Africa: Ghana: unpublished.

Johannesbaptistia De Toni 1934

Plants microscopic, elongate and cylindrical, straight or curved; cells discoid, dividing in one plane perpendicular to the axis of the plant, in a single series in a narrow hyaline gelatinous matrix or sheath.

Johannesbaptistia pellucida W.R. Taylor & Drouet

Taylor and Drouet, in Drouet, 1938.

Filaments to about 1 mm in length; cells discoid or spherical to discoid, from 3-20 μm in diameter, arranged in a single series; gelatinous matrix homogeneous and the sheath around the cells firm or diffluent.

For figures see Drouet and Daily, 1956, p. 180, figs 182-184.

Occurs mixed with other blue-green algae as a thick spongy lining to the floor of shallow salt pans or growing over the surface of *Sesuvium* plants.

Distribution. World: from boreal-antiboreal to tropical regions, more common in the latter. West Africa: Ghana: unpublished.

Order Chamaesiphonales · Family Chamaesiphonaceae

Entophysalis Kützing 1843b

Plants initially forming microscopic growths, later becoming macroscopic and encrusting or cushion-like; solitary cells usually spherical and giving rise by growth and division to a cushion of cells which are spherical, elliptical, cylindrical or polyhedral, often the cells arranged in vaguely ordered radial rows, with the basal cells producing downgrowing uni- or multiseriate filaments penetrating the substratum; endosporangia of diverse form.

Key to the Species

1. Epilithic or growing on various inanimate objects .... *E. deusta*
1. Epibiotic... *E. conferta*

Entophysalis conferta Drouet & Daily

Drouet and Daily, 1948.

Plants epibiotic, microscopic or macroscopic and forming crusts or cushions; cells of the microscopic stage solitary, spherical, ovoid or pyriform, to about 20 μm in diameter; cells of cushion-like layer of the macroscopic stage spherical and spherical to polyhedral after division, from (1-)3-6 μm in diameter; cells of the downgrowing filaments of the macroscopic stage spherical, ovoid, cylindrical or polyhedral, from 3-8 μm in diameter; endosporangia spherical, obovoid, cylindrical, pyriform or tube-like, to about 50 μm in diameter.

For figures see Drouet and Daily, 1956, p. 181, 182, figs 196-215.

Occurs in the lower part of the eulittoral zone on larger algae often growing in tidepools, whilst in the littoral fringe it is sometimes found as a dark blue-green covering to the roots of trees.

Distribution. World: probably cosmopolitan. West Africa: Ghana: Drouet & Daily 1956. Nigeria: Fox 1957. Sierra Leone: Drouet & Daily 1956, John & Lawson 1977a, Lawson 1954b; *Dermocarpa hemisphaeria, D. leibleiniae, D. prasina,* Aleem 1980a.

Entophysalis deusta (Meneghini) Drouet & Daily

Drouet and Daily, 1948; *Coccochloris deusta* Meneghini, 1840.

Plants growing on various inanimate objects; microscopic stage consisting of solitary, spherical, ovoid or pyriform cells, to about 10 μm in diameter; macroscopic stage crustose or as cushion-like layers, with the cells spherical to polyhedral after division and from (1-)3-6 μm in diameter, with the downgrowing filaments from the cushion-like layer uniseriate or less commonly multiseriate and the cells from (1-)2-15 μm in diameter; gelatinous matrix homogeneous or lamellated; endosporangia spherical or ovoid, to about 30 μm in diameter.

For figures see Drouet and Daily, 1956, p. 180, 181, figs 185-194, p. 183, figs 247-250; Drouet, 1981, p. 204, fig. 36.

Occurs as a powdery and greenish to blackish covering to rocks, perforating molluscan shells, or sometimes as a sheet mixed with other blue-green algae growing in the littoral fringe.

Distribution. World: probably cosmopolitan. West Africa: Gambia: John & Lawson 1977b. Ghana: Drouet & Daily 1956, Lawson 1960a; *E. crustacea*, Lawson 1956, 1966. Sierra Leone: Drouet & Daily 1956, John & Lawson 1977a; *E. crustacea*, Lawson 1957a; *E. granulosa*, *Hyella caespitosa*, Aleem 1980a.

Order Nostocales · Family Oscillatoriaceae*

Arthrospira Gomont 1893

Plants consisting of trichomes distinctly divided into cells by cross walls; cells having a homogeneous or granulose protoplasm, with many granules along the cross and side walls; terminal cells often attenuate, rounded or conical, with the outer membrane not thickened; sheath present or absent.

Key to the Species

1. Trichomes with all the cells slightly shorter or slightly longer than broad; terminal cell hemispherical . . . . *A. jenneri*
1. Trichomes with cells other than the terminal cell shorter than broad; terminal cell conical and to six times as long as broad . . . . *A. neapolitana*

Arthrospira jenneri Gomont

Gomont, 1892.

Trichomes from 3-5 μm in diameter and sometimes attenuate towards the tip; cells from 2-4 μm in length, as long as broad or shorter, with the terminal cell cylindrical and becoming hemispherical with age.

*For terms used to describe the shape of the end cell and the outer wall of the apical cell see pl. 58, fig. 7a-k.

For figures see Drouet, 1968, p. 339, figs 84, 85.

Occurs with other blue-green algae as a purplish-black mat in deep rock crevices in the littoral fringe.

Distribution. World: probably cosmopolitan. West Africa: Ghana: unpublished.

Arthrospira neapolitana (Gomont) Drouet

Drouet, 1969; *Oscillatoria neapolitana* Gomont, 1892.

Trichomes from 2-10 μm in diameter and gradually attenuate towards the apex; cells from 1.5-4 μm in length and very much shorter than broad; terminal cell to about 6 times as long as broad, rotund or truncate-conical, becoming longer to acute-conical.

For figures see Drouet, 1968, p. 339, figs 86-88 (as *A. brevis*).

Occurs as a yellow-green or blue-green sheet or scum growing together with other blue-green algae on the mud along the banks of lagoons or on rocks in the littoral fringe where there is freshwater run-off.

Distribution. World: probably cosmopolitan. West Africa: Ghana: *A. brevis,* Drouet 1968; *Oscillatoria brevis,* Lawson 1960a. Nigeria: *A. brevis,* Drouet 1968. Príncipe: *Oscillatoria brevis,* Sampaio 1963. São Tomé: *A. brevis,* Drouet 1968; *Oscillatoria roseirai,* Sampaio 1958, 1962a. Sierra Leone: *Oscillatoria brevis,* Aleem 1980a.

Microcoleus Gomont 1892

Plants consisting of trichomes, often attenuate towards the end; terminal cell truncate or depressed-conical or hemispherical, with an outer membrane becoming thickened with age; cells having the protoplasm homogeneous or granulose, with the cross walls and more rarely the side walls lined with granules.

Key to the Species

1. Cells having the granules along both the cross and side walls .... *M. lyngbyaceus*
1. Cells having the granules confined to the cross walls.............. *M. vaginatus*

Microcoleus lyngbyaceus Gomont

Gomont, 1892.

Trichomes from 3.5-80 μm in diameter; cells from 1.5-8 μm in length, usually shorter than broad; terminal cell rotund and the outer membrane becoming thickened and depressed-rotund, depressed-conical, depressed-hemispherical in shape; cells having dense protoplasm and granules along the side and cross walls.

For figures see Drouet, 1968, p. 341, figs 101-129.

Occurs as separate dark green tufts or as thick mats with other blue-green algae; found in the eulittoral zone on rocks exposed to wave-action, also on mud in sheltered lagoon systems and lining shallow salt pans, purplish-black only when in deep rock crevices. Grows mixed with coarse tufts of *Schizothrix mexicana* occurring on rock platforms at a depth of about 22 m.

Distribution. World: cosmopolitan. West Africa: Benin: John 1977a, John & Lawson 1972a. Cameroun: Drouet 1968, John 1977a; *Lyngbyea majuscula,* Lawson 1955, Steentoft 1967. Côte d'Ivoire: John 1972c, 1977a. Gabon: Drouet 1968, John 1977a, John & Lawson 1974a; *Hydrocoleum lyngbyaceum,* Steentoft 1967; *Lyngbya majuscula,* Lawson 1955, Steentoft 1967, Vickers 1897. Ghana: Drouet 1968, 1974, John 1977a; *Hydrocoleum lyngbyaceum,* Lawson 1956, 1966, Steentoft 1967, Stephenson & Stephenson 1972; *Lyngbyea aestuarii,* Lawson 1960a; *L. confervoides,* Lawson 1956, 1966; *L. semiplena,* Lawson 1960a, 1965, 1966. Liberia: De May *et al.* 1977, John 1972c. Nigeria: Drouet 1968, John 1977a; *Lyngbyea confervoides,* Fox 1957; *L. semiplena,* Fox 1957. Príncipe: Drouet 1968, John 1977a; *Lyngbyea aestuarii,* Sampaio 1963. São Tomé: Drouet 1968, John 1977a; *Hydrocoleum lyngbyaceum* var. *typica* and *rupestre,* Sampaio 1958, 1962a, 1962b, Steentoft 1967; *Lyngbyea majuscula,* Steentoft 1967; *Oscillatoria tenuis,* Henriques 1917. Sierra Leone: Drouet 1968. John & Lawson 1977a; *Hydrocoleum lyngbyaceum,* Aleem 1980a; *Lyngbyea confervoides,* Aleem 1980a, Lawson 1954b, 1957a, Longhurst 1958, Whitton 1968; *L. majuscula,* Aleem 1980a; *L. martensiana,* Aleem 1980a; *Oscillatoria chlorina,* Lawson 1954b, 1957a; *O. margaritifera,* Aleem 1980a. Togo: Drouet 1968, John 1977a, John & Lawson 1972b.

Remarks. The names of ecophenes found in Ghana but which have not been published previously are *Hydrocoleum cantharidosum, H. comoides, H. glutinosum, H. lyngbyaceum, Oscillatoria corallinae* and *O. margaritifera.* This is probably one of the most ubiquitous members of the blue-greens in the region being commonly found growing on soil as well as in freshwater habitats.

Microcoleus vaginatus Gomont

Gomont, 1892.

Trichomes from 2.5-9 μm in diameter; cells from 1-10 μm in length, shorter to longer than broad and the terminal cell conical, hemispherical or truncate-cylindrical, with the outer membrane thickened to form a cone, cup or convex plate; cells having dense protoplasm and granules confined to the cross walls.

For figures see Drouet, 1968, p. 339, figs 89-99.

Occurs associated with a number of other species of blue-green algae as a bluish-grey slime over the surface of barnacles growing in the lower part of the eulittoral zone.

Distribution. World: cosmopolitan. West Africa: Ghana: unpublished. São Tomé: *Phormidium boryanum,* Henriques 1885, 1886a, 1917.

Remarks. This species is known from a number of countries bordering the region but only growing in freshwater or on soil, and so would seem to be of uncommon occurrence in the marine environment.

Oscillatoria Gomont 1893

Plants consisting of trichomes distinctly divided into cells by cross walls; terminal cell truncate, hemispherical, or depressed-conical, having the outer membrane thickened; cells having the protoplasm homogeneous or granular, with granules absent from the side and cross walls.

Key to the Species

1. Cells at least one third as long as broad 2
1. Cells less than one third as long as broad 3

2. Outer wall of the terminal cell having a depressed-conical or hemispherical distal thickening .... *O. submembranacea'*
2. Outer wall of the terminal cell having a convex plate-like thickening .... *O. lutea*, in part

3. Trichomes distinctly attenuate towards the tips .... *O. princeps*
3. Trichomes not tapering or very slightly .... *O. lutea*, in part

Oscillatoria lutea Gomont

Gomont, 1893.

Trichomes from 2.5-10 μm in diameter; cells from 1-7 μm in length and as long or more often a good deal shorter than broad; terminal cell broadly truncate-conical and the outer membrane becoming thickened as a convex plate; cells rarely containing pseudovacuoles.

For figures see Drouet, 1968, p. 337, figs 51-57.

Occurs often as a thick mat or yellowish-green slime growing over rocks or silt-covered boulders in the littoral fringe; often very abundant in tidepools especially when influenced by freshwater run-off.

Distribution. World: probably cosmopolitan. West Africa: Côte d'Ivoire: John 1972c, 1977a. Ghana: Drouet 1968, John 1977a; *Lyngbyea lutea*, Lawson 1956. Sierra Leone: *Lyngbya lutea*, Aleem 1980a.

Oscillatoria princeps Gomont

Gomont, 1893.

Trichomes from 9-90 μm in diameter and briefly and obviously attenuate over several cells; cells from 3-8 μm in length and much shorter than broad; terminal cell becoming truncate-depressed-conical, with the outer membrane eventually thickened and depressed-hemispherical; cells with the protoplasm occasionally containing pseudovacuoles.

For figures see Drouet, 1968, p. 335, figs 48-50.

Occurs with other blue-green algae as floating scums on the surface of tidepools in the upper eulittoral zone and littoral fringe; found also in rock crevices as purplish-black mats.

Distribution. World: probably cosmopolitan. West Africa: Ghana: Drouet 1968.

Oscillatoria submembranacea Gomont

Gomont, 1892.

Trichomes from 2.5-9 μm in diameter and if attenuate then only slightly towards the tip; cells from 3-11 μm in length and longer or slightly shorter than broad; terminal cells truncate-cylindrical or truncate-conical, with the outer membrane becoming thickened and depressed-conical or depressed-hemispherical; cells with the protoplasm only rarely containing pseudovacuoles.

For figures see Drouet, 1968, p. 337, figs 62, 64.

Occurs in the littoral fringe mixed with *Entophysalis deusta* and *Bostrychia* spp. or forming a mat with other blue-green algae on silt-covered boulders.

Distribution. World: probably cosmopolitan. West Africa: Cameroun: unpublished. Gabon: John & Lawson 1972a. Ghana: Drouet 1968; *Symploca atlantica,* Lawson 1956, Steentoft 1967. Nigeria: Drouet 1968; *Symploca atlantica,* Fox 1957, Steentoft 1967. São Tomé: Drouet 1968; *Symploca atlantica,* Steentoft 1967. Sierra Leone: John & Lawson 1977a.

Remarks. *Phormidium submembranaceum* from Ghana is the only unpublished ecophene of this species so far found in the marine environment in the region.

Porphyrosiphon Gomont 1892

Plants consisting of trichomes having conspicuous cross walls and attenuate for several cells towards the tip; terminal cell conical or hemispherical and with the outer membrane not thickened; cells not containing granules along the cross or side walls.

Key to the Species

1. Cells mostly less than one third as long as broad. *P. kurzii*
1. Cells one third as long as broad or longer . . . . *P. notarisii*

Porphyrosiphon kurzii (Gomont) Drouet

Drouet, 1968; *Chthonoblastus kurzii,* Gomont, 1892.

Trichomes from 6-12 μm in diameter; cells from 2-4 μm in length, mostly less than one third as long as broad or rarely longer, with the terminal cell hemispherical and later becoming obtuse- or acute-conical.

For figures see Drouet, 1968, p. 339, fig. 76.

Occurs over a wide range of brackish-water habitats as a dirty greenish layer on rocks in the littoral fringe and on mud in lagoons; occasionally on open shores mixed with other blue-green algae in the algal turf in the lower eulittoral subzone.

Distribution. World: in warm temperate and tropical seas. West Africa: Gambia: John & Lawson 1977b. Ghana: Drouet 1968; *Microcoleus kurzii,* Lawson 1960a, 1965, 1966.

Porphyrosiphon notarisii Bornet & Flahault

Bornet and Flahault, 1886c.

Trichomes from 3-40 μm in diameter; cells from 3-15 μm in length, slightly shorter or longer than broad, with the terminal cell hemispherical and becoming obtuse- or acute-conical.

For figures see Drouet, 1968, p. 337, figs 65-75.

Occurs with other species of blue-green algae as a greenish or bluish-grey slime covering both rocks and barnacles in the lower part of the eulittoral zone; occasionally on mud in lagoons.

Distribution. World: cosmopolitan. West Africa: Cameroun: Drouet 1968. Ghana: Drouet 1968, Lawson 1960a. Nigeria: Drouet 1968. Sierra Leone: Aleem 1980a, John & Lawson 1977a; *Oscillatoria nigro-viridis*, Aleem 1980a.

Remarks. *Oscillatoria nigro-viridis* and *O. subuliformis* are the only two unpublished ecophenes of this species which occur in marine habitats in Ghana.

Schizothrix Gomont 1892

Plants consisting of trichomes, cylindrical throughout or if attenuate then only the terminal cell involved; terminal cell hemispherical or rotund-cylindrical or conical, with the outer membrane always thin; protoplasm homogeneous or granulose, never with granules lining the cross or side walls.

Key to the Species

1. Terminal cell becoming rounded 2
1. Terminal cells becoming conical 4

2. Terminal cells cylindrical and rotund at the ends .... *S. friesii*
2. Terminal cells hemispherical or almost spherical............................ 3

3. Trichomes from 4-60 μm in diameter .... *S. mexicana*
3. Trichomes up to 3.5 μm in diameter.............................. *S. calcicola*

4. Terminal cells long and acuminate.............................. *S. tenerrima*
4. Terminal cells becoming obtusely, acutely or truncately conical 5

5. Trichomes becoming constricted at the cross walls.................. *S. arenaria*
5. Trichomes without constrictions at the cross walls .... *S. rubella*

Schizothrix arenaria Gomont

Gomont, 1892.

Trichomes from 1-6 μm in diameter, becoming constricted at the cross walls; cells from 2-10 μm in length and as long as broad or longer, with the terminal cells acute or obtusely conical, only very occasionally cylindrical-conical.

For figures see Drouet, 1968, p. 335, figs 28-34.

Occurs as a thin covering to rocks in the eulittoral zone or together with other blue-greens as a thick spongy mat on the muddy floors of tidepools, shallow salt pans or lagoons.

Distribution. World: cosmopolitan. West Africa: Cameroun: Drouet 1968; *Microcoleus chthonoplastes,* Pilger 1911; *Sirocoleum guyanense,* Lawson 1955, Stephenson & Stephenson 1972. Gabon: John & Lawson 1974a. Gambia: John & Lawson 1977b. Ghana: Drouet 1968; *Microcoleus chthonoplastes,* Lawson 1960a, 1966. Nigeria; Drouet 1968; *M. chthonoplastes,* Fox 1957. São Tomé: Drouet 1968; *M. chthonoplastes,* Sampaio 1958, 1962a, 1962b. Sierra Leone: Drouet 1968, John & Lawson 1977a; *M. chthonoplastes,* Aleem 1980a, Lawson 1954b, 1957a, Longhurst 1958.

Schizothrix calcicola Gomont

Gomont, 1892.

Trichomes from 0.2-3.5 μm in diameter; cells from 2-6 μm in length and shorter or longer than broad; terminal cell cylindrical and somewhat bulbose, with the thin outer membrane at first quasi-truncate and becoming rotund or very occasionally eccentrically swollen.

For figures see Drouet, 1968, p. 333, figs 8-19.

Occurs as slimy patches on rocks or mixed with other blue-greens and forming thick spongy mats, purplish-black when in deep rock crevices; occasionally as brownish scums floating on the surface of tidepools or else lining the floor of shallow salt pans. Grows on a variety of subtrates in the eulittoral zone and sometimes found in coarse tufts of *Schizothrix mexicana* at a depth of about 22 m.

Distribution. World: probably cosmopolitan. West Africa: Benin: John 1977a, John & Lawson 1972b. Gabon: John & Lawson 1974a. Ghana: Drouet 1968, John 1977a; *Oscillatoria amphibia,* Lawson 1960a; *Plectonema nostocorum,* Lawson 1960a. Gambia: John & Lawson 1977b. Liberia: De May *et al.* 1977, John 1972c, 1977a. Nigeria: Drouet 1968, John 1977a. Sierra Leone: John & Lawson 1977a; *Lyngbya porphyrosiphonia,* Aleem 1980a, Whitton 1968; *L. inflexa, Oscillatoria amphibia, Phormidium fragile, Plectonema terebrans,* Aleem 1980a.

Remarks. Unpublished ecophenes found in Ghana are *Lyngbyea epiphytica* and *Phormidium fragile.*

Schizothrix friesii Gomont

Gomont, 1892.

Trichomes from 2-10 μm in diameter; cells from 3-20 μm in length, slightly shorter or longer than broad; terminal cell long-cylindrical or somewhat attenuate-cylindrical, with the outer wall hemispherical.

For figures see Drouet, 1968, p. 333, figs 23-27.

Occurs associated with other blue-greens as spongy mats growing over the muddy floor of shallow salt pans and on *Sesuvium* plants.

Distribution. World: cosmopolitan. West Africa: Ghana: unpublished.

Schizothrix mexicana Gomont

Gomont, 1892.

Trichomes from 4-65 μm in diameter; cells from 2-10 μm in length, shorter and longer than broad, with the terminal cell subspherical to hemispherical.

For figures see Drouet, 1968, p. 333, figs 20-22.

Occurs with other blue-green algae as a greenish-black crust covering mud and rocks in lagoons and tidepools in the eulittoral zone; occasionally in the sublittoral to depths of about 22 m as coarse purplish tufts up to 20 cm in height.

Distribution. World: widespread in warm temperate and tropical seas. West Africa: Côte d'Ivoire: John 1972c, 1977a. Gabon: John 1977a, John & Lawson 1974a. Ghana: Drouet 1968, John 1977a. Nigeria: Drouet 1968, John 1977a; *Lyngbyea meneghiniana,* Fox 1957. Sierra Leone: John 1977a, John & Lawson 1977a.

Remarks. *Lyngbyea sordida* is the only unpublished ecophene of this species found in marine habitats in the region being known from Ghana.

Schizothrix rubella Gomont

Gomont, 1892.

Trichomes from 1.5-8 μm in diamter; cells from 2-10 μm long and equal or longer or shorter than broad; terminal cell obtusely, acutely or truncately conical or cylindrical-conical, with the outer membrane thin.

For figures see Drouet, 1968, p. 335, figs 35-41.

Occurs together with *Microcoleus lyngbyaceus* growing over shaded logs in the littoral fringe in the brackish-water habitat of an estuary.

Distribution. World: cosmopolitan. West Africa: Gabon: John & Lawson 1974a.

Schizothrix tenerrima (Gomont) Drouet

Drouet, 1968; *Microcoleus tenerrimus* Gomont, 1892.

Trichomes from 1-6 μm in diameter; cells from 3-12 μm in length, to about 4 times as long as broad; terminal cell cylindrical, becoming long-cylindrical and extended out into a hair-like tip.

For figures see Drouet, 1968, p. 335, figs 42-45.

Occurs together with other blue-green algae in brackish-water situations either in the littoral fringe as a mat on stranded logs or mixed with larger algae growing in the eulittoral zone where there is often heavy sedimentation.

Distribution. World: probably cosmopolitan. West Africa: Gabon: John & Lawson 1974a. Gambia: John & Lawson 1977a. São Tomé: *Microcoleus tenerrimus,* Sampaio 1958, 1962a, 1962b. Sierra Leone: John & Lawson 1977a; *Oscillatoria acuminata,* Aleem 1980a.

Spirulina Gomont 1893

Plants consisting of trichomes in which the cross walls between the cells are not readily discernable; trichomes straight or more often curved or spiralling and having the ends hemispherical, with the outer membrane unthickened; sheath absent.

Spirulina subsalsa Gomont

Gomont, 1893.

Trichomes cylindrical and usually spiralled, from 0.4-4 μm in diameter; protoplasm homogeneous or granulose, occasionally containing pseudovacuoles; sheath material absent or the trichome sometimes enclosed in a homogeneous mucus.

For figures see Drouet, 1968, p. 333, figs 1-7.

Occurs with other blue-green algae either as a mat on open rocks in the eulittoral zone or as a thick spongy layer covering the floor of salt pans.

Distribution. World: probably cosmopolitan. West Africa: Ghana: Drouet 1968. Nigeria: Drouet 1968. São Tomé: *S. subtilissima,* Sampaio 1958, 1962a, 1962b. Sierra Leone: Aleem 1980a; *S. labyrinthiformis, S. subtilissima,* Aleem 1980a; *S. major,* Aleem 1980a, Whitton 1968. Togo: John & Lawson 1972b.

Family Nostocaceae

Calothrix Bornet & Flahault 1886a

Plants consisting of erect, more or less parallel and basally attached filaments; trichomes simple or occasionally falsely branched, often drawn out into a hair-like tip; heterocysts basal or sometimes intercalary; sheath homogeneous or lamellated.

Key to the Species

1. Trichomes attenuated gradually to the tips .... *C. parietina*
1. Trichomes abruptly attenuated near the tips..................... *C. crustacea*

Calothrix crustacea Bornet & Flahault

Bornet and Flahault, 1886a.

Trichomes cylindrical or abruptly tapering at one end, both ends or swollen at one end, with or without constrictions at the cross walls, from 3-22 μm in diameter; cells shorter or longer than broad.

For figures see Drouet, 1973, p. 263, figs 67-83.

Occurs as a crust together with other blue-green algae in moderately to very wave-exposed situations growing on rocks or over lithothamnia in the eulittoral zone. Found in amongst species of *Bostrychia* in rock crevices in the littoral fringe and occasionally on shaded estuarine rocks.

Distribution. World: probably cosmopolitan. West Africa: Benin: Drouet 1973, John & Lawson 1972b. Bioko: unpublished. Cameroun: Drouet 1973; *Brachytrichia maculans,* Lawson 1955, Stephenson & Stephenson 1972. Gambia: John & Lawson 1977b. Ghana: Drouet 1973, Lawson 1960a; *Symploca hydnoides,* Lawson 1960a, 1966. Togo: Drouet 1973; *Isactis plana,* John & Lawson 1972b. Sierra Leone: Aleem 1980a, Drouet 1973, John & Lawson 1977a, Whitton 1968; *Calothrix contarenii,* Aleem 1980a, Lawson 1954b, 1957a, Longhurst 1958; *Calothrix scopulorum,* Aleem 1980a, Lawson 1954b, 1957a.

Calothrix parietina Bornet & Flahault

Bornet and Flahault, 1886a.

Trichomes cylindrical or towards the tip long and gradually attenuate, or swollen at one end, with or without constrictions at the cross walls, from 3-24 μm in diameter; cells shorter or longer than broad.

For figures see Drouet, 1973, p. 257, 259, 261, figs 28-60.

Occurs usually as a bluish-grey slime or mat with other blue-green algae or sometimes as globose colonies about 1 mm across, growing in the eulittoral zone on rocks or shells subject to a wide range of exposure to wave-action.

Distribution. World: probably cosmopolitan. West Africa: Benin: Drouet 1973, John & Lawson 1972b. Ghana: Drouet 1973. Togo: John & Lawson 1972b.

Scytonema Bornet & Flahault 1886c

Filaments containing falsely branched trichomes and usually arising laterally between two heterocysts; trichomes straight and single in a sheath; terminal cell at first hemispherical, becoming blunt or almost so; hormogonia and pseudohormogonia terminal and solitary.

Scytonema hofmannii Bornet & Flahault

Bornet and Flahault, 1886c.

Trichomes more or less cylindrical, with indistinct constrictions often present at the cross walls, from 3-30 μm in diameter; cells shorter to longer than broad.

For figures see Drouet, 1973, p. 253, 255, figs 1-27.

Occurs as a blackish crust on the surface of mud in lagoons.

Distribution. World: probably cosmopolitan. West Africa: Cameroun: Drouet 1973. Ghana: Drouet 1973, Lawson 1960a.

Order Stigonematales · Family Nostochopsidaceae

Mastigocoleus Bornet & Flahault 1886c

Filaments consisting of repeatedly and irregularly branched trichomes, often unilateral, and surrounded by a thin sheath; lateral branches generally of two types, one long and ending in a hair, the other short and a few cells (1 to 4) in length; hetrocysts terminal, lateral or rarely intercalary; growing within calcareous material.

Mastigocoleus testarum Bornet & Flahault

Bornet and Flahault, 1886c.

Trichomes interlaced and variously curved, from 3.5-10 μm in diameter, with the cells cylindrical to subcylindrical; heterocysts from 6-18 μm in length and breadth, arising terminally or laterally, rarely intercalary.

For figures see Geitler, 1930-1932, p. 474, fig. 284; Drouet, 1981, p. 203, fig. 14.

Occurs together with other blue-green algae as a bluish-green covering to the pyramidal shells of the small limpet *Siphonaria* found in tidepools in the littoral fringe, also in the shells of *Ostrea tulipa.*

Distribution. World: probably cosmopolitan. West Africa: Sierra Leone: Aleem 1980a, John & Lawson 1977a.

APPENDIX I

Keys to the Genera

All the keys in this book are artificial and are designed to permit identification of well-developed specimens of any of the four major divisions* of marine algae so far known from tropical West Africa. Whenever possible characters of gross morphology and vegetative anatomy are used in the keys to simplify determination (see Appendix II). Reproductive structures have been used mainly in those cases where positive identification cannot be made without reference to them . It is often rare to find fertile plants and special techniques of preparation and staining may be required to show the necessary details. Ideally, identifications that have been made using the keys provided in this book should be checked by referring to authenticated specimens and/or more specialized treatments dealing with the taxon in question. The following monographic works on certain groups will be found especially useful: - Hoek (1963, 1982) and Söderström (1963) for the genus *Cladophora*. Bliding (1963, 1968) for the order Ulvales, Woelkerling (1971, 1973a) for the family Acrochaetiaceae, Clayton (1974) for the order Ectocarpales, Drouet and Daily (1956) and Drouet (1968, 1973, 1978, 1981) for the division Cyanophyta. There are a large numbers of algal genera which are known to be extremely polymorphic and are in need of critical re-appraisal and possibly global revision, these include *Chaetomorpha, Rhizoclonium, Enteromorpha, Dictyota, Padina, Sargassum, Gelidium, Gracilaria, Gymnogongrus, Scinaia* and the lithothamnia; identification of species within such genera can at best be regarded as provisional.

Plants belonging to the four divisions can usually be readily recognised on the shore by the following characters:

Chlorophyta: (Green algae)	plants green as the chlorophyll is unmasked by other pigments.
Phaeophyta: (Brown algae)	plants olive-brown to very dark brown due to other pigments masking the chlorophyll.
Rhodophyta: (Red algae)	plants usually red due to other pigments masking the chlorophyll, but varying widely from dark purple through dirty green to yellow or yellowish-brown under differing environmental conditions, usually retaining some red colouration near the base.
Cyanophyta: (Blue-green algae)	plants usually blue-green in colour, but varying from bright red to purplish-black under different environmental conditions.

*A fifth division, the Xanthophyta, is represented by the genus *Vaucheria*. The green coenocytic filaments of this alga have been collected on several occasions growing in shaded parts of the littoral fringe. As most of the specimens lack organs of sexual reproduction identification to species has not been possible, and so the genus has been omitted from this flora. Also omitted is *Chara zeylanica*, a representative of yet another division, the Charophyta; this alga is sometimes found in brackish-water lagoons.

CHLOROPHYTA

1. Plants consisting of irregularly divided procumbent filaments; minute and epi- or endophytic. . . . . *Entocladia* (p. 73)
1. Plants of another form; epilithic, epibiotic or free-living. 2

   2. Plants composed of simple or branched filaments. 3
   2. Plants not obviously filamentous, pseudoparenchymatous or truly parenchymatous 16

3. Plants umbrella-like, consisting of a stalk and often a terminal whorl of branchlets fused into a disc. . . . . *Acetabularia* (p. 94)
3. Plants not as above. 4

   4. Plants not usually divided by cross wall into distinct cells when in the vegetative state. 5
   4. Plants obviously divided into cells. 7

5. Plants composed of dichotomously or irregularly divided branches constricted at the forks, with branchlets absent. *Boodleopsis* (p. 109)
5. Plants composed of uni- or bilaterally, or irregularly arranged branchlets formed successively from the apex. 6
   6. Branchlets radially arranged and considerably narrower than the main axes . . . . *Trichosolen* (p. 103)
   6. Branchlets uni- or bilaterally arranged along the axes, with often little difference in diameter between the axes and the branchlets. . . *Bryopsis* (p. 100)

7. Filaments simple or bearing short rhizoidal branches. 8
7. Filaments usually many times branched. 9

   8. Filaments often bearing short rhizoidal branches and sometimes having 'knee-like' joints; cells generally cylindrical. . . . . *Rhizoclonium* (p. 89)
   8. Filaments always simple and without 'knee-like' joints; cells cylindrical to barrel-shaped. . . . . *Chaetomorpha* (p. 74)

9. Branchlets normally having a distinct basal cross wall; cell division not segregative; chloroplasts almost invariably distinctly net-like. 10
9. Branchlets commonly without a basal cross wall; cell division segregative; chloroplasts initially net-like and later becoming discoid. 11

   10. Branches arising oppositely or several borne on a single cell arranged in a fan-like manner; compound branches alternating wiith simple ones. . . . . *Willeella* (p. 93)
   10. Branches usually radially arranged and rarely more than three arising from a single cell; no regular alternation of simple and compound branches. . . . . *Cladophora* (p. 82)

11. Basal cell becoming subdivided by a number of irregular walls. . . . . *Siphonocladus* (p. 97)
11. Basal cell persisting. 12

12. Branching of each successive order arising in distinct whorls; cells large and macroscopic. . . . . *Ernodesmis* (p. 97)
12. Branches radially arranged or divided in only one plane, occasionally with some branches irregularly whorled; cells small and microscopic. 13

13. Branches free from one another and the branchlets often arising in unilateral series. . . . . *Cladophoropsis* (p. 96)
13. Branches often attached to one another to occasionally form a net-like structure and the branchlets either radially or bilaterally arranged. 14

14. Branches radially arranged and the branchlets fused to one another in an irregular manner by tentacular cells to form intricately woven, spongy tufts . . . . *Boodlea* (p. 94)
14. Branches divided in one plane and the branchlets fused directly to one another at regular intervals to form a net-like structure. 15

15. Plants having a somewhat pyramidal or triangular outline; branchlets arising terminally on a large stalk-like basal cell. *Struvea* (p. 100)
15. Plants having a somewhat indefinite outline; branchlets and the basal cell similar in size and appearance. . . . . *Microdictyon* (p. 92)

16. Plants consisting of thick spongy branches. . . . . *Codium* (p. 110)
16. Plants not as above. 17

17. Plants having a well-developed creeping base bearing erect branches at intervals . . . . *Caulerpa* (p. 104)
17. Plants not so differentiated. 18

18. Plants many times branched and only rarely simple; thallus hollow and tubular, at least in part. . . . . *Enteromorpha* (p. 63)
18. Plants membranaceous, simple or divided usually into narrow segments; thallus distromatic and with no space between the cell layers. . . *Ulva* (p. 69)

PHAEOPHYTA

1. Plants forming lobed or irregular crusts. 2
1. Plants of another form. 4

2. Crust readily removed from the substratum; thallus parenchymatous with the medulla a single layer of tall cells and the cortex of several layers of cells . . . . *Lobophora,* prostrate form (p. 144)
2. Crust very firmly affixed to the substratum; thallus essentially filamentous consisting of erect filaments arising from one or more basal layers. 3

3. Plants blackish, soft and slimy when wet; erect filaments loosely adherent and readily separated on squashing. . . . . *Basispora* (p. 122)
3. Plants dark brown and leathery; erect filaments firmly adherent and almost impossible to separate from one another. . . . . *Ralfsia* (p. 123)

4. Plants soft, gelatinous and worm-like. . . . . *Levringia* (p. 125)
4. Plants of another form. 5

5. Plants filamentous in structure. 6
5. Plants parenchymatous in structure. 12

6. Filaments mostly multiseriate; growth from a conspicuous apical cell. . . . 7
6. Filaments uniseriate throughout; growth intercalary or trichothallic. 8

7. Lateral branches arising from the upper part of a segment; vegetative reproduction commonly by distinctive propagulae. . . . . *Sphacelaria* (p. 133)
7. Lateral branches straddling the cross wall between two adjacent segments; propagulae not produced. . . . . *Halopteris* (p. 135)

8. Chloroplasts in stellate groups in mature cells; sporangia intercalary or lateral in position. . . . . *Bachelotia* (p. 115)
8. Chloroplasts not as above; sporangia arising terminally or laterally. 9

9. Plants forming minute epiphytic discs consisting of 2 basal cell layers from which arise short erect filaments. . . . . *Hecatonema* (p. 125)
9. Plants not as above. 10

10. Chloroplasts ribbon-shaped, simple or branched and usually a few in each cell. . . . . *Ectocarpus* (p. 116)
10. Chloroplasts more or less discoid or band-shaped and usually many in each cell 11

11. Branches generally arising below the growth zone, often this meristem subtending a long unbranched filament. . . . . *Feldmannia* (p. 118)
11. Branches usually arising above the growth zones which are scattered throughout the plant. . . . . *Giffordia* (p. 120)

12. Plants differentiated into a distinct terete central axis and lateral foliar appendages; air bladders usually present. 13
12. Plants not so differentiated; air bladders absent. 14

13. Foliar appendages filiform or spine-like; air bladders and receptacles formed in the ordinary branches. . . . . *Cystoseira* (p. 154)
13. Foliar appendages leaf-like, with a distinct midrib; air bladders and receptacles lateral and axillary. . . . . *Sargassum* (p. 155)

14. Plants terete, subterete, globose or tubular. 15
14. Plants markedly flattened above and occasionally stipitate below. 18

15. Plants solid throughout. . . . . *Chnoospora* (p. 126)
15. Plants hollow, at least in the upper parts. 16

16. Plants tubular and having terete or slightly flattened branches. . . . . *Rosenvingea* (p. 129)
16. Plants irregularly globose, sometimes becoming flattened and lobed or perforated 17

17. Plant body without apertures, usually thrown into folds. . . *Colpomenia* (p. 129)
pearance. . . . . *Hydroclathrus* (p. 128)
17. Plant body without apertures and usually thrown into folds*Colpomenia* (p. 129)

18. Plants strap-like and many times branched. 19
18. Plants foliaceous or fan-shaped and simple or irregularly cleft. 20

19. Branches having a distinct midrib. . . . . *Dictyopteris* (p. 136)
19. Branches without a midrib. . . . . *Dictyota* (p. 136)

20. Apical or distal margin of the frond incurved. . . . . *Padina* (p. 145)
20. Apical margin of the frond always flat. 21

21. Fronds having distinct concentric zones of hairs on the surface. . . . .
. . . . *Stypopodium* (p. 152)
21. Fronds without zones of hairs. 22

22. Plants fan-shaped in outline and sometimes divided into deltoid to suborbicular lobes; medulla of a single layer of tall cells and cortex of several layers of small cells. . . . . *Lobophora,* erect form (p. 144)
22. Plants somewhat irregular in outline and often divided into cuneate to oblong or linear segments; medulla several cell layers thick and bounded by a single-layered cortex. . . . . *Spatoglossum* (p. 151)

RHODOPHYTA

1. Plants calcified to varying degree. 2
1. Plants uncalcified. 15

2. Plants flexible, either due to uncalcified nodes between heavily calcified segments or to the weak development of calcification throughout the. . . . . . thallus 3
2. Plants inflexible, completely and heavily calcified throughout. 8

3. Plants distinctly and usually regularly segmented; calcification strong at the internodes; conceptacles usually prominent. 4
3. Plants unsegmented or if segmented then only at irregular intervals; calcification light to moderate; conceptacles completely immersed. 7

4. Branching distichous and oppositely pinnate. 5
4. Branching irregular or di- or trichotomous. 6

5. Tetrasporic conceptacles containing 30 to 40 sporangia; cystocarpic conceptacles having a broad fertile area on the floor and a long exit canal. . . . .
. . . . *Corallina* (p. 228)
5. Tetrasporic conceptacles containing 4 to 8 sporangia; cystocarpic conceptacles having a restricted fertile area on the floor and a short exit canal. . . . .
. . . . *Haliptilon* (p. 229)

6. Conceptacles scattered and arising laterally on the segments............. .... *Amphiroa* (p. 224)
6. Conceptacles terminal on the segments.................. *Jania* (p. 230)

7. Branches having an obviously filamentous outer cortex, readily squashed out on a slide after decalcification; cystocarps not surrounded by a pericarp....... .... *Liagora* (p. 182)
7. Branches having a pseudoparenchymatous cortex and not readily disrupted on squashing after decalcification; cystocarps surrounded by a filamentous pericarp.. *Galaxaura* (p. 185)

8. Plants either forming a delicate crust consisting of 1 to 3 layers of cells in the vegetative parts or as a net-work of meandering filaments; never differentiated into distinct strata......................... *Fosliella* (p. 234)
8. Plants always polystromatic and usually differentiated into many distinct strata.. 9

9. Plants usually dark reddish-brown and attached to the substratum by rhizoids; tetrasporangia borne in superficial nemathecia... *Peyssonnelia,* in part (p. 257)
9. Plants often light pink to pinkish-red and attached directly to the substratum; tetrasporangia arising in sunken conceptacles.......................... 10

10. Conceptacles of asexual plants having a number of pores in the roof.... 11
10. Conceptacles of asexual plants having a single pore in the roof......... 12

11. Crust in section having rounded epithallial cells and a noncoaxial hypothallus.. .... *Phymatolithon* (p. 242)
11. Crust in section having angular epithallial cells and a coaxial hypothallus...... .... *Mesophyllum* (p. 240)

12. Cells having secondary pit connections........... *Lithophyllum (p. 236)*
12. Cells without secondary pit connections.......................... 13

13. Heterocysts absent......................... *Pseudolithophyllum* (p. 245)
13. Heterocysts present... 16

14. Hypothallus monostromatic..................... *Hydrolithon* (p. 235)
14. Hypothallus polystromatic.. 15

15. Heterocysts in vertical rows or single................ *Neogoniolithon* (p. 241)
15. Heterocysts in horizontal rows......................... *Porolithon* (p. 243)

16. Plants horizontally expanded and crustose.......................... 17
16. Plants erect, semi-erect or prostrate but never crustose................ 20

17. Plants epiphytic and forming microscopic crusts; cells with no obvious protoplasmic continuity............................... *Erythrocladia* (p. 162)
17. Plants epilithic and forming large macroscopic crusts; cells with obvious protoplasmic continuity.. 18

18. Crust differentiated into distinct strata and attached to the substratum by rhizoids. . . . . *Peyssonnelia,* in part (p. 257)
18. Crust not differentiated into strata and attached directly to the substratum. 19

19. Crust usually 2 cells thick; tetrasporangia borne in superficial nemathecia. . . . . *Rhodophysema* (p. 261)
19. Crust usually several cells thick; tetrasporangia borne in sunken conceptacles. . . . . *Hildenbrandia* (p. 260)

20. Plants rigid and horny, with the main axis covered by short peltate warts arranged in whorls of four. *Thamnoclonium* (p. 256)
20. Plants of another form. 21

21. Plants consisting wholly or in part of one or more fluid-filled cavities. 22
21. Plants otherwise. 26

22. Plants parasitic on the tips of the branches of *Laurencia*. . *Ricardia* (p. 353)
22. Plants free-living. 23

23. Plants consisting of a solid, terete, branched axis bearing a number of vesicles. . . . . *Botryocladia* (p. 266)
23. Plants not so differentiated. 24

24. Plants irregularly constricted into saccate and oblong segments; tetrasporangia tetrahedrally divided. . . . . *Chrysymenia* (p. 267)
24. Plants regularly constricted into cylindrical to barrel-shaped segments; tetrasporangia tetrapartitely divided. 25

25. Branches slightly constricted and a transverse septum at each constriction, with the segments cylindrical to barrel-shaped. . . . . *Champia* (p. 262)
25. Branches deeply constricted and solid at each constriction, with the segments elliptical to barrel-shaped. . . . . *Lomentaria* (p. 264)

26. Cells without obvious protoplasmic continuity between one another. . . . 27
26. Cells with obvious protoplasmic continuity. 29

27. Filaments uniseriate throughout and pseudodichotomously branched. . . . . *Goniotrichum* (p. 161)
27. Filaments often only uniseriate in part and unbranched. 28

28. Plants less than 1 cm in length; filaments uniseriate and if pluriseriate then only towards the base. . . . . *Erythrotrichia* (p. 162)
28. Plants from 2-10 cm in length; filaments uniseriate near the base and pluriseriate above, with the groups of cells arranged in transverse rows. . . . . *Bangia* (p. 163)

29. Plants consisting of a central axis surrounded by an elaborate sac-like network of anastomosing branchlets. . . . . *Dictyurus* (p. 312)
29. Plants of another form. 30

30. Plants largely membranaceous or flattened and somewhat fleshy....... 31
30. Plants not as above... 39

31. Thallus polystromatic throughout.................................. 32
31. Thallus partly monostromatic, leaf-like having a midrib present........... 38

32. Plants appearing parenchymatous throughout....................... 33
32. Plants having an obviously filamentous medulla..................... 34

33. Branches usually less than 1 cm in breadth, often narrowing upwards and the tip sometimes spathulate............................ *Rhodymenia* (p. 270)
33. Branches usually from 2-4 cm in breadth, with the tips acute or rounded....... .... *Gracilaria,* in part (p. 214)

34. Plants of firm consistency; branches with a partial midrib or else having a crenulate margin.............................. *Cryptonemia* (p. 247)
34. Plants of fleshy or gelatinous consistency especially when first collected; branches without a midrib and the margin entire or if crenulate or erose-dentate then the surface usually papillose........................... 35

35. Plants semi-prostrate and dorsiventral, with the overlapping fronds locally fused to one another and attached to the substratum by shortly stalked haptera often arising marginally.............................. *Halichrysis* (p. 268)
35. Plants having the branches free from one another and basally attached...... 36

36. Plants consisting of dichotomously divided cuneate segments; cystocarps borne in surface papillae........................... *Anatheca* (p. 203)
36. Plants simple, irregularly divided or occasionally pinnate, often proliferous; cystocarps immersed in the thallus.......................... 37

37. Fronds often delicate and obovate or suborbicular; cortex consisting of a few layers of cells decreasing in size towards surface.... *Halymenia,* in part (p. 253)
37. Fronds often fleshy, from linear to lanceolate in shape, sometimes with ligulate proliferations; cortex of anticlinal rows of equally sized cells............... .... *Grateloupia,* in part (p. 250)

38. Plants semi-prostrate and having an entire margin; found in the littoral zone usually in brackish-water habitats............... *Caloglossa* (p. 302)
38. Plants erect and often having a margin of small and regular dentations; found only in deep water habitats................. *Hypoglossum* (p. 305)

39. Plants fundamentally monosiphonous................................ 40
39. Plants fundamentally polysiphonous................................ 53
39. Plants otherwise.. 73

40. Plants corticated by a few delicate filaments or a dense mass of interwoven filaments and appearing parenchymatous.......................... 41
40. Plants ecorticate or with an essentially parenchymatous cortication..... 42

41. Branchlets of determinate growth sharply delimited from the branches of indeterminate growth, with the upper branchlets more or less whorled and often terminating in 1 or 2 spine-like cells............... *Wrangelia,* in part (p. 301)

41. Branches not readily delimited as above, with the branches never whorled and generally without spine-like terminal cells...... *Callithamnion,* in part (p. 276)

42. Plants ecorticate..43
42. Plants partially or completely corticated............................52

43. Plants consisting of an axial filament bearing whorls of short lateral branchlets from the upper part of each cell; branchlets close together and completely enclosed by a mucilage sheath.............................*Gulsonia* (p. 294)
43. Plants of another form..44

44. Papilliform cells often borne at the distal end of the upper cells of branchlets..*Dohrniella* (p. 291)
44. Papilliform cells absent..45

45. Plants epiphytic or epilithic and growing as tufts or mats from 1-12 cm in height 46
45. Plants usually epiphytic and often minute, always less than 1 cm in height...49

46. Plants forming soft and bushy tufts, often iridescent; cells of the branches small and microscopic..47
46. Plants forming coarse and often erect tufts; cells large and visible to the naked eye..48

47. Plants markedly delimited into branches and branchlets, with the upper branchlets more or less whorled and often terminating in 1 or 2 spine-like cells....... ....*Wrangelia,* in part (p. 301)
47. Plants not differentiated as above, with the branches not arranged in whorls and generally without spine-like terminal cells.. *Callithamnion,* in part (p. 276)

48. Branches formed laterally and the cells subsphaerical to pyriform, to about 1.5 mm in diameter.............................*Griffithsia* (p. 292)
48. Branches regularly dichotomously divided and the cells cylindrical to pyriform, to about 420(-700) μm in diameter...............*Bornetia* (p. 274)

48. Erect filaments arising from a basal cell, a prostrate system of filaments or a pseudoparenchymatous structure; cells less than 12 μm in diameter........... ....*Audouinella* (p. 166)
49. Erect filaments arising from a prostrate system of filaments; cells always greater than 15 μm in diameter..50

50. Plants attached by elongate holdfasts arising directly from the creeping filaments; sporangia polysporic or tetrasporic (tetrahedrally divided)........ ....*Spermothamnion* (p. 296)
50. Plants attached by rhizoids arising from the basal cell of the branchlets; tetrasporangia only (tetrapartitely divided)..........................51

51. Branchlets spirally arranged on the branches; gland cells absent............. ....*Grallatoria* (p. 291)
51. Branchlets usually arising opposite one another or in whorls on the branches; gland cells present.................................*Antithamnion* (p. 271)

52. Cortication restricted to the nodes of the axial filament; spines absent..... .... *Ceramium* (p. 283)
52. Cortication complete and cortical cells arranged in vertical rows; spines sometimes present at the nodes................... *Centroceras* (p. 282)

53. Plants feather-like, with branchlets distichous and alternate............... 54
53. Plants otherwise.. 55

54. Branchlets arising at intervals of two segments; tetrasporangia one per segment.................................... *Pterosiphonia* (p. 352)
54. Branchlets less closely set together; tetrasporangia several per segment.... .... *Bostrychia,* in part (p. 317)

55. Branches more or less flattened, at least in the ultimate divisions........... 56
55. Branches terete throughout.. 57

56. Branches terminating in 2 or 3 hair-like filaments....... *Taenioma* (p. 308)
56. Branches terminating in a distinct apical cell........ *Platysiphonia* (p. 306)

57. Plants only partly polysiphonous, with at least the ultimate divisions of the branchlets monosiphonous... 58
57. Plants polysiphonous throughout... 64

58. Branchlets monosiphonous and bearing bands of cortical cells confined to the nodes of the axial row of cells...................... *Spyridia* (p. 297)
58. Branchlets monosiphonous and without such cortical bands........... 59

59. Plants radially organized... 60
59. Plants dorsiventrally organized... 63

60. Plants ecorticate................................. *Murrayella* (p. 346)
60. Plants with cortication variously developed......................... 61

61. Cortication consisting of a rhizoid running in a furrow between the pericentral cells... *Lophocladia* (p. 344)
61. Cortication well-developed in the older branches, often rhizoidal and pseudoparenchymatous 62

62. Tetrasporangia arising in the distal portion of little modified branches.... .... *Micropeuce* (p. 314)
62. Tetrasporangia arising in distinct stichidia borne on ramellate branchlets.. .... *Dasya* (p. 310)

63. Branchlets arising alternately from every second segment of a branch; plants of the sublittoral fringe or zone...................... *Heterosiphonia* (p. 314)
63. Branchlets not having such a regular arrangement; plants of the littoral fringe, usually in brackish-water habitats................ *Bostrychia,* in part (p. 317)

64. Branches corticated throughout................................... 65
64. Branches ecorticate... 68

65. Branches beset with numerous, short, stiff and radially arranged branchlets... .... *Digenea* (p. 329)

65. Branches normally stout and branchlets of another form..................66

66. Branchlets somewhat spindle-shaped being contracted towards the base and sometimes tapering at the apex........................*Chondria* (p. 326)
66. Branchlets normally short and spinulose...............................67

67. Branches having from 7 to 9 pericentral cells; branchlets arising in 2 or 4 rows.. ....*Bryothamnion* (p. 324)
67. Branches having 5 pericentral cells; branches and branchlets of determinate growth usually radially arranged....................*Acanthophora* (p. 315)

68. Branches having 3 pericentral cells and each containing a darkly coloured body............................*Asparagopsis* - asexual stage (p. 196)
68. Branches having 4 or more pericentral cells and the cells not containing a conspicuous body..69

69. Plants radially organized and the apices of the branches often straight......70
69. Plants dorsiventrally organized and the apices of the branches often strongly curved...72

70. Tetrasporangia several per segment...........*Bostrychia,* in part (p. 317)
70. Tetrasporangia one per segment...71

71. Plants differentiated into branches and branchlets, with the branchlets often densely clustered; pericentral cells 10 in number...........*Bryocladia* (p. 323)
71. Plants not so differentiated and branches not noticeably clustered; pericentral cells 4 or 10 in number.............................*Polysiphonia* (p. 348)

72. Prostrate branches bearing a characteristic sequence of simple and compound branches; simple erect branches usually straight and sometimes bearing a terminal cluster of hairs..................*Herposiphonia* (p. 330)
72. Prostrate branches having no fixed sequence or arrangement of erect branches; erect branches usually having their apices curved towards the tip of the creeping axis from which they arise, often with a cluster of hairs on the adaxial side of the curve...................*Lophosiphonia* (p. 345)

73. Plants fundamentally multiaxial...74
73. Plants fundamentally uniaxial but appearing parenchymatous especially in the lower parts of the older branches.......................................84

74. Plants erect and clavate, simple or forked and occasionally branches arising in whorls...................................*Corynomorpha* (p. 246)
74. Plants of another form...75

75. Plants soft and gelatinous, worm-like or flattened and somewhat lobed or warty...76
75. Plants otherwise..78

76. Plants consisting of worm-like and di- or trichotomously divided branches. ....*Nemalion* (p. 184)
76. Plants often initially flattened and becoming expanded and sometimes lobed and warty with age...77

77. Gland cells present in the subcortex; nutritive cells not developed in the vicinity of the carposporophyte. . . . . *Nemastoma* (p. 198)
77. Gland cells absent; nutritive cells develop in clusters in the vicinity of the carposporophyte. . . . . *Predaea* (p. 199)

78. Plants prostrate or semi-prostrate and intricate, usually less than 4 cm in height. 79
78. Plants normally erect and rarely intricate, generally greater than 4 cm in height or length. 80

79. Branches constricted at intervals to form terete or somewhat compressed segments; branching di- or trichotomous. . . . . *Catenella* (p. 200)
79. Branches terete throughout and not constricted to form segments; branching irregularly dichotomous and pinnate or secund near tips. . . . . *Gigartina* (p. 223)

80. Branches terete and only slightly compressed just below each dichotomy; cortical cells very compact and not evidently filamentous. . . . . *Halymenia,* in part (p. 253)
80. Branches terete or compressed; cortex and medulla of obvious filamentous construction 81

81. Branches usually somewhat compressed and irregularly or more often pinnately divided. . . . . *Grateloupia,* in part (p. 250)
81. Branches terete throughout and dichotomously or more rarely alternately divided 82

82. Branches sparingly alternately divided and with a number of softer lateral branches present. . . . . *Solieria* (p. 204)
82. Branches regularly dichotomously divided and lateral branches absent. . 83

83. Outer cortical layer consisting of large, swollen hyaline cells and anticlinal rows of assimilatory cells. . . . . *Pseudogloiophloea* (p. 191)
83. Outer cortical layer consisting only of large, swollen assimilatory cells. . . . . *Scinaia* (p. 191)

84. Branches markedly compressed, at least in the upper parts. 85
84. Branches terete or occasionally subterete. 91

85. Branches and branchlets having obtuse apices terminating in a pit containing a tuft of hairs. . . . . *Laurencia,* in part (p. 333)
85. Branchlets and branchlets usually having acute apices and so never terminating in a pit. 86

86. Cortex consisting of anticlinal rows of cells; tertasporangia in swollen nemathecia. . . . . *Gymnogongrus,* in part (p. 219)
86. Cortical cells not arranged in obvious anticlinal rows; tetrasporangia embedded in unmodified cortical tissue or sometimes in stichidia. 87

87. Medulla consisting of large colourless cells; tetrasporangia zonately divided. 88
87. Medulla consisting of colourless cells and small interlacing filaments; tetrasporangia tetrapartitely divided. 89

88. Lateral branches 2 or 3 times pinnately divided and arising almost opposite spine-like branchlets, with the incurved margin of the ultimate pinnae bearing 2 or 3 subulate teeth. . . . . *Plocamium* (p. 212)
88. Lateral branches 1 or 2 times pinnately divided and without spine-like branches or subulate teeth. . . . . *Rhodophyllis* (p. 205)

89. Branches thick and fleshy and usually greater than 5 mm in width; apical cell prominent and surrounded by rudimentary hairs. . . . . *Waldoia* (p. 353)
89. Branches not as above, often thin and less than 5 mm in width; apical cell not prominent and without hairs. 90

90. Thick-walled refractive filaments largely confined to the cortex; cystocarps having 2 pores. . . . . *Gelidium,* in part (p. 175)
90. Thick-walled refractive filaments lying chiefly in the central medullary region; cystocarps having a single pore. . . . . *Pterocladia* (p. 180)

91. Refractive filaments largely confined to the cortex. . . *Gelidium,* in part (p. 175)
91. Refractive filaments absent from the medulla and the cortex. 92

92. Branches and branchlets having obtuse apices terminating in a pit containing a tuft of hairs. . . . . *Laurencia,* in part (p. 333)
92. Branches and branchlets usually having acute apices or if obtuse then without an apical pit. 93

93. Plants soft and plumose, with the older branches often subtubular. . . . . *Asparagopsis* - sexual stage (p. 196)
93. Plants of firmer consistency and never plumose; branches never tubular. 94

94. Plants having spine-like, stellate or sometimes awl-shaped branchlets. . . 95
94. Plants not as above. 96

95. Plants composed of a compact cortex surrounding a much looser tissue. . . . . *Caulacanthus* (p. 203)
95. Plants composed of compact cells throughout. . . . . *Hypnea* (p. 205)

96. Cortical cells arranged in anticlinal rows. . . *Gymnogongrus,* in part (p. 219)
96. Cortical cells not obviously arranged in rows. 97

97. Branches often very variable in width, usually greater than 1 mm in diameter or width; spermatangia borne in vase-shaped depressions in the surface of the thallus. . . . . *Gracilaria,* in part (p. 214)
97. Branches of more or less constant diameter or width, usually less than 1 mm; spermatangia borne directly on the surface. 98

98. Growth from a single apical cell. . . . . *Gelidiella* (p. 172)
98. Growth from a row of apical cells. 99

99. Plants more of less prostrate and intricate, with the filiform branches irregularly divided and often fused to one another. . . . . *Wurdemannia* (p. 181)
99. Plants normally erect and the terete to subterete branches free from one another, sometimes branches arising opposite each other so giving the frond a trifid appearance or if many developing close together then having a palmate appearance. . . . . *Gelidiopsis* (p. 174)

CYANOPHYTA

1. Plants consisting of solitary or groups of trichomes. 2
1. Plants consisting of unicells or colonies of unicells, sometimes forming a pseudofilamentous colony but never as true trichomes. 10

 2. Heterocysts present. 3
 2. Heterocysts absent. 4

3. Terminal cell of trichome initially hemispherical, becoming blunt-conical or cylindrical with the tip rounded. . . . . *Calothrix* (p. 369)
3. Terminal cell of trichome hemispherical, becoming more or less acute-conical. . . . . *Scytonema* (p. 370)

 4. Trichomes branched, with one lateral branch short and the other long and terminating in a hair. . . . . *Mastigocoleus* (p. 370)
 4. Trichomes always simple. 5

5. Trichomes spirally twisted; cells with the cross walls not readily discernable. . . . . *Spirulina* (p. 368)
5. Trichomes not regularly twisted, usually straight; cells with the cross wall clearly visible. 6

 6. A layer of granules absent, or at the most 2 granules found on either side of each cross wall. 7
 6. A layer of several granules present on either side of each cross wall. 9

7. Distal wall of the apical cell becoming thickened. . . . . *Oscillatoria* (p. 363)
7. Distal wall of the apical cell remaining thin. 8

 8. Trichomes not narrowing towards the tip, or if narrowing then only involving the apical cell. . . . . *Schizothrix* (p. 365)
 8. Trichomes attenuate towards the tip and involving several cells. . . . . *Porphyrosiphon* (p. 364)

9. Distal wall of the apical cell becoming thickened. . . . . *Microcoleus* (p. 361)
9. Distal wall of the apical cell remaining thin. . . . . *Arthrospira* (p. 360)

 10. Plants differentiated into a base and an apex, with the cells arranged in vague rows. . . . . *Entophysalis* (p. 359)
 10. Plants not so differentiated. 11

11. Cells ovoid to cylindrical before division, longer than broad, dividing in a plane perpendicular to the long axis of the cell. . . . . *Coccochloris* (p. 357)
11. Cells spherical, ovoid, discoid, cylindrical or pyriform before division, never dividing in a plane perpendicular to the long axis of the cell. 12

 12. Cells arranged in a linear series; cell division in one plane. . . . . *Johannesbaptistia* (p. 358)
 12. Cells irregularly arranged in a gelatinous matrix or in rows; cell division in 2 or 3 planes. 13

13. Cells irregularly distributed in a matrix or arranged in a series of rows; cell division in 3 planes perpendicular to one another. . . . . *Anacystis* (p. 355)
13. Cells always regularly or irregularly arranged; cell division in 2 planes perpendicular to one another. 14

  14. Plants forming a flat or curved plate; cells arranged in a regular series of rows perpendicular to one another. . . . . *Agmenellum* (p. 355)
  14. Plants spherical to ovoid, sometimes tuberculate; cells regularly or irregularly radially arranged. . . . . *Gomphosphaeria* (p. 358)

APPENDIX II

Features Used in Classification and Identification

The algae constitute a very heterogeneous assemblage of plants whose primary classification into a number of divisions is based on criteria such as photosynthetic pigments, assimilatory products or food reserves, flagellation, cell wall components, and certain details of cellular structure. At the lower levels of classification such as at the family, genus and species levels, details of vegetative structure and the organs and processes of reproduction are particularly important. The species treated in this flora include representatives of the four main divisions of largely benthic or attached macroscopic algae found along the tropical West African coast and offshore islands, namely the Chlorophyta (green algae), Phaeophyta (brown algae), Rhodophyta (red algae), and the Cyanophyta (blue-green algae).

Plants as a whole may be divided into two broad groupings based on internal cellular organization. The Procaryota have cells in which the only true membrane is the outer one and hence there is no clear distinction of internal structures or definite organelles; this group includes not only the blue-green algae but also bacteria and viruses. The remainder are included in the Eucaryota which have a limiting membrane about every organelle within the cell; this group includes all the higher plants and all the other algae. The blue-green algae have little internal differentiation when viewed under the light microscope, lacking definite organelles such as chloroplasts and nuclei. The chlorophyll-containing bodies in the eucaryotic algae are termed chloroplasts but the green coloration may be masked by accessory pigments such as phycocyanin and phycoerythrin (red algae), or fucoxanthin (brown algae). The chloroplasts can have diagnostic value at the generic and species level. Their position within the cell, whether located towards the periphery (parietal) or in the centre of the cell (axial), and numbers and shape (cup-shaped, discoid, rod-like, net-like, etc.), may have taxonomic significance. Another cellular character that may be of some importance is the presence or absence and number of bodies often found in the chloroplasts and known as pyrenoids. Such bodies are associated with food reserves and are readily recognised after appropriate staining. The cell wall itself may provide features useful in classification. For example, in the genus *Laurencia* the presence or absence of lens-like thickenings in the walls of the medullary cells or obvious pores between adjacent cells, is of taxonomic importance.

The four divisions under consideration show a wide range in vegetative structure, but usually produce a unicellular stage at some time in their life histories. The attached blue-green algae usually consist either of a well-defined colony of cells embedded in a mucilaginous matrix or of a trichome made up of a linear row of cells which may be sheathed or naked. The shape of the cells, the plane or planes of cell division, and the form of the sheath surrounding groups of cells, are all useful characters for distinguishing genera and species in this division. In those blue-green algae consisting of trichomes the presence or absence of spores or distinctive cells known as heterocysts, and the relative position of these when present, together with the shape of the trichome, are characters of taxonomic importance.

The three other divisions have filamentous forms consisting of simple or branched linear rows of cells, often differentiated into erect and prostrate systems. In the red algae, all thalli have a structure that is essentially filamentous, even the large and elaborate forms being pseudoparenchymatous. There are two basic types of thallus construction in this division and these are used in classification. The uniaxial type has a single main or axial filament, and the multiaxial type has many axial filaments often in a bundle. The simplest case of the former type is a single row of cells (monosiphonous) whilst more elaborate whorls of lateral branches may develop (see *Gulsonia*), which when highly developed may form a continuous outer layer or cortex. The downgrowth of small filaments can bring about the cortication of the central axial filament to varying degrees (see *Wrangelia*). In some cases the original filament or siphon is surrounded by a ring of large cells (polysiphonous), with or without the development of an external cortical layer of smaller cells (see *Polysiphonia, Herposiphonia*). Uniaxial plants consisting of well-developed thalli, whether terete, subterete or obviously membranaceous, may be recognised as such by the presence of a single apical cell. Similarly varying degrees of elaboration occur in the multiaxial types, but each of the axial filaments terminates in its own apical cell. The axial filaments in some cases are relatively distinct and bear laterals arising from the central filaments (see *Liagora, Predaea*). In very elaborate types, secondary filaments develop which obscure the original axial filaments and the laterals that are produced may coalesce to form a more or less compact cortex differentiated into different cell layers (see *Galaxaura, Cryptonemia*). There are some genera (see *Champia, Chrysymenia*) in which the thallus is in the form of a hollow cylinder with the axial filaments running separately along the cavity and their lateral branches coalescing to give rise to a compact cortex. Cell division in more than one plane in some brown and green algae results in a truly parenchymatous construction which in the former division can give rise to large and very elaborate plant bodies. The siphonaceous type of structure is found in certain groups of green algae in which the vegetative thallus is not divided into cells and is multinucleate. The thallus may consist of a single non-cellular unit or it may be made up of many coenocytic and siphoneous filaments forming an elaborate pseudoparenchymatous structure (see *Codium*). A number of morphological features also have diagnostic value such as the type of branching (alternate, irregular, dichotomous), the shape and relative lengths of the branches, the form of the attachment organ (rhizoids, discoid holdfast), the dimensions of the organs, the presence of calcification and sometimes its distribution, and the presence or absence of zones of hairs.

One of the things that makes for difficulties in algal taxonomy is that the algal plant body, unlike that of many of the higher plants, is usually of simple form with few definite structures. Furthermore, these forms themselves can show great plasticity in response to environmental variations. Thus completely unrelated entities may often appear superficially alike. As already indicated, the internal structure may give valuable clues to the real position of many algae but even more useful for this purpose are the reproductive structures. It is generally accepted that in the evolution of plants, reproductive structures have been more conservative than vegetative structures and therefore give a better indication of phyllogenetic relationships. In addition, their shapes and dimensions show less variability than those of vegetative organs and give the taxonomist something more definite to make use of. On the other hand it must be pointed out that taxonomists have often separated plants — especially red algae — on such obscure details of their reproductive processess that the amount of research required to make a determination using such characters virtually rules out their use as a practical possibility. It should also be mentioned that large-scale classification on the basis of reproductive structures has often been carried out on the assumption that groups of plants with similar morphological characteristics must have reproductive processess similar to the very few representatives in which such processess have been thoroughly worked out. Recent investigations have frequently shown that these assumptions were unjustified and that the variety of reproductive processess is greater than originally supposed.

Reproductive processess do not always involve special structures. In some of the unicellular algae, for instance, reproduction may be simply by cell multiplication and in many algae, including even elaborate forms, fragments of the thallus may give rise to new plants. Such reproduction by fragmentation may be a chance occurrence or a more definite part of the reproductive cycle as in the case of some of the blue-green algae which produce short mobile lengths of trichome called hormogonia. Even in the course of such asexual reproductive processess, however, definite structures may sometimes occur as in the case of *Sphacelaria* which has specially modified branches known as propagules, the features of which are quite distinctive and readily permit the separation of species within the genus.

Asexual reproduction is distinguished from vegetative reproduction by the fact that a spore is produced which is released from the parent plant. The spores are produced either in relatively unmodified vegetative cells, or in somewhat distinctive sporangia. In the brown algae two basic types of sporangia are produced depending on whether they contain one or a number of chambers or loculi. The unilocular sporangia have less taxonomic importance than the plurilocular structures which are often termed gametangia although it is not definitely known in most cases whether the liberated spores function as gametes (motile cells which fuse in pairs enabling sexual reproduction to take place) or zoospores (motile cells giving rise directly to new plants asexually). The position, shape, size, and outline of the plurilocular structures provide characters which have been used for distinguishing between taxa (see *Ectocarpus, Giffordia*). The spores in the green and brown algae are usually flagellated but in the red algae they are always non-motile and often produced singly (monospores). Endospores are formed when the entire contents of a cell, or a small portion of the protoplasm, becomes divided into non-motile elements. These spores are characteristic of certain members of the blue-green algae (see Chamaesiphonales). Tetraspores are non-motile spores formed in tetrasporangia, and are usually produced after meosis (the production of four nuclei each containing half the number of chromosomes found in the parent nucleus). They are commonly found in the red algae and possess a number of features that may have taxonomic significance i.e. the type of division of the tetrasporangia (zonate, tetrapartite, tetrahedral), their occurrence on ordinary or specialized branches (stichidia), the possession or otherwise of a cover cell or cells, and as to whether one or more than one arise in a single segment. Tetrasporangia are also to be found in such brown algae as *Dictyota* and *Padina*. Sporangia containing more than four spores may be found in certain red algae and are commonly termed polysporangia.

Sexual reproduction involves the fusion of morphologically similar gametes (isogamy) or dissimilar gametes (anisogamy). The most "advanced" form is oogamy where the large female gamete is non-motile. The nature and position of the male reproductive organs, whether scattered, grouped in sori or within conceptacles, can in some cases have taxonomic importance. The female organ in the red algae consists of the carpogonium, essentially an egg cell with a protuberance of variable length and shape known as the trichogyne. In this division the non-flagellated spermatium becomes attached to the trichogyne and its nucleus fuses with that of the egg cell. The zygote produced is believed to usually give rise by direct division to carpospores in the Bangiophyceae, but in the Florideophyceae development is far more complex. Here the fertilized carpogonium develops into the carposporophyte which shows variations from group to group and has great significance in classification, especially at the family level.

It is not possible to give details here of the variations and complications of the process of development of the carposporophyte, but a number of the recent text books on the algae should be consulted for an account of at least the more general features (e.g. Dixon, 1973; Bold and Wynne, 1977). The usual end result of the process in the Florideophyceae is the production of filamentous outgrowths known as gonimoblasts from the carpogonium, or from other cells sometimes remote from the position of the original carpogonium. These filaments produce carpospores at their apices. In some groups the carposporangium may be enclosed by sterile tissue called the pericarp. The whole structure is referred to as the cystocarp and many of its features may have significance in distinguishing between taxa at different levels. For example, the shape of the cystocarp, the presence or absence of a pore, the shape of the gonimoblast, the presence or absence of connecting filaments between the gonimoblast and the pericarp, whether or not it is immersed in the thallus, are all useful features.

The life history of an alga, which is the sequence of phases through which it passes during its growth, frequently involves the alternation of generations between a gametophyte (sexual) and a sporophyte (asexual) plant. There is considerable variation in the complexity of algal life histories though most commonly the haploid gametes from the sexual parent fuse to form a diploid asexual generation. The diploid sporophyte plant produces spores after meiosis which once again grow into the gametophyte generation. The two generations may be similar in form as in *Ulva* and *Dictyota,* in which case no special taxonomic problems arise, but where the two generations are very different in size and morphology they may be mistaken for entirely separate species and given separate names as has happened in the genus *Galaxaura.* An even more striking example is where the two generations are so different that they have originally been placed in separate genera and only linked together after cultural studies have shown them to be stages in a single life history, as has been found with *Asparagopsis* (sexual stage) and *Falkenbergia* (asexual). Where such links can be definitely established the taxonomist must decide which name has priority. Similar complications may arise where the male and female sexual plants are separate and have morphological differences.

In the Florideophyceae the life histories are very complicated and there are many variations making it difficult to generalize. There are often three phases in the life history of members of this group, namely gametophyte, carposporophyte, and sporophyte. After fertilization the carposporophyte develops parasitically on the gametophyte and produces carpospores which are shed and develop into separate sporophyte plants. On maturity these sporophyte plants produce tetraspores that in turn give rise to the gametophytic plants. In the brown algae the Fucales have a large sporophyte which is often very elaborate. Here no gametophyte is produced and gametes are the only haploid stages in the life history fusing to give rise once again to the sporophyte.

APPENDIX III

Collection, Examination and Preservation of Specimens

Collecting of shore algae is obviously limited by the character and period of the tides. Local tide tables should be consulted beforehand in order to ascertain the time of low water and the period of the month when the tides are lowest (Spring tides). Ideally, it is best to arrive at the shore about 2 hours before the time of the predicted low tide and to work the shore from the high to the low levels as the water recedes. The small tidal range and the more or less continuous heavy wave action action along much of the open coast of West Africa makes collecting often difficult and somewhat hazardous. The most favourable time of the year for collecting in most parts of the region is over the major dry season when the sea is calmer, the climate comparatively stable, and the lower low water in any 24 hours is during the day. This is also the best time for subtidal collecting as the sea is not only comparatively calm but underwater visibility is best due to the low rainfall over the period and hence reduced discharge of turbid lagoon and river water.

A certain amount of simple equipment is necessary for shore collecting. This includes a number of variously sized containers ranging from small specimen tubes to large open-mouthed jars, preferably of polythene so as to avoid breakage. The containers may be carried in a strong polythene bag or a plastic bucket with a fitted lid. General collections of seaweeds including portions of the algal turf, may be placed in small polythene bags whilst small, rare or fertile plants may be put into specimen tubes or jars. A pair of forceps is necessary for collecting many of the very small and more delicate algae. For removing crustose or small prostrate algae a hammer or cold chisel are useful tools for obtaining the substratum along with the adhering specimen. A knife or other sharp instrument can be used for scraping rock surfaces or cutting the holdfasts of larger algae.

A small field-collection notebook is necessary for recording field data. Pertinent information should include the name of the locality, collecting date, the type of habitat, the nature of the substratum, the amount of wave action, the dominant species at various levels on the shore, and other data which might be thought to be useful. It is advisable to immediately label each collection giving it a code number so as to avoid any confusion when sorting back in the laboratory. No water should be added to the containers as algae tend to decay more rapidly when floating in water. The containers should be sealed, and the bucket in which they are carried should contain seawater to minimize the overheating that is liable to occur as collecting proceeds. High temperatures will not only readily damage some of the more delicate plants but will also lead to rapid decay. If the plants have to be kept for any length of time before they are to be examined, then it is advisable to store them in a cold box or a refrigerator.

A diving mask or glass bottomed bucket may be used for looking for subtidal algae which may sometimes be found in some of the larger and deeper tidepools at lower levels on the shore. Shallow water collecting can be achieved by snorkel diving whilst deeper water sampling is only possible by dredging or SCUBA (self-contained underwater breathing apparatus) diving. When collecting underwater the containers may be conveniently carried in a fine mesh draw-string bag. Small containers such as specimen tubes should be filled with seawater so as to prevent them from readily floating away or buoying up the draw-string bag. The wide measuring scale on capillary depth gauges makes them particularly useful for accurately recording the depth of each collection. Information regarding subtidal collections can be recorded *in situ* by writing with a wax crayon or a lead pencil on a pad of roughened plexiglass or on a slate.

An attempt should always be made to collect only those plants which appear to be healthy and mature. A number of specimens of a particular species are always more useful then just a single one which might be atypical and so unrepresentative. Drift specimens may be gathered either when still floating in the sea or after having been cast-up on the beach. These plants may often be partly bleached and decayed, nevertheless, their study might be highly rewarding as they may be species that only grow in deep water.

Miscellaneous collections of seaweeds may be sorted back in the laboratory in shallow basins or white enamelware trays filled with seawater. Large individual plants and groups of smaller ones may be transferred to smaller containers such as basins, glass petri dishes or specimen tubes, remembering to keep the code number given on the shore with the relevant collection. These specimens may now be observed with a lens or under the low power of the microscope and further divided should a number of smaller plants be present in a single collection.

The most satisfactory method for preserving algae permanently is to keep them as dried herbarium mounts. Nevertheless, it is always useful and especially so in the case of unusual or rare plants, to have both dried and pickled collections which might need to be morphologically or cytologically investigated at a later date.

For preparing herbarium mounts a collection of variously sized sheets of good quality paper is essential, preferably a medium weight herbarium paper. This all rag paper is usually supplied as standard sized sheets of 29.5 cm by 45.0 cm. The larger specimens for mounting should be washed immediately prior to drying in freshwater to remove inorganic material as well as salt. Very delicate specimens may burst their cells when placed in freshwater and so are better mounted directly in seawater. Specimens selected for mounting should be supported by an appropriate sized sheet of herbarium paper in a broad, shallow dish or tray of water. The branches should be carefully displayed in characteristic and orderly arrangement using a brush, needle, or water from a pipette. If the specimen is too thick or bushy then a few branches may be removed though not enough to obscure the characteristic habit of the plant. The herbarium paper should itself be supported in the dish by a metal sheet or, if the mounting paper has been cut into a smaller sheet, by the hand. The support should be carefully raised so as to allow the water to drain slowly away without interfering with the arrangement of the specimen.

The mounting sheet bearing the specimen should be placed on a dry blotter in a plant press and the specimen covered with cloth such as muslin, mosquito netting or nylon stockings, alternatively waxed paper may be used to prevent the specimen sticking to the drying paper. The mounting sheet with the specimen or specimens thus covered is now placed between blotters or newspapers and a label bearing the collection code number is also included. The press can now be closed and the drying papers replaced preferably once a day until the specimens are dry. Air circulation through the press may be increased by placing corrugated cardboard at intervals between the drying papers so as to facilitate drying. Various driers described in most works on herbarium techniques may also considerably speed up the drying process.

Most algae stick to herbarium paper naturally but coarse wiry plants may have to be glued or tied down with cotton thread or alternatively placed in small envelopes and appended to herbarium sheets. Very small algae may be mounted on small white cards or thin mica sheets. Coralline algae should be dried with very little pressure applied as they readily disintergrate. They, and the larger calcareous crustose forms, may be kept in boxes or envelopes. Each sheet, envelope or box should bear a label containing the name of the specimen, the date of collection, the locality, the habitat, the collectors name and the person who made the identification. On an herbarium sheet the label should be fixed to the lower left hand corner. Dried specimens are more or less permanent and in the dark the plants retain their colour almost indefinitely and are not usually subject to attacks by animals such as herbarium beetles.

Algae may also be preserved in liquids such as 3- to 4-percent formaldehyde in seawater or 70 percent alcohol. Borax should be added to formaldehyde to make it alkaline otherwise it brings about the gradual dissolution of the calcareous matrix of some algae. Specimens pickled in this way require more attention than dried mounts. To cut down on storage space and minimize attention the pickled specimens may be kept in tubes plugged with cotton wool and these stored in large glass or plastic jars which only require occasional topping-up with water. When collecting far from the home base it may be necessary to transport large numbers of specimens over long distances. This may be conveniently undertaken by placing numbered collections into suitably sized bags of net, muslin or perforated polythene, and packing these tightly into tins containing the liquid preservative, or, more simply, by draining the collections after fixing in the preservative and then putting them in polythene bags sealed with rubber bands.

Most algae can only be identified with any certainty if they are examined and sometimes measured under the microscope. This may sometimes necessitate examining sectioned, squashed, or teased material. Many dried specimens can be quickly sectioned by shaving them with a razor blade or scalpel on a microscope slide, and gathering the resulting sections up in dilute detergent solution or 50 percent alcohol which softens the material and allows the cells to swell and once more take on a life-like appearance. This method of shaving material on a glass slide can also be used with most larger and tougher plants whether freshly collected or preserved in formaldehyde. The freezing microtome is a useful instrument for obtaining in a short time a number of good sections of predetermined thickness. Before calcareous algae can be sectioned they have to be placed in a decalcifying solution such as Perenyi's solution (4 parts of 10 percent nitric acid, 3 parts of 70 percent alcohol, 3 parts of 0.5 percent chromic acid) for 4 to 6 hours in the case of thin portions of thallus and several days if very thick. For most purposes of identification, staining of the material is not necessary but if required details of suitable stains and techniques, as well as methods for preparing permanent mounts of algae, may be found in a number of publications on laboratory techniques such as Johansen (1940) and Purvis *et al.* (1966). A comparatively simple method of making more or less permanent slides is to mount the material in a few drops of 50 percent corn syrup, often known commercially as Karo, diluted with boiled or freshly distilled water and with a small drop of formaldehyde added. The slide may be left overnight to thicken and 80 percent corn syrup is now added. A cover slip may now be careful placed onto the drop of syrup which eventually dries forming a seal. Material mounted in this way should previously be soaked in formaldehyde/seawater.

APPENDIX IV

Glossary

Abaxial - on the side or face away from an axis or branch.
Acicular - needle-shaped.
Aculeate - with coarse teeth or sharp spine-like emergences.
Acute - ending abruptly in a sharp point; describing an angle of less than a right angle.
Adaxial - on the side or face next to an axis or branch.
Aegagropilous - free and more or less spherical.
Akinete - a thick-walled spore derived from a vegetative cell.
Anastomose - to join by cross connections.
Anisogamy - a union of motile gametes consistently different in size.
Anticlinal - perpendicular to the surface of a structure.
Annulate - ring-like.
Antheridium - a gametangium containing motile male gametes in oogamous sexual reproduction.
Antherozoid - a flagellated male gamete.
Apiculate - with a short and sharp point.
Aplanosporangium - a cell or sporangium that contains aplanospores.
Aplanospore - a non-motile spore with a sporangial wall.
Arcuate - bent or curved.
Articulated - segmented or jointed in appearance.
Articulation - the uncalcified joint between the segments in some members of the Corallinaceae.
Assurgent - curving obliquely upwards.
Attenuate - narrow and gradually tapering.
Auxiliary cell - a specialized cell in the red algae (Florideophyceae) that receives the fusion nucleus or its division product.
Axial - relates to the main or morphological axis.
Axil - angle between axis and an organ.
Basionym - the original name of a taxon now used in a new combination.
Benthic - bottom-living; attached to or resting on the substratum.
Biflagellate - with two flagella.
Bispore - one of two spores produced by division of a sporangium.
Brackish-water - water with salinity intermediate between that of oceanic water and true freshwater.
Branchlet - a small secondary branch.
Calcareous - containing large amounts of calcium carbonate.
Calcified - encrusted or impregnated with calcium carbonate.
Carpogonium - the female sex organ in the red algae containing the egg.
Carposporangium - a sporangium produced on a carposporophyte and containing carpospores.

Carpospore - a red algal spore arising as a result of fertilization.
Carposporophyte - a multicellular phase produced following fertilization in some red algae (Florideophyceae) and producing gonimoblast filaments and carposporangia.
Cartilaginous - firm and gristly.
Catenate - end to end as a chain.
Cervicorn - antler-like.
Chloroplast - an organelle containing the green pigment chlorophyll, applied regardless of whatever auxiliary pigments are also present. Formerly such bodies as were coloured other than green were called chromatophores (brown algae) or rhodoplasts (red algae).
Clavate - club-shaped.
Coaxial - parallel with axis, or having a common axis.
Coccoid - unicellular and rounded.
Coenocytic - containing many nuclei and without transverse walls.
Conceptacle - an invagination or cavity within a thallus containing reproductive organs.
Conspecific - belonging to the same species.
Coralline - articulated or jointed, calcareous red alga.
Cordate - heart-shaped.
Cortex - the outermost layer or layers of cells.
Corticate - with a cortex.
Corymb - a raceme with the younger branches shorter towards the top giving a flat-topped cluster.
Costate - with a rib.
Crenate - scalloped or toothed with rounded notches.
Crenulate - finely crenate.
Crustose - crust-like.
Cryptostoma - a sunken cavity containing sterile hairs.
Cuneate - wedge-shaped.
Cymose - with the oldest branch at the apex and growth continued by lateral branches to give a usually obconical outline.
Cystocarp - in the red algae, carposporophyte and carpospores together with any protective envelope or pericarp provided by the parent female plant.
Deciduous - falling off naturally.
Decumbent - prostrate and curving upwards.
Decussate - in pairs alternating at right angles.
Deltoid - shaped like an equilateral triangle.
Dentate - toothed.
Determinate - of limited growth.
Diaphragm - a cellular partition across a hollow thallus.
Dichotomous - bifurcate; forking into two equal or unequal (irregularly dichotomous) branches.
Diffluent - readily becoming liquid.
Dioecious - having male and female organs borne on separate individuals.
Diploid - having twice the basic (haploid) number of chromosomes.

Discoid - disc-like.
Distichous - arranged in two rows or ranks on opposite sides of an axis.
Distromatic - of two cells in thickness.
Divaricate - extremely divergent; widely spreading.
Dorsiventral - in algae, with reference to plants having morphologically distinct upper and lower surfaces.
Ecophene - an ecological growth form.
Ecorticate - without cortication.
Emarginate - indented at the apex.
Endogenous - arising internally.
Endophytic - growing within the tissues or sheath of another plant.
Endosporangium - a cell containing endospores (blue-green algae).
Endozoic - growing within part of an animal.
Entire - with a smooth or continuous margin.
Epibiotic - living on the surface of plants or animals.
Epilithic - living on rock surfaces.
Epithallus - outermost layer or layers of cells in crustose algae.
Epiphytic - living on the surface of another plant.
Epizoic - living on the surface of an animal.
Eulittoral - zone on the shore between the levels of low and high tides (intertidal).
Excrescence - an outgrowth such as a wart.
Exogenous - arising externally; refering in the algae to a branch arising from an outer cell of the thallus.
Exospore - a spore produced by budding from another cell.
Falcate - sickle-shaped.
Fascicle - a dense cluster or bundle of branches or branchlets.
Fastigiate - having the branches erect, parallel and more or less adpressed.
Filament - plant or branch composed of a single row of cells.
Filiform - thread-like.
Flagellum - a whip-like structure on a motile cell.
Flexuous - bent alternately in opposite directions; zigzag.
Foliaceous - leaf-like.
Forcipate - forked and incurved.
Frond - a leaf-like or erect portion of a thallus.
Fusiform - tapering at both ends; spindle-shaped.
Gametangium - a structure believed to produce gametes.
Gamete - a male or female sexual cell.
Gametophyte - the sexual phase in the life history of a plant.
Glabrous - smooth, without hairs.
Gland cell - special, often highly refractive, cell occurring in red alage and may be secretory or storage in function.
Globose - more or less spherical in shape.
Gonimoblast - a filament bearing a carpospore or carpospores or the entire collection of these filaments forming the carposporophyte (Florideophyceae).
Hair - an elongated colourless cell or filament often projecting from the plant.
Haploid - with a single basic set of chromosomes in each nucleus.

Hapteron - an organ of attachment.
Heterocyst - a distinctive thick-walled cell with clear cytoplasm in a trichome (blue-green algae), or an enlarged hyaline cell in an otherwise homogeneous tissue (red algae).
Hormogonium - a short and often motile section of trichome, forming an organ of vegetative reproduction in some blue-green algae.
Hyaline - thin and transparent; colourless.
Hypothallus - the lowermost layer or layers of cells in a crustose alga.
Indusium - a cover to sporangia.
Intercalary - inserted at a point along the length of a structure.
Internode - a portion between two nodes.
Intricate - intertwined, entangled.
Involucre - a structure protecting reproductive organs.
Involucrate - having an involucre.
Isogamous - having morphologically indistinguishable gametes.
Laciniate - cut into deep narrow divisions.
Lamellate - as thin plates.
Lanceolate - lance-shaped; broadest usually about one-third of the distance from the base and tapering to each end.
Lenticulate - like a biconvex lens.
Life history - the sequence of nuclear and morphological phases of an organism.
Ligulate - tongue-shaped; strap-like and short.
Linear - narrow and several times longer than wide with parallel sides.
Littoral fringe - zone on the shore beyond the tidal range and only influenced by wash, splash or spray.
Locule - a cavity or 'cell', usually in a reproductive structure in the algae.
Locular - having cavities or loculi.
Lubricous - smooth, slippery.
Lumen - the central cavity of a cell.
Macroscopic - visible to the unaided eye; discernable without magnification.
Mammilliform - applied to papillate protruberances.
Medulla - the central region of a thallus, usually colourless.
Membranaceous - thin and often semi-transparent.
Meristematic - describing a region where growth takes place.
Micron, micrometre (μm) - a unit of measurement = 0.001 mm.
Microscopic - discernable only with magnification.
Midrib - a thickened region extending longit udinally down the middle of a flattened branch.
Moniliform - like a string of beads.
Monoecious - having male and female organs on the same individual.
Monosiphonous - having a single row of cells.
Monosporangium - a sporangium producing a single spore.
Monospore - a spore produced in a monosporangium.
Monostromatic - composed of a single layer of cells.
Mucronate - abruptly terminating in a short, sharp point.
Multi - prefix meaning many.
Nemathecium - an elevation of the thallus containing reproductive organs.

Node - the junction between two adjacent cells of a monosiphonous filament, the segments of a polysiphonous filament or between articulations in members of the Corallinaceae.
Ob - prefix meaning inversely or contrarily.
Obtuse - blunt or roundet at the end.
Oogamous - sexual reproduction involving the union of a small motile gamete or non-motile spermatium with a non-motile egg.
Orbicular - orb-shaped; flattened and circular.
Organelle - a specialized structure in a cell.
Ovate - egg-shaped.
Palisade - cells elongated at right angles to the surface and united laterally into a compact tissue.
Palmate - flattened and lobed or divided like the palm of a hand.
Panicle - a loose cluster.
Paraphysis - a sterile branch or filament among sporangia or gametangia.
Parasporangium - a vegetative organ producing without meiosis variable numbers of spores.
Parenchyma - a compact tissue composed of thin-walled cells which are not usually much elongated.
Parietal - adjoining the cell wall, outer or peripheral.
Patent - spreading.
Pectinate - branches or branchlets set close together like the teeth of a comb.
Pedicel - a stalk of a reproductive organ.
Pericarp - a sterile envelope around a carposporophyte (Florideophyceae).
Pericentral cell - a cell surrounding the central cell in a polysiphonous branch.
Perithallus - the cells of the middle layer of a crustose alga and usually developing inwardly from the intercalary meristem.
Phycology - the study of the algae.
Pilose - hairy.
Pinna - the primary branch of a pinnately divided thallus.
Pinnate - with the filaments or branches arising on opposite sides of an axis; feather-like.
Pit connection - a discrete lens-shaped plug within the aperture between two adjacent cells.
Placenta - a large cell usually of fused fertile tissue.
Pluri - prefix for many or several.
Polymorphic - of very variable form or appearance.
Polysiphonous - a plant made up of tiers of cells each consisting of a central cell surrounded by pericentral cells.
Polysporangium - a sporangium producing by meiosis more than four spores.
Proliferous - producing many adventitious branches or branchlets.
Propagulum - a specially modified branch for vegetative reproduction.
Pseudo - prefix signifying false.
Pyrenoid - an organelle usually associated with the accumulation of reserve food materials.
Pyriform - pear-shaped; broader at base than at top.
Quadrate - square.

Quadriflagellate - with four flagella.
Quasi - prefix meaning in a sort, as it were.
Racemose - with the youngest and smallest branches nearest the apex, often conical in outline.
Radial - radiating from an axis or centre.
Ramellate - bearing small branches.
Ramulus - a small secondary branch or branchlet.
Receptacle - a fertile portion of a branch on which are borne gametangia or sporangia.
Refracted - abruptly backwardly bent.
Reticulate - net-like.
Revolute - rolled backwards and usually downwards.
Rhizine - a slender, thick-walled and refractive filament.
Rhizoid - a unicellular or filamentous attachment organ.
Rotund - more or less rounded or circular.
Rugose - wrinkled or ridged.
Secund - arranged in a row along one side.
Segment - one division of a plant organ or a portion between successive nodes of an articulated thallus.
Segregative division - a type of cell division in which the protoplasm of a cell divides and each portion becomes surrounded by a new membrane (green algae, in Siphonocladiales).
Septum - cross wall.
Seriate - in series.
Serrate - saw-toothed; with the teeth pointing upwards towards the apex.
Sessile - borne directly on the thallus, without a stalk.
Simple - unbranched.
Sinuate - with a deep wavy margin.
Siphonaceous - relating to green algae, tubular and not divided into cells, without cross walls.
Sorus - a group or cluster of reproductive structures.
Spathulate - spatula-shaped; oblong with the basal end narrowed.
Spermatangium - a male sex organ of the red algae producing non-motile gametes (spermatia).
Sporangium - a spore-producing structure.
Spore - a cellular agent of asexual reproduction.
Sporophyte - the asexual or sporangial stage in the life history of a plant.
Stellate - star-shaped; radiating from a common centre.
Stichidium - a specialized branch producing tetrasporangia (Florideophyceae).
Stipe - a stem-like region below the frond.
Stipitate - having a stipe.
Stoloniferous - with stolons, long and slender horizontal branches.
Strobilus - a cone-like structure.
Sub - prefix denoting under or below or less than, approaching, etc.
Sublittoral - below the lowest level of the tides.
Subtend - to extend under.
Subulate - awl-shaped.

Synonym - a superseded or unused name.
Taxon - a unit in a classification system.
Terete - cylindrical, rounded in cross section.
Tetrahedral - a term applied to tetrasporangia whose contents are divided obliquely so that only three of the four spores are visible in any one view.
Tetrapartite - a term applied to tetrasporangia divide into four spores by two mutually anticlinal divisions, so that all four spores are always visible.
Tetrasporangium - a sporangium producing by meiosis tetraspores.
Tetraspore - a spore produced in a tetrasporangium (red algae).
Thallus - a plant body not differentiated into a true root, stem, leaf or leaves, etc.
Tortuous - winding irregularly.
Trabeculae - slender strands or bars lending rigidity to a thallus.
Trichocyte - a large hair-bearing cell.
Trichogyne - a receptive protuberance or elongation of a female gametangium to which male gametes become attached (red algae).
Trichome - the living portion of the filament in members of the blue-green algae.
Trichothallic growth - a type of growth in which there is an intercalary meristem at the base of a hair or filament (brown algae).
Trifid - slit into three.
Truncate - blunt-ended; cut-off abruptly.
Tuberculate - beset with knobbly projections or irregularly warty outgrowths.
Turbinate - top-shaped.
Uni - prefix meaning one or single.
Uniseriate - arranged in a single row or series.
Utricle - a large and swollen terminal portion of a filament or portion of a tubular thallus.
Variety - a taxon below the level of a species or subspecies.
Vesicle - a swollen membrane-bounded organelle or structure.
Verrucose - warty or wart-like.
Verticillate - with successive whorls of branches or branchlets along an axis.
Virgate - much branched and long, slender and stiff.
Whorled - arranged in a circle around an axis.
Zonate - divided on parallel planes, usually referring to the type of division in the tetrasporangia where the tetraspores lie in one row.
Zoospore - a flagellated asexual spore produced in a zoosporangium
Zygote - a cellular product produced by the union of gametes.

Bibliography

ABBOTT, I.A. (1976) - On the red algal genera *Grallatoria* Howe and *Callithamniella* Feldmann-Mazoyer (Ceramiales). *Br. phycol. J.*, 11: 143-149.

ABBOTT, I.A. & HOLLENBERG, G.J. (1976) - *Marine algae of California.* pp. XII+827. Stanford, California.

ADEY, W.H. (1970) - A revision of the Foslie crustose coralline herbarium. *K. norske Vidensk. Selsk. Skr.*, 1: 1-46.

ADEY, W.H. & LEBEDNIK, P.A. (1967) - Catalog of the Foslie Herbarium. *K. norske Vidensk. Selsk. Mus. Trondheim, Norge.* pp. 1-92 (mimeo.).

ADMIRALTY TIDE TABLES Volume 2. *Atlantic and Indian Oceans including tidal stream tables Parts I & II.* H.M.S.O. pp. XXXIV + 432.

AGARDH, C.A. (1821-1828) - *Species algarum rite cognitae,...*vol. 1, part 1, pp. IXXXVI + 168. 1821a; vol. 1, part 2, pp. 169-531. 1823. *Id.* vol. 2, section 1, pp. IXXVI + 189. 1828. Gryphiae. Note: There is another version of volume 1 besides that issued at Griefswald (Gryphiae). The parts issued at Lund were dated 1820 (part 1) and 1822 (part 2) and had different title pages to those of the Griefswald issues.

AGARDH, C.A. (1821b) - *Icones algarum ineditae.* (1), 20 pls. Lund.

AGARDH, C.A. (1824) - *Systema algarum,...* pp. XXXVIII + 312. Lund.

AGARDH, J.G. (1841) - In historiam algarum symbolae. *Linnaea,* 15: 1-50, 443-457.

AGARDH, J.G. (1842) - *Algae Maris Medeterranei et Adriatici,...*pp. X+164.

AGARDH, J.G. (1848a) - Nya alger från Mexico. *Öfvers. K. Vetensk Akad. Förh.,* 4: 5-17 (1847).

AGARDH, J.G. (1848-1876) - *Species, genera et ordines algarum, seu descriptiones succinctae specierum, generum et ordinum, quibus constituitur,* I. *Species, genera et ordines Fucoidearum..., algas fucoideas complectens.* Vol. 1, pp. VIII+363. 1848b. *Id.,* II. *Species, genera et ordines Floridearum...*vol. 2, part 1, pp. XII + 351. 1851; vol. 2, part 2, pp. 337-720. 1852; vol. 2, part 3, pp. IV + 701-1291. 1863. *Id.,* III. *Epicrisis systematis Floridearum.* Vol. 3, part 1, pp. VII+724. 1876. Lund.

AGARDH, J.G. (1850) - Algologiska Bidrag. *Öfvers. K. Vetensk Akad. Förh.,* 4: 5-17 (1847).

AGARDH, J.G. (1872a) - Bidrag till Florideernes Systematik. *Acta Univ. Lund,* 8: 1-60.

AGARDH, J.G. (1872-1887) - Till Algernes Systematik. Nya Bidrag. I. *Caulerpa,* II. *Zonaria,* III. *Sargassum, Acta Univ. Lund,* 9: 1-45. 1872. Första afd., IV Chordarieae, V. Dictyoteae. *Ibid.,* 17: 1-134, 3 pls. 1881. Tredfe afd., V. Ulvaceae. *Ibid.,* 19: 1-181, 4 pls. 1882-1883. Femte afd., VIII. Siphoneae. *Ibid.,* 23: 1-174, 3 pls. 1887.

AGARDH, J.G. (1889) - Species sargassorum australiae... *K. Svenska Vetensk-Akad. Handl.,* 23: 1-133, 31 pls.

AGARDH, J.G. (1894-1897) - Analecta algologica... Continuatio I. *Acta Univ. Lund,* 29: 1-144, 2 pls. 1894; Id., Continuatio IV. *Ibid.,* 33: 1-106, 2 pls. 1897.

ALEEM, A.A. (1978) - A preliminary list of marine algae from Sierra Leone. *Botanica mar.,* 21: 397-399.

ALEEM, A.A. (1980a) - The Cyanophyta of Sierra Leone (West Africa). *Botanica mar.,* 23: 49-51.

ALEEM, A.A. (1980b) - Distribution and ecology of marine fungi in Sierra Leone (Tropical West Africa). *Botanica mar.,* 23: 679-688.

ALLEN, J.R.L. (1965) - Late Quaternary Niger delta and adjacent areas: Sedimentary environments and lithofacies. *Bull. Am. Ass. Petrol. Geol.,* 49: 547-600.

ALLEN, J.R.L. & WELLS, J.W. (1962) - Holocene coral banks and subsidence in the Niger Delta. *J. Geol.,* 70: 381-397.

AMBRONN, H. (1880) - Über einige Fälle von Bilateralität bei den Florideen. *Bot. Ztg.,* 38: 193-200.

AMOSSÉ, A. (1970) - Diatomées marines et saumâtres du Sénégal et de Côte d'Ivoire. *Bull. Inst. fond. Afr. noire,* sér. A, 32: 289-311.

ANANG, E.R. (1979) - The seasonal cycle of the phytoplankton in the coastal waters of Ghana. *Hydrobiologia,* 62: 33-45.

ARDRÉ, F. & GAYRAL, P. (1961) - Quelques *Grateloupia* de l'Atlantique et du Pacifique. *Revue algol.,* 6: 38-48, 3 pls.

AREGOOD, C.C. & HACKETT, H.E. (1972) - A new *Dictyurus* (Rhodophyceae - Dasyaceae) from the Maldive Islands, Indian Ocean. *J. Elisha Mitchell scient. Soc.,* 87: 91-96 (1971).

ARENS, G., DELTEIL, J.R., VALÉRY, P., DAMOTTE, B., MONTADERT, L. & PATRIAT, P. (1971) - The continental margin off the Ivory Coast and Ghana [pp. 61-78]. In: Delany, F.M. (ed.), ICSU/SCOR Symposium Cambridge 1970: The geology of the East Atlantic continental margin. Part 4 Africa. *Rep. No.* 70/16, *Inst. geol. Sci.,* pp. IV + 209.

ARESCHOUG, J.E. (1843) - Algarum minus rite cognitarum pugillus secundus. *Linnaea,* 17: 257-269.

ARESCHOUG, J.E. (1854) - *Phyceae Extraeuropaeae Exisiccatae quas distribuit John Ehrh. Areschoug.* Fasc. 2, nos. 31-60, Upsaliae.

ASKENASY, E. (1888) - Algen, mit Unterstützung der Herren E. Bornet, A. Grunow, P. Hariot, M. Moebius, O. Nordstedt bearbeitet [pp. 1-58]. In: Engler, A., *Die Forschungsreise S.M.S. 'Gazelle'* ... IV. *Theil Botanik,* pp. XVI+58 + pls I-XII + 16 + 48 + pls I-VIII + 64 + 20 + pls I-III + 49 + pls I-XV. Berlin (1889). Note: For details of this publication along with dates of issue of various portions see Lawson and Price (1969).

ASKENASY, E. (1896) - Enumération des algues des îles du Cap Vert. *Bolm Soc. broteriana,* 13: 150-175.

BAKUN, A., MCLAIN, D.R. & MAYO, F.V. (1973) - *Upwelling studies based on surface observations. Coastal Upwelling Experimental Workshop Abstracts.* Dept. of Oceanography, Florida State University, Tallahasse.

BALAKRISHNAN, M.S. (1962) - Studies on Indian Cryptonemiales - II. *Coryno-morpha* J. Ag. *Phytomorphology,* 12: 77-86.

BALDOCK, R.N. & WOMERSLEY, H.B.S. (1968) - The genus *Bornetia* (Rhodophyta, Ceramiaceae) and its southern Australian representatives, with description of *Involucrana* gen. nov. *Aust. J. Bot.,* 16: 197-216, 4 pls.

BARBEY, C. (1968) - Africa in the physical world seas and coasts in western Africa [pp. 1-2, pl. 1]. In *International Atlas of West Africa.* OAU Scientific, Technical, and Research Commission. 1-42 pls.

BARTON, E.S. (1897) - Welwitsch's African marine algae. *J. Bot., Lond.,* 35: 369-374.

BARTON, E.S. (1901) - Marine algae [pp. 324-328]. In: Anon., *Catalogue of the African plants collected by Dr. Fredrich Welwitsch in 1853-61.* Vol. 2, part 2. *Cryptogamia.* pp. 261-566. London.

BASSINDALE, R. (1961) - On the marine fauna of Ghana. *Proc. zool. Soc. London,* 137: 481-510.

BATTEN, L. (1923) - The genus *Polysiphonia* Grev., a critical revision of the British species, based upon anatomy. *J. Linn. Soc. (Bot.),* 46: 271-311, 4 pls.

BEAUVOIS, A.M.F.J. Palisot de (1805) - *Flore d'Oware et de Benin, en Afrique.* 1, Livraisons 2 & 3. pp. 9-32, 12 pls. Paris.

BÉLANGER, C. (1834) - *Voyagées aux Indies-Orientales... 1825, 1826, 1828 et 1829. Botanique. IIe Partie. Cryptogamie.* pp. 81-192, 7 pls. Paris. Note: For date of publication see Ross (Taxon, 13: 193-196, 1964).

BERNARD, A. (1937) - *Géographie Universelle, XI. Afrique septentrionale et occidentale.* pp. 1-529. Paris.

BERRIT, G.R. (1961, 1962a) - Contribution à la connaissance des variations saisonnières dans le golfe de Guinée. Observations de surface le long des lignes de navigation. Introduction. *Cah. océanogr.,* 13: 715-727. 1961. Id., Étude régionale. *Ibid.,* 14: 633-643. 1962a.

BERRIT, G.R. (1962b) - Contribution à la connaissance des variations saisonnières dans le golfe de Guinée. *Cah. océanogr.,* 14: 719-729.

BERRIT, G.R. (1969) - Les eaux dessalées du Golfe de Guinée [pp. 13-22]. In: *Proceedings of the Symposium on Oceanography and Fisheries Resources of the Tropical Atlantic, 20-28 Oct. 1966, Abidjan.* UNESCO. pp. 1-430. Paris.

BERTHOLD, G. (1882) - Über die Vertheilung der Algen im Golf von Neapel nebst einem Verzeichniss der bisher daselbst beobachteten Arten. *Mitt. zool. Stn Neapel,* 3: 394-536.

BLACKLER, H. (1964) - Some observations on the genus *Colpomenia* (Endicher) Derbès et Solier 1851. *Proc. Int. Seaweed Symp.*, 4: 50-54.

BLACKLER, H. (1967) - The occurrence of *Colpomenia peregrina* (Sauv.) Hamel in the Mediterranean (Phaeophyta: Scytosiphonales). *Blumea*, 15: 5-8.

BLIDING, C. (1963) - A critical survey of European taxa in Ulvales. Part I. *Capsosiphon, Percursaria, Blidingia, Enteromorpha. Op. bot. Soc. bot. Lund*, 8: 1-160.

BLIDING, C. (1968) - A critical survey on European taxa in Ulvales. Part II. *Ulva, Ulvaria, Monostroma, Kornmannia. Bot. Notiser*, 121: 535-629.

BODARD, M. (1966a) - Les *Gracilaria* et *Gracilariopsis* au Sénégal. *Annls Fac. Sci. Univ. Dakar*, 19: 27-55, 3 pls.

BODARD, M. (1966b) - Sur le développement des tétrasporocystes d'*Anatheca montagnei* Schmitz (Solériacées, Gigartinales). *Bull. Inst. fond. Afr. noire*, sér. A, 28: 867-894, 7 pls.

BODARD, M. (1966c) - Première liste des espèces d'algues présentes sur la Pointe de Sarène (Sénégal). *Notes afr.*, 111: 81-89.

BODARD, M. (1967) - Sur le développement des cystocarpes des *Gracilaria* et *Gracilariopsis* au Sénégal. *Bull. Inst. fond. Afr. noire*, sér. A, 29: 869-897.

BODARD, M. (1968) - Les *Hypnea* au Sénégal (Hypnéacées, Gigartinales). *Bull. Inst. fond. Afr. noire*, sér. A, 30: 811-829.

BODARD, M. (1971) - *Halymenia senegalensis*, nov. sp. (Algae), espèce charactéristique de l'intralittoral sénégalais. *Bull. Inst. fond. Afr. noire*, sér. A, 33: 1-19.

BODARD, M. & MOLLION, J. (1974) - La végétation infralittorale de la petite côte sénégalaise. *Bull. Soc. phycol. France*, 19: 193-221

BOLD, H.C. & WYNNE, M.J. (1977) - *Algae: Structure and Reproduction*. pp. XIV+706. New Jersey.

BONNEMAISON, T. (1828) - Essai sur la hydrophytes loculées (ou articulées) de la famille des Épidermées et des Céramiées. *Mém. Mus. Hist. nat., Paris*, 16: 49-148.

BØRGESEN, F. (1909) - Some new or little known West Indian Florideae I. *Bot. Tidsskr.*, 29: 1-19.

BØRGESEN, F. (1910) - Some new or little known West Indian Florideae II. *Bot. Tidsskr.*, 30: 177-207.

BØRGESEN, F. (1911, 1912) - Some Chlorophyceae from the Danish West Indies. *Bot. Tidsskr.*, 31: 127-152. 1911. Id., II, *ibid.*, 32: 241-273. 1912.

BØRGESEN, F. (1913-1920) - The marine algae of the Danish West Indies. I. Chlorophyceae. *Danske bot. Ark.*, 1: 1-158. 1913. Id., II. Phaeophyceae. *Ibid.*, 2: 159-226. 1914. Id., III. Rhodophyceae, Part 1. *Ibid.*, 3: 1-80. 1915; Part 2. *Ibid.*, 3: 81-144. 1916; Part 3. *Ibid.*, 3: 145-240. 1917; Part 4. *Ibid.*, 3: 241-304. 1918; Part 5. *Ibid.*, 3: 305-368. 1919; Part 6. *Ibid.*, 3: 369-504. 1920. Issued separately as follows: 1. Chlorophyceae. I (1): 1-158. 1913. 2 Phaeophyceae. I (2): 159-228. 1914. Rhodophyceae, with addenda to the Chlorophyceae, Phaeophyceae and Rhodophyceae. II: 1-498, 1915-1920. Copenhagen. Note: References to page numbers in the part of the text relating to the brown algae are taken from the latter issues.

BØRGESEN, F. (1925-1930) - Marine algae from the Canary Islands, especially from Teneriffe and Gran Canaria. I. Chlorophyceae. *K. danske Vidensk. Selsk., Biol. Medd.,* 5: 1-123. 1925. Id., II. Phaeophyceae. *Ibid.,* 6: 1-112. 1926. Id., III. Rhodophyceae, Part I. Bangiales and Nemalionales. *Ibid.,* 6: 1-97. 1927; Id., Part II. Cryptonemiales, Gigartinales and Rhodymeniales. Les Mélobésiées par Mme. P. Lemoine. *Ibid.,* 8: 1-97, 4 pls. 1929; Id., Part III. Ceramiales. *Ibid.,* 9: 1-159. 1930a.

BØRGESEN, F. (1930b) - Some Indian green and brown algae especially from the shores of the Presidency of Bombay. *J. Indian bot. Soc.,* 9: 151-174, 2 pls.

BØRGESEN, F. (1931) - Sur *Platysiphonia* nov. gen. et sur les organes mâles et femelles du *Platysiphonia miniata* (Ag) nov. comb. (*Sarcomenia miniata* (Ag) J. Ag.). *Recl Trav. Crypt. dédiés à Louis Mangin.* 1-9.

BØRGESEN, F. (1932) - A revision of Forsskål's algae mentioned in Flora aegypt-iaco-arabica and found in his herbarium in the Botanical Museum of the University of Copenhagen. *Dansk bot. Ark.,* 8: 1-14, 1 pl.

BØRGESEN, F. (1934) - Some Indian Rhodophyceae especially from the shores of the Presidency of Bombay: IV. *Kew Bull.,* 1934: 1-30, 4 pls.

BØRGESEN, F. (1941-1953) - Some marine algae from Mauritius. II. Phaeophyceae. *K. danske Vidensk. Selsk., Biol. Medd.,* 16: 1-81, 8 pls. 1941. Id., III. Rhodophyceae. Part 1. Porphyridiales, Bangiales, Nemalionales. *Ibid.,* 17: 1-64, 2 pls. 1942; Part 2. Gelidiales, Cryptonemiales, Gigartinales. *Ibid.,* 19: 1-85, 1 pl. 1943; Part 4. Ceramiales. *Ibid.,* 19: 1-68. 1945; Id., Additions to the parts previously published, V. *Ibid.,* 21: 1-62, 2 pls. 1953.

BØRGESEN, F. (1950) - A new species of the genus *Predaea. Dansk bot. Ark.,* 14: 1-8.

BORNET, É. (1892) - Les algues de P.K.A. Schousboe récoltées au Maroc et dans la Méditerrannée de 1815 à 1829. *Mém. Soc. natn. Sci. nat. Math. Cherbourg,* 28: 165-376, 3 pls.

BORNET, É. & FLAHAULT, C. (1886) - Revision des Nostocacées hétérocystées contenues dans les principaux herbiers de France. *Annls Sci. nat. (Botanique),* sér. 7, 3: 323-381. 1886a. *Ibid.,* 4: 343-373. 1886b. *Ibid.,* 5: 51-129. 1886c [1887]. *Ibid.,* 7: 177-262. 1886d [1888]. Note: This work was published 1886-1888, though the date has been artificially fixed as 1 January 1886 for nomenclatural purposes.

BORNET, É. & THURET, G. (1876) - *Notes algologiques..* pp. XX+70, 25 pls. Paris.

BORY DE SAINT-VINCENT, J.B.G.M. (1804) - *Voyage dans les quatre principales îles des mers d'Afrique,... (1801 et 1802)... sur la corvette Le Naturaliste,...* 3+Atlas, 58 pls. Paris.

BORY DE SAINT-VINCENT, J.B.G.M. (1825) - *Dictionnaire classique d'histoire naturelle, par Messieurs Audouin, Isid. Bourdon,... et Bory de Saint-Vincent... Ouvrage dirigé par ce dernier collaborateur,...* 7. pp. 1-626. Paris.

BORY DE SAINT-VINCENT, J.B.G.M. (1827-29) - Cryptogamie [pp. 1-301, Atlas 39 pls]. In: Duperry, L., *Voyage autour de monde... sur... "La Coquille" pendant... 1822, 1823, 1824 et 1825.* Paris.

BOUGHEY, A.S. (1957) - Ecological studies of tropical coast-lines. 1. The Gold Coast, West Africa. *J. Ecol.*, 45: 665-687.

BOURRELLY, P. (1957) - Review of F. Drouet and W.A. Daily. Revision of the coccoid Myxophyceae. *Revue algol.*, 2: 279-280.

BOURRELLY, P. (1970a) - *Les algues d'eau douce.* III: *Les algues bleues et rouges, le Eugléniens, Peridiniens et Cryptomonadines.* pp. 1-512. Paris.

BOURRELLY, P. (1970b) - Note sur la famille des Oscillatoriacées. *Schweiz. Z. Hydrol.*, 32: 519-522.

BRAND, F. (1904) - Über die Anheftung der Cladophoraceen und über verschiedene polynesische Formen dieser Familie. *Bot. Zbl.*, 18: 165-193, 2 pls.

BRAND, F. (1911) - Cladophoraceae [pp. 313-316]. In: Engler, A., Beiträge zur Flora von Africa. XXXIX. *Bot. Jb.*, 46: 293-464.

BUCHANAN, J.B. (1954) - The zoogeographical significance of the Madreporaria collected in the Gold Coast, West Africa. *Revue Zool. Bot. afr.*, 49: 84-88.

BUCHANAN, J. B. (1957) - Benthic fauna of the continental edge off Accra, Ghana. *Nature, Lond.*, 179: 634-635.

BUCHANAN, J.B. (1958) - The bottom fauna communities across the continental shelf off Accra, Ghana (Gold Coast). *Proc. zool. Soc. Lond.*, 130: 1-56.

CARPINE, C. (1959) - Aperçu sur les peuplements littoraux [pp. 75-90]. In: Forest, J. (ed.), Campagne de la Calypso dans le Golfe de Guinée et aux îles Príncipe, São Tomé, Annobon (1956). *Annls Inst. océanogr., Monaco*, N.S., 37: 1-244, 6 pls.

CARRUTHER, J.N. (1961) - The Atlantic Ocean - North and South [pp. 1-17]. In: Borgstrom, G. & Heighway, A.J. (eds), *Atlantic Ocean Fisheries.* pp. VIII+336. London.

CHAPMAN, V.J. (1961, 1963) - The marine algae of Jamaica. 1. Myxophyceae and Chlorophyceae. *Bull. Inst. Jamaica, Sci. Ser.,* 12(1): 1-159, 1961. Id., 2. Phaeophyceae and Rhodophyceae. *Ibid.,* 12 (2): 1-201, 1963.

CHAPMAN, V.J. (1971) - What is *Enteromorpha tubulosa* Kütz. *Rev. algol.*, N.S., 10: 133-143.

CHEVALIER, A. (1920) - *Exploration botanique de l'afrique occidentale française. Tome 1. Enumération des plantes récoltées avec une carte botanique, agricole et forestière.* pp. XIII+798. Paris.

CHI-BONNARDEL, R. van (1973) - *The Atlas of Africa.* pp. 1-350, 138 maps.

CHIHARA, M. (1962) - Life cycle of the bonnemaisoniaceous algae in Japan (2). *Sci. Rep. Tokyo Kyoiku Daigaku*, sect. B, 11: 27-54.

CHOU, R.C. (1945, 1947) - Pacific species of *Galaxaura* I. Asexual types. *Pap. Mich. Acad. Sci.*, 30: 35-56, 11 pls. 1945 (1944). Id., II. Sexual types. *Ibid.*, 31: 3-24, 8 pls. 1947 (1945).

CHRISTENSEN, T. (1957) - *Chaetomorpha linum* in the attached state. *Bot. Tidsskr.*, 53: 311-316.

CLAYTON, M.N. (1974) - Studies on the development, life history and taxonomy of the Ectocarpales (Phaeophyta) in Southern Australia. *Aust. J. Bot.*, 22: 743-813.

CLAYTON, M.N. (1976) - A study of variation in Australian species of *Colpomenia* (Phaeophyta, Scytosiphonales). *Phycologia*, 14: 187-195 (1975).

CLIMAP [Project Members] (1976) - The surface of the ice-age earth. *Science, N.Y.*, 191 (4232): 1131-1137.

COLLINS, F.S. (1901) - The algae of Jamaica. *Proc. Amer. Acad. Arts. Sci.*, 37: 229-270.

COLLINS, F.S. (1906) - *Acrochaetium* and *Chantransia* in North America. *Rhodora*, 8: 189-196.

COLLINS, F.S. (1909) - Green algae of North America. *Tufts Coll. Stud. (Scientific series)*, 2: 79-480, 18 pls.

COLLINS, F.S. & HERVEY, A.B. (1917) - The algae of Bermuda. *Proc. Am. Acad. Arts Sci.*, 53: 1-195, 6 pls.

CORDEIRO-MARINO, M. (1978) - Rodoficeas bentonicas marinhas do estado de Santa Catarina. *Rickia*, 7: 1-243.

CRIBB, A.B. (1954-1958) - Records of marine algae from South-eastern Queensland I. *Pap. Dept. Bot. Univ. Qd.*, 3(3): 15-37. 1954. Id., II. *Ibid.*, 3(16): 131-147. 1956. Id., III. *Ibid.*, 3(19): 159-191, 13 pls. 1958.

CROUAN , P.L. & CROUAN, H.M. (1852) - *Algues marines du Finistère, recueillies et publiées par Crouan, Frères, Pharmaciens à Brest... Deuxième volume. Floridées: Crouan Frères.* nos. 113-322. Brest.

CROUAN, P.L. & CROUAN, H.M. (1867) - *Florule du Finistère...*, pp. X+262. Paris.

DANGEARD, P. (1951a) - Deux espèces nouvelles de genre *Chondria* de la région de Dakar. *Botaniste*, 35: 13-20, 1 pl.

DANGEARD, P. (1951b) - Sur les Gélidiacées de Dakar et de Port Etienne. *Botaniste*, 35: 21-24, 1 pl.

DANGEARD, P. (1952) - Algues de la presqu'île du Cap Vert (Dakar) et ses environs. *Botaniste*, 36: 193-329, 8 pls.

DANGEARD, P. (1955) - Remarques sur quelques *Codium*, en particulier le *Codium fragile* (Sur.) Hariot: *Botaniste*, 39: V+XVII, 2 pls.

DANGEARD, P. (1958) - Sur quelques espèces d'*Ulva* de la région de Dakar. *Botaniste*, 42: 163-171, 2 pls.

DAWSON, E.Y. (1944) - The marine algae of the Gulf of California. *Allan Hancock Pacif. Exped.*, 3: 189-453, 47 pls.

DAWSON, E.Y. (1950a) - A review of *Ceramium* along the Pacific Coast of North America with special reference to its Mexican representatives. *Farlowia*, 4: 113-138, 4 pls.

DAWSON, E.Y. (1950b) - Notes on some Pacific Mexican Dictyotaceae. *Bull. Torrey bot. Club*, 77: 83-93.

DAWSON, E.Y. (1953, 1954b) - Marine red algae of Pacific Mexico. I. Bangiales to Corallinaceae Subf. Corallinoideae. *Allan Hancock Pacif. Exped.*, 17: 1-239, 33 pls. 1953. Id., 2. Cryptonemiales (cont.). *Ibid.*, 17: 241-397, 44 pls. 1954b.

DAWSON, E. Y. (1954a) - Marine plants in the vicinity of the Institut Océanographique de Nha Trang, Viêt Nam. *Pacif. Sci.*, 8: 373-481.

DAWSON, E.Y. (1962) - New taxa of benthic green, brown and red algae published since De Toni 1889, 1895, 1924, respectively, as compiled from the Dawson Algal Library. *Contrib. Beaudette Found. Biol. Res.*, pp. 1-105. Santa Ynez, California.

DAWSON, E.Y. (1963) - Marine red algae of Pacific Mexico. VIII. Ceramiales: Dasyaceae, Rhodomelaceae. *Nova Hedwigia*, 5: 401-481.

DECAISNE, M.J. (1841) - Plantes de l'Arabie Heureuse,... *Archs Mus. natn Hist. nat., Paris*, 2: 89-199, 3 pls.

DELILE, A.R. (1813) - Florae Aegyptiaceae illustratio [pp. 49-82]. In: *France (Commission d'Égypte), Description de l'Égypte,... l'éxpedition de l'armée Francaise (1798-1801)... Histoire Naturelle.* Vol. 2. pp. 1-320, 62 pls. Paris.

DE MAY, D., JOHN, D.M. & LAWSON, G.W. (1977) - A contribution to the littoral ecology of Liberia. *Botanica mar.*, 20: 41-46.

DENIZOT, M. (1968) - *Les algues floridées encroutantes (à l'exclusion des Corallinacées).* pp. 1-310. Paris

DERBÈS, A. (1856) - Description d'une nouvelle espèce de floridée, devant former un nouveau genre, et observations sur quelques algues. *Annls Sci. nat. (Botanique)*, sér. 4, 5: 209-220, 1 pl.

DERBÈS, A. & SOLIER, A.J.J. (1851) - Algues. In: Castagne, L., *Supplement au catalogue des plantes qui croissent naturellement aux environs de Marseille.* Aix, pp. 93-121.

DERBÈS, A. & SOLIER, A.J.J. (1856) - Mémoire sur quelques points de la physiologie des algues. *C. r. hebd. Séanc. Acad. Sci., Paris*, Suppt 1: 1-120, 23 pls.

DE ROUVILLE, M.A. (1946) - *Le régime des côtes.* Paris.

DESIKACHARY, T.V. (1959) - Cyanophyta. *Indian Council of Agricultural Research Monograph.* pp. X+686. New York and London.

DE SILVA, M.W.R.N. & BURROWS, E.M. (1973) - An experimental assessment of the status of the species *Enteromorpha intestinalis* (L.) Link and *Enteromorpha compressa* (L.) Grev. *J. mar. biol. Ass. U.K.*, 53: 895-904.

DESMAZIÈRES, J.B.H.J. (1827) - *Plantes cryptogames du nord de la France.* Fasc. V, espèces nos 201-250.

DE TONI, G. (1936) - *Noterelle di nomenclatura algologica.* VII. *Primo elenco di Floridee omonime.* Brescia.

DE TONI, G.B. (1889-1924) - *Sylloge algarum omnium hucusque cognitarum.* I. *Sylloge Chlorophycearum...* pp. 12+CXXXIX+1325. 1889. Id., III. *Sylloge Fucoidearum...* pp. XVI+638. 1895. Id., IV. *Sylloge Floridearum...* Sect. II. pp. 387-776. 1900; Sect. III. pp. 775- 1525. 1903; Sect. IV. pp. 1532-1873. 1905; Sect. V. pp. XI+767. 1924. Patavii.

DÍAZ-PIFERRER, M. (1969) - Corrective note on a previously published paper on the genus *Ceramiella. Caribb. J. Sci.*, 9: 179-180.

DICKINSON, C.I. (1951, 1952) - Marine algae from the Gold Coast. III. *Kew Bull.*, 6: 293-297, 2 pls. 1951. Id., IV. *Ibid.*, 7: 41-43. 1952.

DICKINSON, C.I. & FOOTE, V.J. (1950, 1951) - Marine algae from the Gold Coast I. *Kew Bull.*, 5: 267-272. 1950. Id., II. *Ibid.*, 6: 133-138. 1951.

DIELS, L., MILDBRAED, J. & SCHULZE-MENZ, G.K. (1963) - Vegetationskarte von Africa (herausgegeben von W. Domke). *Willdenowia*, Beih. 1.

DILLWYN, L.W. (1809) - *British confervae;...* pp. 1-87, 109 pls. + 7 suppl. pls. London.

DIXON, P.S. (1958) - The structure and development of the thallus in the British species of *Gelidium* and *Pterocladia. Ann. Bot.*, 22: 353-368.

DIXON, P.S. (1960) - Taxonomic and nomenclatural notes on the Florideae, II. *Bot. Notiser,* 113: 295-319.

DIXON, P.S. (1963) - Variation and speciation in marine Rhodophyta [pp. 51-62]. In: Harding, J.P. & Tebble, N. (eds), Speciation in the sea. *Syst. Ass. Publn,* 5. pp. 1-199. London.

DIXON, P.S. (1964) - *Asparagopsis* in Europe. *Nature, Lond.*, 204: 902.

DIXON, P.S. (1966) - On the form of the thallus in the Florideophyceae [pp. 45-63, 2 pls]. In: Cutter, E.G. (ed.), *Trends in Plant Morphogenesis.* pp. XVI+329. London and Colchester.

DIXON, P.S. (1967) - The typification of *Fucus cartilagineus* L. and F. *corneus* Huds. *Blumea,* 15: 55-62.

DIXON, P.S. (1973) - Biology of the Rhodophyta. *University Reviews in Botany,* 4. pp. XIII+285. Edinburgh.

DIXON, P.S. & IRVINE, L.M. (1977a) - *Seaweeds of the British Isles,* Vol. 1. *Rhodophyta,* Part 1. *Introduction, Nemaliales, Gigartinales.* pp. XI+252. BM(NH), London.

DIXON, P.S. & IRVINE, L.M. (1977b) - Miscellaneous notes on algal taxonomy and nomenclature. IV. *Bot. Notiser,* 130: 137-141.

DIXON, P.S. & PRICE, J.H. (1981) - The genus *Callithamnion* (Rhodophyta: Ceramiaceae) in the British Isles. *Bull. Br. Mus. nat. Hist.,* 9(2): 99-141.

DIZERBO, A.-H. (1974) - La répartition des *Gigartina* (Gigartinales, Gigartinacées) des Massif Armoricain. *Bull. Soc. Phycol. France,* 19: 88-94.

DOODSON, A.T. & WARBURG, H.D. (1941) - *Admiralty manual of tides.* pp. XII+270. H.M.S.O. London.

DRONKERS, J.J. (1964) - *Tidal computations in rivers and coastal waters.* pp. XII+518. Amsterdam.

DROUET, F. (1938) - Notes on Myxophyceae, I-IV. *Bull. Torrey bot. Club,* 65: 285-292.

DROUET, F. (1968) - Revision of the classification of the Oscillatoriaceae. *Academy of Natural Sciences of Philadelphia,* Monograph 15. pp. 1-370. Pennsylvania.

DROUET, F. (1969) - Homonymy in *Arthrospira* Stizenb. (Oscillatoriaceae). *Phytologia,* 18: 339.

DROUET, F. (1973) - *Revision of the Nostocaceae with cylindrical trichomes (formerly Scytonemataceae and Rivulariaceae).* pp. 1-292. New York.

DROUET, F. (1978) - Revision of the Nostocaceae with constricted trichomes. *Beih. Nova Hedwigia,* 57: 1-258.

DROUET, F. (1981) - Revision of the Stigeonemataceae with a summary of the classification of the blue-green algae. *Beih. Nova Hedwigia,* 66: 1-221.

DROUET, F. & DAILY, W.A. (1948) - Nomenclature transfers among coccoid algae. *Lloydia,* 11: 77-79.

DROUET, F. & DAILY, W.A. (1952) - A synopsis of the coccoid Myxophyceae. *Butler Univ. bot. Stud.,* 10: 220-223.

DROUET, F. & DAILY, W.A. (1956) - Revision of the coccoid Myxophyceae. *Butler Univ. bot. Stud.,* 12: 1-218.

DROUET, F. & DAILY, W.A. (1957) - Revision of the coccoid Myxophyceae; additions and corrections. *Trans. Am. microsc. Soc.,* 76: 219-222.

DUBY, J.E. (1830) - *Botanicon gallicum...* Vol. 2, ed. 2. pp. 545-1068 + I-LVIII. Paris

DUCLUZEAU, J.A.P. (1905) - *Essai sur l'histoire naturelle des conferves des environs de Montpellier...* pp. 1-92. Montpellier.

DURAIRATNAM, M. (1961) - Contribution to the study of the marine algae of Ceylon. *Bull. Fish. Res. Stn Ceylon,* 10: 1-181, 32 pls.

EARLE, S.A. (1969) - Phaeophyta of the Eastern Gulf of Mexico. *Phycologia,* 7: 71-254.

EDMUNDS, J. & EDMUNDS, M. (1973) - Preliminary report on the mollusca of the benthic communities off Tema, Ghana. *Malacologia,* 14: 371-376.

EDWARDS, P. (1979) - A cultural assessment of the distribution of *Callithamnion hookeri* (Dillw.) S.F. Gray (Rhodophyta, Ceramiales) in nature. *Phycologia,* 18: 251-263.

EGEROD, L. (1974) - Report of the marine algae collected on the Fifth Thai-Danish Expedition of 1966 Chlorophyceae and Phaeophyceae. *Botanica mar.,* 17: 130-157.

EKMAN, S. (1953) - *Zoogeography of the Sea.* pp. XII+417. London.

ELLIS, J. & SOLANDER, D. (1786) - *The natural history of many curious and uncommon zoophytes, collected from various parts of the globe...* pp. XII+208, 63 pls. London.

FAGADE, S.O. & OLANIYAN, C.I.O. (1974) - Seasonal distribution of the fish fauna of the Lagos Lagoon. *Bull. inst. fond. Afr. noire,* sér. A, 36: 244-252.

FALKENBERG, P. (1879) - Die Meeres-algen des Golfes von Neapel. Nach Beobachtungen in der Zool. Station während der Jahre 1877-78 zusammengestellt. *Mitt. zool. Stn Neapel,* 2: 218-277 (1878).

FALKENBERG, P. (1901) - *Die Rhodomelaceen des Golfes von Neapel und der angrenzenden Meeresabschnitte. Fauna und Flora des Golfes von Neapel.* 26. pp. XVI+754, 24 pls. Berlin.

FAN, K.C. (1956) - Revision of *Calothrix* Ag. *Revue algol.,* N.S., 2: 154-178.

FARNHAM, W.F. (1980) - Studies on aliens in the marine flora of southern England [pp. 875-914]. In: Price, J.H., Irvine, D.E.G. & Farnham, W.F. (eds), *The Shore Environment,* Vol. 2: *Ecosystems.* Systematics Association Special Volume No. 17(b). pp. XX + 323-945 + Iiii-IIc.

FELDMANN, J. (1931) - Contribution à la flore algologique marine de l'Algerie Les Algues de Cherchell. *Bull. Soc. Hist. nat. Afr. N.,* 22: 179-254.

FELDMANN, J. (1938) - Recherches sur la végétation marine de la Méditerranée. La Côte des Albères. *Revue algol.,* 10: 1-339.

FELDMANN, J. (1939) - Les algues marines de la côte des Albères. IV. Rhodophycées. *Revue algol.,* 11: 247-330.

FELDMANN, J. (1942) - Les algues marines de la côte des Albères IV. Rhodophycées (fin). *Trav. algol.,* 1: 29-113, 1 pl.

FELDMANN, J. & FELDMANN, G. (1942) - Recherches sur les Bonnemaisoniacées et leur alternance de générations. *Annls Sci. nat. (Botanique),* sér. 6, 3: 75-175.

FELDMANN, J. & HAMEL, G. (1934) - Observations sur quelques Gélidiacées. *Revue gen. Bot.,* 46: 528-549.

FELDMANN, J. & HAMEL, G. (1936) - Floridées de France. VII. Gélidiales. *Revue algol.,* 9: 85-140, 5 pls.

FELDMANN-MAZOYER, G. (1938) - Sur un nouveau genre de Céramiacées de la Méditerranée. *C. r. hedb. Séanc. Acad. Sci., Paris,* 207: 1119-1121.

FELDMANN-MAZOYER, G. (1941) - *Recherches sur les Céramiacées de la Méditerranée occidentale.* pp. 1-510, 4 pls. Alger.

FLETCHER, R.L. (1980) - *Catalogue of Main Marine Fouling Organisms* Volume 6 *Algae Marine Fouling Algae.* pp. 1-61. Brussels.

FOREST, H. (1968) - The approach of a modern algal taxonomist [pp. 185-199]. In: Jackson, D.F. (ed.), *Algae, Man and the Environment.* pp. VIII + 554. New York.

FORSSKÅL, P. (1775) - *Flora Aegyptiaco-Arabica,...* pp. 1-32 + CXXVI + 219. Havniae.

FOSLIE, M. (1897) - On some Lithothamnia. *K. norske Vidensk. Selsk. Skr., (1): 1-20.*

FOSLIE, M. (1898) - Some new or critical Lithothamnia *K. norske Vidensk. Selsk. Skr.,* (6): 1-19.

FOSLIE, M. (1900a) - New or critical calcareous algae. *K. norske Vidensk. Selsk. Skr.,* (5): 1-34 (1899).

FOSLIE, M. (1900b) - Five new calcareous algae. *K. norske Vidensk. Selsk. Skr.,* (3): 1-6.

FOSLIE, M. (1900c) - Revised systematical survey of the Melobesiae. *K. norske Vidensk. Selsk. Skr.,* (5): 1-22.

FOSLIE, M. (1901) - New Melobesiae. *K. norske Vidensk. Selsk. Skr.,* (6): 1-24 (1900).

FOSLIE, M. (1906-1909) - Algologiske notiser II. *K. norske Vidensk. Selsk. Skr.,* (2): 1-28. 1906a. Id., III. *Ibid.,* (8): 1-34. 1907. (1906). Id., VI. *Ibid.,* (2): 1-63. 1909.

FOSLIE, M. (1906b) - Den botaniske samling. *K. norske Vidensk. Selsk. Mus.* Aarsberetn, (Arb): 1-43 (1905).

FOSLIE, M.H. & PRINTZ, H. (1929). - *Contribution to a monograph of the Lithothamnia.* pp. 1-60, 75 pls. Trondheim. Note: Collected and edited by Henrik Printz after the author's death.

FOX, M. (1957) - A first list of marine algae from Nigeria. *J. Linn. Soc. (Bot.),* 55: 615-631, 1 pl.

GABRIELSON, P.W. & HOMMERSAND, M.H. (1982) - The Atlantic species of *Solieria* (Gigartinales, Rhodophyta) : their morphology and affinities. *J. Phycol.,* 18: 31-45.

GAILLARD, J. (1967) - Étude monographique de *Padina tetrastromatica* (Hauck). *Bull. Inst. fond. Afr. noire,* sér. A, 29: 447-463.

GANESAN, E.K. (1968) - Studies on the marine algal flora of Venezuela. 1. The occurrence of the brown alga *Levringea brasiliensis* (Montagne) Joly in the Caribbean. *Bolm Inst. oceanogr., Univ. Oriente,* 7: 129-136.

GANESAN, E.K. (1974) - Studies on the marine algal flora of Venezuela. V. *Pseu-*

dogloiophloea Halliae. J. Phycol., 10: 415-418.

GANESAN, E.K. & WEST, J.A. (1975) - Culture studies on the marine red alga *Rhodophysema elegans* (Cryptonemiales, Peysonneliaceae). *Phycologia,* 14: 161-166.

GATES, W.L. (1976) - Modeling the ice-age climate. *Science, N.Y.,* 191(4232): 1138-1144.

GAULD, D.T. & BUCHANAN, J.B. (1959) - The principal features of rock shore fauna in Ghana. *Oikos,* 10: 121-132.

GAYRAL, P. (1958) - *La Nature au Maroc* II. *Algues de la Côte Atlantique Marocaine.* pp. 1-524, 3 pls. + errata/addendum slip. Rabat.

GAYRAL, P. (1960) - Sur la présence au Maroc et à Dakar de *Levringea brasiliensis* (Mont.) B. Joly. *Revue algol.,* N.S., 5: 49-54.

GEESINK, R. (1973) - Experimental investigations on marine and freshwater *Bangia* (Rhodophyta) from the Netherlands. *J. exp. mar. Biol. Ecol.,* 11: 239-247.

GEITLER, L. (1930-1932) - Cyanophyceae. In: Rabenhorst, L., *Kryptogamen-Flora von Deutschland, Österreich und Schweiz.* Vol. 14. pp. VI + 1196. Leipzig.

GEITLER, L. (1960) - Schizophyzeen. *Handbuch der Pflanzenanatomie.* 6. pp. VII+131. Berlin.

GERLOFF, J. (1959) - *Bachelotia* (Bornet) Kuckuck ex Hamel oder *Bachelotia* (Bornet) Fox? *Nova Hedwigia,* 1: 37-39.

GHANA, FISHERY RESEARCH UNIT, (1970) - The NCOR report on the mechanism of the upwelling of Ghana's coastal waters. *Mar. Fish. Res. Rep., Tema,* (3): 1-13.

GIRAULT, G. & KIMPE, P. de (1967) - La productivité primaire d'un milieu aquatique lagunaire tropical. *Bull. inst. fond. Afr. noire,* sér. A, 24: 710-734.

GMELIN, S.G. (1768) - *Historia Fucorum...,* pp. 1-6 + VI + 239, 33 pls. Petropoli.

GOMONT, M. (1892, 1893) - Monographia des Oscillatoriées (Nostocacées homocystées... *Annls Sci. nat. (Botanique),* sér. 7, 15: 263-368. 1892. *Ibid.,* sér. 7, 16: 91-264. 1893.

GOODENOUGH, S. & WOODWARD, T.J. (1797) - Observations on the British Fuci, with particular descriptions of each species. *Trans. Linn. Soc.,* 3: 84-235, 4 pls.

GOOR, A.C.J. van (1923) - Die Holländischen Meeresalgen (Rhodophyceae, Phaeophyceae und Chlorophyceae) insbesondere der Umgebung von Helder, des Wattenmeeres und der Zuidersee. *Verh. K. Akad. Wet.,* Tweede Sectie, 23, (2): I-IX +232.

GORDON, M.E. (1972) - Comparative morphology and taxonomy of the Wrangelieae, Spondylothamnieae, and Spermothamnieae (Ceramiaceae, Rhodophyta). *Aust. J. Bot.,* suppl. ser. (4): 1-180.

GRAY, S.F. (1921) - *A Natural Arrangement of British Plants,* ..1. pp. XXVIII +824. London.

GREVILLE, R.K. (1830) - *Algae britannicae,...* pp. LXXXVIII+218, 19pls. Edinburgh and London.

GRUNOW, A. (1868) - Algae [pp. 1-104, 11 pls]. In: Fenzl, E., *Reise der Österreichischen Fregatte Novara um die Erde in den Jahren 1857, 1858, 1859. ... Bot. Theil.* I. *Sporenpflanzen.* pp. 1-261, Wien.

GRUNOW, A. (1915, 1916) - Additamenta ad cognitionem Sargassorum. *Verh. zool-bot. Ges. Wien,* 65: 329-448. 1915. Id., (Fortsetzung). *Ibid.,* 66: 1-48, 136-185. 1916.

GUIRY, M.D. (1977) - Studies on marine algae of the British Isles. 10.The genus *Rhodymenia. Br. phycol. J.,* 12: 385-425.

GUIRY, M.D. & IRVINE, L.M. (1974) - A species of *Cryptonemia* new to Europe. *Br. phycol. J.,* 9: 225-237.

HAMEL, G. 1927 - *Algues de France... Ce recueil publié à 100 exemplaires par la Revue Algologique sous la direction de G. Hamel.* Fasc. 1, nos. 1-50, Dec. 1927. Paris.

HAMEL, G. (1928) - Sur les genre *Acrochaetium* Naeg. et *Rhodochorton* Naeg. *Revue algol.,* 3: 159-210.

HAMEL, G. (1931-1939) - Pheophycées de France. pp. XLVII+432. Paris.

HAMEL, G. & LEMOINE, P. (1953) - Corallinacées de France et d'Afrique du Nord. *Archs Mus. natn Hist. nat., Paris,* sér. 7, 1: 15-136, 23 pls.

HARDY-HALOS, M.-TH. (1968) - Les Ceramiaceae (Rhodophyceae-Florideae) des côtes de Bretagne: 1. - Le genre *Antithamnion* Nägeli. *Revue algol.,* N.S., 9: 152-183.

HARIOT, P. (1895) - Liste des algues recueilles au Congo par M.H. Lecomte. *J. Bot., Paris,* 9: 242-244.

HARIOT, P. (1896) - Contribution à la flore algologique du Gabon et du Congo française. *C. r. Ass. fr. Avanc. Sci.,* 24(2): 641-643 (1895).

HARIOT, P. (1908) - Les algues de San Thomé (côte occidentale d'Afrique). *J. Bot., Paris,* sér. 2, 1: 161-164.

HARTOG, C. den (1959) - *The epilithic algal communities occurring along the coast of the Netherlands.* pp. XI+241. Amsterdam.

HARVEY, W.H. (1833) - Cryptogamia Algae... Div. II. Confervoideae [pp. 322-385]. Div. III. Gloiocladeae [pp. 385-401]. In: Hooker, W.J., *The English Flora of Sir James Edward Smith. Class XXIV. Cryptogamia, ...,* vol. 5 (or Vol. II of Dr. Hooker's British Flora), Part I, ... pp. X + 4 + 432. London.

HARVEY, W.H. (1934) - Notice of a collection of algae, communicated to Dr. Hooker by the late Mrs Charles Telfair, from "Cap Malheureux", in the Mauritius; ... *J. Bot., Hooker,* 1: 147-157, 2 pls.

HARVEY, W.H. (1847) - *Nereis australis,...* pp. VII+124, 50 pls. London.

HARVEY, W.H. (1849) - *Phycologia britannica:...* Vol. 2. pp. I-VI, 120 pls. London.

HARVEY, W.H. (1852-1858) - Nereis boreali-americana;... I. Melanospermeae. *Smithson. Contr. Knowl.*, 3: 1-150, 12 pls. 1852, Id., II. Rhodospermeae. *Ibid.*, 5:1-258, 24 pls. 1853. Id., III. Chlorospermeae, including supplements. *Ibid.*, 10: 1-140, 14 pls. 1858.

HARVEY, W.H. (1859, 1863) - *Phycologia australica;...* Vol. 2. pp. VIII + 1-59 pls. 1859. *Id.*, Vol. 5. pp. LXXIII + 60 pls. 1863. London.

HAUCK, F. (1887) - Über einige von J.M. Hildebrandt im Rothen Meere und Indischen Ocean gesammelte Algen. III. *Hedwigia*, 26: 18-21, 41-45.

HEDGPETH, J.W. (1957) - Marine Biogeography [pp. 359-382]. In: Hedgpeth, J.W. (ed.), Treatise on marine ecology and paleoecology I. Ecology. *Mem. geol. Soc. Am.*, Memoir 67. pp. VIII+1296.

HEEREBOUT, G.R. (1968) - Studies on the Erythropeltidaceae (Rhodophyceae-Bangiophycidae). *Blumea*, 16: 139-157.

HENRIQUES, J. (1885) - Contribução para o estudo da flora d'algumas possessões portuguezas 1 Plantas colhidas por F. Newton na Africa occidental. *Bolm Soc. broteriana*, 3: 129-140.

HENRIQUES, J. (1886a) - Contribução para o estudo da flora d'algumas possessões portuguezas. Plantas colhidas por F. Newton na Africa occidental (dal Boletim da Sociedade Broteriana III-IV p. 129 - Coimbra 1885). Algae [pp. 121, 122]. In: De Toni, G.B. & Levi, D., Contributiones ad phycologiam extra-italicam. *Notarisia*, 1: 117-122.

HENRIQUES, J. (1886b) - Algae [pp. 218-221]. In: Henriques, J., Contribuçoes para o estudo da Flora d'Africa. Flora de S. Thomé. *Bolm soc. broteriana*, 4: 129-221.

HENRIQUES, J. (1887) - Flora de S. Thomé [pp. 381-383]. In: De Toni, G.B. & Levi, D., Contributiones ad phycologiam extra-italicam. *Notarisis*, 2: 375-383.

HENRIQUES, J. (1917) - Catálogo das espéces de animais e plantas até hoje encontradas na ilha de S. Thomé. *Bolm Soc. broteriana*, 27: 138-197.

HERING, K. (1842) - Diagnoses algarum novarum a cl. Dre Ferdinand Krauss in Africâ Australi lectarum. *Ann. Mag. nat. Hist.*, 8: 90-92.

HEYDRICH, F. (1892) - Beiträge zur Kenntnis der Algenflora von Kaiser-Wilhelms-Land (Deutsch-Neu-Guinea). *Ber. dt. bot. Ges.*, 10: 458-485, 3 pls.

HEYDRICH, F. (1897) - Neue Kalkalgen von Deutsch-Neu-Guinea (Kaiser-Wilhelms-Land). *Biblthca bot.*, 41: 1-11, 1 pl.

HIERONYMUS, G. (1895) - Klasse: Chlorophyceae [pp. 21-24]. In: Engler, A., *Deutsch-Ost-Afrika. ... V. Die Pflanzenwelt Ost-Afrikas und der Nachbargebiete. Theil C. Verzeichnis der bis jetzt aus Ost-Afrika bekannt gewordenen Pflanzen.* pp. II+433. Berlin

HILL, M.O. (1973) - Reciprocal averaging: an eigenvector method of ordination. *J. Ecol.*, 61: 237-249.

HOEK, C. van den (1963) - *Revision of the European species of Cladophora.* pp. VII + 248. Leiden.

HOEK, C. van den (1975) - Phytogeographic provinces along the coasts of the northern Atlantic Ocean. *Phycologia,* 14: 317-330.

HOEK, C. van den (1982) - A taxonomic revision of the American species of *Cladophora* (Chlorophyceae) in the North Atlantic Ocean and their geographic distribution. *Verh. K. ned. Akad. Wet.,* Tweede Sectie, 78: 1-236.

HOLLENBERG, G.J. (1968a) - An account of the species of *Polysiphonia* of the central and western tropical Pacific Ocean. 1. Oligosiphonia. *Pacif. Sci.,* 22(1): 56-98.

HOLLENBERG, G.J. (1968b) - An account of the species of the red alga *Herposiphonia* occurring in the central and western tropical Pacific Ocean. *Pacif. Sci.,* 22 (4): 536-559.

HOMMERSAND, M. H. (1963) - The morphology and classification of some Ceramiaceae and Rhodomelaceae. *Univ. Calif. Publ. Bot.,* 35: 165-366, 6 pls.

HOOKER, W.J. (1833) - Cryptogamia Algae ... Div. I. Inarticulatae [pp. 264-322]. In: Hooker, W.J. *The English Flora of Sir James Edward Smith. Class XXIV. Cryptogamia,...* Vol. V. (or Vol. II of Dr. Hooker's British Flora), Part I, ... pp. X+4+432. London.

HOOKER, J.D. (1845-1847) - *The botany of the Antarctic voyage...* I. *Flora Antarctica.* Pt. II, Algae, ... pp. 209-574 + 117 pls. London.

HOOKER, J.D. & HARVEY, W.H. (1845) - Algae antarcticae... *J. Bot., Lond.,* 4: 249-276, 293-298.

HOPPE, H.A. (1969) - Marine algae raw materials [pp. 126-287, 16 pls]. In: Levring, T., Hoppe, H.A. & Schmid, O.J., *Marine algae a survey of research and utilization.* Botanica Marina Handbooks 1. pp. 1-421, 16 pls. Hamburg.

HORNEMANN, J.W. (1819) - *Anniversaria in memoriam Reipublicae Sacrae et Litterariae cum Universae,... De Indole plantarum Guineensium.* pp. 1-27. Hauniae.

HOSPER, J. (1971) - The geology of the Niger Delta area [pp. 121-142]. In: Delany, F.M. (ed.), ICSU/SCOR Symposium Cambridge 1970: The geology of the East Atlantic continental margin. Part 4. Africa. *Rep. No.* 70/16, *Inst. geol. Sci.,* pp. IV+209.

HOUGHTON, R.W. (1973) - Evaporation during the upwelling in Ghanaian coastal waters. *J. phys. oceanogr.,* 3: 487-489.

HOUGHTON, R.W. (1976) - Circulation and hydrographic structure over the Ghana continental shelf during 1974 upwelling. *J. phys. oceanogr.,* 6: 909-924.

HOWE, M.A. (1905) - Phycological studies - II. New Chlorophyceae, new Rhodophyceae, and miscellaneous notes. *Bull. Torrey bot. Club,* 32: 563-586, 7 pls.

HOWE, M.A. (1909) - Phycological studies - IV. The genus *Neomeris* and notes on other Sipnonales. *Bull. Torrey bot. Club,* 36: 75-104, 8 pls.

HOWE, M.A. (1911) - Phycological studies - V. Some marine algae of Lower California, Mexico. *Bull. Torrey bot. Club,* 38: 489-514, 8 pls.

HOWE, M.A. (1914) - The marine algae of Peru. *Mem. Torrey bot. Club,* 15: 1-185, 66 pls.

HOWE, M.A. (1915) - Report on a visit to Porto Rico for collecting marine algae. *Jl N.Y. bot Gdn,* 16: 219-225.

HOWE, M.A. (1920) - Class 2. Algae [pp. 553-618]. In: Britton, N.L. &Millspaugh, C.F., *The Bahama Flora.* pp. VIII+695. New York.

HOWE, M.A. (1934) - Hawaiian algae collected by Dr. Paul C. Galtsoff. *J. Wash. Acad. Sci.,* 24: 32-42.

HOWE, M.A. & TAYLOR, W.R. (1931) - Notes on new or little-known marine algae from Brazil. *Brittonia,* 1: 7-33, 2 pls.

HUDSON, W. (1762, 1778) - *Flora anglica,...* pp. VIII+506. 1762. *Id.,* Ed. 2, 2. pp. 397-690. 1778.

HUVÉ, P. & HUVÉ, H. (1976) - Contribution à la connaissance de l'algue *Halichrysis depressa* (Montagne 1838 in J. Ag. 1851) Bornet 1892 (Rhodophycées, Rhodymeniales). *Phycologia,* 15: 377-392.

INGHAM, M.C. (1970) - Coastal upwelling in the northwestern Gulf of Guinea. *Bull. Mar. Sci.,* 20: 1-34.

JAASUND, E. (1969, 1970) - Marine algae in Tanzania I. *Botanica mar.,* 12: 255-274. 1969. Id., III. *Ibid.,* 13: 65-70. 1970a. Id., IV. *Ibid.,* 13: 71-79. 1970b.

JAASUND, E. (1976) - *Intertidal seaweeds in Tanzania. A field guide.* First edition. pp. 1-160. University of Tromsø.

JACKSON, S.P. (1961) - *Climatological Atlas of Africa.* Commission for technical co-operation in Africa south of the Sahara. Joint Report No. 1. 55 pls. Pretoria.

JARDIN, E. (1851 ?) - *Herborisation sur la côte occidentale d'Afrique pendant les années 1845-1846-1847-1848.* pp. 1-19. Paris. Note: For doubt regarding the year of publication see Lawson and Price (1969).

JARDIN, E. (1891) - *Aperçu sur la flore du Gabon avec quelques observations sur les plantes les plus importantes.* pp. 1-71. Paris. Note. Originally published in *Bull. Soc. linn. Normandie,* sér. 4, 4: 135-203 (1891).

JÉNIK, J. & LAWSON, G.W. (1967) - Observations on water loss of seaweeds in relation to microclimate on a tropical shore (Ghana). *J. Phycol.,* 3: 113-116, 1 pl.

JOHANSEN, D.A. (1940) - *Plant Microtechnique.* pp. XI+523. New York and London.

JOHANSEN, H.W. (1970) - The diagnostic value of reproductive organs in some genera of articulated coralline red algae. *Br. phycol. J.,* 5: 79-86.

JOHN, D.M. (1972a) - A new species of *Botryocladia* (Rhodophyceae) from the Gulf of Guinea. *Phycologia,* 12: 33-36.

JOHN, D.M. (1972b) - *Sphacelaria elliptica* Dick. conspecific with *Sphacelaria brachygonia* Mont. (Phaeophyceae). *Bull. Inst. fond. Afr. noire,* sér. A, 34: 1-4.

JOHN, D.M. (1972c) - The littoral ecology of rocky parts of the north-western shore of the Guinea Coast. *Botanica mar.,* 15: 199-204.

JOHN, D.M. (1974) - New records of *Ascophyllum nodosum* (L.) Le Jol from the warmer parts of the Atlantic Ocean. *J. Phycol.,* 10: 243-244.

JOHN, D.M. (1977a) - The marine algae of Ivory Coast and Cape Palmas in Liberia (Gulf of Guinea). *Revue algol.,* N.S., 11: 303-324 (1976).

JOHN, D.M. (1977b) - A new West African species of *Trichosolen* (Chlorophyceae). *Phycologia,* 16: 407-410.

JOHN, D.M. (1980) - A new species of *Botryocladia* (Rhodophyceae, Rhodymeniales) from Ghana (Tropical West Africa). *Phycologia,* 19: 91-95.

JOHN, D.M. & ASARE, S.O. (1975) - A preliminary study of the variation in yield and properties of phycocolloids from Ghanaian seaweeds. *Marine Biology,* 30: 325-330.

JOHN, D.M. & GRAFT-JOHNSON, K.A.A. de (1977) - Preliminary observations on the growth and reproduction of *Sphacelaria brachygonia* Sauv. (Phaeophyceae) in culture. *Bull. Inst. fond. Afr. noire,* sér. A, 37: 751-760 (1975).

JOHN, D.M. & LAWSON, G.W. (1972a) - Additions to the marine algal flora of Ghana I. *Nova Hedwigia,* 21: 817-841 (1971).

JOHN, D.M. & LAWSON, G.W. (1972b) - The establishment of a marine algal flora in Togo and Dahomey (Gulf of Guinea). *Botanica mar.,* 15: 64-73.

JOHN, D.M. & LAWSON , G.W. (1974a) - Observations on the marine algal ecology of Gabon. *Botanica mar.,* 17: 249-254.

JOHN, D.M. & LAWSON, G.W. (1974b) - *Basispora,* a new genus of the Ralfsiaceae. *Br. phycol. J.,* 9: 285-290.

JOHN, D.M. & LAWSON, G.W. (1977a) - The marine algal flora of the Sierra Leone peninsula. *Botanica mar.,* 20: 127-135.

JOHN, D.M. & LAWSON, G.W. (1977b) - The distribution and phytogeographical status of the marine algal flora of Gambia. *Feddes Reprium,* 88: 287-300.

JOHN, D.M. & POPLE, W. (1973) - The fish grazing of rocky shore algae in the Gulf of Guinea. *J. exp. mar. Biol. Ecol.,* 11: 81-90.

JOHN, D.M., LAWSON, G.W. & PRICE, J.H. (1981) - Preliminary results from a recent survey of the marine algal flora of Angola (southwestern Africa). *Proc. Int. Seaweed Symp.* 8: 367-371.

JOHN, D.M., LIEBERMAN, D. & LIEBERMAN, M. (1977) - A quantitative study of the structure and dynamics of benthic subtidal algal vegetation off Ghana (Tropical West Africa). *J. Ecol.,* 65: 497-521.

JOHN, D.M., LIEBERMAN, D., LIEBERMAN, M. & SWAINE, M.D. (1980) - Strategies of data collection and analysis of subtidal vegetation [pp. 265-283]. In: Price, J.H., Irvine, D.E.G. & Farnham, W.F. (eds), *The Shore Environment* Vol. 1: *Methods*. Systematics Association Special Volume No. 17(a). pp. XX + 321 + IXLI. London & New York.

JOHN, D.M., PRICE, J.H., MAGGS, C. & LAWSON, G.W. (1979) - Seaweeds of the western coast of tropical Africa and adjacent islands: a critical assessment. III. Rhodophyta (Bangiophyceae). *Bull. Br. Mus. nat. Hist.* (Bot.), 7: 69-82.

JOLY, A.B. (1953) - Re-discovery of *Mesogloia Brasiliensis* Montagne. *Bolm Inst. oceanogr., S. Paulo,* 3: 39-47, 1 pl. (1952).

JONES, J.I. (1969) - Planktonic foraminifera as indicator organisms in the eastern Atlantic equatorial current system [pp. 213-230]. In: *Proceeding of the Symposium on Oceanography and Fisheries Resources of the Tropical Atlantic, 20-28 Oct. 1966, Abidjan.* UNESCO. pp. 1-430. Paris.

KAPRAUN, D.F. (1970) - Field and cultural studies of *Ulva* and *Enteromorpha* in the vicinity of Port Aransas, Texas. *Publs Inst. mar. Sci. Univ. Texas,* 15: 205-285.

KEAY, R.W.J. (1959) - *Vegetation map of Africa south of the Tropic of Cancer: explanatory notes by R.W.J. Keay,* A.E.T.F.A.T. pp. 1-24+map. London.

KENSLEY, B. & PENRITH, M.-L. (1973) - The constitution of the intertidal fauna of rocky shores of Moçamedes, Southern Angola. *Cimbebasia,* ser. A, 2: 114-123,5 pls.

KJELLMAN, F.R. (1900) - Om floridé-slägtet *Galaxaura* dess organografi och systematik. *K. svenska Vetensk-Akad. Handl.,* 33: 1-109, 20 pls.

KOMAREK, J. (1973) - Prospects for taxonomic developments [pp. 482-486]. In: Carr, N.G. & Whitton, B.A. (eds), The biology of blue-green algae. *Botanical Monogr.,* 9: 1-686.

KOSTER, J.T. (1955) - The genus *Rhizoclonium* in the Netherlands. *Publ. Staz. zool. Napoli,* 27: 335-357.

KRAFT, G.T. & JOHN, D.M. (1976) - The morphology and ecology of *Nemastoma* and *Predaea* species (Nemastomataceae, Rhodophyta) from Ghana. *Br. Phycol. J.,* 11: 331-344.

KRISHNAMURTHY, V. & JOSHI, H.V. (1969) - The species of *Ulva* L. from Indian waters. *Bot. J. Linn. Soc.,* 62: 123-130, 1 pl.

KUNTH, C.S. (1822) - *Synopsis plantarum, quas in itinere ad plagam aequinoctialem orbis novi collegerunt Al. De. Humboldt et Am. Bonopland.* Part 6, I. *Botanique.* pp. IV+491. Paris.

KÜTZING, F.T. (1833) - Synopsis Diatomearum oder Versuch einer systematischen Zusammenstellung der Diatomeen. *Linnaea,* 8: 529-620.

KÜTZING, F.T. (1836) - *Algarum aquae dulcis Germanicarum, collegit Fridericus Traugott Kützing, Decade* XVI. nos. 151-160. Halis Saxonum.

KÜTZING, F.T. (1841) - Über *Ceramium* Ag. *Linnaea,* 15: 727-746.

KÜTZING, F.T. (1843a) - Über die Systematische Eintheilung der Algen. *Linnaea,* 17: 75-107.

KÜTZING, F.T. (1843b) - *Phycologia generalis...* pp. XXXII+458, 80 pls. Lipsiae.

KÜTZING, F.T. (1845) - *Phycologia Germanica. d., Deutschlands Algen in bündigen Beschreibungen.* pp. X+340. Nordhausen.

KÜTZING, F.T. (1847) - Diagnosen und Bemerkungen zu neuen oder kritischen Algen. *Bot. Ztg.,* 5: 164-167.

KÜTZING, F.T. (1849) - *Species algarum.* pp. VI+922. Leipzig.

KÜTZING, F.T. (1853-1869) - *Tabulae phycologicae oder Abbildungen der Tange.* 3, pp. 1-28, 100 pls. 1853. *Id.,* 5, pp. II+30, 100 pls. 1855. *Id.,* 8, pp. II+48, 100 pls. 1858. *Id.,* 9, pp. VIII+42, 100 pls. 1859. *Id.,* 11, pp. II+32, 100 pls. 1861. *Id.,* 13, pp. I+31, 100 pls. 1863. *Id.,* 14, pp. I+35, 100 pls. 1864. *Id.,* 15, pp. I+36, 100 pls. 1865. *Id.,* 16, pp. I+35, 100 pls. 1866. *Id.,* 18, pp. I+35, 100 pls. 1868. *Id.,* 19. pp. IV+36, 100 pls. 1869. Nordhausen.

KYLIN, H. (1906) - Zur Kenntnis einiger schwedischen *Chantransia*-Arten. In: *Botaniska Studier Tillagnade F.R. Kjellman den 4 November 1906.* pp. 113-126. Upsala.

KYLIN, H. (1932) - Die Florideenordnung Gigartinales. *Acta Univ. Lund,* N.F., 2, 28: 1-88, 28 pls.

KYLIN, H. (1938) - Verzeichnis einiger Rhodophyceen von Südafrika. *Acta Univ. Lund.,* N.F., 2, 34: 1-26.

KYLIN, H. (1944) - Die Rhodophyceen der schwedischen Westküste. *Acta Univ. Lund,* N.F., 2, 40: 1-104, 32 pls.

KYLIN, H. (1956) - *Die Gattungen der Rhodophyceen.* pp. XV+673. Lund.

LABOREL, J. (1974) - West African reef coral An hypothesis on their origin. *Proc. Second. Intern. Coral Reef Symposium.* Vol. 1: 425-443.

LAMOUROUX, J.V.F. (1805) - *Dissertations sur plusiers espèces de Fucus, peu connues ou nouvelles; avec leur description en Latin et en Français.* (1). pp. XXIV+ 83, 36 pls. Agen and Paris.

LAMOUROUX, J.V.F. (1809a) - Observations sur la physiologie des algues marines, et description de cinq nouveaux genres de cette famille. *Nouv. Bull. Sci. Soc. philom. Paris,* 1: 329-333, 1 pl. (May 1809).

LAMOUROUX, J.V.F. (1809b) - Mémoire sur trois nouveaux genres de la famille des algues marines... *J. Bot., Paris,* 2: 129-135, 1 pl. (April 1809).

LAMOUROUX, J.V.F. (1813) - Essai sur les genres de la famille des thallassiophytes non articulées. *Annls Mus. Hist. nat. Paris,* 20: 21-47, 115-139, 267-293, 7 pls.

LAMOUROUX, J.V.F. (1816) - *Histoire des polypiers coralligènes flexibles, vulgairement nommés zoophytes.* pp. LXXXIV + 559, 19 pls. Caen.

LAMOUROUX, J.V.F. (1821) - *Exposition méthodique des genres de l'ordre des polypiers,...* pp. VII+115, 84 pls. Paris.

LAMOUROUX, J.V.F. (1825) - Gélidie [pp. 190-191]. In: Audouin *et al., Dictionnaire classique d'histoire naturelle.* 7. pp. 1-626. Paris.

LANJOUW, J. *et al.* (1966) - *International Code of Botanical Nomenclature adopted by the Tenth International Botanical Congress, Edinburgh, August 1964.* pp. 1-402. Utrecht.

LAWSON, G.W. (1953) - The general features of seaweed zonation on the Gold Coast [pp. 18-19]. In: Black, W.A.P. *et al.* (eds), *Proceedings of the first international seaweed symposium held in Edinburgh 14th-17th July 1952.* pp. VII+129. Inveresk.

LAWSON, G.W. (1954a) - Intertidal zonation in West Afrcia in relation to ocean currents. *Rapp. Commun. int. bot. Congr.,* 8(17): 153-155.

LAWSON, G.W. (1954b) - Seaweeds from Sierra Leone. *Jl W. Afr. Sci. Ass.,* 1: 63-67.

LAWSON, G.W. (1955) - Rocky shore zonation in the British Cameroons. *Jl W. Afr. Sci. Ass.,* 1: 78-88, 1 pl.

LAWSON, G.W. (1956) - Rocky shore zonation on the Gold Coast. *J. Ecol.,* 44: 153-170, 2 pls.

LAWSON, G.W. (1957a) - Some features of the intertidal ecology of Sierra Leone. *Jl W. Afr. Sci. Ass.,* 3: 166-174.

LAWSON, G.W. (1957b) - Seasonal variation of intertidal zonation on the coast of Ghana in relation to tidal factors. *J. Ecol.,* 45: 831-860.

LAWSON, G.W. (1960a) - The Caulerpas of West Africa. *Niger. Fld,* 25: 23-31.

LAWSON, G.W. (1960b) - The genus *Taenioma* in West Africa. *New Phytol.,* 59: 361-366, 1 pl.

LAWSON, G.W. (1960c) - A preliminary check-list of Ghanaian fresh- and brackish-water algae. *Jl W. Afr. Sci. Ass.* 6: 122-136, 1 pl.

LAWSON, G.W. (1965) - Additions to a preliminary check-list of Ghanaian fresh- and brackish-water algae. *Jl W. Afr. Sci. Ass.,* 10: 45-55.

LAWSON, G.W. (1966) - The littoral ecology of West Africa. *Oceanogr. mar. Biol. ann. Rev.,* 4: 405-448, 2 pls.

LAWSON, G.W. (1978) - The distribution of marine algal floras in the tropical and subtropical Atlantic Ocean: a quantitative approach. *Bot. J. Linn. Soc.,* 76: 177-193.

LAWSON, G.W. (1980) - The Nigerian marine flora comes of age. *Niger. Fld Soc., Field notes,* No. 3: 9-12.

LAWSON, G.W. & JOHN, D.M. (1977) - The marine flora of the Cap Blanc peninsula: its distribution and affinities. *Bot. J. Linn. Soc.,* 75: 99-118.

LAWSON, G.W. & PRICE, J.H. (1969) - Seaweeds of the western coast of tropical Africa and adjacent islands: a critical assessment. I. Chlorophyta and Xanthophyta. *Bot. J. Linn. Soc.*, 62: 279-346.

LE JOLIS, A. (1863) - Liste des algues marines de Cherbourg. *Mém. Soc. natn Sci. nat. Math. Cherbourg,* 10: 6-168, 6 pls.

LEMOINE, P. (1911) - Structure anatomique des Mélobésiées. Application à la classification. *Annls. Inst. océanogr. Monaco,* 2: 1-213, 5 pls.

LEMOINE, P. (1917) - Les Mélobésiées des Antilles. Danoises recoltées par M. Børgesen. *Bull. Mus. Hist. nat., Paris,* 23: 133-136.

LEMOINE, P. (1924) - Corallinacées du Maroc, I. *Bull. Soc. Sci. nat. Maroc,* 4: 113-134, 2 pls.

LEMOINE, P. (1964) - Contribution à l'etude des Melobesiées de l'Archipel du Cap Vert. *Proc. Int. Seaweed Symp.,* 4: 234-239.

LEMOINE, P. (1965) - Algues calcaires (Mélobésiées) recueillies par le Professeur Drach (croisière de la Calypso en Mer Rouge, 1952). *Bull. Inst. océanogr. Monaco,* 64: 1-20.

LEVRING, T. (1938) - Verzeichnis einiger Chlorophyceen und Phaeophyceen von Südafrika. *Acta Univ. Lund,* 34: 1-25, 8 pls.

LEVRING, T. (1969) - The vegetation in the sea [pp. 1-46]. In: Levring, T., Hoppe, H.A. & Schmid, O.T., *Marine algae A survey of research and utilization.* Botanica Marina Handbooks, 1. pp. 1-421, 16 pls. Hamburg.

LEWIS, J.R. (1961) - The littoral zone on rocky shores - a biological or physical entity? *Oikos,* 12: 281-301.

LEWIS, J.R. (1964) - *The ecology of rocky shores.* pp. XII+323. London.

LIEBERMAN, M., JOHN, D.M. & LIEBERMAN, D. (1979) - Ecology of subtidal algae on seasonally devastated cobble substrates off Ghana. *Ecology,* 60: 1151-1161.

LIGHTFOOT, J. (1777) - *Flora scotica;...* Vol. 2, pp. 531-1151. London.

LINK, H.F. (1820) - Epistola... de algis aquaticis in genera disponendis [pp. 1-8, 1pl.]. In: Nees von Esenbeck, C.G., *Horae physicae berolinensis, collectae ex symbolis virorium doctorum...* pp. 1-123, 27 pls. Bonnae.

LINNAEUS, C. (1753, 1763) - *Species plantarum,...* Vol. 2. pp. 561-1200 [+I-XXXI]. 1753. *Id.,* Ed. 1, Vol. 2. pp. 785-1684. 1763. Stockholm.

LINNAEUS, C. (1767) - *Systema naturae...* Ed. 12, Vol. 1. pp. 1284-1304. Stockholm.

LINNAEUS, C. (1771) - *Mantissa plantarum altera...* Ed. 2. pp. 143-586. Stockholm.

LODGE, S.M. (1948) - Algal growth in the absence of *Patella* on an experimental strip of foreshore, Port St. Mary, Isle of Man. *Proc. Trans. L'pool biol. Soc.,* 56: 78-83.

LONGHURST, A.R. (1958) - An ecological survey of the West African marine benthos. *Fishery Publs colon. Off. London,* 11: 1-102.

LONGHURST, A.R. (1962) - A review of the oceanography of the Gulf of Guinea. *Bull. Inst. fr. Afr. noire,* sér. A, 24: 633-663.

LONGHURST, A.R. (1964) - The coastal oceanography of western Nigeria. *Bull. Inst. fr. Afr. noire,* sér. A, 26: 337-402.

LUCAS, A.H.S. & PERRIN, F. (1947) - *The seaweeds of South Australia.* II. *The red seaweeds.* pp. 111-458. Adelaide.

LYNGBYE, H.C. (1819) - *Tentamen hydrophytologiae Danicae...* pp. XXXII+248, 70 pls. Hafniae.

MAGNE, F. (1964) - Recherches caryologiques chez les Floridées (Rhodophycées). *Cah. Biol. mar.,* 5: 461-671, 17 pls.

MARCHAL, E. (1960) - Premières observations sur la répartition des organismes de la zone intercotidale de la région de Konakri (Guinée). *Bull. Inst. fr. Afr. noire,* sér. A, 22: 137-141.

MARTENS, G.M. von (1866) - *Die Preussische Expedition nach Ost-Asien. Nach amtlichen Quellen. Botanischer Theil. Die Tange...* pp. 1-152, 8 pls. Berlin.

MARTIN, L. (1971) - The continental margin from Cape Palmas to Lagos: bottom sediments and submarine morphology [pp. 79-96]. In: Delany, F.M. (ed.), ICSU/SCOR Symposium Cambridge 1970: The geology of the East Atlantic continental margin. Part 4 Africa. *Rep. No.* 70/16, *Inst. geol. Sci.*, pp. IV+209.

MARTIUS, K.F.P. von (1828) - *Icones Selectae Plantarum Cryptogamicarum...* fasc. 1. pp. 28, pls. 14. Monochii.

MARTIUS, K.F.P. von (1833) - Ordo primus. Algae Roth [pp. 1-50]. In: Martius, K.F.P. von, Eschweiler, F. & Nees von Esenbeck, C.G., *Flora Brasiliensis...* Vol. 1. Pars prior, Algae, Lichens, Hepaticae. Exposierunt Martius, Eschweiler, Nees von Esenbeck. pp. IV+390. Stuttgartiae and Tubingae. Note. The overall editor of this work was Martius but parts dealing with the lichens and hepaticae were written by Eschweiler, F.G. and Nees von Esenbeck, C.G. respectively. A limited number of copies were published in 1826 and minor changes and additions were made in the re-issue in 1833.

MASAKI, T. (1968) - Studies on the Melobesioideae of Japan. *Mem. Fac. Fish. Hokkaido Univ.,* 16: 1-80, 79 pls.

MAZÉ, H. & SCHRAMM, A. (1870-1877) - *Essai de classification des algues de la Guadeloupe.* pp. XIX+283+III. Basse-Terre.

MAZOYER, G. (1938) - Les Céramiées de l'Afrique du Nord. *Bull. Soc. Hist. nat. Afr. N.,* 29: 317-331.

MCMASTER, R.L., LACHANCE, T.P. & ASHRAF, A. (1970) - Continental shelf geomorphic features off Portuguese Guinea, Guinea, and Sierra Leone (West Africa). *Marine Geol.,* 9: 203-213.

MENEGHINI, G. (1840) - Botanische Notizen. *Flora, Jena,* 23: 510-512. Note. This would appear to be the published form of a document which is often cited in the literature as "Lettera al Dott. Iacob Corinaldi a Pisa" (see Woelkerling, 1973a, for further comments).

MENEGHINI, G. (1841) - Adunanza del di 16 Settembre 1841. *Atti Riun. Sci. ital.,* pp. 417-436.

MENEGHINI, G. (1842) - *Alghe Italiana e Dalmatiche.* (1/5), pp. 1-384, 5 pls. Padua.

MENEGHINI, G. (1843) - Monographia nostochinearum Italicarum addito specimine de Rivulariis,... *Mem. Acad. Sci. Torino,* ser. 2, 5: 1-143, 17 pls.

MICHANEK, G. (1971) - A preliminary appraisal of world seaweed resources. *FAO Fisheries Circular,* No. 128, pp. II+37. Rome.

MICHANEK, G. (1975) - Seaweed resources of the ocean. *FAO Fisheries Technical Paper,* No. 138, pp. V+127. Rome.

MILLS, F.W. (1932) - Some diatoms from Warri, South Nigeria. *Jl R. microsc. Soc.,* 52: 383-394.

MOLINIER, R. & PICARD, J. (1953) - Recherches analytiques sur les peuplements littoraux Méditerranéens se développant sur substrat solide. *Recl. Trav. Stn mar. Endoume,* 9: 1-18.

MONTAGNE, J.F.C. (1836) - Notice sur les plantes cryptogames récemment découvertes en France, ... *Annls Sci. nat. (Botanique),* sér. 2, 6: 321-339.

MONTAGNE, J.F.C. (1839) - Botanique, Cryptogamie. II. Florula Boliviensis stirpes novae et minus cognitae. Algae [pp. 13-39, 7 pls]. In: d'Orbigny, A., *Voyage dans l'Amérique méridionale.* Vol. 7. pp. 1-119.

MONTAGNE, J.F.C. (1839-1841) - Plantes cellulaires [pp. I-XV+1-208]. In: Barker-Webb, P. & Berthelot, S., *Histoire naturelle des Iles Canaries,...* Tome IIIe, 2e partie, *Phytographia Canariensis,* Sectio Ultima. pp. XV+208, 9 pls. Paris (1835-1850). Note. For a detailed consideration of this work see Stearn (1937, *J. Soc. Biblphy nat. Hist.,* 1: 49-63).

MONTAGNE, J.F.C. (1840) - Seconde centurie de plantes cellulaires exotiques nouvelles. Décade VII. *Annls Sci. nat. (Botanique),* sér. 2, 13: 193-207.

MONTAGNE, J.F.C. (1842) - Algae [pp. 1-104, 5 pls]. In: Ramon de la Sagra, *Histoire physique, politique et naturelle de l'Ile de Cuba. Botanique - plantes cellulaires.* pp. X+549. Paris (1838-1842).

MONTAGNE, J.F.C. (1843) - Quatrième centurie de plantes cellulaires exotiques nouvelles. Décade VII. *Annls Sci. nat. (Botanique),* sér. 2, 20: 294-306.

MONTAGNE, J.F.C. (1846-1849) - Phyceae [pp. 1-197, 16 pls]. In: Durieu de Maisonneuve, M.C., *Exploration Scientifique de l'Algérie. Sciences naturelles, botanique.* Partie 1. *Cryptogamie.* pp. 1-631. Paris.

MONTAGNE, J.F.C. (1950a) - Pugillus algarum yememsium quas collegerunt annis 1847-1849, clarr. Arnaud et Vaysiere. *Annls Sci. nat. (Botanique)*, sér. 3, 13: 236-248.

MONTAGNE, J.F.C. (1950b) - Cryptogamia guayanensis, seu plantarum cellularium in Guayana gallica annis 1835-1849 a Cl. Leprieu collectarum enumeratio universalis. *Annls Sci. nat. (Botanique)*, sér. 3, 14: 283-309.

MONTAGNE, J.F.C. (1856) - *Sylloge generum specierumque cryptogamarum quas in variis operibus descriptas iconibusque illustratas...* pp. XXIV+498. Paris.

MORLIERE, A. (1970) - Les saisons marines devant Abidjan. *Doc. Scient. Centre Rech. Oceanogr., Abidjan.* 1: 1-15.

MORLIERE, A. & REBERT, J.P. (1972) - Étude hydrologique du plateau continental ivorien. *Doc. Scient. Centre Rech. Oceanogr., Abidjan.* 3: 1-30.

MÜLLER, O.F. (1778) - *Icones plantarum... ad illustrandum opus... Florae Danicae nomine inscriptum.* 5 (13), pp. 1-8, 59 pls. Hauniae.

MURRAY, G. (1888) - Catalogue of the marine algae of the West Indian region. *J. Bot., Lond.*, 26: 358-363.

NÄGELI, G.A. (1862) - Beiträge zur Morphologie und Systematik der Ceramiaceen. *Sber. bayer Akad. Wiss.*, 2: 297-415, 1 pl.

NAKAMURA, Y. (1965) - Development of zoospores in *Ralfsia*-like thallus, with special reference to the life cycle of the Scytosiphonales. *Bot. Mag., Tokyo,* 78: 109-110.

NASR, A.H. (1947) - Synopsis of the marine algae of the Egyptian Red Sea Coast. *Bull. Fac. Sci. Egypt. Univ.*, 26: 1-155, 14 pls.

NEUMANN, G. (1965) - *Oceanography of the tropical Atlantic. De Ciencias Separata do vol. 37, Supplemento, Dus "Anais da Academia Brasileira".* Rio de Janeiro.

NEWTON, L. (1931) - *A Handbook of the British seaweeds.* pp. XIII+478. London

NIELSEN, R. (1972) - A study of the shell-boring marine algae around the Danish Island Laesø. *Bot. Tidsskr.*, 67: 245-269.

NIENHUIS, P.H. (1975) - *Biosystematics and ecology of Rhizoclonium riparium (Roth) Harv. (Chlorophyceae: Cladophorales) in the estuarine area of the rivers Rhine, Meuse and Scheldt.* Thesis, Rijksuniversiteit te Groningen. Rotterdam.

NIZAMUDDIN, M. (1964) - Studies on the genus *Caulerpa* from Karachi. *Botanica mar.*, 6: 204-223, 7 pls.

NIZAMUDDIN, M. (1969) - Contribution to the marine algae from West Pakistan. 1. Morphology and ecology of siphoneous algae. *Revue algol.*, N.S., 9: 239-274, 7 pls.

NIZAMUDDIN, M. & SAIFULLAH, S.M. (1967) - Studies on marine algae of Karachi: *Dictyopteris* Lamouroux. *Botanica mar.*, 10: 169-179, 5 pls.

NORTON, T.A. & BURROWS, E.M. (1969) - Studies on marine algae of the British Isles. 7. *Saccorhiza polyschides* (Lightf.) Batt. *Br. phycol. J.*, 4: 19-53.

OFORI-ADU, D.W. (1975) - Beach hydrography and the inshore fishery of Ghana for the period 1968-71. *Mar. Fish. Res. Rep. Ghana,* (5): 1-56.

OHMI, H. (1968) - A descriptive review of *Gracilaria* from Ghana, West Africa. *Bull. Fac. Fish. Hokkaido Univ.,* 19: 83-86, 2 pls.

OKAMURA, K. (1897) - On the algae from Ogasawarajima (Bonin Islands). *Bot. Mag., Tokyo,* 11: 1-16.

OKAMURA, K. (1908) - *Icones of Japanese Algae.* I, pp. 179-208, 4 pls. Tokyo.

OKAMURA, K. (1921) - *Icones of Japanese Algae.* IV, pp. 109-125, 5 pls. Tokyo.

OKAMURA, K. (1930) - On the algae from the island Hatidyo. *Rec. Oceanogr. Wk. Japan,* 2: 92-110, 5 pls.

O'KELLY, C.J. & YARISH, C. (1981) - Observations on marine Chaetophoraceae (Chlorophyta). II. On the circumscription of the genus *Entocladia* Reinke. *Phycologia,* 20: 32-45.

PADMAJA, T.D. & DESIKACHARY, T.V. (1967) - Trends in the taxonomy of algae. *Bull. natn Inst. Sci. India,* 34: 338-364.

PAPENFUSS, G.F. (1940a) - Notes on South African marine algae I. *Bot. Notiser,* 1940: 200-226.

PAPENFUSS, G.F. (1940b) - A revision of the South African marine algae in Herbarium Thunberg. *Symb. bot. Upsal.,* 4: 1-17.

PAPENFUSS, G.F. (1943) - Notes on algal nomenclature. II. *Gymnosporus* J. Agardh. *Am. J. Bot.,* 30: 463-468.

PAPENFUSS, G.F. (1945) - Revision of the *Acrochaetium-Rhodochorton* complex of the Red Algae. *Univ. Calif. Publs Bot.,* 18: 299-334.

PAPENFUSS, G.F. (1947) - Generic names proposed for conservation. I. *Madroño,* 9: 8-17.

PAPENFUSS, G.F. (1950) - Review of the genera of algae described by Stackhouse. *Hydrobiologia,* 2: 181-208.

PAPENFUSS, G.F. (1952-1968a) - Notes on South African marine algae, III. *Jl S. Afr. Bot.,* 17: 167-188. 1852. Id., IV. *Ibid.,* 22: 65-77. 1956. Id., V. *Ibid.,* 34: 267-287. 1968a.

PAPENFUSS, G.F. (1961) - The structure and reproduction of *Caloglossa leprieurii. Phycologia,* 1: 8-31.

PAPENFUSS, G.F. (1967) - Notes on algal nomenclature - V. Various Chlorophyceae and Rhodophyceae. *Phykos,* 5: 95-105 (1966).

PAPENFUSS, G.F. (1968b) - A history, catalogue and bibliography of Red Sea benthic algae. *Israel J. Bot.,* 17: 1-118.

PAPENFUSS, G.F. & CHIANG, Y.-M. (1969) - Remarks on the taxonomy of *Galaxaura* (Nemaliales, Chaetangiaceae). *Proc. Int. Seaweed Symp.,* 6: 303-314.

PAPENFUSS, G.F. & EGEROD, L.E. (1957) - Notes on South African marine Chlorophyceae. *Phytomorphology,* 7: 82-93.

PARKE, M. & BURROWS, E.M. (1976) - Chlorophyceae [pp. 566-570]. In: Parke, M. & Dixon, P.S., Check-list of British Marine Algae - Third Revision. *J. mar. biol. Ass. U.K.,* 56: 527-594.

PARKE, M. & DIXON, P.S. (1976) - Check-list of British Marine Algae - Third Revision. *J. mar. biol. Ass. U.K.,* 56: 527-594.

PAULY, D. (1975) - On the ecology of a small West-African lagoon. *Ber. Dtsch. wiss. Komm. Meeresforsch,* 24: 46-62.

PENRITH, H.-L. & KENSLEY, B.F. (1970a) - The constitution of the intertidal fauna of rocky shores of South West Africa. Part I. Lüderitzbucht. *Cimbebasia,* ser. A, 1: 189-239.

PENRITH, M.-L. & KENSLEY, B.[F.] (1970b) - The constitution of the fauna of the intertidal shores of South West Africa. Part II. Rocky Point. *Cimbebasia,* ser. A, 1: 243-268, 8 pls.

PÉRÈS, J.M. & PICARD, J. (1956) - Considérations sur l'étagement des formations benthiques. *Recl. Trav. Stn mar. Endoume,* 18: 15-30.

PERLROTH, I. (1969) - The distribution of water type structure in the first 300 feet of the equatorial Atlantic [pp. 185-191]. In: *Proceedings of the Symposium on Oceanography and Fisheries Resources of the Tropical Atlantic. 20-28 Oct. 1966, Abidjan,* UNESCO. pp. 1-430. Paris.

PHAM-HOÀNG, HÔ (1969) - Rong bîên Việtnam Marine algae of South Vietnam. *Trung-Tâm Hoc-Liêu Xuât-Bẫm,* pp. 1-558.

PICCONE, A. (1884) - *Crociera del Corsaro alle Isole Madera e Canarie del Capitano Enrico d'Albertis* 2. *Alghe.* pp. 1-60, 1 pl. Genova.

PICCONE, A. (1900) - Noterelle ficologiche. XI. Pugillo di alghe dell'isola S. Thiago (Capo Verde). *Atti Soc. Ligust. Sci. nat. geogr.,* 11: 238-239.

PICCONE, A. (1901) - Noterelle ficologiche XI.-Pugillo di alghe dell'isola S. Thiago (Capo Verde). *Nuova Notarisia,* 12: 45-47.

PILGER, R. (1911) - Die Meeresalgen von Kamerun. Nach der Sammlung von C. Ledermann [pp. 294-313, 316-323]. In: Engler, A., Beiträge zur Flora von Afrika. XXXIX. *Bot. Jb.,* 46: 293-464.

PILGER, R. (1919) - Über Corallinaceae von Annobon [pp. 401-435]. In: Engler, A., Beiträge zur Flora von Afrika. XLVII. *Bot. Jb.,* 55: 350-463.

PILGER, R. (1920) - Algae Mildbraedianae Annobonenses [pp. 1-14]. In: Engler, A., Beiträge zur Flora von Afrika. XLVIII. *Bot. Jb.,* 57: 1-301 (1920-1921).

POPLE, W. & MENSAH, M.A. (1971) - Evaporation as the upwelling mechanism in Ghanaian Coastal waters. *Nature Physical Science,* 234: 18-20.

POST, E. (1936) - Systematische und pflanzen-geographische Notizen zur *Bostrychia-Caloglossa*-Assoziation. *Revue algol.,* 9: 1-84.

POST, E. (1955-1959) - Weitere Daten zur Verbreitung des Bostrychietum IV. *Arch. Protistenk,* 100: 351-377, 5 pls. 1955a. Id., VI. *Ibid.,* 102: 84-112. 1957b. Id., VII. *Ibid.,* 103: 489-506, 3 pls. 1959.

POST, E. (1955b) - Weitere Daten zur Verbreitung des Bostrychietum V. *Ber. dt. bot. Ges.,* 68: 205-216.

POST, E. (1957a) - Fruktifikationen und Keimlinge bei *Caloglossa. Hydrobiologia,* 9: 105-125.

POST, E. (1963a) - *Bostrychia radicans* in Süsswasser Westafrikas (P. Fremy in Verehrung). *Revue algol.,* N.S., 6: 270-281.

POST, E. (1963b) - Zur Verbreitung und Ökologie der *Bostrychia-Caloglossa*-Assoziation. *Int. Revue ges. Hydrobiol.,* 48: 47-152.

POST, E. (1965) - *Bostrychia scorpioides* in tropischen Westafrika. *Hydrobiologia,* 26: 301-306.

POST, E. (1966a) - *Caloglossa ogasawaraensis* in Westafrika. *Hydrobiologia,* 27: 317-322.

POST, E. (1966b) - Neues zur Verbreitungsökologie neuseeländischer und mittelamerikanischer *Bostrychia-Caloglossa*-Assoziation. *Revue algol.,* N.S., 8: 127-150.

POST, E. (1968) - Zur Verbreitungsökologie des Bostrychietum. *Hydrobiologia,* 31: 241-316.

POSTEL, E. (1968) - Marine hydrology and biogeography in western Africa [pp. 13-16, pls 18, 19]. In: *International Atlas of West Africa.* OAU Scientific, Technical and Research Commission. 1-42 pls.

POSTELS, A. & RUPRECHT, F. (1840) - *Illustrationes algarum in itinere circa orben jussu Imperatoris Nicolai I. algue auspiciis Navarchi Friderici Lutke...* pp. VI + 28 + IV + 22, 40 pls. Petropoli.

PRICE, J.H. (1978) - Ecological determination of adult form in *Callithamnion:* its taxonomic implications [pp. 203-300]. In: Irvine, D.E.G. & Price, J.H. (eds), *Modern Approaches to the Taxonomy of Red and Brown Algae.* Systematics Association Special Volume No. 10, pp. XII+484. London and New York.

PRICE, J.H., JOHN, D.M. & LAWSON, G.W. (1978) - Seaweeds of the western coast of tropical Africa and adjacent islands: a critical assessment. II. Phaeophyta. *Bull. Br. Mus. nat. Hist.* (Bot.), 6: 87-182.

PRICE, J.H., JOHN, D.M. & LAWSON, G.W. (1983) - Seaweeds of the western coast of tropical Africa and adjacent islands: a critical assessment. III. Rhodophyta (Florideophyceae). *Bull. Br. Mus. nat. Hist.* (Bot.): in press.

PRICE, S.M. (1973) - Studies on *Bachelotia (Pilayella?) antillarum.* I. The occurrence of plurilocular sporangia in culture. *Br. phycol. J.,* 8: 21-29.

PRUD'HOMME VAN REINE, W.F. (1982) - A taxonomic revision of the European Sphacelariaceae (Sphacelariales, Phaeophyceae). *Leiden Botanical Series,* 6: 1-293.

PURVIS, M.J., COLLIER, D.C. & WALLS, D. (1966) - *Laboratory techniques in botany*. 2nd edit., pp. VIII+439. London.

RAO, P.S. (1970) - Systematics of Indian Gelidiales. *Phykos*, 9: 63-78, 2 pls.

RAVANKO, O. (1970) - Morphological, developmental and taxonomic studies in the *Ectocarpus* complex (Phaeophyceae). *Nova Hedwigia*, 20: II+179-252.

RAVEN, P.H. & AXELROD, D.I. (1974) - Angiosperm biogeography and past continental movements. *Ann. Missouri bot. gard.*, 61: 539-673.

REINKE, J. (1879) - Zwei parasitische Algen. *Bot. Ztg.*, 37: 473-478, 1 pl.

REYSSAC, J. (1970) - Phytoplancton et production primaire au large de la Côte d'Ivoire. *Bull. Inst. fond. Afr. noire*, sér. A, 32: 869-981.

RICHARDS, H.M. (1901) - *Ceramothamnion Codii*, a new Rhodophyceous alga. *Bull. Torrey bot. Club*, 28: 257-265, 2 pls.

RICHARDSON, W.D. (1969) - Some observations on the ecology of Trinidad marine algae. *Proc. Int. Seaweed Symp.*, 6: 357-363.

RICHARDSON, W.D. (1975) - The marine algae of Trinidad, West Indies. *Bull. Br. Mus. nat. Hist.* (Bot.), 5: 71-143.

ROBERTS, M. (1967) - Studies on marine algae of the British Isles. 3. The genus *Cystoseira*. *Br. phycol. Bull.*, 3: 345-366.

ROBERTS, M. (1977) - Studies on marine algae of the British Isles. 9. *Cystoseira nodicaulis* (Withering) M. Roberts. *Br. phycol. J.*, 12: 175-199.

RODRIGUES, J.E. de M. (1960) - Revisão das algas de S. Tomé e Príncipe do herbário do Instituto Botânico de Coimbra I - Phaeophyta. *Garcia de Orto*, 8: 583-595, 6 pls.

ROSANOFF, S. (1886) - Recherches anatomiques sur les Mélobésiées. *Mém. Soc. Sci. nat. math. Cherbourg*, 12: 5-112, 7 pls.

ROSENVINGE, K.L. (1909) - The marine algae of Denmark - contributions to their natural history. Pt. 1. Introduction. Rhodophyceae 1. (Bangiales and Nemalionales). *K. danske Vidensk. Selsk. Skr.*, 7: 1-151, 2 pls.

ROTH, A.G. (1797-1806) - *Catalecta botanica quibus plantae novae et minus cognitae describunter atque illustrantur.* 2 (1), pp. VIII+244, 2 pls. 1797. *Id.*, 2 (2), pp. 1-258, 9 pls. 1800. *Id.*, 2 (3), pp. 1-350, 12 pls. 1806. Leipzig.

SAENGER, P. (1974) - Additions and comments on the Rhodomelaceae of Inhaca Island, Moçambique. *Nova Hedwigia*, 24: 19-37.

SAITO, Y. (1967) - Studies on Japanese species of *Laurencia* with special reference to their comparative morphology. *Mem. Fac. Fish. Hokkaido Univ.*, 15: 1-81, 18 pls.

SAITO, Y. & WOMERSLEY, H.B.S. (1974) - The southern Australian species of *Laurencia* (Ceramiales: Rhodophyta). *Aust. J. Bot.*, 22: 815-874.

SAMPAIO, J. (1958, 1963) - Cianófitas de S. Tomé e Príncipe. Estudos Ensaios Docum. *Junta Invest. Ultramar.*, 47: 11-80, 10 pls. 1958. *Ibid.*, 108: 1-52, 6 pls. 1963.

SAMPAIO, J. (1962) - Cianófitas de S. Tomé e Príncipe. *Bolm Soc. broteriana,* 48: 243-259. 3 pls. 1962a. *Ibid.,* 49: 1-74, 10 pls. 1962b.

SAUVAGEAU, C. (1901, 1903) - Remarques sur les Sphacélariacées. *J. Bot., Paris,* 15: 222-255. 1901. *Ibid.,* 17: 378-422. 1922.

SAUVAGEAU, C. (1927) - Sur le *Colpomenia sinuosa* Derb. et Sol. *Bull. Stn biol. Arcachon,* 24: 309-353.

SCHMIDT, O.C. (1929) - Beiträge zur Kenntnis der Meeresalgen der Azoren II. *Hedwigia,* 69: 165-172.

SCHMIDT, O.C. & GERLOFF, J. (1957) - Die marine Vegetation Afrikas in ihren Grundzügen dargestellt. *Willdenowia,* 1: 709-756.

SCHMITZ, F. (1889) - Systematische Übersicht der bisher bekannten Gattungen der Florideen. *Flora, Jena,* 72: 435-456.

SCHMITZ, C.J.F. (1893) - Die Gattung *Lophothalia* J. Ag. *Ber. dt. bot. Ges.,* 11: 212-232.

SCHMITZ, C.J.F. (1895) - Marine Florideen von Deutsch-Ostafrika. *Bot. Jb.,* 21: 137-177.

SCHMITZ, C.J.F. & HAUPTFLEISCH, P. (1896) - Rhodophyllidaceae [pp. 366-382]. In: Engler, A. & Prantl, K. (eds), *Die natürlichen Pflanzenfamilien nebst ihren Gattungen und wichtigeren Arten.* Vol. 1. pp. XII + 544. Leipzig.

SCHNEIDER, C.W. & SEARLES, R.B. (1975) - North Carolina marine algae. IV. Further contribution from the continental shelf including two new species of Rhodophyta. *Nova Hedwigia,* 26: 83-104.

SCHNELL, R. (1950) - Esquisse de la végétation côtiére de la Basse Guinée Française [pp. 201-214]. In: Anon., *Conferência Internacional dos Africanistas ocidentais 2A. Conferência Bissau, 1947,* Vol. 2 *Trabalhos apresentados à 2ª Secçao (Meio Biológico),* 1ª *Parte.* pp. 1-338. Lisboa.

SETCHELL, W.A. (1914) - The *Scinaia* assemblage. *Univ. Calif. Publs Bot.* 6: 75-152, 7 pls.

SETCHELL, W.A. & GARDNER, N.L. (1924) - The marine algae. Expedition of the California Academy of Sciences to the Gulf of California in 1921. *Proc. Calif. Acad. Sci.,* ser. 4, 12: 695-949, 77 pls.

SETCHELL, W.A. & GARDNER, N.L. (1930) - Marine algae of the Revillagigedo Islands Expedition in 1925. *Proc. Cailf. Acad. Sci.,* ser. 4, 19: 109-215, 12 pls.

SETCHELL, W.A. & MASON, L.R. (1943) - *Goniolithon* and *Neogoniolithon. Proc. natn. Acad. Sci. U.S.A.,* 29: 87-92.

SILVA, P.C. (1952) - A review of nomenclatural conservation in the algae from the point of view of the type method. *Univ. Calif. Publs Bot.,* 25: 241-324.

SILVA, P.C. (1960) - *Codium* (Chlorophyta) in the Tropical Western Atlantic. *Nova Hedwigia,* 1: 497-536, 16 pls.

SILVA, P.C. (1980) - Names of classes and families of living algae. *Regnum Vegetabile,* 103: 1-156.

SIMONS, R.H. (1964) - Species of *Plocamium* on the South African coast. *Bothalia,* 7: 183-193.

SKUJA, H. (1956) - Review of F. Drouet and W.A. Daily. Revision of the coccoid Myxophyceae. *Svensk. bot. Tidsskr.,* 50: 550-556.

SLUIMAN, H.J. (1979) - A note on *Bostrychia scorpioides* (Hudson) Montagne ex Kützing and *B. montagnei* Harvey (Rhodomelaceae, Rhodophyta). *Blumea,* 24: 301-305.

SÖDERSTRÖM, J. (1963) - Studies in *Cladophora. Botanica gotob.,* 1: 1-147.

SOLMS-LAUBACH, G. (1881) - II. *Corallina. Flora und Fauna des Golfes v. Neapel und der angrenzenden Meeresabschnitte,* 4. Monographie. pp. 1-172.

SOMMERFELT, S.C. (1826) - *Supplementum Florae Lapponicae, quam edidit... G. Wahlenberg,...* pp. XII+331, 3 pls. Christianiae.

SONDER, O.G. (1854) - Algae [pp. 1-4]. In: Zollinger, H., *Systematisches Verzeichniss der im indischen Archipel in den Jahren 1842-1848 gesammelten sowie der aus Japan empfangenen Pflanzen.* 2. pp. XII+160, 1 pl. Zürich.

SOURIE, R. (1954) - Contribution à l'étude ecologique des côtes rocheuses du Sénégal. *Mém. Inst. fr. Afr. noire,* 38: 1-342.

SOUTH, G.R. (1976) - A check-list of marine algae of eastern Canada - first revision. *J. mar. biol. Ass. U.K.,* 56: 817-843.

SOUTH, G.R. & HOOPER, R.G. (1980) - A catalogue and atlas of the benthic marine algae of the island of Newfoundland. *Mem. Univ. Newfoundland Occ. Pap. Biol.,* 3: 1-136.

SOUTH, R.G. & WHITTICK, A. (1976) - Aspects of the life history of *Rhodophysema elegans* (Rhodophyta, Peyssonneliaceae). *Br. phycol. J.,* 11: 349-354.

SPRENGEL, C. (1807) - Observations botanicae in Flora Halensem [pp. 1-26]. In: *Mantissa Prima, Florae Halensis.* pp. XVI+420, 12 pls. Halae.

SRINIVASAN, K.S. (1967) - Conspectus of *Sargassum* species from Indian territorial waters. *Phykos,* 5: 127-159 (1966).

SRINIVASAN, K.S. (1969) - *Phycologia Indica. Icones of Indian Marine Algae.* 1. pp. XVIII+52. Calcutta.

STACKHOUSE, J. (1795) - *Nereis britannica,...* (1). pp. VIII+30. Bath and London.

STANIER, R.Y., KUNISAWA, R., MANDEL, M. & COHEN—BAZIRE, G. (1971) - Purification and properties of unicellular blue-green algae (Order Chroococcales). *Bact. Rev.,* 35: 171-205.

STEENTOFT NIELSEN, M. (1958) - Common seaweeds at Lagos I. *Niger. Fld,* 23: 34-44.

STEENTOFT, M. (1967) - A revision of the marine algae of São Tomé and Príncipe (Gulf of Guinea). *J. Linn. Soc. (Bot.)*, 60: 99-146, 2 pls.

STEPHENSON, T.A. & STEPHENSON, A. (1949) - The universal features of zonation between tidemarks on rocky coasts. *J. Ecol.*, 38: 289-305, 1 pl.

STEPHENSON, T.A. & STEPHENSON, A. (1972) - *Life between tidemarks on rocky shores.* pp. XII+425+errata page. San Francisco.

SUNESON, S. (1937) - Studien über die Entwicklungsgeschichte der Corallinaceen. *Acta. Univ. Lund,* 33: 1-101, 4 pls.

SVEDELIUS, N.E. (1895) - Algen aus den Ländern der Magellanstrasse und West-patogonien. I. Chlorophyceae [pp. 283-316]. In: *Svenska Expeditionem till Magellansländern.* 3. *Botany.* pp. 1-304. Stockholm.

TANAKA, T. (1941) - The genus *Hypnea* from Japan. *Hokkaido Univ. Inst. Algol. Res., Sci. papers,* 2: 227-250, 2 pls.

TANDY, G. (1944) - Algae [p. 386, Appendix I]. In: Exell, A.W., *Catalogue of the vascular plants S. Tomé (with Príncipe and Annobon).* pp.XI-428. London.

TATEWAKI, M. (1966) - Formation of a crustaceous sporophyte with unilocular sporangia in *Scytosiphon lomentaria. Phycologia,* 6: 62-66.

TAYLOR, W.R. (1928) - The marine algae of Florida with special reference to the Dry Tortugas. *Publns Carnegie Instn Washington,* 379 [*Pap. Tortugas Lab.,* 25]. V+219, 37 pls.

TAYLOR, W.R. (1939) - Algae collected on the Presidential Cruise of 1938. *Smithsonian Miscellaneous Collections,* 98: 1-18, 2 pls.

TAYLOR, W.R. (1943) - Marine algae from Haiti collected by H.H. Bartlett in 1941. *Pap. Mich. Acad. Sci.,* 28: 143-163, 4 pls. (1942).

TAYLOR, W.R. (1945) - Pacific marine algae of the Allan Hancock expeditions to the Galapagos Islands. *Allan Hancock Pacif. Exped.,* 12: IV+528, 100 pls.

TAYLOR, W.R. (1960) - *Marine algae of the eastern tropical and subtropical coasts of the Americas.* Univ. Mich. Stud., Sci. Ser. 21. pp. IX + 870, 80 pls. Ann Arbor.

TAYLOR, W.R. (1962a) - *Marine algae of the northeastern coast of North America.* Ed. 2, 2nd printing, with corrections. Univ. Mich. Stud., Sci. Ser. 13. pp. IX + 509, 60pls. Ann Arbor.

TAYLOR, W.R. (1962b) - Marine algae from the tropical Atlantic Ocean. V. Algae from the Lesser Antilles. *Contr. U.S. natn Herb.,* 36: 43-62, 4 pls.

TAYLOR, W.R. (1966) - Records of Asian and western Pacific marine algae, particularly from Indonesia and the Philippines. *Pacific Sci.,* 20: 342-359.

TAYLOR, W.R. (1969) - Notes on the distribution of West Indian marine algae particularly in the Lesser Antilles... *Contr. Univ. Mich. Herb.,* 9: 125-203, 8 pls.

TAYLOR, W.R. & ARNDT, C.H. (1929) - The marine algae of the south-western peninsula of Hispaniola *Am. J. Bot.,* 16: 651-662.

TAYLOR, W.R., JOLY, A.B. & BERNATOWICZ, A.J. (1953) - The relation of *Dichotomosiphon pusillus* to the algal genus *Boodleopsis*. *Pap. Mich. Acad. Sci.,* 38: 97-107, 3 pls. (1952).

TAYLOR, W.R. & RHYNE, C.F. (1970) - Marine algae of Dominica. *Smithson. Contr. Bot.,* 3: 1-16.

THIVY, F. (1943) - New records of some marine Chaetophoraceae and Chaetosphaeriaceae for North America. *Biol. Bull. Mar. biol. tab. Woods Hole,* 85, 3: 244-264.

THOMPSON, B.W. (1965) - *The climate of Africa.* pp. 1-15, 132 maps. Nairobi.

THURET, G.W. (1855) - Note sur nouveau genre d'algues, de la famille des Floridées. *Mém. Soc. Sci. nat. math. Cherbourg,* 3: 155-160, 2 pls.

THURET, G.W. & BORNET, É. (1878) - *Études phycologiques. Analyses d'algues marines.* pp. III+105, 51 pls. Paris.

TOUPET, C. (1968) - Major climatic elements [pp. 1-3, pls 10-13]. In: *International Atlas of West Africa.* OAU Scientific, Technical, and Research Commission. 1-42 pls.

TOWNSEND, C. & LAWSON, G.W. (1972) - Preliminary results on factors causing zonation in *Enteromorpha* using a tide simulating apparatus. *J. exp. mar. Biol. Ecol.,* 8: 265-276.

TREVISAN, V.B.A. (1845) - *Nomenclator algarum, ou collection des noms imposées aux plantes de la famille des algues.* 1. pp. 1-80. Padua.

TROCHAIN, J. (1940) - Contribution à l'étude de la végétation du Sénégal. *Mém. Inst. fr. Afr. noire,* 2: 1-433, 30 pls.

TSENG, C.K. (1943) - Marine algae of Hong Kong. IV. The genus *Laurencia. Pap. Mich. Acad. Sci.,* 28: 185-208, 4 pls (1942).

TSENG, C.K. (1944) - Marine algae of Hong Kong. VI. The genus *Polysiphonia. Pap. Mich. Acad. Sci.,* 29: 67-82, 4 pls. (1943).

TURNER, D. (1808-1819) - *Fuci, sive plantarum fucorum generi a botanicis...* Vol. 1. pp. 1-164, 71 pls. 1808. *Id.,* Vol. 2. pp. 1-162, 62 pls. 1809. *Id.,* Vol. 3. pp. 1-148, 62 pls. 1811. *Id.,* Vol. 4. pp. 1-153, 61 pls. 1819. London.

VAHL, M. (1802) - Endeel kryptogamiske planter fra St. Croix. *Skr. Naturh.- Selsk. København,* 5: 29-47.

VICKERS, A. (1897?) - Contribution à la flore algologique des Canaries. *Annls Sci. nat. (Botanique),* sér. 8, 4: 293-306. (1896). Note. The date of this publication is difficult to cite with any certainty (see Lawson and Price, 1969).

VICKERS, A. (1905) - Liste des algues marines de la Barbade. *Annls Sci. nat. (Botanique),* sér. 9, 1: 45-66.

VICKERS, A. & SHAW, M.H. (1908) - *Phycologia Barbadensis. Iconographie des algues marines récoltées à l'île Barbade (Antilles).* pp. IX + 44, 87 pls. Paris.

WATTS, J.C.D. (1958) - The hydrology of a tropical West African estuary. *Bull. Inst. fr. Afr. noire,* sér. A, 20: 697-752.

WEBB, J.E. (1958) - The ecology of Lagos lagoon I. The lagoons of the Guinea Coast. *Phil. Trans. R. Soc.,* ser. B, 241: 307-318, 4 pls.

WEBB, J.E. (1960) - *The erosion of Victoria Beach. Its cause and cure.* pp. 1-42. Ibadan.

WEBER-VAN BOSSE, A. (1913) - *Liste des algues du Sibago.* I. *Myxophyceae, Chlorophyceae, Phaeophyceae... Siboga Exped.,* Monogr. 59a. pp. 1-186, 5 pls. Leiden.

WHITTON, B.A. (1968) - Blue-green algae from Sierra Leone. *Nova Hedwigia,* 15: 203-209.

WITHERING, W. (1776) - *A botanical arrangement of all the vegetables naturally growing in Great Britain...* Vol. 2. pp. 285-838. Birmingham and London.

WITHERING, W. (1796) - *An arrangement of British plants;...* Ed. 3, Vol. 4. pp. 1-418. London.

WOELKERLING, W. J. (1971) - Morphology and taxonomy of the *Audouinella* complex (Rhodophyta) in southern Australia. *Aust. J. Bot.,* suppl. ser. 1: 1-91.

WOELKERLING, W.J. (1973a) - The morphology and systematics of the *Audouinella* complex (Acrochaetiaceae, Rhodophyta) in north-eastern United States. *Rhodora,* 75: 529-621.

WOELKERLING, W.J. (1973b) - The *Auouinella* complex (Rhodophyta) in the western Sargasso Sea. *Rhodora,* 75: 78-101.

WOLLASTON, E.M. (1968) - Morphology and taxonomy of southern Australian genera of Crouanieae Schmitz (Ceramiaceae, Rhodophyta). *Aust. J. Bot.,* 16: 217-417, 10 pls.

WOMERSLEY, H.B.S. (1954) - The species of *Macrocystis* with special reference to those on southern Australian coasts. *Univ. Calif. Publns Bot.,* 27: 109-132.

WOMERSLEY, H.B.S. (1967) - A critical survey of the marine algae of southern Australia II. Phaeophyta. *Aust. J. Bot.,* 15: 189-270.

WOMERSLEY, H.B.S. & BAILEY, A. (1970) - Marine algae of the Solomon Islands. *Phil. Trans. R. Soc.,* ser. B, 259: 257-352, 4 pls.

WOMERSLEY, H.B.S. & SHEPLEY, E.A. (1959) - Studies on the *Sarcomenia* group of the Rhodophyta. *Aust. J. Bot.,* 7: 168-223, 5 pls.

WOODWARD, T.J. (1797) - Observations upon the generic character of *Ulva,* with descriptions of some new species. *Trans. Linn. Soc.,* 3: 46-58.

WOOSTER, W.S., BAKUN, A. & MACLAIN, D.R. (1976) - The seasonal upwelling cycle along the eastern boundary of the North Atlantic. *J. Mar. Res.,* 34: 131-141.

WULFEN, F.X. (1789) - Plantae rariores Carinthiaceae [pp. 3-166]. In: Jacquin, N.J., *Collectanea Austriaca ad botanicam, chemiam, et historiam naturalem spectantia,...* Vol. 3. pp. 1-306, 23 pls. Vindobonae.

WULFEN, F.X. (1803) - Cryptogama aquatica. *Arch. Bot., Leipzig,* 3: 1-64, 1 pl.

WYNNE, M.J. (1969) - Life history and systematic studies of some Pacific North American Phaeophyceae (brown algae). *Univ. Calif. Publns Bot.,* 50: 1-88, 24 pls.

WYNNE, M.J. (1972) - Studies on the life forms in nature and in culture of selected brown algae [pp. 133-146]. In: Abbott, I.A. & Kurogi, M. (eds), *Contributions to the systematics of benthic marine algae of the North Pacific.* Japanese Soc. Phycol., pp. XIV+279. Kobe.

WYNNE, M.J. & TAYLOR, W.R. (1973) - The status of *Agardhiella tenera* and *Agardhiella baileyi* (Rhodophyta, Gigartinales). *Hydrobiologia,* 43: 93-107.

YAMADA, Y. (1931) - Notes on *Laurencia,* with special reference to the Japanese species. *Univ. Calif. Publns Bot.,* 16: 185-310.

YENDO, K. (1902) - Corallinae verae Japonicae. *J. Coll. Sci. imp. Univ. Tokyo,* 16: 1-36, 7 pls.

YENDO, K. (1917) - Notes on algae new to Japan. VI. *Bot. Mag., Tokyo,* 31: 75-95.

ZANARDINI, G. (1839) - *Biblioteca italiana,* 96: 134-137.

ZANARDINI, G. (1847) - Notizie intorno alle cellulari marine delle lagune e de litorali di Venezia. *Atti Ist. veneto Sci.,* ser. 1, 5: 1-88, 6 pls.

ZANARDINI, G. (1851) - Algae novae vel minus cognitae in mari rubro a Portiero collectae. *Flora, Jena,* 34: 33-38.

ZANARDINI, G. (1858) - Plantarum in mari rubro prucusque collectarum enumeratio. *Mem. Ist. Veneto Sci. Lett. Arti,* 7: 209-309, 12 pls.

ZANARDINI, G. (1860) - Iconographia phycologica adriatica ossia scelta di ficee nuove o più rare del mare adriatico. *Mem. R. Ist. veneto Sci.,* 10: VIII+176, 40 pls.

ZANARDINI, G. (1863) - Scelta de ficee nuove o più rare del mare adriatico,... *Mem. R. Ist. veneto Sci.,* 11: 269-306, 8 pls.

ZANARDINI, G. (1872) - Phycearum indicarum pugillus. *Mem. R. Ist. veneto Sci.,* 17: 209-309, 12 pls.

Systematic Index

This is an index to every algal name used in the text. Families and higher taxa are given in CAPITALS, recognised genera, species and infraspecific taxa (subspecies, ssp.; variety, var.; form, f.) in roman type (including ecophenes of blue-green algae), and synonyms in *italic* type (*italic* type for the genus of a synonymous species does not necessarily imply that the name is invalid). Page numbers are in **bold type** where a genus or subgeneric taxon is described, in *italic type* where it is mentioned in the legend accompanying its illustration, and in roman type where the taxon is simply noted.

F

G

H

R

S